Exploring Traditional Wild Edible Plants

Wild edible plants are native species that grow and reproduce naturally in their natural habitats without domestication. These plants can serve as a healthier alternative to farmed crops that may be heavily laden with pesticides and other poisonous substances. This book focuses on assessment of the nutritional value, potential health benefits, and mechanisms of action of various wild edible plants. It presents information on nutrients and bioactive ingredients that can have health advantages, including antioxidant properties, antimicrobial, anti-inflammatory, and antidiabetic effects.

Features:
- Comprehensive exploration of potential benefits as well as side effects of wild edible plants.
- Special emphasis placed on diversity, category, and pharmacological values of wild edible plants.
- Challenges regarding the usage of wild edible plants such as overharvesting and habitat destruction, safety, and toxicity.

A volume in the *Exploring Medicinal Plants* series, this book highlights bioactive compounds and therapeutic efficacies of various wild edible plant species. It is useful reading for plant scientists as well as scientists, researchers, and academia interested in the health benefits of wild edible plants.

Exploring Medicinal Plants
Series Editor: Azamal Husen, *Wolaita Sodo University, Ethiopia*

Medicinal plants render a rich source of bioactive compounds used in drug formulation and development; they play a key role in traditional or indigenous health systems. As the demand for herbal medicines increases worldwide, supply is declining as most of the harvest is derived from naturally growing vegetation. Considering global interests and covering several important aspects associated with medicinal plants, the Exploring Medicinal Plants series comprises volumes valuable to academia, practitioners, and researchers interested in medicinal plants. Topics provide information on a range of subjects including diversity, conservation, propagation, cultivation, physiology, molecular biology, growth response under extreme environment, handling, storage, bioactive compounds, secondary metabolites, extraction, therapeutics, mode of action, and healthcare practices.

Led by Azamal Husen, PhD, this series is directed to a broad range of researchers and professionals consisting of topical books exploring information related to medicinal plants. It includes edited volumes, references, and textbooks available for individual print and electronic purchases.

Pharmacological Aspects of Essential Oils
Current and Future Trends
Edited by Sunita Singh, Pankaj Kumar Chaurasia and Shashi Lata Bharati

Promising Antiviral Herbal and Medicinal Plants
Edited by Nadeem Akhtar, Azamal Husen, Vagish Dwibedi, and Santosh Kumar Rath

Antimalarial Medicinal Plants
Edited by Azamal Husen

Medicinal Spice and Condiment Crops
Edited by Azamal Husen

Therapeutic Medicinal Plants in Traditional Persian Medicine
Edited by Roja Rahimi and Roodabeh Bahramsoltani

Medicinal Plants for the Management of Neurodegenerative Diseases
Edited by Jamal Akhtar, Maryam Sarwat and Fouzia Bashir

Nutraceuticals Inspiring the Contemporary Therapy for Lifestyle Diseases
Edited by Mala Trivedi, Sachidanand Singh, Parul Johri, and Pedro López-Sánchez

Plants as Medicine and Aromatics: Uses of Botanicals
Edited by Mohd Kafeel Ahmad Ansari, Mushtaq Ahmad, and Gary Owens

Exploring Traditional Wild Edible Plants
Edited by Vibhor Agarwal, Sachidanand Singh and Rahul Datta

For more information about this series, please visit: www.routledge.com/Exploring-Medicinal-Plants/book-series/CRCEMP

Exploring Traditional Wild Edible Plants

Edited by
Vibhor Agarwal
Sachidanand Singh
Rahul Datta

CRC Press
Taylor & Francis Group
Boca Raton London New York

CRC Press is an imprint of the
Taylor & Francis Group, an **informa** business

First edition published 2025
by CRC Press
2385 NW Executive Center Drive, Suite 320, Boca Raton FL 33431

and by CRC Press
4 Park Square, Milton Park, Abingdon, Oxon, OX14 4RN

CRC Press is an imprint of Taylor & Francis Group, LLC

© 2025 selection and editorial matter, Vibhor Agarwal, Sachidanand Singh and Rahul Datta; individual chapters, the contributors

Library of Congress Cataloging-in-Publication Data
Names: Agarwal, Vibhor, editor. | Singh, Sachidanand, editor. | Datta, Rahul, editor.
Title: Exploring traditional wild edible plants / edited by Vibhor Agarwal, Sachidanand Singh and Rahul Datta.
Description: First edition. | Boca Raton, FL : CRC Press, 2025. |
Series: Exploring medicinal plants | Includes bibliographical references and index. | Summary: "Wild edible plants are native species that grow and reproduce naturally in their natural habitats without domestication. These plants can serve as a healthier alternative to farmed crops that may be heavily laden with pesticides and other poisonous substances. This book focuses on assessment of the nutritional value, potential health benefits, and mechanisms of action of various wild edible plants. It presents information on nutrients and bioactive ingredients that can have health advantages, including antioxidant properties, antimicrobial, anti-inflammatory, and antidiabetic effects. Features: Comprehensive exploration of potential benefits as well as side effects of wild edible plants. Special emphasis is placed on diversity, category, and pharmacologi-cal values of wild edible plants. Discusses challenges regarding the usage of wild edible plants such as overharvesting and habitat destruction, safety, and toxicity. A volume in the Exploring Medicinal Plants series, this book highlights bioactive compounds and therapeutic efficacies of various wild edible plant species. It is useful reading for scientists, researchers, and academia interested in the health benefits of wild edible plants as well as plant scientists"-- Provided by publisher.
Identifiers: LCCN 2024011072 | ISBN 9781032498867 (hbk) | ISBN 9781032498898 (pbk) | ISBN 9781003395935 (ebk)
Subjects: LCSH: Wild plants, Edible. | Medicinal plants.
Classification: LCC QK98.5.A1 E97 2025 | DDC 581.6/32--dc23/eng/20240521
LC record available at https://lccn.loc.gov/2024011072

ISBN: 978-1-032-49886-7 (hbk)
ISBN: 978-1-032-49889-8 (pbk)
ISBN: 978-1-003-39593-5 (ebk)

DOI: 10.1201/9781003395935

Typeset in Times
by SPi Technologies India Pvt Ltd (Straive)

Contents

Preface

This book focuses on the latest developments and applications of wild edible plants with a special emphasis placed on diversity, category, and pharmacological values of wild edible plants. The book particularly facilitates the various studies and knowledge of wild edible plants, focusing on their bioactive components. Many wild and cultivated food plant species that have medicinal, nutraceutical, and pharmaceutical constituents are excellent sources of active phytochemicals with importance in the prevention of different diseases. The book highlights the therapeutic use of herbal medicinal products and supplements, which has increased rapidly over the past few decades. Most of the drugs formulated with herbal compounds are often viewed as side-effect free, balanced, and moderate in curing diseases. However, it is not to ignore that a few wild edible plants have certain toxic substances too and people must be aware of the use of certain drugs that are being produced through synthetic means. In order to cure ailments through natural sources, the therapeutic knowledge of more plants is needed. In this book, we attempt to remove the uncertainty arising out of insufficient scientific proof and documentation about the bioactive compounds and therapeutic efficacy and potential of wild edible plant species.

We have the honor of working with a group of respected researchers in the disciplines of chemistry, botany, pharmacology, pharmacognosy, and biotechnology who have generously contributed their expertise and knowledge to our project. Their inputs have been crucial not only to enhance the book but also to transform this book into a trustworthy and knowledgeable resource for comprehending and utilizing the power of wild edible plants. We, the editors, hope that readers who are curious about the potential of wild edible plants will find this book to be a useful resource. We sincerely believe that by implementing the ideas presented in these pages, we may take significant steps toward a healthy future.

Finally, we express our sincere appreciation to the authors and hardworking staff at Taylor & Francis Group/CRC Press for their constant support and dedication to this undertaking. We are indebted to Dr. Azamal Husen, the series editor, *Exploring Medicinal Plants*, for providing constant support throughout the compilation of the book. We also want to thank the readers for inspiring us to establish this extensive resource because of their curiosity and need for information. Let's explore the incredible potential of wild edible plants to alter our lives as we set off on a journey of empowerment and discovery together.

Editors

Dr. Vibhor Agarwal is currently an Assistant Professor in the School of Science and Mathematics at Emporia State University. Previously, he served as a visiting assistant professor at The College of Wooster and as a postdoctoral researcher at the University of Dayton. He earned a PhD from The Ohio State University and an MSc from Indian School of Mines. He has been the recipient of several prestigious awards such as the Junior Heiskanen Award from OSU Graduate School and Graduate Fellowship from OSU Graduate School. He is also the recipient of Erasmus Mundus scholarship for a semester study abroad program at KTH Stockholm. He has several research papers in international journals of repute in field of GIS and remote sensing application in glaciology and hydrogeology. He has also collaborated on research related to soil sciences, plant science, and food production.

Dr. Sachidanand Singh is currently an associate professor and Head, Department of Biotechnology, School of Energy and Technology at Pandit Deendayal Energy University, India. He has completed B.Tech Biotechnology, M.Tech Bioinformatics, and PhD in Bioinformatics. He has a postdoctoral research experience from the College of Pharmacy, The Ohio State University, USA. His specialization area includes plant biotechnology, systems biology, and drug design. He has more than 12 years of research and academic experience. Dr. Singh has served for a decade in the School of Biotechnology and Health Sciences, Karunya University, Coimbatore, Tamil Nadu, India. He also worked as a consultant for the College of Pharmacy, The Ohio State University in statistical genetics during 2018. He has published more than 38 research articles in peer-reviewed journals and completed one SERB, Government of India funded project and two University research grant projects. Dr. Singh has received several awards in his area of research and for his teaching strategies. He has organized different conferences and workshops and received government funds for the same. He has shared his area of expertise with different colleges and universities as an invited speaker at different conferences and workshops. He has guided 36 B.Tech and 10 M.Tech projects. Presently six PhD students are enrolled under him and two completed PhDs are working in the field of herbal products, drug design, and integration with network biology. Presently he has completed two book proposals on phytopharmaceutical and green technology under the umbrella of Taylor & Francis Group and Springer respectively.

Dr. Rahul Datta is an assistant professor in the field of soil science currently at Mendel University in Brno, Czech Republic. His research has focused largely on increasing crop productivity using "green" and sustainable methods. Some of the most notable problems he has addressed through his research are drought stress, heavy metal toxicity in soil, and climate change greenhouse gas. His research has helped in increasing the quality and productivity of staple crops like wheat, maize, barley, rice, and other crops like mangoes, spinach, and cotton. His work has been cited in excess of 3000 times by researchers from all over the world, including research groups at major national institutions in the United States, such as the USDA. He has reviewed over 400 papers and published around 115 articles in peer-reviewed journals. Additionally, he is an editorial board member of the journals BMC Plant Biology, Sustainability (MDPI), BMC Plant Biology, and Open Agriculture, in which he is overseeing a special issue on current trends in agriculture (MDPI).

Contributors

Milica Aćimović
Institute of Field and Vegetable Crops Novi Sad
 – National Institute of the Republic of Serbia
Novi Sad, Serbia

Duygu Ağagündüz
Department of Nutrition and Dietetics
Gazi University
Ankara, Türkiye

Vibhor Agarwal
School of Science and Mathematics
Emporia State University
Emporia, Kansas, USA

Muhammad Shoaib Amjad
Department of Botany
Women University of Azad Jammu & Kashmir
Bagh, Pakistan
and
Birmingham Institute of Forest Research
University of Birmingham
Birmingham, United Kingdom

Daniel Dias Rufino Arcanjo
Departamento de Biofísica e Fisiologia
Universidade Federal do Piauí
Teresina, Brazil

Robina Aziz
Department of Botany
Government College
Women University
Sialkot, Pakistan

Sara Léa Fortes Barbosa
Departamento de Biofísica e Fisiologia
Universidade Federal do Piauí
Teresina, Brazil

Megha Barot
Research and Development Cell
Parul University
Vadodara, India
and

Department of Environmental Science
Parul Institute of Applied Sciences
Parul University
Vadodara, India

Zakia Binish
Department of Botany
Women University of Azad Jammu & Kashmir
Bagh, Pakistan

Ria Cahyaningsih
Research Center for Plant Conservation,
 Botanic Gardens and Forestry
National Research and Innovation Agency
Kabupaten Bogor, Indonesia

Iolanda Souza do Carmo
Laboratório de Geoquímica Orgânica
Universidade Federal do Piauí
Teresina, Brazil

Özge Cemali
Department of Nutrition and Dietetics
Trakya University
Edirne, Türkiye

Antônia Maria das Graças Lopes Citó
Laboratório de Geoquímica Orgânica
Universidade Federal do Piauí
Teresina, Brazil

Angélica Gomes Coêlho
Departamento de Biofísica e Fisiologia
Universidade Federal do Piauí
Teresina, Brazil

Francisco Valmor Macedo Cunha
Centro Universitário UNINOVAFAPI
Teresina, Brazil

Srijit Das
Department of Human & Clinical Anatomy
College of Medicine & Health Sciences
Sultan Qaboos University
Muscat, Oman

Hiral S. Desai
Department of Biotechnology, Smt. S.S Patel
    Niitan Science and Commerce College
Sankalchand Patel University
Visnagar, India

Asia Farooq
Department of Botany
Women University of Azad Jammu & Kashmir
Bagh, Pakistan

Lilla Nur Firli
Universitas Jember
Jember, Indonesia

Yogesh Godiyal
Department of Pharmacognosy and
    Phytochemistry
School of Pharmaceutical Sciences
Delhi Pharmaceutical Sciences and Research
    University (DPSRU)
New Delhi, India

Allah Bakhsh Gulshan
Department of Botany
Ghazi University
Dera Ghazi Khan, Punjab, Pakistan

Mumtaz Hussain
PARC, Arid Zone Research Institute
Bahawalpur, Pakistan

Sneha Joshi
Department of Pharmaceutical Chemistry
PCTE Group of Institution
Ludhiana, India

Yusof Kamisah
Department of Pharmacology
Faculty of Medicine
Universiti Kebangsaan Malaysia
Kuala Lumpur, Malaysia

Allah Nawaz Khan
Department of Botany
University of Agriculture Faisalabad
Punjab, Pakistan

Noopur Khare
Department of Biotechnology
Bhai Gurdas institute of Engineering and
    Technology
Sangrur, India

Pragati Khare
Department of Pharmacy
Bhagwant University
Ajmer, India

Biljana Lončar
Faculty of Technology Novi Sad
University of Novi Sad
Novi Sad, Serbia

Muhammad Majeed
Department of Botany
University of Gujrat
Gujrat, Pakistan

Wali Muhammad Mangrio
Department of Zoology
Faculty of Natural Sciences
Shah Abdul Latif University
Khairpur, Pakistan

Murad Muhammad
State Key Laboratory of Desert and
    Oasis Ecology
Xinjiang Institute of Ecology and
    Geography
Chinese Academy of Sciences
Urumqi, China

Alice Nabatanzi
Department of Plant Sciences, Microbiology
    and Biotechnology
Makerere University
Kampala, Uganda

Bernardo Melo Neto
Faculdade de Ciências Médicas
Universidade Federal do Piauí
Teresina, Brazil

José de Sousa Lima-Neto
Departamento de Farmácia
Universidade Federal do Piauí
Teresina, Brazil

Dimitrios D. Ntakoulas
Department of Chemistry
National and Kapodistrian University
    of Athens
Athens, Greece

Ari Satia Nugraha
Universitas Jember
Jember, Jawa Timur, Indonesia

Lívio César Cunha Nunes
Programa de Pós Graduação em Ciencias
  Farmaceuticas
Universidade Federal do Piauí
Teresina, Brazil

Ioannis N. Pasias
General Chemical Lab of Research and Analysis
University of Athens
Athens, Greece

Charalampos Proestos
Department of Chemistry
National and Kapodistrian University of Athens
Athens, Greece

Reza Yuridian Purwoko
Research Center for Pre-Clinical and Clinical
  Research
National Research and Innovation Agency
  Republic of Indonesia
Jarkarta, Indonesia

Huma Qureshi
Department of Botany
University of Chakwal
Chakwal, Pakistan

S. Srinivasa Rao
Department of Human & Clinical Anatomy
College of Medicine & Health Sciences
Sultan Qaboos University
Muscat, Oman

Hakim Ali Sahito
Department of Zoology
Faculty of Natural Sciences
Shah Abdul Latif University
Khairpur, Pakistan

Watchara Sangsopha
Chulabhorn Research Institute
Bangkok, Thailand

Brenda Nayranne Gomes dos Santos
Programa de Pós Graduação em Ciencias
  Farmaceutica
Universidade Federal do Piauí
Teresina, Brazil

Sachidanand Singh
Department of Biotechnology
School of Energy and Technology
Pandit Deendayal Energy University
Gandhinagar, India

Hawa Nordin Siti
Faculty of Medicine
Universiti Sultan Zainal Abidin
Terengganu, Malaysia

Devesh Tewari
Department of Pharmacognosy and
  Phytochemistry
School of Pharmaceutical Sciences
Delhi Pharmaceutical Sciences and Research
  University (DPSRU)
New Delhi, India

Konstantina Tsikrika
Department of Chemistry
National and Kapodistrian University
  of Athens
Athens, Greece

Phurpa Wangchuk
College of Public Health, Medical and
  Veterinary Sciences
James Cook University
Smithfield, Queensland, Australia

Zhihong Xu
College of Arts & Sciences
University of Pikeville
Pikeville, Kentucky, USA
Institute of Interventional & Vascular Surgery
Tongji University
Shanghai, China

1 Ethnobotanical Review of Wild Edible Plants

Hiral S. Desai, Sachidanand Singh, Vibhor Agarwal and Megha Barot

1.1 INTRODUCTION

The scientific study of a social group's botanical knowledge and how they use locally growing plants for clothing, food, medicine, and religious rituals is known as ethnobotany. It entails gathering information about how various cultural groups use wild plants and examining their beliefs and understanding about them (Srivastava et al., 2023). The term "Ethnobotany" was coined by J.W. Harshberger to indicate plants used by the aboriginals: "ethno" – study of people and "botany" – study of plants.

Since the dawn of humanity, indigenous peoples have evolved their own localized expertise in managing plant usage and conservation. This sophisticated information, belief structures, and customs are together referred to as indigenous knowledge or traditional knowledge (Fenetahun and Eshetu, 2017). According to this perspective, ethnobotanical studies are helpful in analyzing, documenting, and conveying knowledge about the interactions between biodiversity and human society, including how natural diversity is used and modified by human activity (Martin, 1995). The ethnobotanical approach is crucial because it incorporates local groups in biodiversity protection. This is predicated on the notion that local people, who manage many species about which science knows little, are in charge of the world's healthiest ecosystems (Eshete et al., 2016).

Throughout history, people have utilized wild plants for a variety of purposes, including food, medicine, and industrial uses. They contribute greatly to the cultural history of numerous societies, and their traditional wisdom is frequently passed down orally from one generation to the next generation.

1.2 MEDICINAL WILD EDIBLE PLANTS

The nutritional and therapeutic benefits of medicinal wild plants have made them a priceless resource that have been utilized by many societies throughout history. The plants possess chemotherapeutic, bacteriostatic, and antimicrobial agents (Gangwar et al. 2022). The majority of secondary metabolites produced by plants are a substantial source of microbicides, pesticides, and a variety of medicine (Kumar et al., 2011). WHO estimates that 80 percent of people on Earth, especially in underdeveloped nations, still rely on plant-based medications for their primary medical care (Gahlawat et al. 2014.). As a result, interest in many traditional natural products has increased.

DOI: 10.1201/9781003395935-1

TABLE 1.1

Wild Edible Plants Used as Medicine

Name	Family	Part Used	Medicinal Uses	References
A.indica	Meliaceae	Leaves	Snakebite & scorpion sting	Sharma and Kumar (2011); Saqib and Gul (2018).
A.catechu	Mimosaceae	Bark, flower tops	Gonorrhoea	
T. undulata	Bignoniaceae	Bark, branch	Syphilis	
A.barbedensis	Liliaceae	Leaves	Sexual vitality	
R.communis	Euphorbiaceae	Seeds	Birth control	
Aloe vera	Liliaceae	Central tissue of leaf	Stomach disorder, moisturizing agent	Kumar (2014)
Melilotus officinalis	Fabaceae	Whole plant	Insomnia, painful menstruation, hemorrhoids	
Foeniculum vulgare	Umbelliferae	Fruit	Gastrointestinal disorders, asthma, cough	
Anethum graveolens	Umbelliferae	Whole herb	Leaves used in ulcer, oil used in hyperacidity, flatulence	
Morus alba	Moraceae	Fruit, leaves, stems, bark	Leaves used in treatment of cold, sore throat, eye infection, and nosebleeds, stem used in high bloodpressure, rheumatic pain, spasm Tincture made from bark relieves toothache Fruit used in urinary incontinence, diabetes, dizziness, tinnitus	
Hibiscus rosa-sinensis	Malvaceae	Whole plant	Flower extract used in liver disorders, high blood pressure	
Ocimum basilicum	Labiatae	Leaves, seeds	Leaves used in fevers, abdominal cramps Water boiled with leaves used in sore throatLeaves also used in respiratory disorders Chewing leaves can act as a protection against stress, ulcer, mouth infection, plant also used in reduction of blood cholesterol level	
Cannabis sativa	Cannabaceae	Flowers, leaves, seed	Asthma, cystitis, dysentry, gout, epilepsy, malaria, also used in treatment of anorexia nervosa	
Asparagus officinalis	Lilaceae	Rhizome, young shoots	Boost fertility, help pregnant women against neural tube defects in infants, laxative, sedative, diuretic	

(Continued)

TABLE 1.1 (Continued)
Wild Edible Plants Used as Medicine

Name	Family	Part Used	Medicinal Uses	References
Tragopogon carcifolius Boiss	Asteraceae	Leaf	Chronic cough	Asadbeigi, Mohsen, et al. (2014)
Altheae hirsuta L.	Malvaceae	Root	Pulmonary infection	
Lamium album L. & Origanum vulgare	Lamiaceae	Flowering shoot	wound healing	Asadbeigi, Mohsen, et al. (2014)
ScrophulariadesertiDel. & ScrophulariastriataBoiss	Scrophulariaceae	Leaves and stem	Cut and burn restoration	
Rumexacetosa	Polygonaceae	Leaves	Skin lesions	
Ficus glumosa Delile, Ficus natalensis Hochst, Ficus platyphylla Delile, Ficus saussureana	Moraceae	Steam bark	Used in TB	Obakiro et al. (2020).
Murraya koenigii	Rutaceae	Whole plant	Diabetes mellitus, leucoderma, body aches, kidney pain, vomiting Leaves used in bruises and eruption, night blindness, bite of poisonous animals	Gahlawat et al. (2014).
Acalypha indica L.	Euphorbiaceae	Leaves	Eczema of hands, sole on legs, burning area, ringworm	Anand, Uttpal, et al. (2022)
Eclipta prostrate L.	Asteraceae	Whole plant	Herpes, infective hepatitis	
Cissampelos pareira L.	Menispermaceae	Root	Pruritus, skin disorder	
Wrightia tinctoria R.	Apocynaceae	Leaves	Psoriasis	
Nerium oleander L.	Apocynaceae	Steam bark, fruit	Steam bark used in ear ache Fruit used in ear pain and rash	
Rotheca serrata	Lamiaceae	Leaves	Eyelid inflammation	
Curcuma longa L.	Zingiberaceae	Rhizome	Premature skin wrinkles, excessive melanin secretion	
Centella astica	Apiaceae	Whole plant	Syphilis	

1.3   WILD PLANTS THAT CAN BE USED AS A SUBSTITUTE FOR FOOD

The difference between the increase of the human population and the supply of food is growing significantly every day. Furthermore, a lot of individuals typically ate diets that consisted of cereal, which led to nutrient shortages (Shaheen et al. 2017). As a result, it has proven to be extremely challenging to provide stable and nourishing food sources for future generations. The experts stressed the need to look for some other food sources in order to satisfy the population's need for food and nourishment. Alternative food sources that do not contain conventional meals must be wholesome and easily accessible. This has led to the development of wild edible plants as a prominent food substitute.

1.3.1   How to Identify Wild Plants That Are Edible

1. Find a habitat for wild edible plants.
2. Prior knowledge of wild edible plant species.
3. Always act on your own initiative, even on a little scale.
4. Switch to different locations of the country.
5. Look for berries on trees.
6. Look for nuts underneath the trees.
7. Seek trees that bear fruit.
8. Look for plants close to water sources.
9. Be cautious around poisonous blooms.
10. Find some vines.
11. Keep a lookout for deciduous and conifer leaves.
12. Check unfamiliar wild edible plants for toxicity.
13. Gather abundantly growing plants.
14. Avoid looking around trash.
15. Adequately clean the plant.

Avoid the plant if it possesses any of the following characteristics:
- Sap that is milky or tarnished.
- Beans, bulbs, or seeds in the pods.
- Spines, adequate hairs, or thorns.
- Carrot, parsnip, or parsley like foliage.
- Woody components and leaves with an almond scent.
- Pink, purple, and black-colored spikes on the heads of the grains.

TABLE 1.2
Wild Edible Plants Used as Food

Botanical Name	Family	Used Part	References
Sambucus ebulus L.	Adoxaceae	Fruits	Akbulut (2022)
Cotinus coggygria Scop.	Anacardiaceae	Fruits	
Arum dioscoridis Sm.	Araceae	Leaves, roots	
Carduus nutans L.	Asteraceae	Stem	
Berberis vulgaris L.	Berberidaceae	Fruits	
Nasturtium officinale R.	Brassicaceae	Aerial parts	
Vaccinium myrtillus L.	Ericaceae	Fruits, leaves	
Juglans regia L.	Juglandaceae	Seeds	
Origanum onites L.	Lamiaceae	Flowering steam, leaves	
Cerasus mahaleb L.	Rosaceae	Seeds	
Smilax excels L.	Smilacaceae	Fresh shoot	

(Continued)

TABLE 1.2 (Continued)
Wild Edible Plants Used as Food

Botanical Name	Family	Used Part	References
Amorphophallusbulbilfer (Roxb) BI. *Homalomena aromatic Schott*	Araceae	Stem, leaf, rhizome, flowers, whole plant	Deb, Dipankar, et al. (2013); Łuczaj et al. (2019)
Amomum spp.	Zingiberaceae	Rhizome	
Melocana baccifera Trin.	Poaceae	Shoot	
Murraya koengii ex L.	Rutaceae	Leaf	
Monochoria hastate (L.) Solms.	Pontederiaceae	Whole plant	
Utrica dioica L.	Urticaceae	Leaves used as vegetables	Thakur, Arti, Somvir Singh, and Sunil Puri et al. (2020)
Malva verticillata L.	Malvaceae	Aerial parts, used as vegetables	
Rubus niveus Thumb.	Rosaceae	Fruit	
Diplazium esculentum (Retz) Sw.	Aspidiaceae	Aerial parts, used as vegetables and pickled	
Rhododendrom Sm.	Ericaceae	Flower, used as chutney	
Allium carnatum L. *A.montanum L.* *A. sphaerocephalon L.* *A. ursinum L.A. victorialis L.* *A. vineale L.*	Liliaceae	Ground part and leaves as salad	Jman Redzic (2006); Garekae and Shackleton et al. (2020)
Alchemilla hybrid Rothm *A. plicatual Gaud* *A. xanthochlora Rothm*	Rosaceae	Young leaves as vegetables	
Amaranthus spinosis L.	Amaranthaceae	Aerial parts	Shad, Shah, and Bakht et al. (2013)
Astragalus anisacanthus Boiss	Leguminosae	Young shoots	
Bauhinia variegate L.	Caesalpinaceae	Bark and flower heads	
Oxallis stricta L. *Oxallis corniculata L.*	Oxalidaceae	Aerial parts, flowers and leaves	
Plantago major L.	Plantaginaceae	Whole plant	

1.3.2 BENEFITS OF WILD EDIBLE PLANTS

- Play an important role in supplying the tribal population's nutritional needs.
- Satisfy dietary requirements during food crises, giving rise to the idea of "famine foods" or plants ingested only during times of food crisis.
- Offer supplemental nutritional support.
- Provide an alternate funding source.
- Wild-food dishes can be nutritious meals.
- Wild plants have considerable potential for hybridization and selection, as well as offering a reservoir of genetic resources, which can be used to domesticate novel food crops.

1.3.3 HAZARDS OF WILD EDIBLE PLANTS

The gathering and consumption of wild edible plants carries some risks. Think about the following before gathering wild plants for food:

- Are pesticides or herbicides being used to treat the area?
- Is there a major road nearby or any other pollution sources?
- Is there any toxic variation of your target species?

1.4   MUSHROOMS AS A FOOD SOURCE

The fleshy, spore-bearing fruiting body of a fungus that is often developed above ground, on soil, or on its food supply is known as a mushroom or toadstool.

Mushrooms are employed as medicinal foods because their chemical makeup largely protects against diseases including cancer, atherosclerosis, hypertension, and high cholesterol (Waktola and Temesgen, 2018). Water (90%), protein (2–40%), fat (2–8%), carbs (1–55%), fibre (3–32%), and ash (Barros et al., 2009; Barros et al., 2010c; Barros et al., 2010a) make up the majority of a mushroom's composition. Mushrooms are more nutritious than fish or beef because they include eight essential amino acids, polyunsaturated fatty acids, and just trace amounts of saturated fatty acids (Waktola and Temesgen, 2018). Additionally, they contain a significant number of vitamins, including folic acid, ascorbic acid, 92–144 mg/100g dry weight basis, riboflavin, 6.7–9.0 mg/100g, niacin, 60–67.3 mg/100g, and thiamine, 1.4–2.2 mg/100 g (Kakon et al. 2012). There are more than 2000 different edible varieties of mushrooms, with white button mushrooms the most prominent mushroom in the world. Other commercially important species include enoki (*Flammmulina ostreatus*), shiitake (*Lentinus edodes*), straw (*Volvariella volvacea*), oyster (*Pleurotus ostreatus*), and lentinus (*Lentinus edodes*). Species that are only present during certain times of the year include chanterelles and morels (*Morchella esculenta*) (Feeney et al., 2014).

1.4.1   THE NUTRITIONAL BENEFITS OF MUSHROOMS

1.4.1.1   Carbohydrate

On a dry weight basis, the carbohydrate content of mushrooms accounts for 50 to 65% of the fruiting bodies, according to Florezak et al. (2004). By dry weight, *Coprinus atramentarius* can accommodate 24% of carbohydrates. The predominant free sugar is mannitol, also referred to as mushroom sugar, which makes up about 80% of the total. According to Mc Connell and Esselen (1947), a fresh mushroom has 0.9% mannitol, 0.28% reducing sugar, 0.59% glycogen, and 0.91% hemicellulose. According to Crisan and Sands' (1978) analysis on *Agaricus bisporus*' carbohydrates, raffinose, sucrose, glucose, fructose, and xylose predominate. Mushroom polysaccharides that are water soluble have anticancer properties (Kakon et al. 2012).

TABLE 1.3
Nutritional Composition of Mushrooms

Species	Initial Moisture	Crude Protein Nx4.38	Fat	Carbohydrate Fiber Total	N-free	Ash	Energy	Value (K.cal)
Agaricus bisporus	89.5	26.3	1.8	59.96	49.5	10.4	12.0	328
Volvariella volvacea	88.0	29.5	5.7	60.0	49.6	10.4	4.8	374
Lentinula Modes	90.0	17.5	8.0	67.5	59.5	8.0	7.0	387
Pleurotus ostreatus	73.7	10.5	1.6	81.8	74.3	7.5	6.1	367
Auricularia polytricha	87.1	7.7	0.8	87.6	73.6	14.0	3.9	347
Flaninnilbia velutipes	89.2	17.6	1.9	73.1	69.4	3.7	7.4	378
Pholiota naineko	95.2	20.8	4.2	66.7	60.4	6.3	8.3	372
Trcinclla fuciforinis	19.7	4.6	0.2	94.8	93.4	1.4	0.4	412

Source: Crisan, E. V., and Anne Sands. "Nutritional value." *The biology and cultivation of edible mushrooms* (1978): 137–168. Academic Press, Inc.

1.4.1.2 Protein

The substratum's composition, the pileus' size, the time of harvest, and the type of mushroom all have an impact on the protein content of mushrooms (Waktola and Temesgen 2018). Compared to fruits and vegetables, mushrooms have 20–35% higher protein by dry weight. Mushrooms are also known as white vegetables or "boneless vegetarian meat" (Kakon et al. 2012). According to Haddad and Hayes (1979), the protein content of *A. bisporus* mycelium varied from 32 to 42% on a dry weight basis. According to Verma et al. (1987), mushrooms are highly beneficial for vegetarians since they include several important amino acids that are found in animal proteins.

1.4.1.3 Fats

Mushrooms' fat content is less than their protein and carbohydrate content. Mushroom fat percentage varies by species, ranging from 2.04% in *Suillus granulatus*to 3.66% in *Suillus luteus* and 2.32% in *A. campestris* (Waktola and Temesgen, 2018). The fat fraction of mushrooms is mainly composed of unsaturated fatty acids (Waktola and Temesgen, 2018).

1.4.1.4 Vitamins

In particular, vitamin B and a tiny amount of vitamin C are abundant in mushrooms. Wild mushrooms contain more vitamin D2 than dark, cultivated *Agaricus bisporus* (Waktola and Temesgen, 2018). Mushrooms are very poor in vitamins A, D, and E.

1.4.1.5 Mineral Constituents

The high quantity of efficiently digested mineral components in mushroom fruiting bodies sets them apart from other organisms. K, P, Na, Ca, Mg, and trace elements including Zn, Cu, Fe, and Cd are the major mineral components found in mushrooms. About 56 to 70% of the overall ash content of mushrooms is made up of K, P, Na, and Mg, with potassium making up 45% of the total ash (Kakon et al. 2012). It has been found that mushrooms can accumulate heavy metals such as cadmium, lead, arsenic, copper, nickel, silver, chromium, and mercury. The mineral content is dependent on the species, age, diameter of the fruiting body, and substratum (Kakon et al. 2012).

1.4.1.6 Fiber

Mushrooms are a good source of fiber, which helps to lower cholesterol. Both soluble and insoluble fibers are found in mushrooms. By decreasing the amount of both LDL and total cholesterol and controlling blood sugar levels, soluble fiber helps prevent heart disease (Kakon et al. 2012).

TABLE 1.4

Essential Amino Acids in 100 gm Dry Mushroom

Ess.Amino Acids	Volvereilla volvacea	Pleurotus ostreatus	Agaricus edodes	Pleurotus sanjorcaju	Pleurotus florida	Agaricus bisporus
Histidine	3.8	1.7	2.7	2.2	2.8	2.7
Leucine	4.5	6.8	7.5	7.0	7.5	7.5
Methionine	1.1	1.5	0.9	1.8	3.0	0.9
Isoleucine	3.4	4.2	4.5	4.4	5.2	4.5
Threonine	3.5	4.6	5.5	5.0	6.1	5.5
Valine	5.4	5.1	2.5	5.3	6.9	2.5
Lysine	7.1	5.7	9.1	5.7	9.9	9.1
Tryptophan	1.5	1.2	2.0	5.0	1.1	2.0

Source: Crisan, E. V., and Anne Sands. "Nutritional value." *The biology and cultivation of edible mushrooms* (1978): 137–168. Academic Press, Inc.

1.5 WILD EDIBLE PLANTS FOR DRUGS

Humans have utilized plants as medicine for very long time. There are numerous sources of novel bioactive chemicals today, including plants, fungus, bacteria, and marine species. In particular, in the field of antibacterial, antifungal, antiparasitic, and antiviral medicines, 61% of the 877 novel small molecule chemical compounds created between 1981 and 2002 were derived from natural materials (Ferreira et al. 2016).

Compounds derived from plant sources fall into one of six categories:

- Bioactive substances that can be administered directly as medications, such as digoxin, which is used to treat cardiac issues.
- Bioactive substances having chemical structures that could serve as a precursor to more effective medications, such as the mitotic inhibitor paclitaxel, which is utilized in cancer chemotherapy.
- Chemophores, which are energy-transmitting cells that can be altered to produce drugs.
- Unadulterated phytochemicals that can be used as benchmarks to standardize unprocessed plant material.
- Plant compounds with potential as pharmacological agents.
- Green tea extracts or herbal extracts used as botanical medicines (Katiyar et al. 2012).

Preclinical research using an animal model is typically initiated once in vitro experiments of a particular drug demonstrate bioactivity (Jain et al. 2022). If permitted, research on humans are conducted to ascertain the toxicity, side effects, and other consequences that cannot be seen in animal models. Before a molecule is marketed as a medicine, years or even decades may pass due to the fact that the process for approving a novel drug is hardly ever linear and has various downsides (Paul et al. 2010).

There are several methods for looking for chemicals in plants:

- Arbitrary selection using chemical screening.
- Incidental selection, followed by a single or a combination of biological assays (in vitro tests that screen a large number of plant species in search of novel drugs).
- Monitoring reports of biological activity.
- Continued research on the use of botanicals for medicine.
- Utilizing databases (Fabricant and Farnsworth, 2001).

TABLE 1.5
Natural Product-Based Drugs

Plant Name	Active Principle	Used Part	Application	Reference
Atropa belladonma L. *Callistemon citrinus Curtis*	Tiotropium Nitisinone	Aerial parts	Chronic obstructive pulmonary disease Tyrosinemia	Balunas et al. (2005)
Artemisia annua L.	Artemisinin	Aerial parts	Antimalarial	Phillipson, 2007; Pohl et al., 2018
Betula spp. L.	Betulinic acid	Bark	Melanoma, anticancer, antimalarial, anti HIV, anthelmintic	Balunas and Kinghorn, 2005; Umar et al., 2020
Calophyllum lanigerum W.	Calanolide	Aerial parts	Anti HIV	Balunas and Kinghorn, 2005

(Continued)

TABLE 1.5 (Continued)
Natural Product-Based Drugs

Plant Name	Active Principle	Used Part	Application	Reference
Callistemmon citrinus Curtis	Nitisinone	Aerial parts	Tyrosinemia	Balunas and Kinghorn, 2005; Parasuraman, 2018
Combretum caffrum (Eckl.& Zeyh.) Kuntze	Combretastain A4 phosphate	Aerial parts	Anaplastic thyroid cancer	Ramawat & Mérillon, 2008; Thomford et al., 2018.
Euphorbia peplos L.	Ingenol 3- angelate	Sap	Skin conditions	Ramawat & Mérillon, 2008; Mehrbod et al., 2021
Galega officinalis L.	Guanidine derivatives	Aerial parts	Type 2 diabetes	Ramawat & Mérillon, 2008; Shedoeva et al., 2019.
Podophyllum peltatum L.	Etoposide	Root	Small cell lung cancer, lymphomas	Phillipson, 2007; Kooti et al., 2017.
Plectranthus barbatus Andrews	Colforsin daropate	Aerial parts	Anticancer	Butler, 2005; Kooti et al., 2017.

1.6 CANCER DRUG FROM WILD EDIBLE PLANTS

A complex illness, cancer exhibits aberrant cell proliferation along with the invasion of neighbouring cells and tissues. Smoke, cigarettes, ionizing radiation, heavy metal exposure, environmental stress, microbial infection, food adulteration, and multiple hereditary factors all contribute to the cumulative accumulation of various genetic mutations that lead to cancer (Upadhyay, 2018). Depending on the position and stage of the tumor, several treatment techniques, including immunotherapy, radiation, chemotherapy, tumor surgery, cancer vaccines, stem cell transformation, and others, have been used recently. However, these sorts of therapies and medications also have an impact on healthy body cells that are rapidly reproducing in a normal environment (Al Kaabi et al., 2022). These kinds of side effects support the need for a more natural approach to treating cancer. Herbal medicines have strong immunomodulatory effects and work by boosting both general and specific immunity (Kumar et al., 2011).

TABLE 1.6
Medicinal Plants Having Anticancer Activity

Plant Name	Parts Used	Family
Aphanamixis polystachya	Bark	Meliaceae
Asparagus racemosus	Root	Liliaceae
Acorus calamus	Rhizome	Araceae
Alium cepa	Bulb	Liliaceae
Bauhinia variegata	Root	Caesalpinaceae
Butea monosperma	Bark	Fabaceae
Bacopa monnieri	Whole plant	Scropulariacae
Cajanus cajan	Leaves	Fabaceae
Calotrophis gigantea	Whole plant	Asclepiadaceae
Citrus medica	Root	Rutaceae
Clerodendrum serratum	Root	Verbanaceae
Cissus quadrangularis	Whole plant	Vitaceae

(Continued)

TABLE 1.6   (Continued)
Medicinal Plants Having Anticancer Activity

Plant Name	Parts Used	Family
Clerodendrum viscosum	Leaves	Verbanaceae
Catunaregum spinosa	Bark/fruit	Rubiaceae
Cassia absus	Leaves	Caesalpinaceae
Cassia senna		
Citrullus colocynthis	Root	Cucurbitaceae
Daucus carota	Root	Apiaceae
Flacourtia jangomos	Bark/leaf	Flacourtiaceae
Embelia ribes	Fruit	Myrsinaceae
Kaempferia galanga	Rhizome	Zingiberaceae
Jatropha curcas	Leaves, seeds	Euphorbiaceae
Limonia acidissima	Fruit	Rutaceae
Lanatn camara	Whole plant	Verbanaceae
Mimosa puduca	Whole plant	Mimosaceae
Macrotyloma uniflorum	seed	Fabaceae
Nicotina tabacum	Leaves	Solanaceae
Rhinacanthus	Whole plant	Acanthaceae
Operculin turoethum	Root	Convolvulaceae
Tylopora indica	Root, leaf	Asclepiadaceae
Salvadora persica	Bark, shoot, leaf, fruit	Salvadoraceae
Vitex trifolia	Leaf	Verbanaceae
Vernonia cinerea	Whole plant	Asteraceae
Xanthium strumarium	Root	Compositae
Zanthoxylum armatum	Bark, fruit	Rutceae

Source: Dhanamani, M.D.S.L., Devi, S.L. and Kannan, S., 2011. Ethnomedicinal plants for cancer therapy – A review. *Hygeia JD Med*, *3*(1), pp.1–10.

1.7   NUTRITIONAL AND ANTINUTRITIONAL VALUE OF WILD EDIBLE PLANTS

As a key factor in determining health, labor productivity, and mental development, nutrition is a crucial basic necessity (Vadivel and Janardhanan, 2005). Due to population growth and the resulting decrease in the availability of food with high nutritional value, food insecurity and malnutrition have increased significantly in most developing and underdeveloped countries (Vadivel and Janardhanan, 2005). Wild edible plants are a good source of nutrients and most people living in rural areas consume it frequently. However, the biggest issue with using these types of plants for nourishment is the presence of poisonous and antinutritional elements (Guil et al., 1997). So, evaluating the nutritional factors of diverse wild edible plants is gaining more and more attention.

TABLE 1.7

Proximate Composition (g/100 g) and Energetic Value (kcal/100 g) of Some WEPs

Plant spp.	Fat	Ash	Moisture	Energy	Protein	Dietry Fiber	Carbohydrates	References
A. nodiflorum	0.10 (0.07–0.14)	1.7 (1.0–3.3)	92.0 (90.0–94.0)	15	1.6 (1.1–2.8)	2.7 (1.9–3.4)	1.2 (0.7–2.1)	García-Herrera et al. (2014)
A. ampeloprasum	0.18 (0.12–0.23)	0.8 (0.5–1.0)	78.3 (76.0–8.15)	79	1.7 (1.2–2.0)	4.2 (3.6–4.7)	16.6 (12.0–20.9)	(García-Herrera et al. 2014, 2013)
A. azurea	0.15 (0.07–0.14)	1.9 (1.8–2.1)	91.2 (88.9–92.7)	18	1.9 (1.1–2.8)	3.9 (3.5–4.4)	1.3 (0.9–1.8)	García-Herrera et al. (2014)
B. officinalis	0.16 (0.13–0.19)	2.4 (2.2–2.5)	86.9 (86.5–87.3)	44	1.2 (1.0–1.4)	5.9 (3.9–9.5)	9.5 (9.2–9.7)	Perreira et al. (2011); Ayeni and Oyeyemi (2021)
C. juncea	0.35 (0.09–0.79)	2.3 (1.4–4.4)	84.6 (65.9–89.7)	30	2.1 (1.9–6.1)	7.8 (4.1–13.4)	2.7 (1.5–9.7)	García-Herrera et al. (2014), Ranfa et al. (2014)
C. intybus	0.13	1.8 (1.7–2.1)	86.4 (84.8–87.9)	33	2.9 (1.5–4.3)	6.1 (5.1–6.7)	3.5 (1.8–4.7)	García-Herrera et al. (2014)
F. vulgare	0.17 (0.08–0.23)	1.9 (1.7–2.3)	86.7 (85.1–90.1)	27	2.1 (0.6–3.8)	5.1 (3.5–6.2)	3.1 (1.4–4.9)	García-Herrera et al. (2014), Trichopoulou et al. (2000); Kumar et al. (2021)
G. headeracea	1.18 (0.95–1.41)	3.47	73.0 (65.0–81.1)	100	1.3	5.1 (3.5–6.2)	21.0 (20.8–21.2)	Barros et al. (2011a); Kumar and Shiddamallayya (2021)
H. lupus	0.20 (0.11–0.26)	5.2 (4.3–6.4)	85.5 (85.2–93.2)	31	4.3 (3.1–5.1)	5.2 (4.3–6.4)	1.6 (1.4–1.8)	Garcia Herrera (2014); Tadele Menigistie (2018)
M. pulegium	0.90 (0.81–0.99)	2.4 (2.4–2.5)	59.5 (50.3–68.7)	157	2.9 (2.7–3.1)	2.4 (2.4–2.5)	5.2 (4.3–6.4)	Tadele Menigistie et al. (2018); Fernandes et al. (2010)
M. sylvestris	0.65 (0.56–0.74)	3.2 (3.1–3.3)	76.3 (75.8–76.8)	85	2.9 (2.7–3.1)	5.2 (4.3–6.4)	16.9 (16.8–17.1)	Barros et al. (2010c); Kumar and Shiddamallayya (2021)
M.fontana	1.66 (0.15–2.15)	1.1 (0.7–1.5)	92.2 (88.9–95.2)	34	1.6 (0.6–2.0)	4.4 (4.0–5.4)	2.0 (0.5–3.3)	García-Herrera et al. (2014), Pereira et al. (2011), Tardio et al. (2011); Adegboyega et al., 2020
N.officinale	0.17 (0.14–0.20)	1.0 (0.9–1.1)	92.2 (91.1–93.2)	24	1.6 (0.9–2.2)	1.8 (1.5–2.0)	4.1 (3.6–4.5)	Pereira et al. (2011), Pinela et al. (2016), Li and Xu (2022)
O. vulgare	2.81 (2.48–3.14)	2.9 (2.8–2.9)	51.8 (46.7–56.9)	195	2.3 (2.2–2.3)	1.8 (1.5–2.0)	40.2 (39.9–40.5)	Barros et al. (2011b)
P. tridentatum	1.05 (0.85–1.25)	0.93	60.8 (60.6–61.0)	158	6.2 (6.0–6.5)	0.9	31.0 (30.7–31.3)	Pinela et al. (2011)
P. rhoeas	0.25 (0.15–0.38)	3.1 (1.9–5.2)	88.1 (68.5–91.0)	37	3.7 (1.5–5.9)	5.5 (2.7–11.1)	3.5 (2.9–5.3)	García-Herrera et al. (2014), Trichopoulou et al. (2000); Padhan and Panda (2020)
P. oleracea	0.39 (0.30–0.44)	1.8	91.8 (91.1–92.3)	25	2.5	0.9	2.7	Cowan et al. (1963), Oliveira et al. (2009); Datta et al. (2019)
R. induratus	0.39 (0.37–0.41)	1.0 (0.9–1.1)	90.1 (90.0–90.3)	37	2.1 (1.3–2.9)	0.9	6.2 (5.5–6.9)	Pereira et al. (2011); Obakiro et al. (2020)
R. ulmifolius	1.05 (0.85–1.25)	0.93	49.19	158	6.2 (6.0–6.5)	0.9	31.0 (30.7–31.3)	Barros et al. (2010e); Ray et al. (2021)
R. acetosella	0.26 (0.23–0.29)	1.2 (1.1–1.3)	89.1 (88.1–90.1)	40	0.9 (0.7–1.1)	0.9	8.6 (8.4–8.8)	Pereira et al. (2011); Seal (2020)
S. marianum	0.o1	1.5 (1.0–1.9)	93.4 (92.9–93.8)	10	0.6 (0.5–0.8)	2.6 (2.3–2.9)	1.1 (0.5–1.7)	García-Herrera et al. (2014)
S. vulgare	0.63 (0.35–0.80)	0.8 (0.3–3.8)	87.1 (80.4–88.5)	33	3.1 (1.9–3.6)	3.6 (2.6–7.0)	2.9 (1.0–3.9)	García-Herrera et al. (2014); Seal (2020)
T. masichina	3.80 (3.70–3.90)	2.7 (2.6–2.8)	54.7 (47.6–61.7)	189	2.2 (2.2–2.3)	7.0 (5.4–8.7)	36.6 (36.6–36.7)	Barros et al. (2010); Ray et al. (2021)
T. obovatum	0.22 (0.19–0.27)	2.1 (1.8–2.5)	83.3 (79.2–86.7)	29	1.6 (1.0–2.1)	7.0 (5.4–8.7)	3.3 (1.6–5.4)	García-Herrera et al. (2014); Seal (2020)
T. communis	0.24 (0.10–0.51)	1.4 (0.9–2.4)	86.2 (83.2–89.0)	36	3.2 (2.5–3.8)	4.7 (3.5–6.0)	4.1 (1.9–11.6)	García-Herrera et al. (2014), Martins et al. (2011); Obidiegwu et al. (2020)

TABLE 1.8

Composition in Individual Sugars (mg/100 g) of WEPs

Plant spp.	Sucrose	Fructose	Glucose	Reference
F. vugare	4710 (4560–4860)	350 (290–410)	4710 (4560–4860)	Barros et al. (2013)
B. officinalis	200 (183–2170)	18 (15–22)	76 (68–84)	Pereira et al. (2011)
G. hederacea	400 (340–460)	150 (140–160)	80 (60–100)	Barros et al. (2011b)
M. fontana	45 (18–79)	36 (28–44)	48 (47–49)	Pereira et al. (2011)
M. sylvestris	941 (934–948)	431 (377–486)	747 (645–848)	Barros et al. (2010c)
M. pulegium	1872 (1759–1986)	969 (924–1013)	1366 (1277–1455)	Fernandes et al. (2010)
N. officinale	45 (18–79)	85 (59–113)	63 (50–72)	Pereira et al. (2011), Pinela et al. (2016)
O. vulgare	300 (299–301)	190 (180–200)	580 (570–590)	Barros et al. (2011b)
P. tridentatum	227 (216–239)	1368 (1325–1411)	466 (447–486)	Pinela et al. (2011)
P. oleracea	151 (75–271)	259 (118–352)	86 (52–138)	Petropoulos et al. (2015)
R. acetosella	23 (15–31)	65 (65–66)	80 (79–81)	Barros et al. (2010d)
R. ulmifolius	432 (427–437)	1692 (1677–1707)	1885 (1870–1900)	Barros et al. (2010d)
T. pulegioides	555 (544–565)	115 (115–116)	173 (157–188)	Fernandes et al. (2010)
T. communis	116 (108–124)	640 (618–661)	301 (277–324)	Martins et al. (2011)
T. mastichina	20 (19–21)	450 (440–460)	970 (860–1080)	Barros et al. (2011b)

Notes: The mean value is presented with, in parentheses, the range of variability of the literature data.

TABLE 1.9
Composition in Mineral Elements (mg/100 g) of WEPs

Plant spp.	Trace Elements				Macrominerals				Reference
	Zn	Fe	Cu	Mn	Mg	K	Na	Ca	
A. nodiflorum	0.50 (0.42–0.70)	1.80 (0.80–3.10)	0.08 (0.04–0.15)	0.29 (0.17–0.34)	28 (16–45)	165 (105–225)	244 (45–288)	152 (64–246)	García-Herrera et al. (2014); Seal et al. (2020)
A. ampeloprasum	0.75 (0.03–1.67)	0.60 (0.20–0.92)	0.11 (0.05–0.22)	0.11 (0.06–0.15)	14 (9–16)	439 (147–533)	55 (44–67)	70 (30–82)	García-Herrera et al. (2014), García-Herrera et al. (2013); Duguma (2020)
A. azurea	0.43 (0.32–0.86)	1.90 (0.60–2.70)	0.13 (0.09–0.28)	0.24 (0.15–0.41)	25 (16–45)	563 (488–1172)	29 (14–37)	158 (126–219)	Borelli et al., 2022; García-Herrera et al. (2014)
B. officinalis	0.35 (0.35–0.36)	2.79 (2.14–3.45)	0.05 (0.04–0.05)	1.62 (1.40–1.84)	20 (17–23)	214 (198–230)	88 (66–109)	232 (219–245)	Pawera et al. (2020)
B. maritima	0.88 (0.64–1.26)	2.24 (1.42–3.97)	0.21 (0.09–0.35)	0.82 (0.57–1.23)	73 (13–136)	1223 (597–2356)	207 (45–288)	96 (19–250)	Seal et al., 2020; García-Herrera et al. (2014), Guil et al. (1997)
C. juncea	1.63 (0.53–3.81)	3.97 (1.47–6.57)	0.43 (0.12–0.90)	0.97 (0.57–1.48)	53 (3–100)	859 (433–1277)	24 (4–58)	273 (22–472)	Datta et al., 2019; García-Herrera et al. (2014)
C. intybus	0.36 (0.08–0.51)	1.03 (0.41–2.00)	0.11 (0.06–0.21)	0.21 (0.09–0.47)	24 (10–34)	481 (104–1085)	76 (37–170)	121 (45–168)	Pereira et al., 2020; García-Herrera et al. (2014), Ranfa et al. (2015)
H. lupulus	1.13 (0.71–1.51)	0.91 (0.37–1.32)	0.14 (0.10–0.17)	0.36 (0.18–0.55)	1.13 (0.71–1.51)	0.91 (0.37–1.32)	29 (25–32)	89 (50–134)	Rana et al., 2019; García-Herrera et al. (2014), Renna et al. (2015)
F. vulgare	0.36 (0.25–0.57)	1.04 (0.07–2.35)	0.05 (0.01–0.11)	0.62 (0.03–0.97)	0.36 (0.25–0.57)	1.04 (0.07–0.97)	0.05 (0.01–0.11)	0.62 (0.03–0.97)	Mahfood Ali and Farouk Abdelsalam, 2020; García-Herrera et al. (2014), Renna et al. (2015), Trichopoulou et al. (2000)
M. fontana	0.38 (0.27–0.56)	1.30 (0.60–1.68)	0.05 (0.04–0.06)	1.07 (0.66–1.93)	0.38 (0.27–0.56)	1.30 (0.60–1.68)	0.05 (0.04–0.06)	1.07 (0.66–1.93)	Tardío et al. (2011); Peduruhewa et al. (2021)
M. sylvestris	1.58 (0.04–2.67)	3.61 (0.76–6.29)	0.21 (0.10–0.33)	0.49 (0.20–0.76)	283 (30–368)	652 (547–836)	93 (68–128)	240 (165–301)	Bianco et al. (1998), Guill-Guerrero and Rodriguez-Garcia (1999), Sanchez- Mata and Tardio (2016)
N. officinale	0.09 (0.27–0.56)	1.65 (2.2–3.1)	0.05 (0.04–0.06)	1.07 (0.66–1.93)	25 (15–34)	276 (265–310)	12 (61–88)	175 (170–180)	Souci et al., 2008; Peduruhewa et al., 2021

(Continued)

TABLE 1.9 (Continued)
Composition in Mineral Elements (mg/100 g) of WEPs

Plant spp.	Trace Elements				Macrominerals				Reference
	Zn	Fe	Cu	Mn	Mg	K	Na	Ca	
P. oleracea	0.46 (0.21–0.88)	2.16 (0.16–6.82)	0.27 (0.16–0.39)	2.16 (0.16–6.82)	143 (56–276)	540 (298–705)	21 (7–42)	160 (51–234)	Seal et al. (2020), Petropoulos et al. (2015), Renna et al. (2015)
P. rhoes	1.43 (0.20–3.19)	5.52 (0.90–16.33)	0.34 (0.09–1.07)	0.67 (0.39–1.06)	33 (9–74)	711 (188–1673)	42 (3–77)	200 (50–545)	García-Herrera et al. (2014), Duguma (2020), Trichopoulou et al. (2000)
S. marianum	0.26 (0.21–0.35)	0.50 (0.47–0.55)	0.08 (0.01–0.17)	0.10 (0.03–0.21)	17 (10–23)	718 (432–1300)	81 (25–128)	132 (42–171)	Suwardi et al. (2022), García-Herrera et al. (2014)
S. vulgaris	0.57 (0.30–1.21)	0.80 (0.50–1.10)	0.07 (0.04–0.11)	0.75 (0.59–0.92)	68 (13–107)	986 (619–1583)	29 (18–53)	93 (22–144)	Seal et al. (2020), García-Herrera et al. (2014)
T. communis	0.50 (0.22–0.90)	3.57 (2.58–4.18)	0.15 (0.08–0.22)	0.33 (0.15–0.53)	18 (2–35)	566 (375–685)	35 (5–62)	117 (16–269)	Duguma (2020), García-Herrera et al. (2014)

Notes: The mean value is presented with, in parentheses, the range of variability of the literature data.

TABLE 1.10

Composition in Folates (microgram/100 g), Ascorbic Acid (mg/100 g), Tocopherols (mg/100 g), Total Phenolics (mg GAE/g Methanolic Extract) and Total Flavonoids (mg CE/g Methanolic Extract) of WEPs

Plant spp.	Bioactive Non-Nutrients		Vitamin E	Vitamin C	Vitamin B$_9$		Reference
	Total Flavonoids	Total Phenolics	Alpha-Tocopherol	Ascorbic Acid	Total Folates	Total Tocopherol	
A. nodiflorum	45.48 (43.87–47.09)	80.47 (76.06–84.88)	0.23 (0.28–0.44)	8.90 (4.91–12.89)	125.2 (97.0–153.4)	0.27 (0.39–0.57)	Talang et al. (2023), Morales et al. (2015, 2012)
A. ampeloprasum	0.86 (0.81–0.91)	5.70 (5.08–6.32)	0.03 (0.02–0.04)	4.15 (1.58–7.89)	145.1 (80.0–170.2)	0.05 (0.04–0.06)	Gracia-Herrera et al. (2014, 2013), Maikhuri et al. (2021), Sanchez-mata et al. (2012)
A. azurea	84.81 (80.78–88.84)	148.62 (146.62–150.62)	0.36 (0.28–0.44)	0.67 (0.53–0.81)	278.2 (256.6–299.7)	0.48 (0.39–0.57)	Kaur and Roy (2021), Morales et al. (2015b), Morales et al., 2014)
B. officinalis	302.1 (249.8–354.4)	0.51 (0.49–0.53)	0.51 (0.49–0.53)	10.13 (8.37–11.63)	302.1 (249.8–354.4)	0.63 (0.48–0.78)	Pereira et al. (2013), Pereira et al., 2011); Mawunu et al. (2020)
B. maritima	21.55 (20.68)	61.91 (54.4–69.42)	1.51 (1.08–1.22)	1.26 (0–2.74)	302.1 (249.8–354.4)	1.51 (1.42–1.60)	Singh and Gairola (2023), Morales et al. (2015b), Morales et al. (2014), Sánchez-Mata et al. (2012)
C. juncea	7.43 (7.151–7.71)	37.66 (35.26–40.06)	0.57 (0.39–0.75)	1.96 (0.56–3.75)	90.2 (74.5–106.0)	0.74 (0.53–0.95)	Ray et al. (2021), Morales et al., 2015b, 2014), Sánchez-Mata et al. (2012)
C. intybus	73.68 (73.02–74.34)	31.35 (30.35–32.35)	0.99 (0.88–1.10)	4.41 (0.92–7.28)	253.5 (244.5–262.5)	2.98 (2.79–3.17)	Kaur and Roy (2021), Morales et al. (2015b), Morales et al., 2014
F. vulgare	9.72 (9.02–10.42)	42.16 (41.18–43.14)	0.43 (0.30–0.56)	13.81 (10.49–17.30)	271.6 (256.2–287.0)	0.57 (0.44–0.70)	Kumar and Shiddamallayya (2021), Barros et al. (2009)
G. hederacea	95.02 (92.29–97.75)	196.61 (190.61–202.61)	73.53 (71.59–75.47)	4.55 (4.49–4.61)	271.6 (256.2–287.0)	99.64 (98.10–101.18)	Stark et al. (2020), Barros et al. (2010d)
H. lupulus	9.56 (8.91–10.21)	55.83 (54.49–57.17)	0.58 (0.52–0.64)	12.79 (10.04–16.15)	143.9 (108.4–179.4)	1.83 (1.76–1.90)	Islary et al. (2017), Sánchez-Mata et al. (2012)
M. fontana	21.28 (16.05–26.89)	61.50 (45.85–82.58)	1.45 (0.72–2.43)	15.43 (0–28.96)	41.8 (41.7–41.9)	1.85 (0.97–3.06)	Morales et al. (2015b), Morales et al. (2012), Pereira et al. (2013), Pereira et al. (2011), Tardío et al. (2011)

(Continued)

TABLE 1.10 (Continued)
Composition in Folates (microgram/100 g), Ascorbic Acid (mg/100 g), Tocopherols (mg/100 g), Total Phenolics (mg GAE/g Methanolic Extract) and Total Flavonoids (mg CE/g Methanolic Extract) of WEPs

Plant spp.	Bioactive Non-Nutrients		Vitamin E	Vitamin C	Vitamin B$_9$		Reference
	Total Flavonoids	Total Phenolics	Alpha-Tocopherol	Ascorbic Acid	Total Folates	Total Tocopherol	
M. sylvestris	210.81 (202.82–218.80)	386.45 (377.91–394.99)	19.84 (19.37–20.31)	4.03 (2.84–5.22)	143.9 (108.4–179.4)	25.24 (24.52–25.97)	Adamu et al. (2022), Barros et al. (2010c)
M. pulegium	139.85 (138.58–141.12)	331.69 (315.06–348.32)	28.18 (23.54–32.82)	3.20 (3.13–3.27)	143.9 (108.4–179.4)	36.36 (30.37–42.35)	Biswas et al. (2022), Fernandes et al. (2010)
N. officinale	35.17 (31.81–38.53)	50.42 (47.65–53.19)	1.01 (0.47–1.52)	0.59(0–1.17)	127.9 (117.1–138.7)	1.17 (0.51–1.82)	Fajardo et al. (2015), Singh and Gairola (2023), Pinela et al. (2016)
O. vulgare	224.15 (223.19–225.11)	368.58 (350.40–386.76)	4.90 (4.69–5.11)	4.11 (0–8.48)	127.9 (117.1–138.7)	6.10 (5.85–6.36)	Kaur and Roy (2021), Pereira et al. (2013)
P. oleracea	1.76 (0.12–5.30)	12.89 (7.65–20.1)	12.2 (1.07–1.19)	26.6 (11.86–16.36)	12 (151.0–153.6)	6.10 (5.85–6.36)	Ray et al. (2021), Petropoulos et al. (2016, 2015)
P. rhoeas	25.86 (22.34–29.38)	12.00 (11.54–12.46)	1.13 (1.07–1.19)	13.71 (11.86–16.36)	152.3 (151.0–153.6)	1.87 (1.68–2.06)	Morales et al. (2015b), Morales et al. (2014), Yimer Forsido, Addis and Ayelign (2023)
R. acetosella	37.91 (34.55–41.27)	141.58 (137.91–145.25)	5.70 (4.70–6.70)	45.91 (19.21–72.23)	12 (151.0–153.6)	10.77 (8.93–12.61)	Adamu et al. (2022), Pereira et al. (2013)
R. ulmifolius	172.45 (16903–175.87)	257.89 (254.61–261.17)	3.03 (2.93–3.13)	87.85 (86.89–88.81)	12 (151.0–153.6)	6.24 (6.13–6.35)	Barros et al. (2010d); de Carvalho and Conte-Junior (2021)
S. marianum	1.13 (0.86–1.40)	3.72 (3.36–4.08)	0.04 (1.31–1.65)	0.47 (0–1.23)	41.7 (11.7–71.6)	0.15 (1.51–1.89)	Morales et al. (2015b), Morales et al. (2012), Kaur and Roy (2021)
S. vulgaris	21.65 (16.12–27.18)	26.72 (25.09–2835)	1.48 (1.31–1.65)	22.02 (16.05–29.44)	267.7 (214.4–320.9)	1.70 (1.51–1.89)	Sánchez-Mata et al. (2012); Kaur and Roy (2021)
T. pulegioides	128.24 (122.24–134.24)	210.49 (189.33–231.65)	6.61 (6.18–7.04)	3.11 (3.05–3.17)	110.7 (91.0–130.4)	7.06 (6.58–7.54)	Kumar and Shiddamallayya (2021), Fernandes et al. (2010)
T.mastichina	83.85 (82.43–85.27)	165.29 (164.18–166.40)	0.16 (0.13–0.19)	2.92(tr–5.93)	110.7 (91.0–130.4)	1.88 (1.79–1.97)	Kewlani et al. (2023), Barros et al. (2010d), Pereira et al. (2013)
T. communis	79.67 (7.89–162.00)	404.26 (45.44–788.00)	2.63 (0.16–5.20)	42.33 (23.55–53.36)	38.1 (37.2–39.0)	4.57 (0.45–8.80)	Ray et al. (2021), Morales et al. (2015b), Morales et al. (2012), Pereira et al. (2013)

TABLE 1.11

Composition in Main Fatty Acids and Categories (in Relative Percentage) of Selected WEPs

Plant spp.	Categories			Individual Fatty Acids				References
	PUFA	MUFA	SFA	Alpha-Linolenic Acid	Palmitic Acid	Linoleic Acid	Oleic Acid	
A. nodiflorum	70.7 (68.7–72.7)	5.4 (4.9–5.8)	23.9 (22.4–25.5)	43.5 (43.4–43.5)	16.3 (15.3–17.3)	24.6 (23.8–25.4)	3.3 (3.3–3.4)	Duguma (2020), Morales et al. (2012)
A. ampeloprasum	54.2 (53.9–54.5)	7.6 (7.2–8.1)	38.2 (37.6–38.9)	nd	26.4 (26.1–26.7)	53.5 (53.2–53.7)	7.4 (7.0–7.8)	García-Herrera et al. (2014), Kaur and Roy (2021)
A. azurea	80.3 (79.4–81.1)	3.2 (3.1–3.2)	16.6 (15.8–17.4)	64.7 (64.5–65.0)	10.5 (9.8–11.1)	12.2 (12.1–12.3)	2.2 (–)	Kumar and Shiddamallayya (2021), Morales et al. (2012)
B. officinalis	45.4 (43.5–47.3)	3.3 (2.8–3.8)	51.3 (50.4–52.3)	12.3 (12.1–12.5)	12.0 (11.3–12.7)	9.5 (8.3–10.8)	2.1 (1.9–2.3)	Seal et al. (2020), Pereira et al. (2011)
B. maritima	79.3 (–)	4.0 (–)	16.7 (–)	57.8 (–)	11.0 (10.9–11.2)	21.3 (21.2–21.3)	3.5 (–)	Padhan and Panda (2020), Morales et al. (2012)
C. juncea	76.5 (6.5–76.6)	2.1 (–)	21.4 (–)	56.3 (56.1–56.4)	13.0 (12.5–13.4)	19.9 (19.8–20.1)	1.9 (–)	Ray et al. (2021), Morales et al. (2012)
C. intybus	82.1 (82.0–82.2)	2.0 (–)	16.0 (15.4–16.5)	60.5 (60.0–60.9)	10.6 (10.0–11.3)	21.1 (21.1–21.2)	1.6 (1.6–1.7)	Ray et al. (2021), Morales et al. (2012)
F. vulgare	73.0 (72.6–73.3)	3.1 (3.0–3.2)	24.0 (23.7–24.2)	35.5 (35.0–36.1)	17.4 (17.1–17.7)	37.0 (36.9–37.1)	2.1 (2.1–2.2)	Kumar and Shiddamallayya (2021), Morales et al. (2012)
G. hederacea	37.1 (36.8–37.3)	36.2 (35.9–36.4)	26.8 (26.3–27.3)	27.9 (27.7–28.1)	12.2 (12.0–12.5)	8.2 (8.1–8.2)	35.1 (34.9–35.4)	Yimer et al. (2023), Barros et al. (2011b)
H. lupulus	69.0 (68.2–69.9)	5.6 (5.5–5.7)	37.5 (37.2–37.7)	38.2 (38.1–38.2)	19.5 (18.9–20.1)	29.7 (28.9–30.6)	1.9 (1.8–2.0)	Adamu et al. (2022), Morales et al. (2012)
M. fontana	71.1 (67.1–75.2)	5.7 (4.0–7.5)	23.1 (20.8–25.4)	51.6 (47.6–55.6)	14.3 (11.3–17.2)	18.5 (18.2–18.7)	4.2 (2.4–6.1)	Ayeni and Oyeyemi (2021), Morales et al. (2012)
M. sylvestris	80.0 (79.5–80.5)	3.7 (3.3–4.0)	16.3 (15.4–17.2)	67.8 (66.8–68.8)	9.8 (8.7–10.9)	12.0 (11.5–12.4)	3.3 (2.9–3.7)	Adamu et al. (2022), Barros et al. (2010c)
M. pulegium	55.5 (55.0–56.0)	6.8 (6.6–7.0)	37.6 (36.8–38.50)	37.0 (36.7–37.4)	14.8 (14.7–14.9)	16.3 (15.9–16.6)	5.8 (5.6–6.0)	Singh and Gairola (2023), Fernandes et al. (2010)
N. officinale	81.2 (80.7–81.7)	2.1 (–)	16.7 (16.2–17.2)	68.4 (68.2–68.7)	13.2 (12.9–13.5)	11.8 (11.7–12.0)	0.7 (–)	Duguma (2020), Pereira et al. (2011)
O. vulgare	85.8 (85.6–86.0)	5.3 (–)	8.9 (8.8–9.1)	62.3 (62.3–62.4)	5.0 (4.9–5.1)	23.2 (23.1–23.4)	5.1 (–)	Stark et al. (2020), Barros et al. (2011b)

(Continued)

TABLE 1.11 (Continued)
Composition in Main Fatty Acids and Categories (in Relative Percentage) of Selected WEPs

Plant spp.	Categories				Individual Fatty Acids					References
	PUFA	MUFA	SFA	Alpha-Linolenic Acid	Palmitic Acid	Linoleic Acid	Oleic Acid			
P. oleracea	52.4 (48.6–56.1)	15.1 (14.1–16.3)	32.5 (29.3–37.1)	23.6 (17.9–28.4)	24.7 (23.4–26.9)	28.8 (25.1–32.9)	12.4 (9.7–15.1)			Fajardo et al. (2015), Singh and Gairola (2023), Petropoulos et al. (2015)
P. rhoeas	81.9 (81.6–82.1)	1.8 (–)	16.4 (16.2–16.6)	65.0 (64.9–65.1)	9.7(9.3–10.1)	16.5 (–)	1.4 9 (–)			Padhan and Panda (2020), Morales et al. (2012)
R. acetosella	72.6 (71.5–73.7)	8.0 (7.9–8.0)	19.5 (18.3–20.6)	51.3 (49.9–52.8)	11.2 (10.5–12.0)	20.2 (19.7–20.7)	3.4 (3.1–3.8)			Kewlani et al. (2023), Pereira et al. (2011)
R. ulmifolius	56.1 (56.0–56.3)	4.6 (–)	39.2 (39.0–39.3)	39.6 (39.5–39.6)	12.0 (–)	16.0 (16.0–16.1)	4.3 (–)			Ray et al. (2021), Barros et al. (2010d)
S. marianum	52.6 (51.7–53.5)	3.9 (3.8–4.0)	43.5 (42.6–44.5)	21.6 (21.4–21.9)	28.7 (27.1–30.3)	31.0 (30.4–31.6)	3.7 (3.8–4.0)			Morales et al. (2012)
S. vulgaris	67.5 (66.5–68.5)	4.4 (4.2–4.5)	28.2 (27.3–29.0)	49.6 (44.6–54.5)	15.8 (13.4–18.2)	22.3 (21.9–24.1)	2.6 (2.4–2.7)			Ayeni and Oyeyemi (2021), Alarcon et al. (2006)
T. pulegioides	49.7 (48.9–50.4)	12.8 (12.8–12.9)	37.5 (36.7–38.3)	36.7 (36.0–36.5)	16.7 (14.5–16.9)	13.0 (12.5–13.5)	11.4 (11.3–11.5)			de Carvalho and Conte-Junior (2021), Fernandes et al. (2010)
T. communis	75.1 (69.9–80.2)	4.5 (0.9–8.2)	20.4 (18.8–22.0)	29.4 (27.5–31.3)	16.0 (14.9–17.0)	42.2 (41.7–42.3)	6.0 (4.6–7.5)			Biswas et al. (2022), Morales et al. (2012)
T. mastichina	60.0 (59.6–60.4)	10.7 (10.4–11.0)	29.3 (29.2–29.4)	45.7 (45.1–46.2)	10.2 (10.0–10.4)	11.8 (11.8–11.9)	9.8 (9.6–10.0)			Singh and Gairola (2023), Barros et al. (2011b)

Notes: The mean value is presented with, in parentheses, the range of variability of the literature data.

TABLE 1.12

Proximate Composition (g/100g) and Energetic Value (kcal/100g) of Some WEPs

Plant spp.	Potentially Toxic Compounds	Parts	References
A. azurea	Pyrrolizidine alkaloids: lycopsamine, laburnine, and acetyllaburnine	Unspecified parts, seeds (oil)	EFSA-European Food Safety Authority (2009), Roeder (1999); Seal, Chaudhuri, Pillai, Chakrabarti, Auddy and Mondal (2020)
B. officinalis	Pyrrolizidine alkaloids: amabiline	Seeds (oil)	EFSA-European Food Safety Authority (2009), EMA (2014). Vacillotto et al. (2013)
F. vulgare	Phenylpropanoids: trans-anethole and estaole (2.3–4.9%) Phenylpropanoid: estragole (0.8–>80%) Phenylpropanoid: estragole (from 11.9 to 56.1% in unripe seeds to 61.8% in ripe seed)	Aerial parts	Seal et al. (2020), EFSA (2012), EFSA-European Food Safety Authority (2009)
G. hederacea	Pyrrolidine alkaloids: hederacines A and B Monoterpene etheroxide: 1,8-cineole (eucalyptol, 1.9–4.6%)	Aerial part, flowering aerial part	EFSA (2012), Kumarasamy et al. (2003); Pearce et al. (2019)
H. lupulus	Prenylflavonoids: 8-prenylnaringenin (hopein), xanthohumol, isoxanthohumol	Inflorescence	Saleem et al. (2022), EFSA (2009)
M. pulegium	Mnocyclic monoterpene ketone: pulegone (71.3–90%); bicyclic monoterpenes: menthofuran and thujones; monoterpene etheroxide; 1,8-cineole (eucalyptol)	Aerial part	Karami et al. (2021), Kristanc and Kreft (2016), Teixeira et al. (2012)
O. vulgare	Bicyclic monoterpene: beta-thujone (0–0.6%); monoterpene etheroxide: 1.8-cineole (eucalyptol, 0–6.5%)	Aerial part	EFSA (2009); Seal, Chaudhuri, Pillai, Chakrabarti, Auddy and Mondal (2020)
P. rhoeas	Bicyclic monoterpene: beta-thujone (0–0.6%); monoterpene etheroxide: 1.8-cineole (eucalyptol, 0–6.5%)	Whole plant	EFSA (2009), Tardio et al. (2011)
S. marianum	Triterpenoid saponins: A, B, and C	Flowering top and seed	Aberoumand (2012), EFSA (2009)
S. vulgaris	Triterpenoid saponins: A, B, and C	Root	Rasalanavho et al. (2019), Glensk et al. (1999)
T. communis	Steroidal saponins: dioscin and gracillin; phenanthrene derivatives	Rhizome	Antoniadis et al. (2021), Kovacs et al. (2008), Réthy et al. (2006)

1.8   CONCLUSION

The chapter concludes with an ethnobotanical review of wild edible plants that emphasizes the cultural significance and priceless knowledge ingrained in the relationship between communities and the diversity of plants around them. Investigating the customs, techniques, and therapeutic uses of these wild edibles enhances our knowledge of regional ecosystems and highlights the close relationship that exists between humans and the natural world.

REFERENCES

Aberoumand, A., 2012. Screening of phytochemical compounds and toxic proteinaceous protease inhibitor in some lesser-known food based plants and their effects and potential applications in food. *International Journal of Food Science and Nutrition Engineering*, 2(3), pp.16–20.

Adamu, E., Asfaw, Z., Demissew, S. and Baye, K., 2022. Antioxidant activity and anti-nutritional factors of selected wild edible plants collected from Northeastern Ethiopia. *Foods*, 11(15), p.2291.

Adegboyega, T. T., Abberton, M. T., AbdelGadir, A. H., Dianda, M., Maziya-Dixon, B., Oyatomi, O. A., ... and Babalola, O. O. 2020. Evaluation of nutritional and antinutritional properties of African yam bean (Sphenostylis stenocarpa (Hochst ex. A. Rich.) Harms.) seeds. *Journal of Food Quality*, 2020, pp.1–11.

Akbulut, S., 2022. Importance of edible wild plants in world food security: The case of Turkey. *Int J Agric Sc Food Technol*, 8(3), pp.209–213.

Al Kaabi, S.K.S., Senthilkumar, A., Kizhakkayil, J., Alyafei, M.A.M., Kurup, S.S., Al Dhaheri, A.S. and Jaleel, A. 2022. Anticancer activity of Moringa peregrina (Forssk.) Fiori.: A Native plant in traditional herbal medicine of the United Arab Emirates. *Horticulturae*, 8(1), p.37.

Alarcón, R., Ortiz, L. T., and García, P. 2006a. Nutrient and fatty acid composition of wild edible bladder campion populations [Silene vulgaris (Moench.) Garcke]. *International Journal of Food Science & Technology*, 41(10).

Alarcón, R., Ortiz, L. T., and García, P. 2006b. Nutrient and fatty acid composition of wild edible bladder campion populations [Silene vulgaris (Moench.) Garcke]. *International Journal of Food Science & Technology*, 41(10).

Anand, U., Tudu, C.K., Nandy, S., Sunita, K., Tripathi, V., Loake, G.J., Dey, A. and Próćków, J. 2022. Ethnodermatological use of medicinal plants in India: From ayurvedic formulations to clinical perspectives–A review. *Journal of Ethnopharmacology*, 284, p.1147.

Antoniadis, V., Shaheen, S.M., Stärk, H.J., Wennrich, R., Levizou, E., Merbach, I. and Rinklebe, J. 2021. Phytoremediation potential of twelve wild plant species for toxic elements in a contaminated soil. *Environment International*, 146, p.106233.

Asadbeigi, M., Mohammadi, T., Rafieian-Kopaei, M., Saki, K., Bahmani, M. and Delfan, M. 2014. Traditional effects of medicinal plants in the treatment of respiratory diseases and disorders: An ethnobotanical study in the Urmia. *Asian Pacific Journal of Tropical Medicine*, 7, pp.S364–S368.

Ayeni, M. J., and Oyeyemi, S. D. 2021. Studies on the Nutritional, Mineral Composition, Mineral Ratio and Anti-nutritional Molar Ratio of Six Underutilized Wild Edible Vegetables in Ado-Ekiti, Ekiti State, Nigeria. *Annual Research & Review in Biology*, 36(12), pp.95–110.

Balunas, M.J. and Kinghorn, A.D. 2005. Drug discovery from medicinal plants. *Life sciences*, 78(5), pp.431–441.

Barros, L., Cabrita, L., Vilas-Boas, M., Carvalho, A. M., and Ferreira, I. C. F. R. 2011a Chemical, biochemical and electrochemical assays to evaluate phytochemicals and antioxidant activity of wild plants. *Food Chemistry*, 127, pp.1600–1608.

Barros, L., Carvalho, A.M. and Ferreira, I.C. 2010c. Leaves, flowers, immature fruits and leafy flowered stems of Malva sylvestris: A comparative study of the nutraceutical potential and composition. *Food and Chemical Toxicology*, 48(6), pp.1466–1472.

Barros, L., Carvalho, A. M., Ferreira, I. C. F. R. 2010d. Leaves, flowers, immature fruits and leafy flowered stems of Malvasylvestris: A comparative study of the nutraceutical potential and composition. *Food ChemToxicol*, 48, pp.1466–1472.

Barros, L., Carvalho, A.M. and Ferreira, I.C. 2011b. From famine plants to tasty and fragrant spices: Three Lamiaceae of general dietary relevance in traditional cuisine of Trás-os-Montes (Portugal). *LWT-Food Science and Technology*, 44(2), pp.543–548.

Barros, L., Carvalho, A.M., and Ferreira, I. C. F. R. 2011c. Comparing the Composition and Bioactivity of CrataegusMonogyna Flowers and Fruits used in Folk Medicine. *Phytochemical Analysis*, 22, pp.181–188.

Barros, L., Carvalho, A.M., Morais, J.S. and Ferreira, I.C. 2010a. Strawberry-tree, blackthorn and rose fruits: Detailed characterisation in nutrients and phytochemicals with antioxidant properties. *Food Chemistry*, *120*(1), pp.247–254.

Barros, L., Heleno, S.A., Carvalho, A.M. and Ferreira, I.C. 2009. Systematic evaluation of the antioxidant potential of different parts of Foeniculum vulgare Mill. from Portugal. *Food and Chemical Toxicology*, *47*(10), pp.2458–2464.

Barros, L., Heleno, S.A., Carvalho, A.M. and Ferreira, I.C. 2010b. Lamiaceae often used in Portuguese folk medicine as a source of powerful antioxidants: Vitamins and phenolics. *LWT-Food Science and Technology*, *43*(3), pp.544–550.

Barros, Lillian, Pereira, Eliana, Calhelha, Ricardo C., Dueñas, Montserrat, Carvalho, Ana Maria, Santos-Buelga, Celestino, and Ferreira, Isabel CFR. 2013. Bioactivity and chemical characterization in hydrophilic and lipophilic compounds of *Chenopodium ambrosioides* L. *Journal of Functional Foods*5(4),1732–1740.

Barroso, M.R., Barros, L., Dueñas, M., Carvalho, A.M., Santos-Buelga, C., Fernandes, I.P., Barreiro, M.F. and Ferreira, I.C. 2014. Exploring the antioxidant potential of *Helichrysum stoechas* (L.) Moench phenolic compounds for cosmetic applications: Chemical characterization, microencapsulation and incorporation into a moisturizer. *Industrial Crops and Products*, *53*, pp.330–336.

Biswas, S.C., Bora, A., Mudoi, P., Misra, T.K. and Das, S. 2022. Evaluation of nutritional value, antioxidant activity, and phenolic content of protium serratum engl and artocarpus chama buch.-ham, wild edible fruits available in Tripura, a North–Eastern state of India. *Current Nutrition & Food Science*, *18*(6), pp.589–596.

Borelli, T., Güzelsoy, N.A., Hunter, D., Tan, A., Karabak, S., Uçurum, H.Ö., Çavuş, F., Ay, S.T., Adanacıoğlu, N., Özbek, K. and Özen, B. 2022. Assessment of the nutritional value of selected wild food plants in Türkiye and their promotion for improved nutrition. *Sustainability*, *14*(17), p.11015.

Butler, M.S. 2005. Natural products to drugs: Natural product derived compounds in clinical trials. *Natural Product Reports*, *22*(2), pp.162–195.

Chanda, S. 2014. Importance of pharmacognostic study of medicinal plants: An overview. *Journal of Pharmacognosy and Phytochemistry*, *2*(5), pp.69–73.

Cowan, J.W., Sakr, A.H., Shadarevian, S.B. and Sabry, Z.I. 1963. Composition of edible wild plants of Lebanon. *Journal of the Science of Food and Agriculture*, *14*(7), pp.484–488.

Crisan, E. V., and Sands, Anne 1978. Nutritional value. *The Biologyand Cultivation of Edible Mushrooms*,137–168. Academic Press, Inc.

Datta, S., Sinha, B. K., Bhattacharjee, S., and Seal, T. 2019. Nutritional composition, mineral content, antioxidant activity and quantitative estimation of water soluble vitamins and phenolics by RP-HPLC in some lesser used wild edible plants. *Heliyon*, *5*(3).

de Carvalho, A.P.A. and Conte-Junior, C.A. 2021. Health benefits of phytochemicals from Brazilian native foods and plants: Antioxidant, antimicrobial, anti-cancer, and risk factors of metabolic/endocrine disorders control. *Trends in Food Science & Technology*, *111*, pp.534–548.

de Cortes Sánchez-Mata, M., and Tardío, J. (Eds.). 2016. *Mediterranean wild edible plants: Ethnobotany and food composition tables*. Springer.

Dhanamani, M.D.S.L., Devi, S.L. and Kannan, S. 2011. Ethnomedicinal plants for cancer therapy–a review. *Hygeia JD Med*, *3*(1), pp.1–10.

Domínguez-Castro, F., Vaquero, J.M., Rodrigo, F.S., Farrona, A.M.M., Gallego, M.C., García-Herrera, R., Barriendos, M. and Sanchez-Lorenzo, A. 2014. Early Spanish meteorological records (1780–1850). *International Journal of Climatology*, *34*(3), pp.593–603.

Duguma, H.T. 2020. Wild edible plant nutritional contribution and consumer perception in Ethiopia. *International Journal of Food Science*. DOI: 10.1155/2020/2958623

EFSA-European Food Safety Authority 2009. Compendium of botanicals that have been reported to contain toxic, addictive, psychotropic or other substances of concern on request of EFSA. *EFSA Journal*, *7*(9), p.100.

Eshete, M. A., Asfaw, Z., & Kelbessa, E. (2016). A review on taxonomic and use diversity of the family Amaranthaceae in Ethiopia. *Journal of Medicinal Plants Sciences*, *4*(2), 185–194.

European Food Safety Authority. 2012. Compendium of botanicals reported to contain naturally occuring substances of possible concern for human health when used in food and food supplements. *EFSA Journal*, *10*(5), p. 2663.

European Medicines Agency 2014. Public statement on the use of herbal medicinal products containing toxic, unsaturated pyrrolizidine alkaloids (PAs).

Fabricant, D.S. and Farnsworth, N.R. 2001. The value of plants used in traditional medicine for drug discovery. *Environmental Health Perspectives*, *109*(suppl 1), pp.69–75.

Feeney, M. J., Miller, A. M. and Roupas, P. 2014. Mushrooms—Biologically distinct and nutritionally unique: Exploring a "third food kingdom". *Nutrition Today*, 49(6), pp.301–307.

Fenetahun, Y. and Eshetu, G. 2017. A review on ethnobotanical studies of medicinal plants use by agro-pastoral communities in, Ethiopia. *Journal of Medicinal Plants*, 5(1), pp.33–44.

Fernandes, Â.S., Barros, L., Carvalho, A.M. and Ferreira, I.C.F.R.,2010. Lipophilic and hydrophilic anti-oxidants, lipid peroxidation inhibition and radical scavenging activity of two Lamiaceae food plants. *European Journal of Lipid Science and Technology*, 112(10), pp.1115–1121.

Ferreira, I.C., Morales, P. and Barros, L. eds. 2016. *Wild plants, mushrooms and nuts: Functional food properties and applications*. John Wiley & Sons.

Florezak, J., Karmnska, A. and Wedzisz, A. 2004. Comparision of the chemical contents of the selected wild growing mushrooms. *Bromatologia i Chemia Toksykologiczna*, 37(4), pp.365–371.

Gahlawat, D.K., Jakhar, S. and Dahiya, P. 2014. Murraya koenigii (L.) Spreng: An ethnobotanical, phytochemical and pharmacological review. *Journal of Pharmacognosy and Phytochemistry*, 3(3), pp.109–119.

Gangwar, P., Johri, P., Datta, R., Singh, S. and Trivedi, M. 2022. Overview on Phytopharmaceuticals and Biotechnology of Herbal Plants. In Sachidanand Singh, Rahul Datta, Parul Johri, Mala Trivedi (eds.) *Phytopharmaceuticals and Biotechnology of Herbal Plants* (pp. 1–17).CRC Press.

García-Herrera, P., Sánchez-Mata, M.C., Cámara, M., Fernández-Ruiz, V., Díez-Marqués, C., Molina, M. and Tardío, J. 2014. Nutrient composition of six wild edible Mediterranean Asteraceae plants of dietary interest. *Journal of Food Composition and Analysis*, 34(2), pp.163–170.

García-Herrera, P., Sánchez-Mata, M. C., Cámara, M., Tardío, J. and Olmedilla-Alonso, B. 2013. Carotenoid content of wild edible young shoots traditionally consumed in Spain (Asparagus acutifolius L., Humulus lupulus L., Bryonia dioica Jacq. and Tamus communis L.). *Journal of the Science of Food and Agriculture*, 93(7), pp.1692–1698.

Garekae, H. and Shackleton, C. M. 2020. Foraging wild food in urban spaces: The contribution of wild foods to urban dietary diversity in South Africa. *Sustainability*, 12(2), p. 678.

Glensk, M., Wray, V., Nimtz, M. and Schöpke, T. 1999. Silenosides A– C, triterpenoid saponins from Silene vulgaris. *Journal of Natural Products*, 62(5), pp.717–721.

Guil, J.L., Rodríguez-Garcí, I. and Torija, E. 1997. Nutritional and toxic factors in selected wild edible plants. *Plant Foods for Human Nutrition*, 51, pp.99–107.

Guill-Guerrero, J. and Rodriguez-Garcia, I. 1999. Lipids classes, fatty acids and carotenes of the leaves of six edible wild plants. *European Food Research and Technology*, 209, 3q–316.

Haddad, N. A. and Hayes, W. A. 1979. Nutritional factors and the composition of Agaricus bisporus mycelium. *Mushroom Science*.

Islary, A., Sarmah, J. and Basumatary, S. 2017. Nutritional value, phytochemicals and antioxidant properties of two wild edible fruits (Eugenia operculata Roxb. and Antidesma bunius L.) from Assam, North-East India. *Mediterranean Journal of Nutrition and Metabolism*, 10(1), pp.29–40.

Jain, T., Srivastava, K., Singh, R., Ansari, M.J., Singh, S. and Dutta, P.K. 2022. Role of nanotechnology in the field of phytopharmaceuticals for the delivery of herbal drugs. In *Phytopharmaceuticals and Biotechnology of Herbal Plants* (pp. 317–330).

Jman Redzic, S. 2006. Wild edible plants and their traditional use in the human nutrition in Bosnia-Herzegovina. *Ecology of Food and Nutrition*, 45(3), pp.189–232.

Kakon, A.J., Choudhury, M.B.K. and Saha, S. 2012. Mushroom is an ideal food supplement. *Journal of Dhaka National Medical College & Hospital*, 18(1), pp.58–62.

Karami, H., Shariatifar, N., Nazmara, S., Moazzen, M., Mahmoodi, B. and Mousavi Khaneghah, A. 2021. The concentration and probabilistic health risk of potentially toxic elements (PTEs) in edible mushrooms (wild and cultivated) samples collected from different cities of Iran. *Biological Trace Element Research*, 199, pp.389–400.

Katiyar, C., Gupta, A., Kanjilal, S. and Katiyar, S. 2012. Drug discovery from plant sources: An integrated approach. *Ayu*, 33(1), p.10.

Kaur, S. and Roy, A. 2021. A Review on the nutritional aspects of wild edible plants. *Current Traditional Medicine*, 7(4), pp.552–563.

Kewlani, P., Tiwari, D., Rawat, S. and Bhatt, I.D. 2023. Pharmacological and phytochemical potential of Rubus ellipticus: A wild edible with multiple health benefits. *Journal of Pharmacy and Pharmacology*, 75(2), pp.143–161.

Kooti, W., Servatyari, K., Behzadifar, M., Asadi-Samani, M., Sadeghi, F., Nouri, B. and Zare Marzouni, H. 2017. Effective medicinal plant in cancer treatment, part 2: Review study. *Journal of Evidence-Based Complementary & Alternative Medicine*, 22(4), pp.982–995.

Kovács, A., Vasas, A. and Hohmann, J. 2008. Natural phenanthrenes and their biological activity. *Phytochemistry*, *69*(5), pp.1084–1110.

Kristanc, L. and Kreft, S. 2016. European medicinal and edible plants associated with subacute and chronic toxicity part II: Plants with hepato-, neuro-, nephro-and immunotoxic effects. *Food and Chemical Toxicology*, *92*, pp.38–49.

Kumar, S., Jawaid, T. and Dubey, S.D. 2011. Therapeutic plants of Ayurveda; a review on anticancer. *Pharmacognosy Journal*, *3*(23), pp.1–11.

Kumar, G.M. and Shiddamallayya, N. 2021. Nutritional and anti-nutritional analysis of wild edible plants in Hassan district of Karnataka, India. *Indian Journal of Natural Products and Resources (IJNPR)[Formerly Natural Product Radiance (NPR)]*, *12*(2), pp.281–290.

Kumar, R., & Bharati, K. A. (2014). *Ethnomedicines of Tharu tribes of Dudhwa national park*. India.

Kumar, S. and Yadav, J. P. 2014. Ethnobotanical and pharmacological properties of Aloe vera: A review. *J Med Plants Res*, *48*(8), pp.1387–98.

Kumarasamy, Y., Cox, P.J., Jaspars, M., Nahar, L. and Sarker, S.D. 2003. Isolation, structure elucidation and biological activity of hederacine A and B, two unique alkaloids from Glechoma hederaceae. *Tetrahedron*, *59*(34), pp.6403–6407.

Li, C. and Xu, S. 2022. Edible mushroom industry in China: Current state and perspectives. *Applied Microbiology and Biotechnology*, *106*(11), pp.3949–3955.

Łuczaj, Ł., Jug-Dujaković, M., Dolina, K., Jeričević, M. and Vitasović-Kosić, I. 2019. The ethnobotany and biogeography of wild vegetables in the Adriatic islands. *Journal of Ethnobiology and Ethnomedicine*, *15*, pp.1–17.

Mahfood Ali, G. and Farouk Abdelsalam, F. 2020. Antioxidant Activity, Antinutritional Factors and Technological Studies on Raw and Germinated Barley Grains (Hordeum vulgare. L). *Alexandria Journal of Agricultural Sciences*, *65*(5), pp.329–343.

Maikhuri, R.K., Parshwan, D.S., Kewlani, P., Negi, V.S., Rawat, S. and Rawat, L.S. 2021. Nutritional composition of seed kernel and oil of wild edible plant species from Western Himalaya, India. *International Journal of Fruit Science*, *21*(1), pp.609–618.

Martins, D., Barros, L., Carvalho, A. M. and Ferreira, I. C. 2011. Nutritional and in vitro antioxidant properties of edible wild greens in Iberian Peninsula traditional diet. *Food Chemistry*, *125*(2), pp.488–494.

Mawunu, M., Pedro, M., Lautenschläger, T., Biduayi, F.M., Kapepula, P.M., Ngbolua, K.N., Luyeye, F.L. and Luyindula, N. 2020. Nutritional value of two underutilized wild plant leaves consumed as food in northern Angola: Mondia whitei and Pyrenacantha klaineana. *European Journal of Nutrition & Food Safety*, *12*(8), pp.116–127.

Mc Connel, J. E. and Esselen, W. B. 1947. Carbohydrate in cultivated mushrooms. *Food Research*, *12*, pp.118–121.

Mehrbod, P., Safari, H., Mollai, Z., Fotouhi, F., Mirfakhraei, Y., Entezari, H., … Tofighi, Z. 2021. Potential antiviral effects of some native Iranian medicinal plants extracts and fractions against influenza A virus. *BMC Complementary Medicine and Therapies*, *21*, pp.1–12.

Morales, P., Carvalho, A.M., Sánchez-Mata, M.C., Cámara, M., Molina, M. and Ferreira, I.C. 2012. Tocopherol composition and antioxidant activity of Spanish wild vegetables. *Genetic Resources and Crop Evolution*, *59*, pp.851–863.

Morales, P., Fernández-Ruiz, V., Sánchez-Mata, M.C., Cámara, M. and Tardío, J. 2015. Optimization and application of FL-HPLC for folates analysis in 20 species of Mediterranean wild vegetables. *Food Analytical Methods*, *8*, pp.302–311.

Morales, P., Ferreira, I.C., Carvalho, A.M., Sánchez-Mata, M.C., Cámara, M., Fernández-Ruiz, V., Pardo-de-Santayana, M. and Tardío, J. 2014. Mediterranean non-cultivated vegetables as dietary sources of compounds with antioxidant and biological activity. *LWT-Food Science and Technology*, *55*(1), pp.389–396.

Obakiro, S.B., Kiprop, A., Kowino, I., Kigondu, E., Odero, M.P., Omara, T. and Bunalema, L. 2020. Ethnobotany, ethnopharmacology, and phytochemistry of traditional medicinal plants used in the management of symptoms of tuberculosis in East Africa: A systematic review. *Tropical Medicine and Health*, *48*(1), pp.1–21.

Obidiegwu, J. E., Lyons, J. B. and Chilaka, C. A. 2020. The Dioscorea Genus (Yam)—An appraisal of nutritional and therapeutic potentials. *Foods*, *9*(9), p.1304.

Oliveira, I., Valentão, P., Lopes, R., Andrade, P. B., Bento, A. and Pereira, J. A. 2009. Phytochemical characterization and radical scavenging activity of Portulacaoleracea L. leaves and stems. *Microchemical Journal*, *92*(2), pp.129–134.

Padhan, B. and Panda, D. 2020. Potential of neglected and underutilized yams (Dioscorea spp.) for improving nutritional security and health benefits. *Frontiers in Pharmacology*, *11*, p.506039.

Parasuraman, S. 2018. Herbal drug discovery: Challenges and perspectives. *Current Pharmacogenomics and Personalized Medicine (Formerly Current Pharmacogenomics)*, *16*(1), pp.63–68.

Paul, M. and Ma, J. K. 2010. Plant-made immunogens and effective delivery strategies. *Expert Review of Vaccines*, *9*(8), pp.821–833.

Pawera, L., Khomsan, A., Zuhud, E.A., Hunter, D., Ickowitz, A. and Polesny, Z. 2020. Wild food plants and trends in their use: From knowledge and perceptions to drivers of change in West Sumatra, Indonesia. *Foods*, *9*(9), p.1240.

Pearce, J. M., Khaksari, M. and Denkenberger, D. 2019. Preliminary automated determination of edibility of alternative foods: Non-targeted screening for toxins in red maple leaf concentrate. *Plants*, *8*(5), p.110.

Peduruhewa, P.S., Jayathunge, K.G.L.R. and Liyanage, R. 2021. Potential of underutilized wild edible plants as the food for the future–a review. *Journal of Food Security*, *9*(4), pp.136–147.

Pereira, C., Barros, L., Carvalho, A.M. and Ferreira, I.C. 2011. Nutritional composition and bioactive properties of commonly consumed wild greens: Potential sources for new trends in modern diets. *Food Research International*, *44*(9), pp.2634–2640.

Pereira, C., Barros, L., Carvalho, A. M. and Ferreira, I. C. 2013. Use of UFLC-PDA for the analysis of organic acids in thirty-five species of food and medicinal plants. *Food Analytical Methods*, *6*, pp.1337–1344.

Pereira, A.G., Fraga-Corral, M., García-Oliveira, P., Jimenez-Lopez, C., Lourenço-Lopes, C., Carpena, M., Otero, P., Gullón, P., Prieto, M.A. and Simal-Gandara, J. 2020. Culinary and nutritional value of edible wild plants from northern Spain rich in phenolic compounds with potential health benefits. *Food &Function*, *11*(10), pp.8493–8515.

Petropoulos, S. A., Karkanis, A., Fernandes, Â., Barros, L., Ferreira, I. C., Ntatsi, G., … Khah, E. 2015. Chemical composition and yield of six genotypes of common purslane (Portulaca oleracea L.): An alternative source of omega-3 fatty acids. *Plant Foods for Human Nutrition*, *70*, pp.420–426.

Petropoulos, S., Karkanis, A., Martins, N. and Ferreira, I. C. 2016. Phytochemical composition and bioactive compounds of common purslane (Portulaca oleracea L.) as affected by crop management practices. *Trends in Food Science & Technology*, *55*, pp.1–10.

Phillipson, J.D. 2007. Phytochemistry and pharmacognosy. *Phytochemistry*, *68*(22–24), pp.2960–2972.

Pinela, J., Barreira, J.C., Barros, L., Antonio, A.L., Carvalho, A.M., Oliveira, M.B.P. and Ferreira, I.C. 2016. Postharvest quality changes in fresh-cut watercress stored under conventional and inert gas-enriched modified atmosphere packaging. *Postharvest Biology and Technology*, *112*, pp.55–63.

Pinela, J., Barros, L., Carvalho, A.M. and Ferreira, I.C. 2011. Influence of the drying method in the antioxidant potential and chemical composition of four shrubby flowering plants from the tribe Genisteae (Fabaceae). *Food and Chemical Toxicology*, *49*(11), pp.2983–2989.

Pinela, J., Prieto, M.A., Antonio, A.L., Carvalho, A.M., Oliveira, M.B.P., Barros, L. and Ferreira, I.C. 2017. Ellagitannin-rich bioactive extracts of Tuberaria lignosa: Insights into the radiation-induced effects in the recovery of high added-value compounds. *Food & Function*, *8*(7), pp.2485–2499.

Ramawat, K.G. and Mérillon, J.M. 2008. *Bioactive molecules and medicinal plants* (pp. 22). Berlin: Springer.

Rana, Z.H., Alam, M.K. and Akhtaruzzaman, M. 2019. Nutritional composition, total phenolic content, antioxidant and α-amylase inhibitory activities of different fractions of selected wild edible plants. *Antioxidants*, *8*(7), p.203.

Ranfa, A., Maurizi, A., Romano, B. and Bodesmo, M. 2014. The importance of traditional uses and nutraceutical aspects of some edible wild plants in human nutrition: The case of Umbria (central Italy). *Plant Biosystems-An International Journal Dealing with all Aspects of Plant Biology*, *148*(2), pp.297–306.

Ranfa, A., Orlandi, F., Maurizi, A. and Bodesmo, M. 2015. Ethnobotanical knowledge and nutritional properties of two edible wild plants from Central Italy: Tordylium apulum L. and Urospermum dalechampii (L.) FW Schmid. *Journal of Applied Botany and Food Quality*, *88*(1).

Rasalanavho, M., Moodley, R. and Jonnalagadda, S.B. 2019. Elemental distribution including toxic elements in edible and inedible wild growing mushrooms from South Africa. *Environmental Science and Pollution Research*, *26*, pp.7913–7925.

Ray, A., Ray, R. and Sreevidya, E.A. 2021. Corrigendum: How many wild edible plants do we eat—Their diversity, use, and implications for sustainable food system: An exploratory analysis in India. *Frontiers in Sustainable Food Systems*, *5*, p.667541.

Renna, M., Cocozza, C., Gonnella, M., Abdelrahman, H. and Santamaria, P. 2015. Elemental characterization of wild edible plants from countryside and urban areas. *Food Chemistry*, *177*, pp.29

Réthy, B., Kovács, A., Zupkó, I., Forgo, P., Vasas, A., Falkay, G. and Hohmann, J. 2006. Cytotoxic phenanthrenes from the rhizomes of Tamus communis. *Planta Medica*, 72(08), pp.767–770.

Roeder, E. 1999. Analysis of pyrrolizidine alkaloids. *Current Organic Chemistry*, 3(6), pp.557–576.

Saha, A. K., Acharya, S. and Roy, A. 2012. Antioxidant level of wild edible mushroom: Pleurotus djamor (Fr.) Boedijn. *Journal of Agricultural Technology*, 8(4), pp.1343–1351.

Saleem, H., Khurshid, U., Tousif, M.I., Anwar, S., Ali, N.A.A., Mahomoodally, M.F. and Ahemad, N. 2022. A comprehensive review on the botany, traditional uses, phytochemistry, pharmacology and toxicity of Anagallis arvensis (L).: A wild edible medicinal food plant. *Food Bioscience*, 52, p.102328.

Sánchez-Mata, M.C., Cabrera Loera, R.D., Morales, P., Fernández-Ruiz, V., Cámara, M., Díez Marqués, C., Pardo-de-Santayana, M. and Tardío, J. 2012. Wild vegetables of the Mediterranean area as valuable sources of bioactive compounds. *Genetic Resources and Crop Evolution*, 59.

Saqib, A. A. and Gul, S. 2018. Traditional knowledge of medicinal herbs among indigenous communities in Maidan Valley, Lower Dir, Pakistan. *Bulletin of Environment, Pharmacology and Life Sciences*, 7, pp.1–23.

Seal, T., Chaudhuri, K., Pillai, B., Chakrabarti, S., Auddy, B. and Mondal, T. 2020. Wild-edible plants of Meghalaya State in India: Nutritional, minerals, antinutritional, vitamin content and toxicity studies. *Pharmacognosy Magazine*, 16(Suppl 1), pp.S142–S151.

Shad, A.A., Shah, H.U. and Bakht, J. 2013. Ethnobotanical assessment and nutritive potential of wild food plants. *Journal of Animal and Plant Sciences*, 23(1), pp.92–97.

Shaheen, S., Ahmad, M. and Haroon, N. 2017. *Edible Wild Plants: An alternative approach to food security* (p. 03). Springer International Publishing.

Sharma, H. and Kumar, A. 2011. Ethnobotanical studies on medicinal plants of Rajasthan (India): A review. *Journal of Medicinal Plants Research*, 5(7), pp.1107–1112.

Shedoeva, A., Leavesley, D., Upton, Z. and Fan, C. 2019. Wound healing and the use of medicinal plants. *Evidence-Based Complementary and Alternative Medicine*, 2019.

Shenstone, E., Lippman, Z. and Van Eck, J. 2020. A review of nutritional properties and health benefits of Physalis species. *Plant Foods for Human Nutrition*, 75, pp.316–325.

Singh, K. and Gairola, S. 2023. Nutritional potential of wild edible rose hips in India for food security. In Ajay Kumar, Pardeep Singh, Suruchi Singh, Bhupinder Singh (eds.) *Wild Food Plants for Zero Hunger and Resilient Agriculture* (pp. 163–179). Singapore: Springer Nature Singapore.

Souci, S.W., Fachmann, W. and Kraut, K. 2008. Food Composition and Nutrition Tables, 7th revised and completed edition. *Technology & Engineering.MedPharm*. 1300 pages.

Srivastava, K., Singh, S., Singh, A., Jain, T., Datta, R. and Kohli, A. 2023. Effect of temperature (cold and hot) stress on medicinal plants. In *Medicinal plants: Their response to abiotic stress* (pp. 153–168). Singapore: Springer Nature Singapore.

Stark, P.B., Miller, D., Carlson, T.J. and Rasmussen de Vasquez, K. 2020. Correction: Open-source food: Nutrition, toxicology, and availability of wild edible greens in the East Bay. *PloS one*, 15(9), p.e0239794.

Sutradhar, B., Deb, D., Majumdar, K. and Datta, B.K. 2015. Traditional dye yielding plants of Tripura, Northeast India. *Biodiversitas Journal of Biological Diversity*, 16(2), 121–127.

Suwardi, A. B., Navia, Z. I., Harmawan, T., Syamsuardi, S. and Mukhtar, E. 2020. Ethnobotany and conservation of indigenous edible fruit plants in South Aceh, Indonesia. *Biodiversitas Journal of Biological Diversity*, 21(5), pp.1850–1860.

Talang, H., Yanthan, A., Rathi, R.S., Pradheep, K., Longkumer, S., Imsong, B., Singh, L.H., Assumi, R.S., Devi, M.B., Kumar, A. and Ahlawat, S.P. 2023. Nutritional evaluation of some potential wild edible plants of North Eastern region of India. *Frontiers in Nutrition*, 10, p.205.

Tardío, J., Molina, M., Aceituno-Mata, L., Pardo-de-Santayana, M., Morales, R., Fernández-Ruiz, V., Morales, P., García, P., Cámara, M. and Sánchez-Mata, M.C. 2011. *Montia fontana L.* (Portulacaceae), an interesting wild vegetable traditionally consumed in the Iberian Peninsula. *Genetic Resources and Crop Evolution*, 58, pp.1105–1118.

Teixeira, B.et al.2012. European pennyroyal (Mentha pulegium) from Portugal: Chemical composition of essential oil and antioxidant and antimicrobial properties of extracts and essential oil. *Industrial Crops and Products*, 36(1), pp.81–87.

Thakur, A., Singh, S. and Puri, S. 2020. Exploration of wild edible plants used as food by Gaddis-a tribal community of the Western Himalaya. *The Scientific World Journal*, 2020. https://doi.org/10.1155/2020/6280153.

Thomford, N. E., Senthebane, D. A., Rowe, A., Munro, D., Seele, P., Maroyi, A. and Dzobo, K. 2018. Natural products for drug discovery in the 21st century: Innovations for novel drug discovery. *International Journal of Molecular Sciences*, 19(6), p.1578.

Trichopoulou, A., Vasilopoulou, E., Hollman, P., Chamalides, C., Foufa, E., Kaloudis, T., … Theophilou, D. 2000. Nutritional composition and flavonoid content of edible wild greens and green pies: A potential rich source of antioxidant nutrients in the Mediterranean diet. *Food Chemistry*, *70*(3), pp.319–323.

Umar, M. F., Ahmad, F., Saeed, H., Usmani, S. A., Owais, M. and Rafatullah, M. 2020. Bio-mediated synthesis of reduced graphene oxide nanoparticles from chenopodium album: Their antimicrobial and anticancer activities. *Nanomaterials*, *10*(6), p.1096.

Upadhyay, R.K. 2018. Plant pigments as dietary anticancer agents. *International Journal of Green Pharmacy (IJGP)*, *12*(01). https://doi.org/10.22377/ijgp.v12i01.1604

Vacillotto, G., Favretto, D., Seraglia, R., Pagiotti, R., Traldi, P. and Mattoli, L. 2013. A rapid and highly specific method to evaluate the presence of pyrrolizidine alkaloids in Borago officinalis seed oil. *Journal of Mass Spectrometry*, *48*(10), pp.1078–1082.

Vadivel, V. and Janardhanan, K. 2005. Nutritional and antinutritional characteristics of seven South Indian wild legumes. *Plant Foods for Human Nutrition*, *60*, pp.69–75.

Verma, R. N., Singh, G. B., & Bilgrami, K. S. (1987). Fleshy fungal flora of NEH India-I. Manipur and Meghalaya. *Indian Mushroom Sciences*, *2*, 414–421.

Verma, R. K. 2014. An ethnobotanical study of plants used for the treatment of livestock diseases in Tikamgarh District of Bundelkhand, Central India. *Asian Pacific Journal of Tropical Biomedicine*, *4*, pp.S460–S467.

Waktola, G. and Temesgen, T. 2018. Application of mushroom as food and medicine. *Advances in Biotechnology and Microbiology*, *113*, pp.1–4.

Yimer, A., Forsido, S. F., Addis, G. and Ayelign, A. 2023. Nutritional composition of some wild edible plants consumed in Southwest Ethiopia. *Heliyon*, *9*(6).

2 Modern Approaches in Identifying Medicinal Wild Edible Plants

Sneha Joshi, Yogesh Godiyal and Devesh Tewari

2.1 INTRODUCTION

Forests have played a central role for providing food, fuel, and fodder as well as medicine to communities. Even though humans are primarily dependent on agriculture for food, millions of people, including rural populations and forest-dwelling communities, rely on wild plant resources for food and medicine. Indigenous people still follow many cultures and traditions associated with nature. For instance, Vietnamese people commonly use Chinese herbal medicine and other folk practices to cure various diseases, which is often termed as "Southern medicine". Additionally, Indian Ayurveda is also considered to be an important method to treat various ailments (Lele et al., 2021).

Wild edible plants (WEP) are those species that grow and reproduce naturally without being cultivated. This plays a central role in empowering local markets by reducing the distance between consumer and producer, hence decreasing the dependency on globalized value chains. It is important to note that the current global food system is sufficient for the survival of mankind, yet many still experience hunger. Even though most of society counts on agricultural crops for everyday existence, the utilization of WEPs has waned completely. As per the reports of the Food and Agriculture Organization (FAO), about one billion people consume wild food in their diet. It is important to mention that WEPs have a salient role in poverty eradication, diversification of agriculture, generation of income resources, security of food availability, and alleviation of malnutrition. The high nutritive value of many WEPs, may decline the problem of food insecurity.

Furthermore, the rise in processed food consumption has led to many fatalistic effects on human health. Certain factors such as climate change and under or over nutrition along with global threat are reasons for the urgent requirement of a healthier and sustainable food system. Wild plants can be substituted as a healthy alternative for plants that are cultivated with pesticides and other chemicals. Wild species are a great source of flavors, dietary supplements, and bioactive compounds. The knowledge of WEPs has a huge impact on agriculture to increase the crop availability where extreme weather conditions such as high temperatures or rain can be tolerated (Motti, 2022).

Classification: Wild edible plants can be classified into nine categories. The further subdivision on vegetables is as follows: cooked (Veg C), raw (Veg R), and preserved. The cooked (Veg C) are generally those which can be consumed only after cooking, grinding, boiling, or mixing with yogurt, whereas the preparation conserved in mustard oil and consumed afterwards as food is considered preserved. Those consumed directly only after washing are categorized as raw. A similar classification is applied for fruits as raw (Fr R) and processed (Fr P). Additional categories are spices (Sp), beverages (Bv), and medicinal plants (Med) (Bhatia et al., 2018).

Wild edible plants play a key role in food security in rural communities as well as providing a proper nutritious diet around the world. Despite the fact that wild vegetables are a good source of many micro and macro nutrients, their phytochemical content is not properly studied. Consider a few examples of wild edible plants such as *Maianthemum atropurpureum*, which grows in high altitude mountains and belongs to family asparagaceae. Although it is consumed in salads and other

DOI: 10.1201/9781003395935-2

dishes, it has excellent medicinal properties in annealing, detoxicating, and low blood pressure. The rhizome part of this medicinal plant can be used to treat menstrual disturbance, kidney disease, rheumatism, and others. Apart from that several nucleosides, flavonoids and saponins have also been isolated from *M. atropurpureum* (Xu et al., 2021).

Psidium guineense a small tree or shrub which generally grows in the region of South and Central America and is widely distributed in Brazil. Despite the fact that the species does not have high nutritional content, it is rich in fiber and other micronutrients such as calcium, magnesium, and zinc. Along with that *P. guineense* has a high content of organic acids, that is, citric acid. Another tree native to South and Central America is *Genipa americana*. The fruit has high levels of calcium, phophorous, and sugar. However, the variation in contents is generally observed as deterioration of fruit occurs quite quickly after the harvest. One of the striking features of *G. americana* is the antineoplastic properties, especially in cancer cells in bladders. In addition, a blue-colored dye can also be extracted from the green fruit of *G. americana*, which is a promising substitute to synthetic dye (de Medeiros et al., 2021).

Indigenous knowledge of wild edible plants is very important, as many plants which are cultivated and used today by civilized societies were initially recognized and developed through the same method. Regrettably, due to the arrival of processed foods, the usage of indigenous food plants is also losing familiarity in rural areas. Hence the declining knowledge on wild edible plants is an alarming warning to speed detailed research work on their nutritive value. Consumption of wild plants is very primitive; however, the level of dependency varies significantly from one community to another. In many traditional cultures people are dependent on WEPs as a staple food for providing a balanced diet to fulfill their nutritional intake. WEPs can also be a source of fiber, energy and micronutrients along with other phytochemicals which includes phenol, flavones, tannins, terpenoids, polysaccharides, steroids, and so on (Dhole, 2021).

Recent scientific research also focuses on filling the void between growing population and food production; hence the researchers also value these traditional foods as they have high nutritive value. The food prepared from these wild species are high in macro- and micronutrient content along with close-to-community culture (Deb et al., 2014).

Proximate analysis on some wild edible plants shows that the nutritional constituents are superior to most of the domesticated varieties of plants. In addition, many wild plants have a high content of vitamin B_2, vitamin C, and protein, which are generally lacking in a conventional human diet. Wild plants contain secondary metabolites as well, which include polyphenols, polysaccharides, and terpenoids. Due to the presence of rich antioxidants and fiber, wild plants can cure multiple disorders including CVS problems, urinary tract problems, inflammation, and digestive disorders. Market assessment on wild plants shows that they are not only utilized for food and medicinal purposes but also used as timber, construction, agriculture tools, fuel, and so on. Consumption of wild edible plants would be beneficial for the tribal population to improve nutrition because they have limited access to anthropogenically produced plants. Certain wild leafy vegetables are great sources of carotenoids, which include vitamin A. If the leaves are consumed along with fats, they maintain a constant vitamin A supply year round (Chauhan et al., 2018).

Before civilization man had a huge influence on wild edible plants due to the nutritive value along with medicinal importance. Nearly 45,000 species of wild plants are known, out of which 9,500 species have ethnobotanical importance along with 7,500 species for medicinal use. About 3,900 species are used by tribal people as food material, among which 145 species constitute roots and tubers while the other 521 species include leafy vegetables. The conventional cultivated vegetables require a lot of supervision as compared to wild plants. Due to social and cultural changes the traditional knowledge of wild edible plants is declining day by day. Therefore, there is an urgent need to document and revivify the knowledge on wild edible plants and preserve it for future generations (Naik et al., 2017).

Some WEPs are used for the preparation of both alcoholic and nonalcoholic beverages, for instance herbal tea (tisanes) and liqueurs. However, medicinal beverages are only taken for a few days to treat specific conditions. Wild plants used as folk medicines are prepared by infusion and decoction techniques to treat several ailments. For digestive problems *Beta maritima*, *Borago officinalis*, and *Humulus lupulus* are used; respiratory problem treatments include *Mentha pulegium*, *Helichrysum stoechas*, and *Genista tridentate*. Other used as disinfectants or for anti-inflammatory effects are *Origanum vulgare*, *Glecoma hederacea*, *Anchusa azurea*, and others. (Pinela et al., 2017).

2.2 ROLE AND IMPORTANCE OF PLANT IDENTIFICATION

Flora and fauna play an important role in preserving the environmental balance. Plants are crucial in sustaining biodiversity, although human civilization development disturbs the ecological balance. Biodiversity preservation at the central level comes with lot of challenges such as drastic climate change, pollution, deforestation, emission of greenhouse gases, and so on. It is one of the biggest responsibilities to search for methods and various techniques to serve our biodiversity. A foremost step toward this is conservation, which requires plant identification to quantify biodiversity. Identification of plants is not only the responsibility of ecologists and botanists, but also of a larger section of society that includes professionals from landscape architect, forestry farmers, and eco tourists.

Plant identification simply means associating it to a scientific name. In other words, assigning that plant to a group called taxon; certain rules are followed to set a name for taxon. The whole process of applying scientific names and delimiting taxa is called taxonomy (systematics) (Bojamma & Chandrasekar, 2019).

2.3 CHALLENGES IN THE IDENTIFICATION OF WILD EDIBLE PLANTS

A lot of challenges are encountered by researchers to provide a reliable and efficient automated method for wild plant species identification. First, distinguishing the plant species into various categories as class, family, species, and genus is a complex task. Another problem faced by researchers is morphological variation in plants belonging to same species due to differences in geographical location, climatic factors, and genetic mutation. In addition to that some species show great similarity even though they belong to different species. Furthermore, variation in image can also cause difficulty in plant identification as the entities are 3D in nature and images are collected purely in 2D projections. Even the quality of image plays a vital role in correct identification because poor quality image leads to wrong plant species authentication.

Challenges found in wild edible plant identification are: (i) It is limited to local people who understand and recognize medicinal plants' usage as well as identification. (ii) Delimited number of books/guides which are difficult to be used in field. (iii) Few authorized institutes and facilities for wild plant identification. (iv) Time consuming and expensive process along with limited number of taxonomists in authorized institutes (Herdiyeni & Adisantoso, 2011) (Figure 2.1).

The morphological features present in plant species that are generally used for identification cannot be used for powdered or other processed plant material. This calls for an urgent requirement of a commercial tool for detection, substitution, and authentication of wild species. The commercial methods used for wild plant species authentication are physical, chemical, or biochemical methods, other methods include immunoassay as well as DNA-based molecular tool. However, these methods require a trained person for taxonomical examination which is a very tedious process. On the other hand, the classical method of wild plant identification involves micro- and macroscopic characters, organoleptic methods, and chemical profile. These methods are not highly successful as they are affected by physiological and storage conditions. For identification and authentication of a large number of organisms, several genome-based methods have been developed during the last decade, but no single method provides a comprehensive solution for plant identification. It is worth mentioning that the advancement in molecular techniques has enabled the scientist to use simple and rapid cost-effective DNA analysis. However, contamination in products by any means leads to compromised therapeutic activity with failure of DNA barcoding to identify and authenticate the wild edible plant material (Mishra et al., 2016).

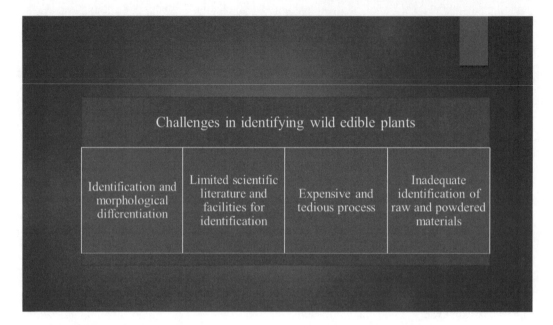

FIGURE 2.1   Challenges in identification of WEPs.

2.4   TRADITIONAL IDENTIFICATION METHOD

The basic and primary objective of systematics is identification. It involves both classification and nomenclature. Prior to the emergence of modern approaches, anatomical and morphological description were predominantly used for plant identification. The conventional approach is declining due to lack of high-level expertise and decreased effectiveness because of incomplete morphological keys. Another mean of authentication was to ascertain the physicochemical properties of individual drugs or proprietary medicine by differentiating them with the standard value of Indian pharmacopoeia along with the botanical aspect supplemented by pharmaceutical chemistry for quality assurance of the drug. Various phytochemical methods were used such as high-performance liquid chromatography, mass spectroscopy, nuclear magnetic resonance spectroscopy, and thin layer chromatography, as well as Fourier transform infrared spectroscopy. In certain circumstances such as differing temperature, light, humidity, soil composition, and storage conditions the phytochemical profiles may vary (Herdiyeni & Adisantoso, 2011).

a. **Expert determination:** This is the best identification method in terms of accuracy and reliability. The expert prepares monographs, synopses, and other groups of questions that include the expert's concept of taxa. Even though the method is precise, it requires a lot of the expert's time, which results in delayed identification.

b. **Taxonomic identification:** The science of classifying and naming the plants is called plant taxonomy; it is also known as systematics, which determines how different organisms are related to each other. The primary goals of plant taxonomy are identification, characterisation, classification, and, lastly, nomenclature. On the other hand, the word systematics originates from the word "*systema*" which means systematic arrangement of organisms. Organisms are sectioned according to different taxonomic categories depending on similarities and specific features, which includes a hierarchical order of: Kingdom, Phylum, Class, Order, Family, Genus, and Species. Identification simply means to recognize an unknown specimen with already available taxon and ascribing it a correct position in the classification. This can be achieved by checking a herbarium and comparing it with identified specimens stored in a herbarium. Alternately it can also be sent to an expert in the field who can guide with the identification. In addition, identification can also accomplished by various literatures such as monographs, floras, and use of identification keys.

c. **Recognition:** This approach is basically dependent on expert determination; it is based on the past experiences of the identifier with a group of plants. This method is not applicable to the larger section plant species.

d. **Comparison:** It is the least abstract approach in the identification process, as comparison of an unknown with a named specimen is done by various means such as photographs or other illustrations. This approach is feasible when identified specimens are available and taxa are manageable, but it gets out of hand for diverse taxonomical groups. Even though the method is reliable, it is time consuming and virtually not possible due to the insufficiency of required material.

e. **Construction and identification keys:** The word key is generally used in English as a solution to something; however, in the process of plant identification it means set of rules for constructing dichotomous keys. This is a pathway of characters that leads to plant identification and a well-constructed key limit the number of steps for identification. It is by far the most widely used method as it is less time consuming and does not require any experienced personnel in comparison and recognition. In a conventional sense, keys are a type of taxonomical literature consisting of numerous contrasting or contradictory statements required by an identifier to make comparisons and decisions for the material to be identified (Walter & Winterton, 2006).

2.5 MODERN APPROACHES IN IDENTIFYING WEPS

Biodiversity conservation has led to the increased demand for routine species identification while the percentage of expert taxonomists is limited and declining.

Digital media is paramount for everyday life as many applications have been developed through various devices for plant identification. These applications use pattern recognition to automatically identify plants by using digital image recognition technique. The program commonly uses images of a particular feature of the plant, which can be either leaves or blossoms, for identification. All the programs upload the picture on a server for analysis, which performs the procedure followed by the species identification.

Apart from digital media, taxonomists are searching for more efficient methods of wild edible plant identification. One such method can be artificial intelligence, which, along with digital image processing, will lead to a revolution in modern identification methods. Taxonomic identification of wild edible plants is being carried out by various molecular techniques that are also responsible for quality control of herbal drugs. The lack of morphological information has led to the development of molecular and biochemical techniques for identification. The recent finding reveals that plant identification is precisely done by using molecular markers.

2.5.1 DNA-Based Method for WEPs Identification

DNA-based methods are preferred over other methods as they are more efficient, less time consuming, require a smaller amount of plant material, and are less expensive. Currently DNA barcoding has proved to be an advantageous tool for accurate and rapid plant identification using standardized DNA markers. However, a few methods are not efficient enough to identify plants in a mixture of several plant species. During the last decade various genome-based methods have been developed for authentication and identification but no single method can be universally accepted as it is unable to meet a comprehensive solution for the many problems concerning plant identification.

Authentication at DNA level is more reliable than other methods (protein or RNA), as it is stable and is not affected by external factors. To overcome the difficulties of traditional taxonomy, DNA barcoding is one of the most successful identification techniques among other genome-based approaches.

In the DNA barcoding technique one must standardize a region of DNA for identification, termed as a DNA barcode. This barcode consists of a small part of the genome which can be easily obtained. This technique is the oversimplified version of a complex problem which will authenticate the plant material. It is quite challenging to perform DNA barcoding for medicinal plants both in generating barcode as well as analyzing the data. Despite the obstacles being faced there is an increase in the number of barcoding studies in plant instant identification. Apart from its authentication it has applications in conservation impact assessment, biodiversity monitoring, forensic, botany, and more. The success and failure of any plant authentication generally depends on how barcoding is combined with other tools (Howard et al., 2012).

2.6   COMPUTER VISION AND MACHINE LEARNING

Wild plant identification increasingly relies on machine learning algorithms, which use images for taxonomic classification. These methods are very powerful for identification of rich and morphologically diverse species. Such rapid increase in machine learning can elevate the capacity of wildlife species identification without burdening the taxonomy experts. The classification done by these automated images will remove the need for manual cross-checking by taxonomic experts (Fujisawa et al., 2021).

Computer vision is generally concerned with replicating human vision using software and hardware. It requires the combined knowledge of electrical engineering, computer science, mathematics, biology, and physiology as well as cognitive science. It basically requires complete understanding from all the fields to stimulate the operation. Image-based recognition techniques studies show how to interpret reconstruct and understand a 3D scene from its converted 2D image. It is done in three steps which are as follows:

   (i) **Low level:** In this step feature extraction is done, that is, corner, edge, or optical flow.
  (ii) **Intermediate level:** In this step object recognition and interpretation of 3D image is carried out.
 (iii) **High level:** This step involves further progress of that interpreted data in intermediate level.

In the current scenario there is an increased interest in using an automatic process for species identification. The accessibility and prevalence of technologies such as mobile devices, digital cameras, and so forth has led to image processing and pattern recognition as the idea of automated species identification becomes closer to reality, driving the recent search of taxonomists for an efficient identification method using image and pattern recognition technique. Digital image processing generally uses algorithms and procedures for image compression, enhancements, analysis, mapping, and so on. These methods are widely accepted for wild species identification as users can take a picture of a plant and directly analyze it with the help of an application. With the use of computer-aided plant identification even a nonprofessional person can take part in the process. The image classification process is divided as:

   (i) **Image Acquisition:** In this step a whole image of a plant is taken so that analysis can be performed.
  (ii) **Pre-processing:** This step enhances image data to suppress the distortions for further processing. It is a subprocess that receives an image as input and modifies an image as output.
 (iii) **Feature extraction and description:** This step involves taking measurements, either geometric or other. Features are described as numbers, which is a characteristic of plant properties.
 (iv) **Classification:** All the extracted features are combined, which are further classified (Wäldchen & Mäder, 2018).

2.6.1 AUTOMATED TAXON IDENTIFICATION

Environmental balance is preserved by the local flora and fauna in the region; despite enormous human development disrupting the ecosystem stability, plants play an important role in maintaining biodiversity. It is one of the biggest responsibilities to conserve the ecosystem by rightly identifying plants to quantify biodiversity. A computerized automated technique is helpful for plant identification as it can conserve natural reserves, botanical gardens, forests, endangered species, and so on. Automated identification is accurate as well as less tedious than manual identification. The objective of scientists is to develop a computerized system of plant identification that reduces the errors stemming from normal vision. For plant recognition various features are available such as flower, leaf, root, fruit, and so on. A number of studies only utilize leaf shape for shape recognition techniques, which will help to model and contour the shape of leaves. More features like texture and color will be considered to improve recognition accuracy. Previously, shape-based recognition used geometrical parameters like breadth, width, length, area, and perimeter to identify, resulting in a 95% accuracy rate. Currently the amalgamation of certain features such as size, shape, and texture of leaves has resulted in progression of plant identification.

It is important to mention that various steps have been made to conserve biodiversity due to the alarming rate of lost species. Plant identification is not only the job of ecologists and botanists but of a large section of society that includes forestry, farmers, landscape artists, ecotourists, conservationists, and many more. Both traditional and conventional methods of plant identification are time consuming as well as resulting in false identification. Biodiversity conservation includes study of climate change, records of endangered species, weed control activities, and is dependent on plant identification techniques. Due to the increasing demands of correct identification the experts in the field are declining as it requires rigorous training and years of experience. Several specialties including scientists, researchers, and botanists have come up with automated plant identification systems. Advancement in technology such as computer image processing and pattern recognition as well as algorithms have helped in plant taxonomy identification (Sun et al., 2017).

2.6.2 DYNAMIC POLYCLAVE

Polyclaves are another type of key which is an alternative to dichotomous key that is becoming immensely popular because of the ease to computerize them. Polyclaves allow selection of characteristics used in specimen identification. Generally, a printed data table, matrix, or chart mentioning the useful characteristics is used as a polyclave, thereby listing all the possible taxa on scratch paper. For identification of larger groups this diagnostic table is more powerful than equivalent keys and requires less space to print. Polyclaves do not require to be automated but might be computerized when a large information base is involved. It is advantageous to use an online computer program for the batch processing system as it allows the user to submit additional data during the procedure, thereby resulting in a conversation between computer and user. It is important to mention that numeric coding is done for lengthy answers as it reduces the chances of error as well as saves printing time.

A polyclave algorithm includes three steps: (i) asking the user for more than one characteristic of the specimen, (ii) eliminating all inconsistent possibilities, and (iii) printing the result of elimination, either an identification or other. Taxon identification can start with a data matrix at any taxonomic hierarchy, however continuation to a subsequent matrix is possible following a family, genus, or any other higher level. Immediately after each identification, other diagnostic characteristics of taxon are listed.

Advantages of Polyclave Keys: Polyclave are easy to use, and they allow the user for multi entry, that is, it can be started from anywhere. Furthermore, the user can work in any direction as they are order free. They are faster as well as easily computerized, which is less tedious than paper versions (Nguyen, 2015) (Figure 2.2).

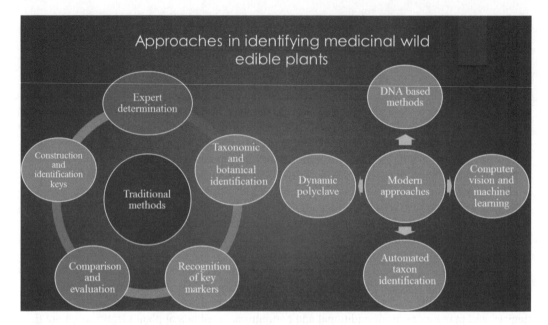

FIGURE 2.2   Approaches in identifying medicinal wild edible plants.

2.7 CONCLUSION AND FUTURE PERSPECTIVE

Numerous studies have highlighted the significance of wild flora consumed largely by indigenous populations and that play an important role in their household diets. They have great potential to serve as food and nutritional sources, yet their utilization remains hindered due to lack of proper identification and species documentation. Proper identification of the families and plants suitable for employment as a source of food is essential for their popularization, with an aim to popularize underused families. These plants with their enormous value and diverse flavor and aroma can also be employed for a healthy and beneficial diet. Development of proper channels to document the traditional knowledge available on the use of WEPs should be strengthened. Conservation of the wild flora is also an important takeaway lesson that can only be ascertained using proper identification tools. Enormous efforts are needed to specify the parameters required for proper identification with an aim for incorporation of highly nutritious flora into diets, thereby strengthening the global value chain.

REFERENCES

Bhatia, H., Sharma, Y. P., Manhas, R. K., & Kumar, K. (2018). Traditionally used wild edible plants of district Udhampur, J&K, India. *Journal of Ethnobiology and Ethnomedicine, 14*(1), 73. https://doi.org/10.1186/s13002-018-0272-1

Bojamma A. M., & Chandrasekar, B. (2019). A study on the machine learning techniques for automated plant species identification: current trends and challenges. *International Journal of Information Technology, 13*. https://doi.org/10.1007/s41870-019-00379-7

Chauhan, S. H., Yadav, S., Takahashi, T., Łuczaj, Ł., D'Cruz, L., & Okada, K. (2018). Consumption patterns of wild edibles by the Vasavas: a case study from Gujarat, India. *Journal of Ethnobiology and Ethnomedicine, 14*(1), 1–20. https://doi.org/10.1186/s13002-018-0254-3

de Medeiros, P. M., dos Santos, G. M. C., Barbosa, D. M., Gomes, L. C. A., da Santos, É. M. C., & da Silva, R. R. V. (2021). Local knowledge as a tool for prospecting wild food plants: experiences in northeastern Brazil. *Scientific Reports, 11*(1), 1–14. https://doi.org/10.1038/s41598-020-79835-5

Deb, D., Sarkar, A., Barma, B. D., Datta, B. K., & Majumdar, K. (2014). Wild edible plants and their use in traditional recipes of tripura, northeast India. *Middle - East Journal of Scientific Research, 19*(2), 277–285. https://doi.org/10.5829/idosi.mejsr.2014.19.2.12222

Dhole, P. A. (2021). Wild edible plantswith its socio-economic importance used by tribes of Gaya District, Bihar. *International Journal of Advanced Research in Biological Sciences*, 8(1), 51–58. https://doi. org/10.22192/ijarbs

Fujisawa, T., Noguerales, V., Meramveliotakis, E., Papadopoulou, A., & Vogler, A. P. (2021). Image-based taxonomic classification of bulk biodiversity samples using deep learning and domain adaptation. *BioRxiv*, 2021.12.22.473797. https://doi.org/10.1101/2021.12.22.473797

Herdiyeni, Y., & Adisantoso, J. (2011). *Computer Vision for Plant Identification*. https://api.semanticscholar. org/CorpusID:17619541

Howard, C., Socratous, E., Williams, S., Graham, E., Fowler, M. R., Scott, N. W., Bremner, P. D., & Slater, A. (2012). PlantID - DNA-based identification of multiple medicinal plants in complex mixtures. *Chinese Medicine*, 7(1), 18. https://doi.org/10.1186/1749-8546-7-18

Lele, Y., Thorve, B., Tomar, S., & Parasnis, A. (2021). Wild Edible Plants of Nutritional and Medicinal Significance to the Tribes of Palghar, Maharashtra, India. *Journal of Ecological Society*, 32–33(1). https://doi.org/10.54081/jes.027/03

Mishra, P., Kumar, A., Nagireddy, A., Mani, D. N., Shukla, A. K., Tiwari, R., & Sundaresan, V. (2016). DNA barcoding: an efficient tool to overcome authentication challenges in the herbal market. *Plant Biotechnology Journal*, 14(1), 8–21. https://doi.org/10.1111/pbi.12419

Motti, R. (2022). Wild Edible Plants: A Challenge for Future Diet and Health. In *Plants* (Vol. 11, Issue 3). https://doi.org/10.3390/plants11030344

Naik, R., Borkar, S. D., Bhat, S., & Acharya, R. (2017). Therapeutic potential of wild edible vegetables – A Review. *Journal of Ayurveda and Integrated Medical Sciences (JAIMS)*, 2(s6). https://doi.org/10.21760/ jaims.v2i06.10930

Nguyen, S. (2015). Biokeys – An integrated system for working with database and polyclave identification keys of various taxonomic levels. *Journal of Science and Technology*, 53. https://doi.org/10.15625/0866 -708X/53/2/3684

Pinela, J., Carvalho, A. M., & Ferreira, I. C. F. R. (2017). Wild edible plants: Nutritional and toxicological characteristics, retrieval strategies and importance for today's society. *Food and Chemical Toxicology: An International Journal Published for the British Industrial Biological Research Association*, 110, 165–188. https://doi.org/10.1016/j.fct.2017.10.020

Sun, Y., Liu, Y., Wang, G., & Zhang, H. (2017). Deep learning for plant identification in natural environment. *Computational Intelligence and Neuroscience*, 2017, 7361042. https://doi.org/10.1155/2017/7361042

Wäldchen, J., & Mäder, P. (2018). Plant Species Identification Using Computer Vision Techniques: A Systematic Literature Review. In *Archives of Computational Methods in Engineering* (Vol. 25, Issue 2). Springer Netherlands. https://doi.org/10.1007/s11831-016-9206-z

Walter, D. E., & Winterton, S. (2006). Keys and the crisis in taxonomy: Extinction or reinvention? *Annual Review of Entomology*, 52(1), 193–208. https://doi.org/10.1146/annurev.ento.51.110104.151054

Xu, L., Wang, Y., Ji, Y., Li, P., Cao, W., Wu, S., Kennelly, E., & Long, C. (2021). Nutraceutical study on Maianthemum atropurpureum, a Wild Medicinal Food Plant in Northwest Yunnan, China. *Frontiers in Pharmacology*, 12(July), 1–10. https://doi.org/10.3389/fphar.2021.710487

3 Efficacy versus Toxicity of Wild Edible Plants

An Inherent Contradiction

Muhammad Shoaib Amjad, Asia Farooq, Huma Qureshi and Zakia Binish

3.1 INTRODUCTION

The main resource used by humans to sustain their life on Earth is plants. All aspects of life can benefit from them. Their applications are not just limited to written works; abundant information is still needed in conventional daily living. Wild edible plants (WEPs) are all non-cultivated plant species that naturally grow on farms, uncultivated land, and in forests that have edible components (Demel & Abeje, 2004). According to Borelli et al. (2020), "plant species that grow and reproduce spontaneously in their natural habitation without being cultivated" is another way to describe wild edible plants. They are referring to non-domesticated plant species that are taken from the environment by the local populations and used as food (Ju et al., 2013). WEPs develop organically and have an impact on communities' ability to sustain themselves without the help of humans. Their use predated the widespread use of agriculture in ancient civilizations (Ojelel et al., 2019). WEPs play a crucial role in sustaining the daily global food basket in all parts of the planet. They are the rural population's only source of food, especially during hunger and drought (Tebkew et al., 2018). The utilization of wild edible plants has been validated by 'State of the World's Biodiversity for Food and Agriculture' (SOWBFA) in numerous nations (FAO, 2019).

According to study, 1,955 different kinds of wild edible plants contribute to food security and nutrition in 69 out of 91 different nations (Ulian et al., 2020). They also help to make diets more varied and healthier. The majority of the time, populations all over the world gather wild edible plants from a variety of ecosystems, including woods, cultivable fields, and even anthropogenically degraded areas like wasteland and roadside (FAO, 2019). Additionally, agricultural fields and hillsides are where they are gathered. Their diversity and accessibility change with the seasons. Locals are knowledgeable about the seasonal availability and best times to pick several species of wild food plants (Menendez-Baceta et al., 2012). Various people around the world harvest wild edible plant species for nutrition, food, and revenue generating. They are the likely source of sustenance since, often, they are more nutrient-dense than conventional crops (Ojelel et al., 2019).

3.2 APPLICATIONS OF WILD EDIBLE PLANTS (FOOD, FEED AND MEDICINE)

3.2.1 FOOD

More than 7,000 species of WEPs have been utilized as food at some point in human history, according to ethnobotanical research. Millions of individuals regularly include WEPs in their diets (Carvalho & Barata, 2016). The genetic and cultural heritage of various regions worldwide includes WEPs. These are sources of nutrients and health-promoting substances, and they have gained significant importance during times of scarcity and hunger, especially in rural and suburban areas (Pinela et al., 2017). One alternative to wholesome and nutritious cuisine is WEPs. They play a crucial role in promoting global food security in every region of the world. By providing them with food and

DOI: 10.1201/9781003395935-3

meeting their nutritional needs, these edible plant species have significantly improved the health of underprivileged communities (Duguma, 2020). Numerous researchers examined the significance of WEPs in the food chain, particularly during the dry and wet seasons or when there is severe food insecurity (Svanberg, 2012; Powell et al., 2014). Locally harvested WEPs are essential to ensuring global food security because they are vital to the human food chain and boost the economic well-being of the indigenous population (Ju et al., 2013).

One of the most important contributions of wild edible plants is their use in traditional societies' food security by giving them access to diverse and alternative food sources (Abbasi et al., 2013). People have been gathering wild food plants from the beginning of time for their own sustenance and for traditional food systems. Low yields do not diminish their contribution to food security, food sovereignty, or even the well-being of their vulnerable households; rather, they increase it (Borelli et al., 2020). Many Ethiopians from the countryside often rely on eating wild plants to survive when there is a food shortage (Fentahun & Hager, 2009). Around the world, various WEPs are consumed as food. The components of those most frequently used were fruits, leaves, and gums. According to one of the writers, fruits are the most often used components of WEPs (Anbessa, 2016).

3.2.2 Feed

Food from WEPs is essential and dependable for both humans and animals. Southwest China is home to about 240 varieties of natural plants that the Asian elephants use as food (Jiang et al., 2019). Wild edible fruits were exploited as cattle feed, according to a study. *Rosa* spp. and *Sorbus aria* produce fleshy fruits that are utilized as fodder. Similar to this, rose fruit is used as goat feed (Pascual & Herrero, 2017). The health of ruminants is greatly hampered by the fatal effects of heavy metals. Additionally, feed production is quite low in many locations due to water scarcity. Therefore, the majority of ruminant animals in that area eat wild plants (Khan et al., 2021). In the area, *Asphodelus albus* Mill. is regularly picked and used to feed pigs and clean pastures (Pascual & Herrero 2017). Along with these plants, oak acorns are also eaten as food by both people and animals in the areas where these species are found (Menendez-Baceta et al., 2012).

3.2.3 Medicine

Wild edible plants provide considerable pharmacological benefits in addition to being valuable in terms of nutrition (Balemie and Kebebew, 2006, Guarrera and Savo, 2013). Different WEPs have been reported for their therapeutic benefits throughout the world (Vitalini et al., 2013). Due to their therapeutic properties, some edible plant species are occasionally also consumed. Different plants are frequently employed in folk phytotherapy as herbal remedies for a variety of diseases (Motti, 2021). Due to their abundance of fiber components and antioxidants, wild fruits also have therapeutic potential for a variety of diseases, including those of the urinary and digestive systems, cardiovascular issues, diabetes, and inflammatory conditions (Alissa & Ferns, 2017). Around 64% of people worldwide depend on traditional medicine to keep their health in check (Phondani et al., 2016). Folk medicine is the sum of all experiments and knowledge used to identify, prevent, and treat illnesses. Traditional medicine only uses observations and experiences that have been passed down verbally and written from generation to generation (WHO, 2000). Through trial-and-error experimentation by ancient people who struggled with health issues such as severe injuries and the misery of pain and sickness, the healing abilities of WEPs have flourished over time (Flatie et al., 2009). The large rural populace in many underdeveloped nations of the world may easily get affordable medicines through WEPs. Additionally, they contain nutrients including vitamins, proteins, and carbs, which have been discovered to be abundant in them. In addition, various health-promoting chemicals, including phenolic compounds, are found in WEPs (Seifu et al., 2017; Sir Elkhatim et al., 2018). Recent reports refer to a variety of wild edible plants as "functional foods" due to their nutritional content. These plants give us a nutritious diet and can help with disease prevention (Pieroni et al., 2018).

Additionally, WEPs have a good fatty acid profile for human diets. Fatty acids are useful in reducing the number of diseases (autoimmune, cardiovascular, and inflammatory), as well as in the prevention of cancer, since the consumption of these vital nutrients is linked to good vision and brain functions (Gemedo-Dalle et al., 2005). One of the most important factors influencing metabolic homeostasis, which includes poor glucose and lipid metabolism, is diet (Gujjala et al., 2017).

3.2.4 THE ROLE OF WILD EDIBLE PLANTS IN INCREASING NUTRITIONAL INTAKE

Wild edible plants contribute significantly to a community's intake of nutrients. They are either utilized as emergency food during famines or are used during periods of seasonal food shortage to cover a food gap and offer necessary nourishment in normal times (Duguma, 2020). The use of wild vegetables, which contain vitamins A and C, iron, thiamine, calcium, niacin, and riboflavin, is crucial for overall micronutrient intake (Ogle et al., 2001). According to research, underutilized green leafy vegetables are full of readily available minerals like ascorbic acid, beta-carotene, iron, and calcium that aid to prevent micronutrient malnutrition (Gupta et al., 2005). A significant legume is the *Vigna unguiculata* (L.) Walp. (Cowpea), a member of the Fabaceae family. In many developing nations, it serves as a cheap source of nutritional protein for people whose daily consumption of protein from other sources is insufficient (Shaheen et al., 2017a, 2017b). Numerous wild fruits and vegetables are significant sources of dietary minerals. They should be regarded as a good source of minerals for humans, and their intake should be increased both to preserve the cultural heritage of traditional foodways and to enhance the nutritional value of the modern human diet (García-Herrera & de Cortes Sánchez-Mata, 2016). When it comes to nutrition, WEPs are sources of a variety of micronutrients that are important for human health and development but are frequently lacking in the staple foods of underdeveloped nations (Leakey, 2005). Given that the world's population is expected to exceed 10 billion people by the year 2050, biodiversity is a critical component in addressing the dual concerns of excess calorie and nutrient intake (obesity) and deficiency of nutritional intake (hunger) that humanity is increasingly confronting today (Ulian et al., 2020). A list of chosen plants utilized as food, medicine, and food supplements is shown in Table 3.1.

3.3 NUTRITIONAL COMPOSITION AND VARIATION OF SELECTED WILD PLANTS

The significance of wild edible plants as a source of nutrients and suppliers to human dietary needs was highlighted by numerous studies conducted in the recent years (Sánchez-Mata et al., 2016). A wide variety of wild foods, such as leafy vegetables, nuts, fruits, flowers, yams, tubers, and berries provide us with a complete diet. These vegetables are essential for the survival of those who live in forested environments (Deshpande et al., 2019). They can consequently have a big impact as a crucial component of people's diets in particular regions of the world and offer more dietary variety for those who rely on them (Motti, 2021).

Rural residents' diets are improved by wild edible plant species because they can include micronutrients (minerals and vitamins) that are superior to those found in cultivated plants (Msuya et al., 2010). High-nutrient WEPs, such as those containing vitamins B2 and C and proteins, can be utilized as an alternative to conventional plant-based meals (Duguma, 2020). Wild plants are a strong possibility for nutrient-rich food because they include a variety of secondary metabolites, such as terpenoids, polysaccharides, and polyphenols, which may have health-promoting properties (García-Herrera et al., 2014, Bélanger & Pilling, 2019). According to studies, they are a significant source of bioactive substances like vitamins, vital fatty acids, and complex carbohydrates (Cornara et al., 2009; Pieroni et al., 2018). The amount of nutrients in wild leafy vegetables is probably comparable to that of domesticated plant species (Cakir, 2017). Mineral nutrition is essential for the growth of health and plays a key part in organism existence. WEPs readily contain these minerals (Mahadkar

TABLE 3.1
Plants Traditionally Used as Food, Medicine, and Nutrition

Scientific Name	Plant Type	Edible Part	Food Category	Traditionally Known Function
Alkekengi officinarum Moench	Shrub	Fruit	Snack	Removing swelling
Allium chrysanthum Regel	Perennial herb	Leaf	Vegetable	–
Allium macranthum Baker	Perennial herb	Bulb	Vegetable	–
Amaranthus tricolor L.	Perennial herb	Bud	Vegetable	–
Amorphophallus konjac K. Koch	Perennial herb	Tuber	Vegetable	–
Amorphophallus variabilis Blume	Perennial herb	Tuber	Snack	Insecticidal
Anredera cordifolia (Ten.) Steenis	perennial grassy vine	Leaf	Vegetable	–
Aralia elata (Miq.) Seem.	Small Tree	Bud	Vegetable	–
Artemisia argyi H.Lév. & Vaniot	Perennial herb	Bud/ Leaf	Vegetable	–
Artemisia indica Willd.	Perennial herb	Bud/Leaf	Vegetable	–
Artemisia lavandulifolia DC.	Perennial herb	Bud/ Leaf	Vegetable	–
Asparagus cochinchinensis (Lour.) Merr.	Perennial herb	Root tuber	_	Nourishing
Aster indicus L.	Perennial herb	Bud	Vegetable	Clearing fire
Aster trinervius subsp. Ageratoides (Turcz.) Grierson	Perennial herb	Bud	Vegetable	–
Atropa bella donna L.	Perennial herb	Fruit	_	–
Basella alba L.	Perennial	Leaf	Vegetable	–
Callipteris esculenta (Retz.) J.Sm.	Fern	Bud	Vegetable	–
Campanumoea javanica Blume	Perennial herb	Root	_	Nourishing
Capsella bursa-pastoris Medik.	Annual herb	Bud	Vegetable	–
Carya cathayensis Sarg	Tree	Fruit/Bud	Vegetable/ nut	Nourishing
Castanea mollissima Blume	Tree	Seeds	Snack	Nourishing
Cestrum inclusum Urb.	Perennial herb	Bud	Vegetable	Cosmetic effect
Chaenomeles japonica (Thunb.) Lindl.	Small Tree	Fruit	Snack	–
Chaenomeles speciosa (Sweet) Nakai	Small Tree	Fruit	Snack	–
Chenopodium album L.	Annual herb	Bud	Vegetable	–
Chimonobambusa quadrangularis (Franceschi) Makino	Small Tree	Bud	Vegetable	–
Cirsium arvense var. *integrifolium* Wimm. & Grab.	Perennial herb	Bud	Vegetable	–
Codonopsis pilosula var. glaberrima (Nannf.) Tsoong	Perennial herb	Root	_	Nourishing
Codonopsis tubulosa subsp. tubulosa	Perennial herb	Root	_	Nourishing
Collybia albuminosa (Berk.) Petch	Fungus	Fruiting body	Vegetable	–
Colocasia esculenta (L.) Schott	Perennial herb	rhizome	Vegetable	–
Crassocephalum crepidioides (Benth.) S.Moore	Perennial herb	Bud	Vegetable	–
Cryptotaenia japonica Hassk.	Annual herb	Bud	Vegetable	–
Chrysanthemum indicum L.	Perennial herb	Inflorescence	_	–
Dendrocalamus tsiangii (McClure) L.C.Chia & H.L.Fung	Tree	Bud	Vegetable	–
Diapensia purpurea Diels	Perennial herb	Root	_	Nourishing
Dioscorea japonica Thunb.	Perennial herb	Root	Vegetable	Tonifying

(Continued)

TABLE 3.1   (Continued)
Plants Traditionally Used as Food, Medicine, and Nutrition

Scientific Name	Plant Type	Edible Part	Food Category	Traditionally Known Function
Dioscorea nipponica Makino	Perennial twining herb	Root	Vegetable	Tonifying
Dioscorea glabra Roxb.	Perennial herb	Root	Vegetable	Tonifying
Diospyros kaki L.f.	Tree	Fruit	Snack	–
Diospyros lotus Lour.	Small Tree	Fruit	Snack	–
Duchesnea filipendula Focke	Perennial herb	Fruit	Snack	–
Elaeagnus pungens Thunb.	Shrub	Fruit	Snack	–
Fagopyrum acutatum Mansf. ex K.Hammer	Perennial herb	Bud	Vegetable	–
Ficus carica L.	Tree	Fruit	Snack	–
Ficus ti-koua Bureau	Shrub	Fruit	Snack	–
Fragaria vesca Pursh	Perennial herb	Fruit	Snack	–
Gastrodia elata Blume	Perennial herb	Tuber	Supplements	Treatment migraine
Ginkgo biloba L.	Tree	Seeds	Vegetable	Nourishing
Laphangium affine (D.Don) Tzvelev	Annual herb	Bud	Vegetable	–
Gonostegia triandra Miq.	Shrub	Fruit	Snack	–
Gynostemma pentaphyllum (Thunb.) Makino	Perennial twining herb	Leaf	Tea	Nourishing
Hemerocallis citrina Baroni	Perennial herb	Bud	Vegetable	–
Houttuynia cordata Thunb.	Perennial herb	Bud/Tender root	Vegetable	–
Hovenia acerba Lindl.	Tree	Fruit	Fruits	–
Hydrocotyle moschata G.Forst.	Perennial herb	Leaf	Vegetable	–
Ilex kudingcha C.J.Tseng	Tree	Leaf	Tea	–
Indocalamus tessellatus (Munro) Keng f.	Perennial herb	Leaf	Auxiliary food	–
Ixeris chinensis (Thunb. ex Thunb.) Nakai	Perennial herb	Bud/Leaf	Vegetable	–
Ixeris polycephala Cass.	Perennial herb	Bud/Leaf	Vegetable	–
Laportea bulbifera (Siebold & Zucc.) Wedd.	Perennial herb	Bud	Vegetable	Treatment skin disease (Itching)
Lilium japonicum Thunb. ex Houtt.	Perennial herb	Bulb	Vegetable	Nourishing
Lithocarpus litseifolius Chun	Tree	Leaf	Tea	expelling Summer-heat
Litsea pungens Hemsl.	Tree	Seeds	Condiment	–
Lonicera macrantha Spreng.	Perennial vine	Flower	Tea/Medicinal materials	–
Lophatherum gracile Brongn.	Perennial herb	Aboveground part	Tea	–
Melastoma dodecandrum Roxb.	Shrub	Fruit	Snack	–
Mentha canadensis L.	Perennial herb	Leaf	Seasoning vegetable	Expelling Summer-heat
Mentha suaveolens Ehrh.	Perennial herb	Leaf	Seasoning vegetable	–
Myrica rubra (Lour.) Siebold & Zucc.	Tree	Fruit	Fruits/Wines	–
Morus alba L.	Tree	Fruit/Leaf	Snack	Tonification
Nicandra physalodes (L.) Gaertn.	Annual herb	Seeds	–	–
Oenanthe javanica DC.	Annual herb	Bud	Vegetable	–
Opuntia stricta (Haw.) Haw.	Succulent shrub	Fruit/Fleshy leaf	Snacks / Vegetable	–
Osmunda lancea Thunb.	Fern	Bud	Vegetable	–

(Continued)

TABLE 3.1 (Continued)
Plants Traditionally Used as Food, Medicine, and Nutrition

Scientific Name	Plant Type	Edible Part	Food Category	Traditionally Known Function
Oxalis corniculata L.	Perennial herb	Root tuber	Snack	–
Phragmites australis (Cav.) Trin. ex Steud.	Perennial herb	Leaf	–	–
Phyllostachys edulis (Carrière) J.Houz.	Tree	Bud	Vegetable	–
Pinellia ternata (Thunb.) Makino	Perennial herb	Tuber	Snack	Insecticidal
Pinus thunbergii Parl.	Tree	Seeds	Snack	–
Pinus tabuliformis Carrière	Tree	Seeds	Snack	–
Platycladus orientalis (L.) Franco	Tree	Shoot	–	–
Reynoutria multiflora (Thunb.) Moldenke	Perennial herb	Root tuber	Snack/Health foods	Nourishing
Polygonatum sibiricum F.Delaroche	Perennial herb	Root tuber	–	Nourishing
Portulaca oleracea L.	Annual herb	Whole grass	Vegetable	Treatment diarrhea
Premna microphylla Turcz.	Shrub	Leaf	Vegetable	–
Pyracantha fortuneana (Maxim.) H.L.Li	Small Tree	Fruit/Leaf	Snack/Medicinal materials	–
Reynoutria japonica Houtt.	Perennial herb	Bud	Vegetable	–
Rhododendron simsii Planch.	Shrub	Flower	Snack/Vegetable	–
Robinia pseudoacacia L.	Tree	Flower	Vegetable	–
Rubus cuneifolius Pursh	Perennial shrub	Fruit	Snack	Nourishing
Rubus parvifolius L	Perennial shrub	Fruit	Snack	–
Schisandra chinensis (Turcz.) Baill.	Woody vine	Fruit	Snack	–
Sedum emarginatum Migo	Perennial herb	Bud	Vegetable	–
Sedum fui G.D.Rowley	Perennial succulent herb	Stem/Leaf	Vegetable	Nourishing
Sedum sarmentosum Bunge	Perennial herb	Bud	Vegetable	–
Smilax aspera L.	Perennial vine	Bud	Vegetable	–
Smilax walteri Pursh	Perennial vine	Bud	Vegetable	–
Solanum nigrum L.	Annual herb	Bud/Fruit	Vegetable/Snack	–
Sonchus brachyotus DC.	Perennial herb	Bud	Vegetable	–
Sonchus wightianus DC.	Perennial herb	Bud/Leaf	Vegetable	–
Stachys aegyptiaca Pers.	Perennial herb	Root tuber	Vegetable	–
Aster subulatus Michx.	Perennial herb	Bud	Vegetable	–
Taxus wallichiana Zucc.	Tree	Fruit	Medicinal materials	Anticancer
Tetradium ruticarpum (A. Jussieu) T. G. Hartley	Small Tree	Fruit	Condiment	Activating blood circulation
Toona sinensis (A.Juss.) M.Roem.	Tree	Bud	Vegetable	–
Urtica fissa E.Pritz.	Perennial herb	Bud/Leaf	Vegetable	–
Vaccinium bracteatum Thunb.	Small Tree	Fruit	Snack	–
Vicia nigricans var. gigantea (Hook.) Broich	Annual herb	Bud/Leaf	Vegetable	–
Viola betonicifolia Sm.	Perennial herb	Bud	Vegetable	–
Vitex negundo L.	Shrub	Stem/Leaf	Auxiliary food	–
Vitis amurensis Rupr.	Perennial vine	Fruit	Snack	–
Youngia japonica (L.) DC.	Perennial herb	Bud/Leaf	Vegetable	–
Zanthoxylum simulans Hance	Shrub	Fruit	Condiment	–
Zingiber striolatum Diels	Perennial herb	Bud	Vegetable	–

et al., 2012). By delivering vitamins, vital fatty acids, fiber, and minerals, WEPs and traditional vegetables improve the nutritional quality as well as the color and flavor of rural diets (Yildirim et al., 2001). They are frequently recognized as functional foods because of their undeniably favorable effects on human health and because they have the highest concentrations of flavonoids, fiber, antioxidants, vitamins, microelements, and phenols when compared to farmed crops. Rural residents are also thinking about growing their own vegetables because those may contain fewer pesticides and other toxins. As a result, WEPs may be excellent sources of exotic flavors, colors, dietary supplements, and bioactive substances (Motti et al., 2020).

3.4   DIETARY AND TOXICOLOGICAL ASPECTS OF A FEW WILD FOOD PLANTS

In addition to providing basic sustenance, certain wild edible plants have physiologically active components that can improve health (Pinela et al., 2017), whereas others may contain substances that could be hazardous to people (Pinela et al., 2017). WEPs have been crucial in helping people in underdeveloped nations get the food they need. Due to population growth, a shortage of arable land and limited access to food sources, poor populations frequently gather wild edible plant species (Tapan et al., 2017). If consumed in excess, some wild foods can have negative side effects (Harris & Mohammed, 2003). Foods made from herbs are not necessarily safer. However, some plant species have components that have the potential to be genotoxic and carcinogenic. Numerous botanical preparations contain a variety of chemicals that have the potential to cause persistent poisoning. In this regard, substances from the pyrrolizidine alkaloid and alkenyl benzene groups are of particular significance to human health (Luka and Samo, 2016). It is also known that many secondary compounds from plants can damage DNA by intercalation or covalent bonding (Wink, 2008). While overusing homoeopathic medicines might have negative effects, widespread usage of medicinal plants does not result in intoxication. Medication residues may contribute to the development of difficult-to-treat microorganisms that are medication resistant (Samudra and Shinde, 2021).

Plant foods contain antinutritional compounds such polyphenols, phytic acid, tannins, and oxalate, which are known to have negative impacts on a person's diet by preventing the absorption of iron and zinc (ul Haq et al., 2019, Hurrell et al., 1999). When nitrates attach to haemoglobin, the amount of oxygen delivered to tissues is reduced. Additionally, they have the capacity to interact with amines to produce carcinogenic nitrosamines (Guil et al., 1997). Oxalic acid creates an insoluble calcium oxalate and calcium complex in plants. The most typical sort of kidney stones in people are caused by this salt (Pinela et al., 2017). Some plant species can seriously intoxicate people. A poisonous plant is a species that contains toxic substances in all or some of its sections, which are primarily fatal to humans or other animals. The poisonous substances are typically organic chemicals rather than minerals, and they frequently cause symptoms when ingested or touched (Elouardia et al., 2022). Examples of a few wild edible plants with potentially harmful substances are included in Table 3.2.

3.5   AN ASSESSMENT OF WILD FLORA'S SAFETY

In order to balance the nutritional needs of the current generation, hybrid varieties and genetically modified crops are being developed for commercial reasons. In these circumstances, we must look for alternatives to satisfy human needs. A few wild medicinal plants serve as the primitive form of today's crops. Protecting them and responsibly using them will ensure a brighter future (Deshpande et al., 2019). According to government assessments and conservation activities, WEPs have always been a crucial component of nutrition and food security. There is evidence of widespread and regular consumption of wild food plants worldwide, primarily by rural and indigenous groups but also in urban areas. The availability of these biological resources in the wild is being constrained by issues related to overexploitation, urbanization, changes in land use, and climate, as well as by the loss

TABLE 3.2

Wild Edible Plants with Potentially Toxic Compounds

Plant Species	Potentially Toxic Compound
Anchusa ochroleuca M.Bieb.	Acetyllaburnine, Pyrrolizidine alkaloids: lycopsamine, laburnine
Borago officinalis L.	Amabiline: Pyrrolizidine alkaloid
Bryonia dioica Sessé & Moc.	Cucurbitacin glycosides: bryodulcoside, Cucurbitacins B, D, E, I, J, K, L, S; bryoside and bryonoside; dihydrocucurbitacins B, E; cucurbitacin saponins: brydioside A, B and C; tetrahydrocucurbitacin I, Brydiofin
Chenopodium ambrosioides L.	Phenylpropanoid: safrole; Peroxygenated monoterpene: ascaridole
Foeniculum vulgare Mill.	Phenylpropanoid: estragole; Phenylpropanoids: trans-anethole, estragole
Glechoma hederacea L.	Monoterpene etheroxide: Pyrrolidine; 1,8-cineole (eucalyptol) alkaloids: hederacines A, B
Humulus lupulus L.	Prenylflavonoids: 8-prenylnaringenin (hopein), isoxanthohumol, xanthohumol,
Origanum vulgare L.	Monoterpene etheroxide: 1,8-cineole, Bicyclic monoterpene: beta-thujone;
Tamus communis L.	Phenanthrene derivatives; Steroidal saponins; dioscin, gracillin

of traditional knowledge associated with their use. Several governmental efforts that specifically attempt to protect them and ensure their sustainable use are now in existence. Part of this can be attributed to the fact that market- and non-market–related constraints inhibit policymakers and other important parties from realizing the untapped potential of WEPs (Borelli et al., 2020).

Due to unauthorized harvesting in surrounding conservation zones, market competitors may suppress this information given the abundance of non-literal marketplaces for WEPs (Petersen et al., 2012). It is usually difficult to find information on the quantity collected, pricing, production costs, and distribution over time and space. Increased profits can frequently lead to excessive use of wild food plants, which has negative consequences on the community as a whole (Shin et al., 2018). However, as the gap between the growing human population and the available food sources widens, certain wild edible plant species must be protected from heedless exploitation. Concerns over the safety of wild foods have also recently focused on their therapeutic properties (Schippmann et al., 2002). To demonstrate the sustainable management approach and harvesting rates, and monitor the ecological effects of rising applications in order to avoid this risk, collaborative research is needed (Hudson et al., 2020).

3.6   WILL FUTURE FOOD COME FROM WILD PLANTS?

Numerous forests that contain WEPs are currently being cleared by local communities. Anthropogenic activities and the rising human population rate are to blame for this. The primary causes of WEP destruction include overgrazing, overharvesting, agricultural land development, wildfires, and the gathering of fuel wood. The primary cause of the destruction of wild edible plants is the growth of agricultural land. As a result, it's essential to preserve wild plants that can be used as food. For the low-income and malnourished populations of developing countries, finding good and nourishing food sources is currently a major concern. Due to food shortage and high cost, it has become a critical problem to provide low-cost and alternative sources of healthful and nourishing food. It has been acknowledged that underutilized wild edible plants (UWEP) are a practical and long-term solution to food poverty. A WEPs-based diet should be increased to ensure food security in the future (Peduruhewa et al., 2021).

Some major present-day concerns include food scarcity, pesticide residue, drought, and nutrition crises. The rate of population growth worldwide is alarmingly high, but due to various factors, including industrialization and civilization, the amount of land set aside for agriculture is shrinking daily. Since wild veggies don't require any specific growing conditions, they can assist us in overcoming the food crisis. Future dietary and health issues may arise, and wild plant species that are edible will allow us to address these issues (Deshpande et al., 2019). The shifting of both geographical ranges and phenological changes in ripening periods is one of the main implications that could be dangerous for the use of WEPs. This can lead to a divergence in the traditional knowledge and methods used by the communities that now gather them (Anderson et al., 2018).

Numerous human and environmental factors can affect WEPs. The main difficult factors influencing the decline in WEPs' growth and production are agriculture and drought (Amente, 2017). Even though there is an abundance of food available to humanity thanks to the present global food system, many individuals still risk going hungry or lack access to a diet that is healthy. On the other side, eating meals that are more heavily processed can be detrimental to one's health. Therefore, wild edible plants can provide a variety of nutritious possibilities for individuals who rely on them and can be a significant element of peoples' meals in some regions of the world (Motti, 2021). Additionally, it is anticipated that in the near future, awareness of the usage of innovative plant materials will increase steadily due to the projected growth in the human population, increasing globalization of trade, and ethnobotanical study. As a result, the demand for WEPs will rise, which will eventually endanger the local species in various parts of the world. Additionally, the price differential between cultivated and wild plants will encourage unsustainable gathering practices in some areas, especially in economically depressed ones where people are unaware of the laws protecting wild plants (Pardo-De-Santayana et al., 2005).

3.7 CONCLUSION

Wild edible plants are an essential part of human existence. They provide a significant contribution to the regional food system. Additionally, WEPs are essential in meeting human nutritional needs. In addition to this, they are also utilized for therapeutic and some cultural practices. The worst scenario, though, is that the current generation's indigenous expertise of how to gather and use WEPs is vanishing at an alarming rate. Therefore, some policies or strategies should be put into place in order to promote food security and community healthcare as well as to safeguard traditionally utilized plants, knowledge, and cultural heritage related to them. Additionally, greater research will support sustainable food production and bio-conservation techniques.

REFERENCES

Abbasi, Arshad Mehmood, Mir Ajab Khan, Nadeem Khan, and Munir H. Shah. "Ethnobotanical survey of medicinally important wild edible fruits species used by tribal communities of Lesser Himalayas-Pakistan." *Journal of Ethnopharmacology* 148, no. 2 (2013): 528–536.

Alissa, Eman M., and Gordon A. Ferns. "Dietary fruits and vegetables and cardiovascular diseases risk." *Critical Reviews in Food Science and Nutrition* 57, no. 9 (2017): 1950–1962.

Amente, D. Ayele. "Ethnobotanical survey of wild edible plants and their contribution for food security used by Gumuz people in Kamash Woreda; Benishangul Gumuz Regional State; Ethiopia." *Journal of Food and Nutrition Sciences* 5, no. 6 (2017): 217–224.

Anbessa, Baressa. "Ethnobotanical study of wild edible plants in Bule Hora Woreda, Southern Ethiopia." *African Journal of Basic & Applied Sciences* 8, no. 4 (2016): 198–207.

Anderson, Darya, James D. Ford, and Robert G. Way. "The impacts of climate and social changes on cloudberry (Bakeapple) picking: A case study from southeastern Labrador." *Human Ecology* 46 (2018): 849–863.

Balemie, Kebu, and Fassil Kebebew. "Ethnobotanical study of wild edible plants in Derashe and Kucha Districts, South Ethiopia." *Journal of Ethnobiology and Ethnomedicine* 2 (2006): 1–9.

Bélanger, Julie, and Dafydd Pilling. The state of the world's biodiversity for food and agriculture. Food and Agriculture Organization of the United Nations (FAO), (2019). https://openknowledge.fao.org/server/api/core/bitstreams/50b79369-9249-4486-ac07-9098d07df60a/content

Borelli, Teresa, Danny Hunter, Bronwen Powell, Tiziana Ulian, Efisio Mattana, Céline Termote, Lukas Pawera et al. "Born to eat wild: An integrated conservation approach to secure wild food plants for food security and nutrition." *Plants* 9, no. 10 (2020): 1299.

Cakir, Ernaz Altundag. "Traditional knowledge of wild edible plants of Igdir Province (East Anatolia, Turkey)." *Acta Societatis Botanicorum Poloniae* 86, no. 4 (2017).

Carvalho, Ana Maria, and Ana Maria Barata. "The consumption of wild edible plants." *Wild plants, mushrooms and nuts: Functional food properties and applications* (2016): 159–198.

Cornara, L., A. La Rocca, S. Marsili, and M. G. Mariotti. "Traditional uses of plants in the Eastern Riviera (Liguria, Italy)." *Journal of Ethnopharmacology* 125, no. 1 (2009): 16–30.

Demel, T., Abeje, E. (2004). Status of indigenous fruits in Ethiopia. Review and appraisal on the status of indigenous fruits in eastern Africa. In: Chikamai B, Eyog-Matig O, editors. A synthesis report for IPGRI-SAFORGEN.

Deshpande, Swapnaja, Uday Pawar, and Rajendra Kumbhar. "Exploration and documentation of wild food plants from Satara district, Maharashtra (India)." *International Journal of Food Science and Nutrition* 4, no. 1 (2019): 95–101.

Duguma, Haile Tesfaye. "Wild edible plant nutritional contribution and consumer perception in Ethiopia." *International Journal of Food Science* 2020 (2020) 2958623. https://doi.org/10.1155/2020/2958623

Elouardi, Mohamed, Touriya Zair, Jamal Mabrouki, Ghizlane Fattah, Mohammed Benchrifa, Najat Qisse, and Mohammed Alaoui El Belghiti. "A review of botanical, biogeographical phytochemical and toxicological aspects of the toxic plants in Morocco." *Toxicologie Analytique et Clinique* 34, no. 4 (2022): 215–228.

FAO (2019). *The State of the World's Biodiversity for Food and Agriculture*; FAO: Rome, Italy.

Fentahun, Mengistu Tiruneh, and Herbert Hager. "Exploiting locally available resources for food and nutritional security enhancement: wild fruits diversity, potential and state of exploitation in the Amhara region of Ethiopia." *Food Security* 1 (2009): 207–219.

Flatie, Teferi, Teferi Gedif, Kaleab Asres, and Tsige Gebre-Mariam. "Ethnomedical survey of Berta ethnic group Assosa Zone, Benishangul-Gumuz regional state, mid-west Ethiopia." *Journal of Ethnobiology and Ethnomedicine* 5, no. 1 (2009): 1–11.

García-Herrera, P., M. C. Sánchez-Mata, M. Cámara, V. Fernández-Ruiz, C. Díez-Marqués, M. Molina, and J. Tardío. "Nutrient composition of six wild edible Mediterranean Asteraceae plants of dietary interest." *Journal of Food Composition and Analysis* 34, no. 2 (2014): 163–170.

García-Herrera, Patricia, and María de Cortes Sánchez-Mata. "The contribution of wild plants to dietary intakes of micronutrients (II): Mineral Elements." In *Mediterranean wild edible plants: Ethnobotany and food composition tables* (2016): 141–171.

Gemedo-Dalle, T., Brigitte L. Maass, and Johannes Isselstein. "Plant biodiversity and ethnobotany of Borana pastoralists in southern Oromia, Ethiopia." *Economic Botany* 59, no. 1 (2005): 43–65.

Guarrera, P. M., and V. Savo. "Perceived health properties of wild and cultivated food plants in local and popular traditions of Italy: A review." *Journal of Ethnopharmacology* 146, no. 3 (2013): 659–680.

Guil, J. L., I. Rodríguez-Garcí, and E. Torija. "Nutritional and toxic factors in selected wild edible plants." *Plant Foods for Human Nutrition* 51 (1997): 99–107.

Gujjala, Sudhakara, Mallaiah Putakala, Srinivasulu Nukala, Manjunatha Bangeppagari, Ramaswamy Rajendran, and Saralakumari Desireddy. "Modulatory effects of Caralluma fimbriata extract against high-fat diet induced abnormalities in carbohydrate metabolism in Wistar rats." *Biomedicine & Pharmacotherapy* 92 (2017): 1062–1072.

Gupta, Sheetal, A. Jyothi Lakshmi, M. N. Manjunath, and Jamuna Prakash. "Analysis of nutrient and antinutrient content of underutilized green leafy vegetables." LWT-Food *Science and Technology* 38, no. 4 (2005): 339–345.

Harris, Frances MA, and Salisu Mohammed. "Relying on nature: wild foods in northern Nigeria." *AMBIO: A Journal of the Human Environment* 32, no. 1 (2003): 24–29.

Hudson, Alex, William Milliken, Jonathan Timberlake, Peter Giovannini, Valdemar Fijamo, Joao Massunde, Hercilia Chipanga, Milagre Nivunga, and Tiziana Ulian. "Natural plant resources for sustainable development: Insights from community use in the Chimanimani trans-frontier conservation area, Mozambique." *Human Ecology* 48 (2020): 55–67.

Hurrell, Richard F., Manju Reddy, and James D. Cook. "Inhibition of non-haem iron absorption in man by polyphenolic-containing beverages." *British Journal of Nutrition* 81, no. 4 (1999): 289–295.

Jiang, Z., Li, Z., Bao, M., and Chen, M. The statistics and analysis of foraging plants species eaten by Asian elephant (*Elephas maximus*) in China. *Acta Theriologica Sinica* 39, no. 5 (2019): 514–30. https://doi.org/10.16829/j.slxb.150237

Ju, Yan, Jingxian Zhuo, Bo Liu, and Chunlin Long. "Eating from the wild: diversity of wild edible plants used by Tibetans in Shangri-la region, Yunnan, China." *Journal of Ethnobiology and Ethnomedicine* 9, no. 1 (2013): 1–22.

Khan, Zafar Iqbal, Ilker Ugulu, A. S. M. A. Zafar, Naunain Mehmood, Humayun Bashir, Kafeel Ahmad, and Madiha Sana. "Biomonitoring of heavy metals accumulation in wild plants growing at soon valley, Khushab, Pakistan." *Pakistan Journal of Botany* 53, no. 1 (2021): 247–252.

Leakey, Roger. "Domestication potential of Marula (Sclerocarya birrea subsp. caffra) in South Africa and Namibia: 3. Multiple trait selection." *Agroforestry Systems* 64 (2005): 51–59.

Luka, K., and Samo, K. (2016). "European medicinal and edible plants associated with subacute and chronic toxicity part I: Plants with carcinogenic, teratogenic and endocrine-disrupting effects." *Food and Chemical Toxicology*, 92: 150–164.

Mahadkar, Shivprasad, Sujata Valvi, and Varsha Rathod. "Nutritional assessment of some selected wild edible plants as a good source of mineral." *Asian Journal of Plant Science and Research* 2, no. 4 (2012): 468–472.

Menendez-Baceta, Gorka, Laura Aceituno-Mata, Javier Tardío, Victoria Reyes-García, and Manuel Pardo-de-Santayana. "Wild edible plants traditionally gathered in Gorbeialdea (Biscay, Basque Country)." *Genetic Resources and Crop Evolution* 59, no. 7 (2012): 1329–1347.

Motti, R., G. Bonanomi, V. Lanzotti, and R. Sacchi. "The contribution of wild edible plants to the Mediterranean Diet: An ethnobotanical case study along the coast of Campania (Southern Italy)." *Economic Botany* 74 (2020): 249–272.

Motti, Riccardo. "Wild plants used as herbs and spices in Italy: An ethnobotanical review." *Plants* 10, no. 3 (2021): 563.

Msuya, Tuli S., Jafari R. Kideghesho, and Theobald C. E. Mosha. "Availability, preference, and consumption of indigenous forest foods in the Eastern Arc Mountains, Tanzania." *Ecology of Food and Nutrition* 49, no. 3 (2010): 208–227.

Ogle, B. M., Hung, P. H., and Tuyet, H. T. Significance of wild vegetables in micronutrient intakes of women in Vietnam: an analysis of food variety. *Asia Pacific Journal of Clinical Nutrition*, 10 no. 1 (2001): 21–30.

Ojelel, Samuel, Patrick Mucunguzi, Esther Katuura, Esezah K. Kakudidi, Mary Namaganda, and James Kalema. "Wild edible plants used by communities in and around selected forest reserves of Teso-Karamoja region, Uganda." *Journal of Ethnobiology and Ethnomedicine* 15 (2019): 1–14.

Pardo-De-Santayana, Manuel, Javier Tardío, and Ramón Morales. "The gathering and consumption of wild edible plants in the Campoo (Cantabria, Spain)." *International Journal of Food Sciences and Nutrition* 56, no. 7 (2005): 529–542.

Pascual, Juan Cruz, and Baudilio Herrero. "Wild food plants gathered in the upper Pisuerga river basin, Palencia, Spain." *Botany Letters* 164, no. 3 (2017): 263–272.

Peduruhewa, P. S., K. G. L. R. Jayathunge, and R. Liyanage. "Potential of underutilized wild edible plants as the food for the future–a review." *Journal of Food Security* 9, no. 4 (2021): 136–147.

Petersen, L. M., E. J. Moll, R. Collins, and Marc T. Hockings. "Development of a compendium of local, wild-harvested species used in the informal economy trade, Cape Town, South Africa." *Ecology and Society* 17, no. 2 (2012). http://www.jstor.org/stable/26269045

Phondani, Prakash C., Indra D. Bhatt, Vikram S. Negi, Bhagwati P. Kothyari, Arvind Bhatt, and Rakesh K. Maikhuri. "Promoting medicinal plants cultivation as a tool for biodiversity conservation and livelihood enhancement in Indian Himalaya." *Journal of Asia-Pacific Biodiversity* 9, no. 1 (2016): 39–46.

Pieroni, Andrea, Renata Sõukand, Hawraz Ibrahim M. Amin, Hawre Zahir, and Toomas Kukk. "Celebrating multi-religious co-existence in Central Kurdistan: The bio-culturally diverse traditional gathering of wild vegetables among Yazidis, Assyrians, and Muslim Kurds." *Human Ecology* 46 (2018): 217–227.

Pinela, José, Ana Maria Carvalho, and Isabel CFR Ferreira. "Wild edible plants: Nutritional and toxicological characteristics, retrieval strategies and importance for today's society." *Food and Chemical Toxicology* 110 (2017): 165–188.

Powell, Bronwen, Abderrahim Ouarghidi, Timothy Johns, Mohamed Ibn Tattou, and Pablo Eyzaguirre. "Wild leafy vegetable use and knowledge across multiple sites in Morocco: a case study for transmission of local knowledge?." *Journal of Ethnobiology and Ethnomedicine* 10 (2014): 1–11.

Samudra, S. M., and H. P. Shinde. "Studies on ethnomedicinal plant diversity at Daund Tehsil, Pune, Maharashtra." *International Research Journal of Plant Science* 12, no. 1 (2021): 1–13.

Sánchez-Mata, María de Cortes, María Cruz Matallana-González, and Patricia Morales. "The contribution of wild plants to dietary intakes of micronutrients (I): Vitamins." In: Sánchez-Mata, M., Tardío, J. (eds) *Mediterranean wild edible plants: Ethnobotany and food composition tables* (2016): 111–139. Springer, New York, NY. https://doi.org/10.1007/978-1-4939-3329-7_6

Schippmann, Uwe, Danna J. Leaman, and Anthony B. Cunningham. "Impact of cultivation and gathering of medicinal plants on biodiversity: global trends and issues." In *Biodiversity and the ecosystem approach in agriculture, forestry and fisheries, Proceedings of the Satellite Event on the Occasion of the Ninth*

Regular Session of the Commission on Genetic Resources for Food and Agriculture, Rome, Italy, 12–13 October 2002; Inter-Departmental Working Group on Biological Diversity for Food and Agriculture: Rome, Italy, (2002); Volume 676, pp. 1–21.

Seifu, T., B. Mehari, M. Atlabachew, and B. Chandravanshi. "Polyphenolic content and antioxidant activity of leaves of Urtica simensis grown in Ethiopia." *Latin American Applied Research* 47, no. 1 (2017): 35–40.

Shaheen, Shabnum, Mushtaq Ahmad, and Nidaa Haroon. *Edible Wild Plants: An alternative approach to food security*. Cham, Switzerland: Springer International Publishing, (2017a).

Shaheen, Shabnum, Mushtaq Ahmad, and Nidaa Haroon. "Nutritional contents and analysis of edible wild plants." In *Edible Wild Plants: An alternative approach to food security*. Cham, Switzerland: Springer International Publishing (2017b): 127–133. https://doi.org/10.1007/978-3-319-63037-3_5

Shin, Thant, Kazumi Fujikawa, Aung Zaw Moe, and Hiroshi Uchiyama. "Traditional knowledge of wild edible plants with special emphasis on medicinal uses in Southern Shan State, Myanmar." *Journal of Ethnobiology and Ethnomedicine* 14, no. 1 (2018): 1–13.

Sir Elkhatim, Khitma A., Randa AA Elagib, and Amro B. Hassan. "Content of phenolic compounds and vitamin C and antioxidant activity in wasted parts of Sudanese citrus fruits." *Food Science & Nutrition* 6, no. 5 (2018): 1214–1219.

Svanberg, Ingvar. "The use of wild plants as food in pre-industrial Sweden." *Acta Societatis Botanicorum Poloniae* 81, no. 4 (2012): 317–327.

Tapan, Seal, Pillai Basundhara, and Chaudhuri Kausik. "Nutritional potential of five unexplored wild edible plants consumed by the tribal people of Arunachal Pradesh state in India." *International Journal of Food Science and Nutrition* 2, no. 2 (2017): 101–105.

Tebkew, Mekuanent, Yohannis Gebremariam, Tadesse Mucheye, Asmamaw Alemu, Amsalu Abich, and Dagim Fikir. "Uses of wild edible plants in Quara district, northwest Ethiopia: implication for forest management." *Agriculture & Food Security* 7, no. 1 (2018): 1–14.

ul Haq, Zahoor, Shujaul Mulk Khan, Javed Iqbal, Abdul Razzaq, and Mazhar Iqbal. "Phyto-medicinal studies in district lower Dir Hindukush range Khyber Pakhtunkhwa Pakistan." *Pakistan Journal of Weed Science Research* 25, no. 3 (2019): 235.

Ulian, Tiziana, Mauricio Diazgranados, Samuel Pironon, Stefano Padulosi, Udayangani Liu, Lee Davies, Melanie-Jayne R. Howes et al. "Unlocking plant resources to support food security and promote sustainable agriculture." *Plants, People, Planet* 2, no. 5 (2020): 421–445.

Vitalini, Sara, Marcello Iriti, Cristina Puricelli, Davide Ciuchi, Alessandro Segale, and Gelsomina Fico. "Traditional knowledge on medicinal and food plants used in Val San Giacomo (Sondrio, Italy)—An alpine ethnobotanical study." *Journal of Ethnopharmacology* 145, no. 2 (2013): 517–529.

Wink, Michael. "Plant secondary metabolism: diversity, function and its evolution." *Natural Product Communications* 3, no. 8 (2008): 1934578X0800300801.

World Health Organization. General guidelines for methodologies on research and evaluation of traditional medicine. No. WHO/EDM/TRM/2000.1. World Health Organization, (2000).

Yildirim, Ertan, Atilla Dursun, and Metin Turan. "Determination of the nutrition contents of the wild plants used as vegetables in Upper Coruh Valley." *Turkish Journal of Botany* 25, no. 6 (2001): 367–371.

4 Current Research in Phytonutrient Potential of Wild Edible Plants

Dimitrios D. Ntakoulas, Konstantina Tsikrika, Ioannis N. Pasias and Charalampos Proestos

4.1 INTRODUCTION

Wild edible plants (WEPs) have been used in the past for their nutritional as well as medicinal properties around the world (e.g., China, Jordan, Egypt, and Greece) (Alu'datt et al. 2018). In fact, WEPs still constitute a major part of the diet in rural deprived and tribal communities in the form of food and nutritional supplement (Seal, Chaudhuri, and Pillai 2017), as their high content in nutrients make them valuable during scarcity periods, while they provide protection for minor health conditions (Torija-Isasa and Matallana-González 2016). Besides, the nutritional qualities of WEPs are similar and sometimes even superior to those of cultivated varieties (Ebert 2014), while they are stronger in genetics than the latter (Shaheen, Ahmad, and Haroon 2017). They also constitute supplementary sources of vitamins and minerals that are vital components of the body in order to maintain its homoeostasis (Datta et al. 2019).

According to the Food and Agriculture Organization (FAO) wild plants are defined as "self-maintained plants that grow spontaneously in natural or semi-natural ecosystems and can exist autonomously of direct human activity" (Heywood 1999). WEPs comprise a vast selection of plant organisms, namely annual and perennial herbs, vines, forbs, grasses, sedges and rushes, broadleaved and needle-like or scale-like leaved shrubs, ferns, and trees. WEPs also include roots and underground storage organs, stems, shoots, leaves, sprouts, fruits and cones, flowers, seeds and nuts, galls, bark, nectar, and gum, alongside lichens, fronds, and algae (Ferreira, Morales, and Barros 2016).

Ever since the industrial revolution, there have been many alterations in lifestyle, leading to an extensive cultivation of a small number of crops (e.g., wheat, maize, and rice), as well as a decreased contact with nature, contributing to the present underuse of WEPs (Łuczaj et al. 2012). However, many people still depend on WEPs to provide at least some of their everyday nutritional needs (Pinela, Carvalho, and Ferreira 2017). Furthermore, WEPs have developed into a commercial crop with a rising market value, owing to their high nutrient content and the absence of pesticide, fungicide, herbicide, and fertilizer residues (Satter et al. 2016; Shaheen, Ahmad, and Haroon, 2017). They are also considered to be appealing gastronomic sources for contemporary culinary encounters as numerous restaurants use a lot more locally produced, farmed, and foraged foods as ingredients on their menus (Ferreira, Morales, and Barros 2016). Additionally, there is growing interest in their possible applications in the food and drug industries as they are rich in bioactive phytonutrients.

Phytonutrients are a group of chemical compounds, naturally occurring in plants with plenty of dietary functions and health benefits for humans (Leitzmann 2016). Phytonutrients, also called phytochemicals, include carotenoids, isoprenoids, indoles, glycosylates, organosulfur compounds, phytosterols, polyphenols, and saponins (Monjotin et al. 2022a; A. Kumar et al. 2023). These compounds are products of the secondary metabolism of plants as a reaction to environmental factors such as light irradiation and temperature, as well as soil, water, salinity, and fertility, and they can exhibit either defensive or disease protective characteristics (Li Yang et al. 2018). Although they are not

DOI: 10.1201/9781003395935-4

essential nutrients, they possess important qualities such as antioxidant, antimicrobial, and anticancer activities; modulation of detoxification enzymes; immune system stimulation; and reduction of platelet aggregation as well as regulation of hormone metabolism (Shiny, Saxena, and Gupta 2013).

Numerous factors involved in the metabolism of each phytonutrient; the digestive capacity and genotype of each person, the membrane transporters, enzymes, and gut microbiota as well as the type of the phytochemical itself (Rathaur and Johar Kaid 2019). Consequently, a small amount of phytonutrients can be immediately absorbed by the human body after oral delivery, while metabolism mostly occurs in the liver and intestine (Kan et al. 2022). On the other hand, the extraction and purification of phytonutrients from WEPs typically necessitate a complicated and prolonged procedure, and even so, the outcome is usually very low. As a result, researchers face a difficult task in discovering workable solutions to speed up the process and increase the yield of phytonutrients (Li Yang et al. 2018).

4.2 PHYTONUTRIENT TYPES AND MODE OF ACTION

Based on their molecular compositions and properties, phytonutrients are typically divided into seven broad groups. These groups include polyphenols, carotenoids, phytosterols, isoprenoids, saponins, and certain polysaccharides (Huang et al. 2016; Kumar et al. 2023). Different subgroups are created within each category by further categorization as per biogenesis or biochemical origin. Over 5000 different alimentary phytonutrients have been isolated in various plant foods. However, the chemical structure and/or biological functions of these substances in humans are still unclear (Liu 2013).

Ascorbic acid, or vitamin C, was the first phytochemical molecule to be identified. It is formed throughout aerobic metabolism and quickly reacts with superoxide, singlet oxygen, ozone, and H_2O_2 to counteract their toxic effects, while it is also implicated in the regeneration of carotenoids and vitamin E (Forni et al. 2019).

Phenolic compounds, carotenoids, and glycosylates from frequently consumed fruits and vegetables are some of the most researched bioactive phytonutrients. These phytonutrients have demonstrated antioxidant, anti-inflammation, anti-carcinogenic, anti-mutagenic, and antibacterial activities (Doughari 2012) occurring from the interaction of phytonutrients with proteins, enzymes, or receptors implicated in precursor-product reactions, signal transduction, and/or cellular redox status (Marrocco, Altieri, and Peluso 2017). Subsequently a series of signaling pathways are being set off, resulting in different outcomes, including altering the expression or activity of inflammatory mediators, improving insulin sensitivity, increasing antioxidant defense, and/or DNA repair mechanisms, disruption of the bacterial or cell cycle, and so on (Si and Liu 2014; Krüger et al. 2014).

4.2.1 PHENOLIC COMPOUNDS

Phenolic compounds comprise a great variety of molecules (approximately 8000 different structures), which include phenolic acids, lignans, lignins, stilbenes, tannins, and flavonoids. The presence of at least one phenol ring is responsible for exhibiting antioxidant activity, by substituting the hydrogen with hydroxyl, methyl, or acetyl groups. Tyrosine and phenylalanine are the amino acids that comprise the precursors of the phenolic compounds that are formed from cinnamic acid (Alu'datt et al. 2018).

4.2.1.1 Phenolic Acids

Phenolic acids are phenols that contain a carboxylic acid. They are categorized in relation to the number and position of the hydroxyl and methoxyl groups bonded to the aromatic rings. Phenolic acids that have shikimic acid as a biogenetic precursor are mainly present in the bound form in plants. These are classified into three main groups, namely hydroxybenzoic acid derivatives (if the functional group of the carboxylic acid is attached to the phenol ring), hydroxycinnamic acid derivatives (when the carboxylic acid functional group and the phenol ring are separated by two doubly bonded carbons) and depsides, with hydroxycinnamic acids being the most common in plants. The most common hydroxycinnamic acids are ferulic, p-coumaric, caffeic, sinapic acid, and others, which occur in foods as esters with quinic acid or sugar components of the food matrix (Tsikrika, O'Brien, and Rai 2019), while vanillic, protocatechuic, p-hydroxy benzoic acid, and so on, that is, the derivatives of hydroxy benzoic acid, are predominantly found as glucosides (Seal 2016). Chlorogenic acid (5-caffeoylquinic acid, 5-CQA) is an ester of caffeic acid with quinic acid and it is produced within the shikimic acid pathway through aerobic respiration (L.-N. Wang et al. 2016).

4.2.1.2   Flavonoids

Flavonoids have a flavone skeleton C6-C3-C6 consisting of 15 carbons and two benzene rings (A and B) connected by a heterocycle pyrene ring (C) that contains oxygen (Dias, Pinto, and Silva 2021). The antioxidant activity is greatly influenced by the position of the B-ring on the C-ring along with the number and position of hydroxy groups on the catechol group of the B-ring (D'Amelia et al. 2018). The latter functional hydroxy groups give electrons via resonance to stabilize free radicals and facilitate antioxidant defense. Flavanols, flavonols, flavones, flavanones, isoflavones, and anthocyanins are the six main classes of flavonoids based on their structure; there is a double bond between C-2 and C-3 in flavonols and flavones, whereas the flavones miss the hydroxyl group at C-3 (Šamec et al. 2021).

Quercetin, kaempferol, myricetin, apigenin, isorhamnetin, luteolin, and their glucosides are ubiquitous in plants. Compounds like catechin, epicatechin, gallocatechin, epigallocatechin, gallic acid esters, and epigallocatechin gallate belong to the subgroup of flavanols, which possess two aromatic rings (Alu'datt et al. 2018; Guan et al. 2021). Anthocyanins are glycosylated polyphenolic compounds of flavonoids, further categorized into pelargonidins, cyanidins, delphinidins, peonidins, petunidins, and malvidins, as per the number and position of the hydroxyl and methoxyl groups on the flavan nucleus (Y. Liu et al. 2018; Tahara 2007).

4.2.2   CAROTENOIDS

Carotenoids are organic pigments of yellow, red, and orange colour, naturally found in the chloroplasts and chromoplasts of photosynthetic organisms, e.g., plants, algae, certain fungi and bacteria (A. Kumar et al. 2023). These phytonutrients possess a polyisoprenoid structure of 40 carbon atoms in a long-conjugated chain of double bonds in the center of the molecule (Valcarcel et al. 2015). Carotenoids are indispensable for photosynthesis and photoprotection in plants as they constitute light harvesting pigments and structural elements of photosystems. They also serve as precursors for the biosynthesis of the phytohormones strigolactones (SLs) and abscisic acid (ABA) (Sun et al. 2022). Furthermore, carotenoid derivatives are involved in plant growth and development as they act as signaling molecules while they also moderate plant responses to environmental cues (Tian 2015; Havaux 2014). In addition to having a key role in plants, carotenoids are vital dietary components for humans, greatly influencing human nutrition and health; α-, and β-carotene, and β-cryptoxanthin are Vitamin A precursors, while lutein, lycopene, and fucoxanthins are strong antioxidants. Consequently, carotenoids are associated with the delay of the onset of some chronic diseases, for example, cancer, cardiovascular diseases, and age-related eye diseases (Sun et al. 2018; A. Kumar et al. 2023).

4.2.3   TERPENOIDS

Terpenoids, also referred to as isoprenoids, are a broad and diverse class of plant secondary metabolites that are formed from isoprene units with five carbons (A. Kumar et al. 2023). The majority of terpenoids are multicyclic compounds with distinct functional groups and fundamental carbon skeletons from one another. Terpenoids are involved in direct and indirect plant defense, pollinator attraction, and various interactions between plants and their environment (Wei et al. 2023). These types of natural lipids are thought to make up the largest group of naturally occurring secondary metabolites as they are present in all classes of living things. Terpenes are abundant in nature, primarily in plants where they are found in essential oils (Koche, Shirsat, and Kawale 2016). They are organic compounds with two or more isoprenes, arranged in a specific pattern. Limonene is the most prevalent monoterpene found in aromatic plants and fruits and is responsible for the lemon-like flavor and aroma (Shahbazi and Shavisi 2021; A. Kumar et al. 2023). Myrcene is another abundant monoterpene in plants and it has two isomers; α-, and β-myrcene. The latter is the most common isomer and is present in the essential oils of plants, for example, hops, lemongrass, verbena, and bay, as well as in citrus fruits (Surendran et al. 2021).

4.2.4 PHYTOSTEROLS

Phytosterols refer to plant sterols and stanols that control numerous plant biological functions. They are steroid alcohols containing a tetracyclic cyclopenta-α-phenanthrene ring and a side chain bonded to the C-17 atom of the ring (Piironen and Lampi 2014). Phytosterols are abundant in olive oil and the oils of various seeds and nuts such as corn, sunflower, sesame, peanuts, macadamia nuts, almonds, and beans (Kumar et al. 2023). Sitosterol is typically the most prevalent plant sterol, though stigmasterol and campesterol can also frequently be found in significant amounts, depending on the specific source being used. Brassicasterol, avenasterol, and spinasterol are less common plant sterols (Bot 2019).

4.2.5 SAPONINS

Saponins are naturally occurring glycosides with surface activity that have a distinctive foaming property caused by the combination of their lipophilic sapogenin and hydrophilic sugar parts. They are primarily produced by plants, but also by some bacteria and smaller marine animals (Desai, Desai, and Kaur 2009). Saponins are divided into steroidal saponins and triterpenoids as per the nature of the aglycone. Dammaranes, tirucallanes, and oleanane are a few examples of the saponins. Dammarane is a tetracyclic triterpene occurring in sapogenins, which forms triterpenoid saponins, whereas oleanane and tirucallane are a triterpenoid and a tetracyclic triterpenoid, respectively (A. Kumar et al. 2023).

4.2.6 POLYSACCHARIDES

Polysaccharides are monomer sugar units connected by glycosidic bond. They can appear in the form of nondigestible substances such as cellulose, beta-glucan, pectin, hemicelluloses, resistant starch, lignin, and so on, which are referred to as dietary fiber, or they can be energy-storing substances like starch and glycogen (Gidley and Yakubov 2019). Although dietary fibers are indigestible and have a limited capacity for absorption in the human small intestine, they can be partially or completely fermentable in the large intestine (Yegin et al. 2020). All plant foods are good sources of dietary fiber; corn, oats, barley, wheat, green beans, and many more are considered as rich sources (A. Kumar et al. 2023).

4.3 PHYTONUTRIENTS IN WILD EDIBLE PLANTS

As mentioned earlier, WEPs are rich in phytonutrients such as carotenoids, phenolic compounds, and other non-nutrient compounds with a demonstrated antioxidant activity linked to health advantages, for example, low cardiovascular disease risk factors and cancer incidence while also acting as a preventative measure against a variety of other chronic illnesses. Among botanical groups, the concentration of these substances in wild plants varies greatly, and it is also controlled by the growth environment, moisture, and other elements (Torija-Isasa and Matallana-González 2016).

This section summarizes the current knowledge about the content of these bioactive compounds and their contribution in WEPs.

Tables 4.1 and 4.2 contains the main phenolic compounds and the respective amounts determined in various WEPs. Alaca et al. (2022) studied some of the most commonly consumed WEPs grown in Eastern Anatolia *viz. Arum conophalloides* Kotschy ex Schott var. *conophalloides* (Khari), *Gundelia tournefortii* L. var. *Tournefortii* (Kenger), *Eremurus spectabilis Bieb.* (Çiriş), *Tragopogon longirostris* Bisch. Ex Schultz Bip. (Yemlik), and *Falcaria vulgaris* Bernh. (Kaz ayağı), *Rumex tuberosus* L. subsp. *horizontalis* (Koch.) Rech. (Evelik), *Rheum ribes* L. (Uşgun), *Chaeropbyllum macropodum* Boiss. (Mendi), *Cichorium intybus* L. (Çatlanguş), *Chenopodium album* L. (Pazı), *Coriandrum sativum* L. (Kişniş), and *Plantago lanceolata* L. (Yılan Dili). According to the authors, the most abundant phenolic acids in the aforementioned WEPs are found to be chlorogenic acid and gallic acids. The flavonoids rutin, quercetin, kaempferol, and luteolin were also found in the tested samples with luteolin being present in almost all the studied plants with the exception of *C. sativum*. High total carotenoid content was also detected in most of the studied samples.

Datta et al. (2019) analyzed six WEPs commonly consumed in India, namely *Ipomoea aquatica*, *Achyranthes aspera*, *Aasystasia ganjetica*, *Enhydra fluctuans*, *Oldenlandia corymbose*, and *Amaranthus viridis* and identified the phenolic acids protocatechuic acid, gallic acid, chlorogenic acid, vanillic acid, *p*-hydroxy benzoic acid, caffeic acid, gentisic acid, syringic acid, *p*-coumaric acid, ferulic acid, sinapic acid, salicylic acid, and ellagic acid, as well as the flavonoids catechin, quercetin, rutin, naringin, myricetin, naringenin, apigenin, and kaempferol. Gallic acid was only found in *A. viridis*, while chlorogenic acid content was detected in both *I. aquatica* and *A. viridis*. *E.fluctuans* contained a good amount of ellagic acid, which can explain its traditional anti-inflammatory use.

Viburnum foetidum, *Houttuynia cordata*, and *Perilla ocimoides* from India were analyzed for free phenolic acids (Seal, Pillai, and Chaudhuri 2016). Specifically, the content of the phenolic acids protocatechuic, gallic, chlorogenic, gentisic, p-hydroxy benzoic, caffeic, vanillic, p-coumaric, syringic, p-coumaric, ferulic, sinapic, salicylic, and ellagic was estimated in those three WEPs. Results revealed that the levels of gentisic acid, chlorogenic acid, syringic acid, ellagic acid, and *p*-coumaric acid were higher in the methanolic than in the 80 % water-ethanol extract of *V. foetidum*. Gallic acid, ferulic acid, and salicylic acid were also found in the methanol extract of *V. foetidum* whereas sinapic acid was only detected in the 80% aq. ethanol extract. High levels of protocatechuic acid were present in the methanol extract of *V. foetidum*, as compared to the 80% aq. ethanol extract of the WEP. The opposite was reported for the *P. ocimoides* extracts; *higher* levels of protocatechuic acid and syringic acid were found in the 80% aq. ethanol extract when compared to that of methanol. The latter extract of *P. ocimoides* had a good amount of the phenolic acids caffeic, vanillic, and ellagic, which were not present in the 80% aq. ethanol extract of the plant. *H. cordata* methanol and 80% aq. ethanol extracts contained good levels of protocatechuic acid, syringic acid and p-coumaric acid, whereas chlorogenic acid and caffeic acid were only found in the methanol extract of the plant. On the contrary, vanillic acid and gallic acid were only present in the *H. cordata* 80% water-ethanol extract.

The same group of authors (Seal, Chaudhuri, and Pillai 2017) investigated the phenolic content of the wild edible fruits of *Cucumis hardwickii* Royle. The authors stated the presence of gallic acid, chlorogenic acid, *p*-hydroxy benzoic acid, caffeic acid, protocatechuic acid, vanillic acid, *p*-coumaric acid, syringic acid, ferulic acid, sinapic acid, ellagic acid, rutin, naringin, and kaempferol.

Earlier, Seal (2016) reported good levels of free phenolic acids in the WEPs *Sonchus arvensis* and *Oenanthe linearis*, collected from N.E. India. High content of ascorbic acid and certain flavonoids *viz*. catechin, rutin, quercetin, myrecetin, apigenin, and kaempferol were also found. Quercetin and kaempferol were detected in the chloroform extract of *S. arvensis* whereas caffeic acid, apigenin, and quercetin were present in that of *O. linearis*. On the other hand, gallic acid, syringic acid, caffeic acid, quercetin catechin, and kaempferol were found in the methanol extract of *O. linearis* while only quercetin was detected in *S. arvensis*. The 80% aq. ethanol extract of both WEPs contained ascorbic acid, gallic acid, catechin, and kaempferol, while low levels of caffeic acid ferulic acid,

TABLE 4.1

Phenolic Acid Content in WEPs

Wild Edible Plants	Gallic Acid	Protocatechuic Acid	Gentisic Acid	p-Hydroxy benzoic Acid	Chlorogenic Acid	Vanillic Acid	Caffeic Acid	Syringic Acid	p-Coumaric Acid	Ferulic Acid	Sinapic Acid	Salicylic Acid	Rosmarinic Acid	Units	References
							Phenolic Acids								
Ipomoea aquatica	ND	ND	ND	0.033 ± 0.003	1.827 ± 0.001	0.360 ± 0.002	ND	0.404 ± 0.001	0.374 ± 0.01	ND	0.055 ± 0.001	ND	NR	mg/100g dry plant	Datta et al., 2019
Achyranthes aspera	ND	ND	ND	ND	ND	0.690 ± 0.006	ND	ND	ND	0.194 ± 0.004	ND	ND	NR		
Aasystasia ganjetica	ND	ND	ND	ND	ND	ND	ND	ND	ND	0.249 ± 0.007	ND	1.518 ± 0.012	NR		
Enhydra fluctuans	ND	ND	ND	ND	ND	ND	ND	ND	ND	0.375 ± 0.008	ND	ND	NR		
Oldenlandia corumbosa	ND	ND	ND	ND	ND	0.500 ± 0.002	ND	ND	ND	0.281 ± 0.002	0.267 ± 0.002	ND	NR		
Amaranthus viridis	0.145 ± 0.001	ND	ND	ND	3.807 ± 0.001	ND	ND	6.065 ± 0.002	ND	0.104 ± 0.002	ND	ND	NR		
Beta vulgaris L var. cicla	NR	NR	NR	NR	NR	NR	NR	NR	3.53	NR	NR	NR	1.02	mg/g extract	Mzoughi et al., 2019
Viburnum foetidum	0.38 ± 0.007	65.08 ± 0.04	5.09 ± 0.03	ND	0.83 ± 0.0004	ND	ND	3.19 ± 0.008	0.29 ± 0.003	0.27 ± 0.001	0.137 ± 0.001	0.21 ± 0.006	NR	mg/gm dry extract	Seal, Chaudhuri and Pillai, 2016
Houttuynia cordata	0.052 ± 0.002	2.91 ± 0.02	ND	ND	0.84 ± 0.02	0.062 ± 0.001	0.34 ± 0.003	0.638 ± 0.001	0.026 ± 0.001	ND	ND	ND	NR		
Perilla ocimoides	ND	2.18 ± 0.012	ND	0.296 ± 0.003	ND	0.027 ± 0.001	0.203 ± 0.001	1.98 ± 0.013	ND	ND	ND	ND	NR		
Cucumis hardwickii	1.03 ± 0.05	0.32 ± 0.003	ND	0.42 ± 0.001	0.78 ± 0.005	0.93 ± 0.002	0.14 ± 0.001	1.05 ± 0.07	0.04 ± 0.0001	0.06 ± 0.0002	0.07 ± 0.003	ND	NR		Seal, Chaudhuri and Pillai, 2017
Sonchus arvensis	0.281 ± 0.05	NR	NR	NR	NR	NR	ND	ND	ND	ND	NR	NR	NR	mg/gm	Seal, 2016
Oenanthe linearis	0.628 ± 0.001	NR	NR	NR	NR	NR	0.096 ± 0.001	0.046 ± 0.002	0.013 ± 0.003	0.023 ± 0.002	NR	NR	NR		

(Continued)

TABLE 4.1 (Continued)
Phenolic Acid Content in WEPs

Wild Edible Plants	Gallic Acid	Proto catechuic Acid	Gentisic Acid	p-Hydroxy benzoic Acid	Chloro genic Acid	Vanillic Acid	Caffeic Acid	Syringic Acid	p-Coumaric Acid	Ferulic Acid	Sinapic Acid	Salicylic Acid	Rosmarinic Acid	Units	References
									Phenolic Acids						
Blumea lacera (Burm. f.) DC.	1.89 ± 0.12	NR	NR	NR	NR	9.40 ± 0.70	5.56 ± 0.31	NR	3.09 ± 0.18	NR	NR	NR	NR	mg/100 g DW	Alam et al., 2020
Erythrina variegata L.	3.25 ± 0.75	NR	NR	NR	NR	8.06 ± 0.82	9.28 ± 1.50	NR	nd	NR	NR	NR	NR		
Berberis aristata	2.85 ± 0.02	NR	NR	NR	NR	0.89 ± 0.05	nd	NR	18.41 ± 0.31	NR	NR	NR	NR		
Sesbania sesban (L.) *Merr*	4.12 ± 0.52	NR	NR	NR	NR	2.22 ± 0.32	1.50 ± 0.05	NR	nd	NR	NR	NR	NR		
Hygrophilla schulli	6.83 ± 1.14	NR	NR	NR	NR	4.86 ± 0.92	3.63 ± 0.03	NR	8.69 ± 0.89	NR	NR	NR	NR		
Origanum vulgare L	NR	NR	NR	NR	NR	NR	50.0	NR	214.8	NR	NR	NR	256.7	mg/100 g DW	Alu'datt et al., 2018
Rosmarinus officinalis L	ND	NR	NR	NR	NR	2.0	40.0	2.0	19.0	NR	NR	NR	1289.4		
Salvia officinalis L.	ND	NR	NR	NR	NR	2.3	121.5	NR	40.0	NR	NR	NR	2186.1		
Thymus vulgaris L.	37.5	NR	NR	NR	NR	NR	54.8	5.0	55.9	NR	NR	NR	681.1		
Mentha spicata	NR	NR	NR	NR	NR	NR	4.8	NR	NR	NR	NR	NR	242.6		
Zingiber	NR	NR	NR	NR	NR	NR	15.5	NR	NR	NR	NR	NR	NR		
Coriandrum sativum L.	2.6	NR	NR	NR	0.4	1.4	22.2	3.3	NR	NR	NR	NR	NR		
Petroselinum crispum	NR	NR	NR	NR	502.4	NR	103.7	14.1	NR	NR	NR	NR	NR		

(Continued)

TABLE 4.1 (Continued)
Phenolic Acid Content in WEPs

Wild Edible Plants		Phenolic Acids													
	Gallic Acid	Proto catechuic Acid	Gentisic Acid	p-Hydroxy benzoic Acid	Chloro genic Acid	Vanillic Acid	Caffeic Acid	Syringic Acid	p-Coumaric Acid	Ferulic Acid	Sinapic Acid	Salicylic Acid	Rosmarinic Acid	Units	References
Chaerophyllum macropodum	ND	NR	NR	NR	18.01 ± 2.38	NR	NR	ND	ND	12.68 ± 0.05	NR	NR	NR	mg/kg DW	Alaca et al., 2022
Eremurus spectabilis	35.77 ± 14.06	NR	NR	NR	ND	NR	NR	ND	ND	ND	NR	NR	NR		
Gundelia toournefortii L.	ND	NR	NR	NR	388.30 ± 23.13	NR	NR	ND	4.09 ± 0.04	ND	NR	NR	NR		
Cichorium intybus L	ND	NR	NR	NR	122.82 ± 78.42	NR	NR	ND	ND	ND	NR	NR	NR		
Chenopodium album L	78.45 ± 23.5	NR	NR	NR	ND	NR	NR	ND	ND	ND	NR	NR	NR		
Coriandrum sativum L	17.99 ± 0.21	NR	NR	NR	22.40 ± 5.57	NR	NR	6.93 ± 0.04	ND	ND	NR	NR	NR		
Tragopogon longirostris	18.08 ± 1.52	NR	NR	NR	1058.81 ± 12.3	NR	NR	ND	ND	ND	NR	NR	NR		
Falcaria vulgaris	ND	NR	NR	NR	388.20 ± 188.93	NR	NR	ND	4.06 ± 0.10	ND	NR	NR	NR		
Rumex tuberosus L.	27.44 ± 1.75	NR	NR	NR	20.80 ± 2.99	NR	NR	ND	ND	12.59 ± 0.09	NR	NR	NR		
Rheum ribes L.	132.06 ± 53.66	NR	NR	NR	ND	NR	NR	ND	ND	ND	NR	NR	NR		
Plantago lanceolata L.	ND	NR	NR	NR	ND	NR	NR	ND	ND	ND	NR	NR	NR		
Arum conophalloides	ND	NR	NR	NR	485.34 ± 22.56	NR	NR	ND	ND	ND	NR	NR	NR		

ND: Not detected
NR: Not reported

TABLE 4.2
Flavonoid Content in WEPs

Wild Edible Plants	Flavonoids									Units	References
	Naringin	Rutin	Myricetin	Quercetin	Naringenin	Apigenin	Kaempferol	Catechin	Epicatechin		
Ipomoea aquatica	ND	0.727 ± 0.001	1.787 ± 0.002	ND	ND	ND	ND	ND	NR	mg/100g dry plant	Datta et al., 2019
Achyranthes aspera	ND	0.058 ± 0.005	ND	0.147 ± 0.006	ND	3.592 ± 0.061	0.489 ± 0.006	ND	NR		
Aasystasia ganjetica	ND	ND	0.414 ± 0.008	1.239 ± 0.005	ND	0.466 ± 0.012	0.196 ± 0.001	9.447 ± 0.009	NR		
Enhydra fluctuans	ND	ND	ND	0.184 ± 0.003	ND	0.194 ± 0.003	0.244 ± 0.005	ND	NR		
Oldenlandia corumbosa	ND	ND	ND	0.091 ± 0.002	ND	0.121 ± 0.002	0.098 ± 0.001	ND	NR		
Amaranthus viridis	ND	ND	ND	ND	ND	4.231 ± 0.002	ND	ND	NR	mg/g extract	Mzoughi et al., 2019
Beta vulgaris L. var. cicla	NR	NR	NR	NR	NR	NR	NR	NR	NR	mg/gm dry extract	Seal, Chaudhuri and Pillai, 2016
Viburnum foetidum	NR	NR	NR	NR	NR	NR	NR	NR	NR		
Houttuynia cordata	NR	NR	NR	NR	NR	NR	NR	NR	NR		
Perilla ocimoides	NR	NR	NR	NR	NR	NR	NR	NR	NR		
Cucumis hardwickii	0.10 ± 0.002	0.51 ± 0.04	ND	ND	ND	ND	1.93 ± 0.06	ND	NR		Seal, Chaudhuri and Pillai, 2017
Sonchus arvensis	NR	ND	0.110 ± 0.004	0.058 ± 0.0004		ND	0.157 ± 0.002	0.136 ± 0.004	NR	mg/gm	Seal, 2016
Oenanthe linearis	NR	0.073 ± 0.005	ND	0.065 ± 0.003		0.029 ± 0.002	0.248 ± 0.002	0.379 ± 0.004	NR		

(Continued)

TABLE 4.2 (Continued)
Flavonoid Content in WEPs

Wild Edible Plants	Naringin	Rutin	Myricetin	Quercetin	Naringenin	Apigenin	Kaempferol	Catechin	Epicatechin	Units	References
					Flavonoids						
Blumea lacera (Burm. f.) DC.	NR	26.33 ± 0.77	1.24 ± 0.18	ND	NR	NR	ND	12.76 ± 1.22	5.45 ± 0.60	mg/100 g DW	Alam et al., 2020
Erythrina variegata L.	NR	24.67 ± 1.10	4.57 ± 0.28	10.15 ± 1.05	NR	NR	1.51 ± 0.11a	76.54 ± 3.42	8.67 ± 0.94		
Berberis aristata	NR	64.23 ± 2.87	ND	1.19 ± 0.03	NR	NR	nd	7.78 ± 1.18	0.95 ± 0.01		
Sesbania sesban (L.) Merr	NR	0.85 ± 0.03	40.75 ± 2.45	4.16 ± 0.80	NR	NR	1.19 ± 0.31a	15.55 ± 2.25	1.59 ± 0.10		
Hygrophilla schulli	NR	4.23 ± 0.64	ND	ND	NR	NR	ND	53.05 ± 4.45	8.25 ± 1.03		
Origanum vulgare L	NR	NR	NR	NR	NR	NR	63.9	NR	NR	mg/100 g DW	Alu'datt et al., 2018
Rosmarinus officinalis L	NR	NR	NR	397.0	NR	1.1	NR	254.9	NR		
Salvia officinalis L.	NR	NR	NR	77.9	NR	3.5	63.9	257.1	NR		
Thymus vulgaris L.	NR	NR	NR	NR	NR	4.0	ND	ND	0.3		
Mentha spicata	NR	NR	NR	NR	NR	NR	NR	NR	NR		
Zingiber	NR	NR	NR	NR	NR	NR	NR	NR	NR		
Coriandrum sativum L.	NR	NR	NR	2.3	NR	NR		1.0 NR	16.2		
Petroselinum crispum	NR	NR	NR	16.3	NR	60.0	NR	NR	NR		

(Continued)

TABLE 4.2 (Continued)
Flavonoid Content in WEPs

Wild Edible Plants	Flavonoids									Units	References
	Naringin	Rutin	Myricetin	Quercetin	Naringenin	Apigenin	Kaempferol	Catechin	Epicatechin		
Chaerophyllum macropodum	NR	17.70 ± 2.81	NR	ND	NR	NR	ND	3.95 ± 0.12	NR	mg/kg DW	Alaca et al., 2022
Eremurus spectabilis	NR	ND	NR	ND	NR	NR	ND	ND	NR		
Gundelia toournefortii L.	NR	ND	NR	56.78 ± 26.14	NR	NR	ND	ND	NR		
Cichorium intybus L	NR	ND	NR	434.91 ± 109.43	NR	NR	ND	ND	NR		
Chenopodium album L	NR	1329.07 ± 367.58	NR	ND	NR	NR	ND	ND	NR		
Coriandrum sativum L	NR	357.86 ± 189.99	NR	221.16 ± 13.31	NR	NR	ND	29.35 ± 3.60	NR		
Tragopogon longirostris	NR	ND	NR	ND	NR	NR	ND	ND	NR		
Falcaria vulgaris	NR	ND	NR	ND	NR	NR	ND	ND	NR		
Rumex tuberosus L.	NR	495.83 ± 222.40	NR	3347.71 ± 374.24	NR	NR	2309.37 ± 67.16	ND	NR		
Rheum ribes L.	NR	137.06 ± 85.33	NR	26.05 ± 0.50	NR	NR	ND	55.37 ± 18.89	NR		
Plantago lanceolata L.	NR	ND	NR	ND	NR	NR	ND	23.55 ± 7.04	NR		
Arum conophalloides	NR	ND	NR	ND	NR	NR	56.62 ± 0.69	ND	NR		

ND:Not detected
NR:Not reported

p-coumaric acid, rutin, and quercetin were also present in *O. linearis*. A significant content of ascorbic acid and gallic acid was also reported for the 1% water-acetic acid extract of both WEPs.

Blumea lacera (Kukurshunga), *Berberis aristata* (Daruharudra), *Hygrophilla schulli* (Kulakhara), *Sesbania sesban* (Dhoinche), and *Erythrina variegata* (Mandargach) from Bangladesh were investigated by Alam et al. (2020). The authors found a wide variety of polyphenols, namely gallic acid, caffeic acid, vanillic acid, p-coumaric acid, ellagic acid, catechin, epicatechin, rutin, myricetin, quercetin, and kaempferol. All of the plants that were tested contained gallic acid, vanillic acid, catechin, and rutin, while the majority of the tested phenolic compounds were present in most of the studied plant samples. *E. variegata* contained the highest levels of caffeic acid, ellagic acid, catechin, epicatechin, quercetin, and kaempferol, whereas *H. schulli* and *B. lacera* had the highest amount of gallic acid and vanillic acid, respectively. On the other hand, the highest content on p-coumaric acid and rutin was found in *B. aristata* while myricetin levels were highest in *S. sesban*.

Mzoughi et al. (2019) reported that Tunisian wild Swiss chard (*Beta vulgaris* L. var. *cicla*) leaves had significant levels of p-coumaric acid, rosmarinic acid, myricitrin, flavonoids, chlorophyll, β-carotene, and lycopene and some volatile compounds such as (E)-anethole, octanoic acid and decanal. In accord, Rocha et al. (2017) detected the aforementioned phenolic acids in *Carpobrotus edulis* L. alongside the hydrocinnamic acids caffeic and ferulic, while the hydrobezoic acids gallic, vanillic p-hydroxybenzoic, and salicylic were found as well. The flavonoids epicatechin, quercetin, and apigenin were also identified. On the other hand, considerable levels of chlorogenic acid were detected in *Leea macrophylla* (a.k.a. Hastikarna palasa) root tubers (Joshi et al. 2016).

Suksathan et al. (2021) studied the phytonutrient content of four Thai wild flowers, namely *Bauhinia variegata* L., *Shorea roxburghii* G. Don, *Gmelina arborea* Roxb, and *Viburnum inopinatum* Craib, which can be used as sources of novel tea formulations. They found gallic acid in both ethanolic and aquatic extracts of all samples, whereas caffeic acid was detected in ethanolic extracts of *B. variegata*, *G. arborea*, and aquatic extracts of *B. variegate*. Vanillin was present in both ethanol and water extracts of *B. variegata* and *G. arborea*, while tannic acid was only found in the aquatic extract of *S. roxburghii*. Both extracts of *V. inopinatum* contained luteolin and quercetin. The latter was also present in the ethanolic extract of *G. arborea* and water extract of *B. variegata*. Rutin hydrate was detected in both extracts of *B. variegata* and *G. arborea*, while all the studied WEPs contained cyanidin chloride.

Monari et al. (2021) analyzed 12 commonly consumed WEPs in Italy, namely *Salvia pratensis* L., *Crepis vesicaria* subsp., *Taraxacum officinale* Weber, *Taraxacifolia* (Thuill) Thell, *Urtica dioica* L., *Sonchus* spp., *Cichorium intybus* L., *Foeniculum vulgare* Mill., *Beta vulgaris* L., *Cichorium intybus* L., *Sambucus nigra* L., *Sonchus* spp., and *Asparagus acutifolius* L. The authors detected 17 different phenolic compounds in the studied plants. Caffeic acid, *p*-coumaric acid, *trans*-ferulic acid, chlorogenic acid, vanillin, piceatannol, *cis*-resveratroloside, *trans*-resveratroloside, *trans*-resveratrol, and *trans*-piceid, as well as luteolin, luteolin-7-glucoside, rutin, quercetin, catechin, naringenin, and epicatechin were identified. It was reported that the luteolin, luteolin-7-glucoside and rutin content found was approximately 100 to 1000 times higher than the average content of the other tested compounds.

Recently, Saleem et al. (2020) reviewed the properties of *Anagallis arvensis* (L.), a WEP widely distributed throughout the world. According to the authors, *A. arvensis* contains high levels of bioactive compounds, and as a result it exerts various biological activities, for example, anti-bacterial, diuretic, antioxidant, antimicrobial, antiviral, antifungal, and many more. High total phenolic content and total flavonoid content in *A. arvensis* has been previously reported too (Saleem et al. 2020). Similar levels of total phenolic content and total flavonoid content were also found in *Adenia viridiflora* Craib, a native WEP of Thailand, Cambodia, and Vietnam (Wannasaksri et al. 2021).

Allium rubellum, Berberis chitria, Capsella bursa-pastoris, Stellaria aquatica, and *Rheum emodi*, underutilized WEPs indigenous in Himalaya were examined by A. Thakur et al. (2022). The phytonutrient content varied greatly among the studied WEPs. The highest phenolic, flavonoid, and terpenoid levels were found in *B. chitria* whereas the lowest were detected in *R. emodi*. On the other hand, the latter WEP exhibited the highest levels of ascorbic acid, while the lowest were in *C. bursa-pastoris*. *A. rubellum* had the highest carotenoid content, whereas *S. aquatica* contained the highest tocopherol content.

A wide range of total phenolic content and total flavonoid content was detected in extracts of *Sphenoclea zeylanica, Cardamine hirsute, Natsiatum herpeticum, Sphaerantus peguensis, Melothria perpusilla*, and *Persicaria chinensis*, WEPs widely consumed by the Bodos, a tribe in Assam, India. *M. perpusilla* exhibited the highest phenolic and flavonoid content, whereas the lowest level of phenolics and flavonoids were found in *N. herpeticum* and *S. zeylanica*, respectively (Basumatary and Narzary 2017). Earlier Narzary, Islary, and Basumatary (2016) analyzed 11 WEPs from Assam, India, namely *Blumea lanceolaria* (Roxb.) Druce, *Tetrastigma angustifolium* (Roxb.), *Oenanthe javanica* (Blume) DC., *Drymaria cordata* (L.) Willd. ex Schult., *Cryptolepis sinensis* (Lour) Merr., *Stellaria media* (L.), *Antidesma acidum* Retz., *Eryngium foetidum* L., *Lippia javanica* (Burm.f.) Spreng., *Polygonum perfoliatum* L., and *Enhydra fluctuans* Lour. *E. fluctuans* had the highest total phenolic content whereas *C. sinensis* the lowest. The highest total flavonoid content was found in *P. Perfoliatum* whereas the lowest was in *S. media*. Swargiary et al. (2016) studied *Clerodendrum viscosum, Eryngium foetidum, Lippia javanica*, and *Murraya koenigii*, also popular WEPs in N.E. India. *C. viscosum* contained higher levels of total phenolic content, total flavonoid content, and vitamin C followed by *M. koenigii, L. javanica*, and *E. foetidum*.

4.4　THE ROLE OF PHYTONUTRIENTS IN THE PREVENTION OF DISEASES

Many chronic and degenerative diseases, for example, atherosclerosis, diabetes, cancer, immunosuppression, and neurological diseases have been linked to oxidative stress (Leyane, Jere, and Houreld 2022). Oxidative stress can be defined as an imbalance between pro-oxidants and antioxidants with a concurrent redox circuitry disorder and macromolecular destruction (Jones 2008; Patel 2016; Sies 2020). Various biochemical processes resulting in the formation of reactive species, exposure to damaging agents such as environmental pollutants and radiations, or low capacities of endogenous antioxidant systems may lead to oxidative stress (Marrocco, Altieri and Peluso, 2017).

The term reactive species concerns reactive radical and no-radical oxygen, and nitrogen derivatives known as reactive oxygen species (ROS) and reactive nitrogen species (RNS), respectively (Mandal et al. 2022). These radicals can be formed within the cells by losing or accepting a single electron, hence acting as oxidants or reductants. They are highly reactive atoms or molecules containing one or more unpaired electron(s) in their external shell and can be formed when oxygen interacts with certain molecules (Chandrasekaran, Idelchik, and Melendez 2017).

Phytonutrients have a significant role in inhibiting DNA damage caused by oxidative stress, through their antioxidant properties. Additionally, they prevent cells from going through molecular alterations that lead to carcinogenesis by modulating a number of oxidative stress-related signaling pathways (Chikara et al. 2018) while they slow the growth of malignant cells, lower tissue inflammation levels, and suppress angiogenesis (Majnooni et al. 2023; Bunea et al. 2013).

Ascorbic acid (Vitamin C) exerts strong antioxidant activity by scavenging free radicals in the plasma, thus protecting cells against ROS-induced oxidative damage (Baenas et al. 2019). It is also involved in the gastrointestinal absorption of iron, while it constitutes a co-factor in the synthesis of catecholamines, cholesterol, L-carnitine, amino acids, and some peptide hormones and is an essential dietary component for the synthesis of collagen. Failure to produce collagen can lead to scurvy, a pathological condition brought on by a deficiency in vitamin C that results in blood vessel fragility, connective tissue damage, and, ultimately, death (Grosso et al. 2013; Dasgupta and Klein 2014).

Phenolics have recently drawn global attention owing to their antioxidant/nutraceutical properties, against various oxidative damage disorders. (Proestos et al. 2013). The biological activities of phenolics include antimutagenic, anti-inflammatory, antibacterial and antiviral activity; apoptotic actions; and anti-biotic properties (Mattila and Hellström 2007).

Yan et al. (2020), have reported that chlorogenic acid can decrease blood sugar levels and exhibit an antidiabetic action, while the consumption of plants containing chlorogenic acid is linked with a reduced risk of liver cirrhosis and liver cancer (Nwafor et al. 2022). Numerous biological properties, for example, antioxidant, antimicrobial, virucidal, antimutagenic, antialgal, antiestrogenic, anti-inflammatory, hypoglycemic, anti-platelet aggregating, and nematocidal have been ascribed to p-hydroxy benzoic acid (Manuja et al. 2013), while vanillic acid has shown liver and kidney protective action (Alamri et al. 2022). A hepatoprotective activity alongside anticancer and antiproliferative attributes have also been reported for syringic acid (Periyannan and Veerasamy 2018), whereas a strong antioxidant activity together with lower levels of the carcinogenic nitrosamines in the stomach have been described for p-coumaric acid (Karthikeyan, Devadasu, and Srinivasa Babu 2015). Ferulic acid and its structural analogue sinapic acid are well known for their high antioxidant activities, while they also exhibit antimicrobial, anti-inflammatory, anticancer, and antianxiety activities (Stompor-Gorący and Machaczka 2021). Ellagic acid possesses antioxidant and antiproliferative activity and acts as a chemopreventive agent to the role of ellagic acid, a dilactone of hexahydroxydiphenic acid, in malignant diseases as a result of its outstanding antioxidant and antiproliferative properties (Lu et al. 2019).

Due to their metabolism into vitamin A (retinol), carotenoids play a significant role in human health while they exert strong antioxidant properties (Cámara, Fernández-Ruiz, and Ruiz-Rodríguez 2016). The carotenoids α-carotene, β-carotene, and β-cryptoxanthin are the dietary precursors of vitamin A and they are indispensable for the immune system and vision. Lutein and zeaxanthin in human eyes are responsible for the filtering of the high-energy wavelengths of blue light, thereby reducing retinal oxidative stresses. Hence, the function of the latter carotenoids has been associated with reducing the onset of age-related eye diseases such as age-related macular degenerative disease and cataracts (Sauer, Li, and Bernstein 2019).

Terpenoids are also good antioxidants that also possess antioxidant, anti-inflammatory, anticancer, antiseptic, antiplasmodial, astringent, digestive, and many additional properties (Cox-Georgian et al. 2019). Saponins have been associated with antiviral, antifungal, hypoglycemic, and hypolipidemic activity (X. Wang et al. 2023). Numerous health benefits have been ascribed to phytosterols; high antioxidant activity, reduction in LDL cholesterol, support for prostate health, and hair growth among them (Kumar et al. 2023). A regular dietary fiber intake improves insulin sensitivity, supports a healthy gut microbiome, and prevents inflammation, hypertension, hypercholesterolemia, hyperlipidemia, obesity, cardiovascular diseases, and various types of cancer (Veronese et al. 2018; Yegin et al. 2020).

4.5 METHODS TO IMPROVE THE BIOACCESSIBILITY OF PHYTONUTRIENTS

Wild edible plants contain many sorts of phytonutrients like proteins, phenolic acids, flavonoids, dietary fibers, peptides, and polyunsaturated fatty acids that show numerous positive impacts on human health (N. Thakur et al. 2020). However, to achieve these innumerable positive effects phytonutrients must be bioavailable, which means that they must be efficiently absorbed from the gut and distributed to the appropriate part of the body via the circulatory system (Fernández-García et al. 2012; Actis-Goretta et al. 2013).

Bioaccessibility is related to the fraction of a compound that is delivered from a food matrix to the luminal content during the process of a meal digestion, which next can be absorbed through the small intestine or can undergo biotransformation by the gut microbiota. The term bioactivity refers to the action of the absorbed compound in various metabolic routes, which result in biological impacts on the human body. The term bioavailability is related to the extent to which a substance that finishes the pathway that passes by the digestive tract, is absorbed, and arrives at the target tissues in order to be stored or perform its bioactivity (Rodrigues et al. 2022).

Many phytonutrients are chemically degraded during their exposure to various environmental changes like heat, light, oxygen, pH changes, and prooxidants (Hu et al. 2023). Therefore, the remaining concentration of phytonutrients in the plants at the consumption stage is significantly reduced. In addition, many phytonutrients have limited bioavailability and bioactivity during plant digestion because of their restricted stability and due to their restrained absorption characteristics and solubility (Kamiloglu et al. 2021; Hayes et al. 2021). Moreover, some phytonutrients exist in plants in a glycoside form, which makes difficult their absorption from the upper gastrointestinal tract (Teng and Chen, 2019).

The impact of the food matrix and the type of processing operations utilized on the bioavailability, bioaccessibility, and bioactivity of phytonutrients are the key determinants of these properties. For instance, the consumption of hydrophobic phytonutrients in combination with lipids increases their bioaccessibility, because it makes it easier for them to dissolve and move through the gastrointestinal fluids (Yao et al. 2021). On the contrary, the consumption of the phytonutrients in combination with dietary fibers decreases their bioaccessibility by preventing their release from the plant matrix (Hu et al. 2023). However, in specific circumstances, the presence of certain dietary fibers like pectin boosts the bioaccessibility of phytonutrients (Wellala et al. 2022). Also, the processing operations used affect the bioavailability and bioaccessibility of phytonutrients in intricate ways. For example, thermal processing can increase or decrease phytonutrient bioavailability depending on how well it affects the release of those nutrients from the plant matrix (Hu et al. 2023). It has been demonstrated that thermal processing in tandem with other methods, for instance vacuum (Burca-Busaga et al. 2021), microwave (Li et al. 2022), or ultrasound (Ramirez-Melo et al. 2022), might lessen the detrimental effects of heating on phytonutrients. Nonthermal techniques including high-pressure processing, freeze-drying, bioprocessing, and pulsed electric field have also been shown to improve the bioaccessibility of phytonutrients in foods (Wellala et al. 2022; Sęczyk, Ozdemir, and Kołodziej 2022; Tomé-Sánchez et al. 2021; Ribas-Agustí et al. 2018).

There have been attempts up to this point to increase the bioavailability of phytonutrients by modifying their physical or chemical structure. Phytonutrient stability and bioaccessibility can be enhanced by chemical alterations. However, due to the safety risks of chemically modifying phytonutrients, physical methods are frequently chosen. The most usual method in the latter category is the encapsulation of phytonutrients into particle-based carriers (Hu et al. 2023). These carriers are commonly gathered from lipids (X.-M. Li et al. 2020) or edible biopolymers (Hao et al. 2022). Encapsulation of phytonutrients in these carriers improve their stability, dispersibility, bioavailability, and bioaccessibility (He et al. 2022; Hu et al. 2023).

The biological benefit of phytonutrients has been evaluated using a variety of models, including cellular absorption and investigations on live animals or humans as well as *in vitro* simulations of digestion. Utilizing in vivo human or animal studies, the bioavailability and bioaccessibility of phytonutrients is most representatively investigated (Kapoor et al. 2021; Zou et al. 2021; Hu et al. 2023).

4.6 THE BIOLOGICAL ACTION OF PHYTONUTRIENTS

The orally ingested phytonutrients undergo various processes, such as digestion, deliverance, and secretion. In order to release phytonutrients from ingested plant matrices, different physical, chemical, and biological processes are taking place within the gastrointestinal tract. For example, the reduction of particle size using mechanical forces; the disintegration promotion by acidic gastric fluids, proteins, fats, and carbohydrates; the hydrolyzation by digestive enzymes; and the solubilization and transportation of lipids by bile salts. After the digestion process, the phytonutrients must be absorbed from the cells that exist in the gastrointestinal tract through various active or passive mechanisms. The lymphatic system or portal vein is subsequently used to transport the phytonutrients into the bloodstream. Through the circulatory system they are then delivered to the body's organs and tissues. The human body excretes phytonutrients that are not used through the urine system or feces (Hu et al. 2023).

4.6.1 Factors Influencing the Phytonutrient Bioaccessibility

Food processing and food matrix are the key elements affecting phytonutrient bioaccessibility. The food matrix is involved in the discharge of phytonutrients into the gastrointestinal tract as it may preserve the phytonutrients included in it against the chemical degradation caused by enzymes or acids found in the digestive tract. The phytonutrients do not exhibit their beneficial effects if they are not delivered at the correct point in the gastrointestinal tract. Through a variety of processes, the presence of lipids in foods enhances the bioaccessibility of phytonutrients that are hydrophobic (Yao et al. 2021). Initially the phytonutrients can be protected from gastrointestinal degradation by being embodied into a lipid phase. In addition, digestible lipids contribute to the creation of mixed micelles, which can make hydrophobic phytonutrients more soluble (Hu et al. 2023).

The bioaccessibility of polyphenols is decreased when polyphenols interact with proteins with non-covalent bonds and form insoluble protein-polyphenol complexes (Kamiloglu et al. 2021). However, the presence of proteins may positively affect the bioaccessibility of hydrophobic phytonutrients and cellular uptake in the small intestine (Pohl 2007). The molecular characteristics of phytonutrients that exist in foods affect their bioaccessibility. The presence of minerals in different plant matrices can change the absorption and bioaccessibility of polyphenols by creating polyphenol-metal complexes. The atomic and molecular structures of the minerals and polyphenols determine the type of complexes that are produced (Pohl 2007).

WEPs may undergo various processing operations that affect the bioaccessibility and bioavailability of phytonutrients. The bioavailability of phytonutrients in foods can be affected by thermal processes, that is, blanching, cooking, pasteurization, or sterilization (Domínguez-Fernández et al. 2021). Thermolabile phytonutrients may undergo chemical degradation as a result of heat. However, the bioavailability of phytonutrients can be increased through thermal processing, by making the food matrix softer. For example, microwaving results in an increment in phytonutrients release but not in their bioaccessibility (Li et al. 2022). On the other hand, heat combined with ultrasound (thermo-sonication) is more effective than traditional sterilization as it can increase the release of phytonutrients while decreasing their chemical degradation (Ramírez-Melo et al. 2022). Vacuum-assisted thermal processing is another technique used to preserve the biological activity as well as the nutritional value of foods containing high amounts of phytonutrients (Hu et al. 2023; Burca-Busaga et al., 2021).

Nonthermal processing is an alternative to conventional thermal techniques that safeguards the bioactivity of phytonutrients in foods. Freeze-drying of foods typically aids in phytonutrient retention. Nevertheless, several investigations claim that freeze-drying reduces the bioaccessibility of phytonutrients compared to traditional convection drying (Sęczyk, Ozdemir, and Kołodziej 2022). In order to preserve the phytonutrients in foods during processing, high-pressure processing and pulsed electric field are also performed as an alternative to conventional thermal methods (Gómez-Maqueo et al. 2021; Hu et al. 2023; Ribas-Agustí et al. 2018).

4.7 METHODS THAT IMPROVE THE BIOACCESSIBILITY OF PHYTONUTRIENTS

The main methods used to enhance the bioavailability and bioaccessibility of phytonutrients are chemical modification and encapsulation techniques such as biopolymer-based particles, emulsions, and lipid-based nanostructures.

The chemical structure of phytonutrients affects their stability, solubility, bioaccessibility, and bioavailability. Exposure to environmental conditions like heat, pH, light, oxygen, and pro-oxidants might lead to phytonutrient decomposition and subsequently a decrease in their bioactivity and bioavailability as well as in their bioaccessibility (Chen et al. 2022; Hao et al. 2022; Hu et al. 2023). Carotenoids such as astaxanthin have greater bioavailability and thermal stability in ester form compared to free form (Zhou et al. 2019). Saturated fatty acid astaxanthin esters with long chains show

greater stability (Lu Yang et al. 2021). Furthermore, esterification enhances the solubility of phytonutrients in food matrices and promotes their bioaccessibility. The solubility of phytosterols is low
both in water and oil, resulting in a low bioaccessibility. However, a phytosterol ester with enhanced
oil solubility is created when oleic acid is used to esterify phytosterols using a microwave-assisted
method. Chemical modification of phytonutrients results in improved biological properties but is
not frequently utilized in food industries due to the cost and time needed to ensure that the new
compounds are safe and comply with regulatory criteria. (Hu et al. 2023).

Encapsulation has emerged as a very important technique in the food industry as it can improve
the bioavailability of phytonutrients and their ability to resist to external environmental conditions,
adjust their release profiles, and cover any unpleasant odors or tastes while promoting their incorporation into food matrices (Hu et al. 2023; Nowak et al. 2019). Proteins, carbohydrates, and lipids are frequently employed to make encapsulation systems. In the food sector, emulsion-based,
biopolymer-based, and other lipid-based systems have been used as encapsulation technologies
(C. Wang et al. 2022; Lin et al. 2021; Hu et al. 2023).

To enhance the functional performance of phytonutrients, biopolymer-based particles have been
used to encapsulate them; the chemical stability of lutein and its bioaccessibility under simulated
gastrointestinal conditions were improved by encapsulating the latter carotenoid in composite microparticles made from sodium alginate and sodium caseinate (Hu et al. 2023; Hao et al. 2022). The
encapsulation of icaritin 28-fold into pectin nanoparticles improved its bioaccessibility (Chen et al.
2022), and the encapsulation of curcumin 6-fold into zein/sophorolipid nanoparticles improved its
bioaccessibility as well. The bioaccessibility of phytonutrients is increased when these are encapsulated within biopolymer nanoparticles, since the percentage of solubilized mixed micelles has
increased (Hu et al. 2023). The encapsulation of phloretin into gliadin/sodium carboxymethyl cellulose nanoparticles led to a sustained regulated delivery of the substance (He et al. 2022). Also in
some cases, the encapsulation of some phytonutrients within biopolymer particles can cover their
unpleasant odor and taste (Hu et al. 2023; Favaro-Trindade et al. 2021; Silva et al. 2022).

Emulsions are also employed as phytonutrient carriers, owing to their ability to encapsulate both
hydrophilic and lipophilic phytonutrients, since they can contain both water and oil phases. In most
cases, lipophilic phytonutrients are encapsulated using oil-in-water emulsions, which are made up of
coated oil droplets scattered in water. The nonpolar cores of the oil droplets contain lipophilic phytonutrients, which makes them more water-soluble and prevents them from degrading. In addition,
mixed micelles are produced by the lipid phase's ingestion, which have the ability to dissolve and
carry the phytonutrients throughout the digestive system (Hu et al. 2023). For instance, the bioaccessibility of β-carotene is increased and its chemical degradation during storage is prevented by being
encapsulated in emulsions (X.-M. Li et al. 2020). The bioavailability of astaxanthin is increased by
six times when it is contained in emulsions as opposed to when it is free. The amount and type of oil
that exists in emulsions affects the capacity of the mixed micelles generated following the digestion
of lipids to dissolve, and this in turn influences the bioaccessibility of phytonutrients. The bioaccessibility is higher when the content of lipids is higher because in this case exists a larger number
of mixed micelles to help the phytonutrients dissolve (Yao et al. 2021). The stability, loading, and
bioaccessibility of phytonutrients can also be affected by the form of oil used to create an emulsion.
For example, triglycerides with large chains give higher bioaccessibility to hydrophobic phytonutrients than triglycerides with medium or short chains. This is happening because triglycerides with
long chains produce mixed micelles with lipophilic regions, which are capable of hosting long-chain
phytonutrients. The bioavailability of phytonutrients is also affected by the polarity of the latter due
to their impact on polarity of the emulsions and mixed micelles (Hu et al. 2023; Yao et al. 2021).

The goal of many studies is the creation of emulsions without surfactants in order to prevent
possible risks to human health from using these surfactants in foods. In Pickering emulsions, for
instance, biopolymer-based nanoparticles, which are made of polysaccharides and/or proteins,

replace the molecular surfactants (Hu et al. 2020). Jo et al. (2021) extended the amount of time that resveratrol remained in the gut, utilizing Pickering emulsions coated with chitosan and waxy maize starch nanocrystals. In another study, enhanced rice starch grains containing octenyl succinic anhydride (OSA) were used to stabilize resveratrol-loaded Pickering emulsions even at partial coverage conditions (Matos et al. 2021). Pickering emulsions containing complexes of gallic acid, dextran, and bovine serum albumin were used to increase the chemical stability of capsorubin (Zhang et al. 2021). The chemical modification of polymers improves the ability of phytonutrients to execute their functions in emulsions (Pan et al. 2021).

Emulsions with a high internal phase are also used to enhance the performance of phytonutrients. These emulsions are created when the close packing limit (about 74%) is exceeded by the scattered phase fraction. Curcumin bioaccessibility has been reported to enhance when high interior phase emulsions, which are stabilized by sugar beet chitosan/tannic acid/pectin complexes, are used (Hu et al. 2023; Miao et al. 2021). In another study, β-carotene and puerarin were encapsulated into meat protein stabilized with high internal phase emulsions and their bioaccessibility was enhanced in simulating the digestive system environment (C. Li et al. 2022).

Many other lipid-based nanostructure types, like solid lipid nanocarriers and liposomes, have been studied for their capacity to encapsulate and deliver phytonutrients. The spherical vesicles known as liposomes consist of one or more phospholipid bilayers, that approximately resembles the structure of cell membranes. In various locations on the carriers, liposomes can encapsulate lipophilic or hydrophilic molecules. Lipophilic phytonutrients may become entrapped in the nonpolar tails of the phospholipids and hydrophilic phytonutrients may become entrapped in the core of the aqueous phase or in the space among the polar head groups of phospholipids. Cell membranes are typically composed of a phospholipid bilayer that can enhance the passage of phytonutrient-loaded liposomes via enterocytes and in this way to increase their bioavailability and bioaccessibility (Hu et al. 2023; Lee 2020). Other lipid-based delivery techniques for encapsulating phytonutrients include nanostructured lipid carriers and solid lipid nanoparticles (McClements 2012). Nanostructured lipid carriers fabricated from olein and palm stearin are used for the encapsulation of β-carotene and enhance the carotenoid bioaccessibility and antioxidant activity (Rohmah, Rahmadi, and Raharjo 2022).

4.8 CLINICAL STUDIES ASSESSING THE BENEFITS OF PHYTONUTRIENTS ON HUMAN HEALTH

Many clinical studies show that a high intake of plant products results in a decrease in mortality and the chance of contracting chronic diseases, which suggests that phytonutrients play a protective role in maintaining human health (Poiroux-Gonord et al. 2010; Aune 2019). In addition to their antioxidant effects, phytonutrients also possess anti-inflammatory and anticarcinogenic properties that benefit human health (R. H. Liu 2013; Kris-Etherton et al. 2002). In this section we will report the existing clinical studies that prove the benefits that arise for human health from the introduction of phytonutrients in the daily diet. We will especially refer to the categories of phytonutrients existing in WEPs.

4.8.1 THERAPEUTIC APPLICATIONS OF PHYTONUTRIENTS

4.8.1.1 Stress and Sleep

Phytonutrients show a pharmacological effect either in problems related to stress or in sleep disorders. Scholey et al. (2012) investigated how flavonoids affect stress levels in humans. The evaluation of the electroencephalogram of the volunteers revealed a reduction in self-assessed stress as well as an increase in self-assessed calmness. The authors speculate that the action mechanism may be explained by a possible effect on the synthesis of nitric oxide, related to the regulation of cerebral vascular permeability. The impact of flavonoids, particularly isoflavones, on the quantity and quality of 38 postmenopausal women's sleep was also investigated (Hachul et al. 2011). After receiving either 80 mg of isoflavones daily for four months or a placebo, a sleep study was conducted utilizing questionnaires and polysomnography. After the completion of treatment, the results indicated an increase in sleep quality and a decrease in the number of insomnia cases.

4.8.1.2 Digestive Health

Phytonutrients present also a positive effect on "digestive health", which refers to a therapeutic field that encompasses both chronic liver disorders as well as a number of common symptoms and conditions like nausea, diarrhea, and constipation (Monjotin et al. 2022b). An 8-week treatment with a flavonoid-rich extract enhanced stool quality while abdominal annoyance was greatly lessened as compared to placebo, and the duration of transit time through the colon was shortened (Baek et al. 2016). The potential of certain flavonoids for stimulating chloride channels and serotonin transmission, which results in the release of water, mucus, and electrolytes in the colon, is thought to be the mechanism of action, according to the authors. Dryden et al. (2013) investigated the role of flavonoids in treating ulcerative colitis and have shown that the rate of pathological remission was significantly increased after 56 days of administration of a flavonoid-rich extract. Flavonoids and particularly epigallocatechin gallate were thought to be responsible for the reported therapeutic benefit, owing to the capacity of the latter phytonutrient to inhibit IkB kinase, hence preventing NF-kB's activation and nuclear translocation, which is a key regulator of inflammation. Additionally, flavonoids are used to treat liver diseases.

Barsalani et al. (2012) examined the impact of isoflavone dietary supplementation on γ-glutamyltransferase levels and on fatty liver index, among 54 post-menopausal overweight or obese women. The results indicated that after 6 months of supplementation with isoflavones, both fatty liver index and γ-glutamyltransferase levels were significantly improved. These results were attributed to isoflavone capacity to reduce the oxidative stress present in liver diseases. Cheraghpour et al. (2019) investigated how therapy with hesperidin affected the symptoms of nonalcoholic fatty liver disorder. The results revealed that after 12 weeks of treatment with 1g/day hesperidin, hepatic enzymes, and hepatic steatosis as well as inflammatory indicators were significantly improved.

4.8.1.3 Joints and Bones

Many products existing in the market can be used to treat joint disorders such as rheumatoid arthritis and osteoarthritis. In this form of joint disorders, several phytonutrients exhibit therapeutic properties (Monjotin et al. 2022b). Gambacciani et al. (1997) investigated potential benefits of the combination of a low dosage estrogen substitute and isoflavone derivative ipriflavone on postmenopausal women's metabolism and mineral content of bones. The authors concluded that after two years of daily treatment with 600 mg of ipriflavone, the density of the vertebral bones increased. Law et al. (2016) investigated the impact of a daily treatment with a juice enriched with flavonoids and phenolic acids on patients with osteoporosis. The results showed that after daily treatment for two months with 100 mL of juice containing flavonoids and phenolic acids, patients with osteoporosis showed positive modulation in bone loss and significant improvements in oxidation markers. Flavonoids' antioxidant properties and their capacity to inhibit the development of progenitors in osteoclasts have been associated with these results. (Monjotin et al. 2022b). Javadi et al. (2017) investigated how quercetin affected inflammatory variables as well as clinical symptoms in female patients suffering from rheumatoid arthritis by administering 500 mg of quercetin every day for eight weeks to patients with rheumatoid arthritis and the results revealed a significant reduction in morning pain and in the pain following activity. These results have been associated with the suppression of NF-kB pathway as well as to the release of anti-inflammatory cytokines.

Wattanathorn et al. (2018) investigated how phenolic acid-rich extract affected the bone turnover markers of perimenopausal and menopausal women, and the results showed that after eight weeks of daily administration of 1.5 g of an extract high in phenolic acids, there was a significant increase in the bone formation-related indicators such as alkaline phosphatase and osteocalcin and decrease in the markers involved in bone resorption (β-carboxy-terminal collagen cross-links). Connelly et al. (2014) investigated results from administering an infusion containing high amount of rosmarinic acid twice daily for 16 weeks to 46 individuals diagnosed with osteoarthritis. The results demonstrated a substantial decrease in pain compared to baseline in the group supplied with rosmarinic acid. In addition, after a 16-week supplementation of rosmarinic acid, life quality was reported to have improved.

4.8.2 Cognitive Function

Several studies have examined how flavonoids can help to treat cognitive problems (Mojnooin et al. 2023). Alharbi et al. (2016) investigated how flavonoid ingestion affected immediate cognitive function. Twenty-four individuals took part in this study, by drinking 240 mL of a beverage that contained 272 mg of flavonoids. The results revealed that six hours after ingestion their verbal abilities and reflexes had significantly improved when compared to the control group. The authors claimed that the flavonoids' antioxidant properties, which enhance NO bioavailability and boost cerebral vascular flow, were responsible for the positive results. Gratton et al. (2020) demonstrated that dietary flavanols have the ability to enhance cognitive function as well as the oxygenation of the cerebral cortex in healthy people. Lamport et al. (2016) investigated the impact of consuming 777 mg of flavonoids per day for 12 weeks on cognitive function and found that driving ability and spatial memory significantly improved. Mastroiacovo et al. (2015) examined the impact on cognitive function of 90 older individuals, after daily ingestion of a beverage high in flavonoids (993 mg/day) for eight weeks. The results demonstrated progress in mental health assessments, especially the trail-making and verbal fluency tests. These outcomes have been attributed to flavonoids' neuroprotective and antioxidant effects related with their ability to enhance cerebral perfusion, through influence on NO-dependent endothelial activity. Phenolic acids have also favorable impacts on cognitive function and agility (Mojnooin et al. 2023). Saitou et al. (2018) investigated the impact on cognitive performance after supplementation for 16 weeks with 300 mg chlorogenic acid in 38 volunteers having memory issues. The results demonstrated that several mental processes, such as executive function and attention, were considerably improved after the supplementation with chlorogenic acid. The results obtained may be explained by the capacity of phenolic acids to influence the synthesis of transthyretin and apolipoprotein A1, two indicators of early cognitive impairment. Falcone et al. (2018) demonstrated that after daily treatment of 142 volunteers for 7–90 days, with an extract rich in phenolic acids, a positive effect on reactive agility was observed. This result has been linked with the phenolic acids' capacity to reduce inflammation in the brain.

4.9 ANALYTICAL TECHNIQUES USED FOR THE DETERMINATION OF PHYTONUTRIENTS IN WILD EDIBLE PLANTS

The phytonutrients of WEPs are complex and monomeric substances which first must be extracted and isolated before being identified, determined, and subjected to bioactivity testing. The last decades innovative extraction, separation, and structural characterization methods and technologies have emerged, which accelerate phytonutrient extraction and analysis. The quantity of phytonutrients separated from WEPs is usually small, thus performing structural research using traditional techniques and methods is frequently challenging. For this reason, the main method utilized is spectral analysis. In the following we will mention the analytical techniques and methods that are used for determining phytonutrients' quality and quantity. Extraction is the first step in the determination of phytonutrients in WEPs and is the necessary process before their isolation. The goal of extraction is to maximize the obtaining of the target phytonutrients while preventing or minimizing the dissolution of undesired ingredients. Phytonutrient separation is the procedure of distinguishing these components of plant extracts or useful parts one at a time and transforming them into monomeric molecules, using chemical and physical processes. Traditional separation techniques include precipitation, crystallization, solvent extraction, salting-out, fractional distillation, and dialysis. Moreover, contemporary separation techniques, such as column chromatography, gas chromatography, high performance liquid chromatography and ultrafiltration significantly contribute to the separation of phytonutrients.

The structural studies on phytonutrients are conducted primarily through spectral analysis. In spectral analysis a small amount of sample is used to get structural information by analyzing various spectra. Furthermore, the structural determination of phytonutrients is attained using ultraviolet-visible spectra, infrared spectra, mass spectroscopy, and nuclear magnetic resonance spectroscopy.

4.9.1 EXTRACTION OF PHYTONUTRIENTS

4.9.1.1 Solvent Extraction Methods

Solvent extraction is a common analytical technique that is used for the extraction of plant material. The selection of the proper solvent is the primary objective in order to effectively extract the desired plant material. Throughout the extraction, in the first step the solvent diffuses into the cell membrane and in the second step dissolves the solutes, which results in the creation of difference between extracellular and intracellular concentration. In the third step the solvent, which is enriched with the solutes that were extracted, spreads out of the plant cells (Feng et al. 2020).

The selection of the suitable solvent varies depending on the plant type, the part of plant that we have to extract, the nature of phytonutrient, and the availability of the solvent. Using a solvent with the appropriate polarity is the key point to attain effectively the goal of the extraction. When extracting polar substances, polar solvents including ethanol, methanol, and water are typically utilized, while nonpolar substances are extracted using nonpolar solvents like hexane and dichloromethane (Abubakar and Haque 2020).

Water is a strong polar solvent that is inexpensive and nontoxic. It is utilized for the extraction of phytonutrients that have high polarities like proteins, tannins, amino acids, saccharides, inorganic salts, alkaloid salts, organic acid salts, and glycosides. The extraction of chemical substances with water has the disadvantage that the aqueous extract is difficult to preserve because it is easy to get moldy. In addition, the boiling point of water is high, and a considerable amount of time must be spent to concentrate the water extract. Additionally, the water extract includes various impurities like pectins, proteins, mucilages, tannins, and inorganic ions, making it difficult to extract the target substances (Feng et al. 2020).

Ethanol and methanol are most commonly used organic polar solvents. Ethanol in different concentrations has the ability to extract chemical substances according to their properties. In addition, ethanol is relatively cheap, nontoxic, and is concentrated easily. Furthermore, hydrolysis of glycosides in ethanol extract is difficult, and ethanol extract does not mold easily. Methanol has similar properties to ethanol but the boiling point of ethanol is lower. Methanol is a toxic solvent and we must pay great attention when using it. The hydrophobic organic solvents benzene, petroleum ether, ether, chloroform, dichloroform, and ethyl acetate are not water miscible. Hydrophobic compounds such lactones, lipids, volatile oils, chlorophyll, phytosterols, certain aglycones, and a few alkaloids are extracted using them. These solvents can be concentrated easily and have low boiling points. Moreover, their disadvantages are toxicity, large loss, flammability, high price, and their difficulty to penetrate the tissues of plant cells (Feng et al. 2020).

Immersion method is a method used for dissolving phytonutrients at temperatures below 80ºC. This method is used for the extraction of phytonutrients with sensitivity at high temperatures.

Plants that contain high amount of pectins, starches, mucilages, or gums, can be extracted using this method as well. To apply this method, plant pieces or powder are loaded in a container and then appropriate solvents are introduced to it, immersing the material for a specified amount of time. The rate of dissolution can be sped up by shaking or intermittent stirring while the process is going on. This method has two drawbacks: a low extraction ratio and easily moldable aqueous extracts (Feng et al. 2020).

A traditional extraction technique used in plant processing is percolation extraction. After adding coarse plant particles to a percolation tank, a continuous flow of extraction solvent is added, and after 24 to 48 hours, percolation extract is simultaneously collected. Afterwards, additional solvent needs to be introduced periodically at the highest point of the percolation device. The advantages of the method are that the equipment is simple, it finds application in the extraction of a wide variety of plants, and plants that are sensitive to high temperature can be easily extracted with this method. The main disadvantages of the method are excessive solvent consumption, extended extraction times, and excessive consumption of energy (Wang, Qu, and Gong 2020).

Reflux extraction is a solid-liquid extraction method that lasts for a set amount of time at a defined temperature with continuous solvent evaporation along with condensation without solvent loss. It is utilized to extract lipophilic phytonutrients, like terpenoids, anthraquinoids, and steroids. This method is widespread in the herbal industry because it is inexpensive, effective, and simple to utilize (Chua, Latiff, and Mohamad 2016). In addition, this method cannot be applied for the extraction of thermal unstable phytonutrients because of the long duration of heating.

Decoction is an extraction method that is used particularly for thermally stable and water soluble phytonutrients. In this instance, the raw plant is heated within an open-type evaporator using a specific amount of water for a set amount of time. It is a method that can be easily applied and the different phytonutrients contained are extracted to different degrees. The disadvantages of the method include the impossibility of extracting plants with many starches and that it cannot be applied to thermally unstable compounds or to volatile compounds. When decoction extraction is finished, the concentrated extract is either filtered or strained and used immediately or after additional processing (Manousi, Sarakatsianos, and Samanidou 2019; Feng et al. 2020).

Ultrasound-assisted extraction utilizes solvents and ultrasonic energy to extract phytonutrients from plant matrices. Ultrasounds are sonic waves with a frequency above ~16 kHz (Tsikrika, Chu, et al. 2022) and it is considered as a 'green technology', as it saves energy and water while reducing carbon and water footprints and posing few chemical and physical dangers (Tsikrika, Lemos, et al. 2022). These waves are composed of multiple cycles of compression and rarefaction and can propagate across a liquid, solid, or gaseous media and cause the molecules to displace from their initial positions. When the intensity is high, cavitation bubbles are produced because the force that pulls apart molecules during rarefaction is stronger than the force that attracts them together. A hot spot and extremely harsh local conditions are produced when these bubbles get bigger during the coalescence and then collapse during the compression phase. The temperature can rise to 5000 K and the pressure may approach 1000 atm. These hot spots accelerate the biochemical reactions; the plant cellular matrix might experience phenomena like fragmentation, pore development, localized erosion, enhanced absorption, and shear force as a result of the sound waves and the collapsing cavitation bubbles. Shockwaves are produced by the collapsed cavitation bubbles, and fragmentation to the structure of cells is brought on by the rapid inter-particle interactions (K. Kumar, Srivastav, and Sharanagat 2021). The bioactive compounds dissolve in the solvent as a result of the fast fragmentation due to an increase in surface area, a decrease in the size of particles, and increased mass transfer rates in the solid matrix's border layer. Ultrasound causes localized damage to plant tissues called corrosion. The collapse of cavitation bubbles on the outer layer of plant tissues can potentially result in corrosion. In addition, this corrosion makes the solvent interaction easier and boosts the extraction yield. The pore formation that occurs during cavitation in the cell membranes, known as 'sonoporation', provokes the release of phytonutrients existing in the plant cell. Furthermore, shear force and turbulence are brought on by the formation and collapse of cavitation bubbles into the fluid, causing the cell walls to break down and release the phytonutrients. Ultrasound also raises the absorption of water in the pomace and improves the phytonutrients' ability to use water as a solvent itself. An increase in the index of swelling of the plant matrix, brought on by ultrasound, promotes solute transport and desorption, leading to more efficient extraction (K. Kumar, Srivastav, and Sharanagat 2021).

Refluxing method is an extraction method used to extract phytonutrients through heating and refluxing. This method uses a special reflux device to save solvents as well as to reduce potential toxicity to the operators. Refluxing method is used to extract hydrophobic phytonutrients like anthraquinoids, terpenoids, and steroids. It is a complex and remarkably effective method of extraction, which is not applied to extract thermal unstable phytonutrients because it demands a long heating time (Feng et al. 2020).

Supercritical fluid extraction method is carried out with plant material and occurs above the applied solvent's critical temperature and pressure. A compound's critical temperature is the temperature at and above which vapor of the compound cannot be converted into a liquid phase, no matter how much pressure is applied. The greatest pressure at which a compound's liquid phase can change into its gas phase through a rise in temperature is known as a compound's critical pressure. It is a fast, clean, and very efficient technique for extracting plant phytonutrients. Supercritical fluid extraction seeks to increase yield with less danger of contamination by using environmentally benign CO_2 gas as an extraction medium, which acts like a liquid solvent. CO_2 gas is converted to a liquid solvent once the critical temperature of 31°C and the critical pressure of 7.4 MPa have been surpassed. In addition, CO_2 has many other positive aspects, like odorless, not being toxic, inflammable, low cost, and chemical stability, which make it the most widely used solvent in supercritical fluid extraction. CO_2 is a nonpolar compound that is utilized to extract lipophilic compounds. It has low dissolvability compared to strong polar compounds and for this reason entrainers are added to make CO_2 more soluble when polar compounds are extracted. The most frequently used entrainers are ethanol, methanol, ethyl acetate, acetone, water, and so on (Rohilla et al. 2022; Feng et al. 2020).

4.9.2 Sublimation Method

The process of sublimation involves heating solid materials without melting them to produce gas. Some phytonutrients show sublimation qualities and could be extracted using the method previously mentioned. Examples are organic acids, coumarins, and alkaloids. Nevertheless, natural products are easy to carbonize due to the prolonged heating duration. The volatile tar-like compounds frequently accompany thermal decomposition and attach to sublimates in most cases. The method frequently gives modest yields and cannot be used for production on a large scale (Feng et al. 2020).

4.9.3 Steam Distillation Method

Steam distillation is the method utilized to extract volatile molecules with elevated boiling points from inactive and complicated matrices, whether they are in solid or liquid form, utilizing superheated or saturated steam. These compounds' vapor pressures are about 100°C and their boiling points are usually higher than 100°C. Steam distillation vaporizes or liberates the volatile compounds from the plants by using water and/or steam as an extraction solvent. The substances are delivered to the steam, where they are diffused after absorbing heat from the steam and becoming vaporized. After cooling and condensing the resultant vapor phase, the organic phase and water are separated depending on their immiscibility. The following two products are generated by this process; hydrosol and volatile oil. The volatile oil remains in the upper phase of the decanter, while the hydrosol (water plus a few hydrolyzed substances) remains in the bottom phase. It is mainly used to extract volatile oils, phenolic substances of small molecules from plants and some alkaloids (Prado et al. 2015; Feng et al. 2020). It is primarily employed for extracting phenolic compounds of very small molecules, volatile oils, and certain alkaloids from plants.

4.9.3.1 Sample Clean Up

The extracts obtained by the aforementioned methods can be analyzed qualitatively and quantitatively to determine phytonutrients. In some cases, depending on the analytical technique to be applied, additional purification may be required. Filtering removes plant unwanted components and purifies the extract. Solid phase extraction or column chromatography may be applied to create further fractions of the extracts prior to instrumental analysis (Tsao and Li 2013).

A method often used to purify the extract is hydrolysis, particularly if we have polyphenol analysis. Polyphenols, a subclass of flavonoids, are frequently heavily glycosylated both via interaction with various sugars and at various aglycone positions. The extract polyphenol profile is simplified by hydrolysis, and in this way a better separation is achieved. Hydrolysis of glycosylated substances is performed under enzymatic basic or acidic conditions. Since different hydrolysis conditions might produce differing hydrolysis products, special care must be given while choosing the most suitable procedure. In addition, hydrolysis is used to release some forms of polyphenols which differently exist as complexes. Another class of phytonutrients that require hydrolysis to provide bioactive oligomers or monomers are polymeric procyanidins. (Tsao and Li 2013).

An effective clean-up process must remove the interfering substances from the extract and be abile to give high recoveries for the substances of interest. Furthermore, there are cases where the phytonutrient of interest has similar physicochemical structure to the interfering compounds, so we have a co-elution of these substances. Additional challenges existing during the sample clean-up process are that some phytonutrients may bound on the solid phase extraction filters and that the hydrolysis might not end precisely when cleavage of the ester or glycosidic bonds takes place, but it might result in the breakdown of some compounds' primary structural components when they are relatively unstable under alkaline or acidic conditions (Nuutila, Kammiovirta, and Oksman-Caldentey 2002; Tsao and Li 2013).

4.10 STRUCTURAL IDENTIFICATION AND QUANTITATIVE DETERMINATION OF PHYTONUTRIENTS

It is very important to identify and quantitatively determine the phytonutrients contained in WEPs, because this may give us the opportunity to conduct further studies on their bioactivities, synthesis, and in vivo metabolism. The quantity of phytonutrients isolated from WEPs is usually low and it is difficult to conduct structural studies using conventional chemical techniques like derivative synthesis and chemical degradation. For this reason, a method is employed that uses the least amount of sample and provides structural data by measuring different spectra (Feng et al. 2020).

4.10.1 PURITY DETERMINATION

In the first step purity has to be defined, because structural identification depends on it; thin layer chromatography is the technique that is most frequently utilized for the determination the purity of substances. A sample can be considered pure substance if it produces a single spot when exposed to three separate developing agents. Sometimes, both reverse phase and normal phase chromatographic techniques are needed. High pressure liquid chromatography and gas chromatography are also frequently used to determine the purity of phytonutrients. Gas chromatography is utilized to analyze volatile substances while high performance liquid chromatography is utilized to analyze both volatile and non-volatile compounds (Feng et al. 2020).

4.10.2 STRUCTURAL DETERMINATION

The structural determination of phytonutrients is achieved using ultraviolet-visible spectra (UV-Vis), infrared spectra, mass spectrometry, and nuclear magnetic resonance (NMR) spectroscopy.

Since aromatic compounds are strong chromophores in the UV range, UV-visible spectroscopy is employed for qualitative and quantitative determination. When molecules absorb electromagnetic waves with wavelengths between 200 and 800 nm, the result is a type of electron transition spectrum known as UV-Vis spectrum. The compounds that contain conjugated double bonds, aromatic compounds, and compounds containing α, β-unsaturated carbonyl groups (ketones, aldehydes, esters and acids) exhibit significant UV spectral absorption due to n $\rightarrow$ π^* or π $\rightarrow$ π^* transitions. The UV-Vis spectroscopy may identify phenolic substances, such as tannins, anthocyanins, and phenols. The UV-Vis spectroscopy can be employed to calculate the total phenolic extract at the wavelength of 280 nm for flavonoids, at the wavelength of 320 nm for flavones, at the wavelength of 360 nm for phenolic acids, and at the wavelength of 520 nm for anthocyanins. This method is cost effective and does not demand too much time compared to other techniques. Only a portion of a compound's structural data, not all of it, can be provided by the UV spectrum and for this reason can be used only as an auxiliary method for structural determination. It is an important tool to figure out the structures of phytonutrients having conjugated substructures (Altemimi et al. 2017; Feng et al. 2020).

Infrared (IR) spectroscopy is an absorption method that is frequently used in both qualitative and quantitative analysis and deals with the infrared region of the electromagnetic spectrum. Electromagnetic radiation that can alter the rotating and vibrating states of covalent bonds in organic compounds is detected in the infrared area of the electromagnetic spectrum. The infrared spectrum of an organic substance is used to identify unknown compounds by comparing it to spectrum libraries. When a substance is exposed to infrared light, internal vibrational changes occur, which are connected to infrared absorption of the compound. The type and the class of bonds (C–C, C=C, C≡C, C=O, C–O, N–H and O–H) show different vibrational frequencies. Compounds with functional groups like carbonyl, hydroxyl, amino, and aromatic rings absorb in the area over 1250 cm^{-1}. Compounds with single bonds absorb in the area 1250 cm^{-1} to 625 cm^{-1}. IR spectroscopy can be used to determine the types and the functional groups of aromatic ring substitution. For instance, an important difference appears in IR spectrum for 25S and 25R spirostanol saponins between the region 960 cm^{-1} and 900 cm^{-1}. Fourier transform infrared spectroscopy is the type of infrared spectroscopy that is most frequently used and is an effective analytical method that offers nondestructive and rapid investigation to the fingerprints of plant extracts (Altemimi et al. 2017; Feng et al. 2020).

In mass spectrometry (MS), organic compounds undergo bombardment by either electrons or lasers. These bombardments change the molecules into charged ions, which are subsequently attracted to the collector by magnetic and electric fields. The plot of an ion relative abundance as a function of mass-to-charge ratio is called a mass spectrum (Altemimi et al. 2017). MS can define formulas, fragment structures, and weights of compounds. Using MS can determine the relative molecular mass of a compound accurately, and then the locations where the molecule has been fragmented can then be used to determine an exact molecular formula (Christophoridou et al. 2005; Altemimi et al. 2017). In the discipline of structural analysis, molecular ion peaks are used to determine molecular weights, and high-resolution mass spectrometry can be used to determine molecular formulas. One method for figuring out chemical structures is to combine fragment ion and molecular ion peaks. Tandem mass spectrometry could be used for the isolation and analysis of mixed ions again. Mass spectrometry could be classified into electrospray ionization mass spectrometry, electron impact mass spectrometry, field desorption mass spectrometry, chemical ionization mass spectrometry, matrix-assisted laser desorption mass spectrometry, fast atom bombardment mass spectrometry, and so on (Feng et al. 2020).

NMR is an efficient and widely employed analytical method used to define chemical structures. The basic principle behind nuclear magnetic resonance spectroscopy is that nuclei of atoms have magnetic properties that can be used to give information about the chemical structure of a compound. It depends on the interaction between an atomic nucleus with radiofrequency radiation that is applied externally. A net energy exchange takes place throughout this interaction, changing the inherent nuclear spin, a property of the atomic nuclei. Apart from the chemical structure, NMR spectroscopy can be used to determine phase changes, configurational and conformational alterations, diffusion potential, and solubility (Singh and Singh 2022). Carbon spectrum and hydrogen spectrum are mainly used. In nuclear magnetic resonance spectroscopy, molecules of the compound are subjected to electromagnetic waves generated by a magnetic field. When nuclei of atoms with magnetic distance absorb a specific amount of energy, energy level transitions then take place. Plotting the absorption strength against the frequencies of the absorption peaks results in the creation of the NMR spectrum (Feng et al. 2020). An NMR spectrum contains an abundance of data about the molecules' structures. Initially the chemical shift values can be used to determine which chemical groups are present in a given molecule. Secondly, signal area in ^{1H}-NMR spectra varies according to the number of atomic nuclei that are responsible for that signal, something that does not apply in the case of ^{13C}-NMR spectra. Finally, for the purpose of determining the structural characteristics of macromolecules, the monitoring of signals resulting from an effect known as the nuclear Overhauser effect is crucial, because it results from the interaction of nuclear spins that are distant in the molecular structure but close in space. An NMR spectrum can provide structural details regarding the kind, number, and arrangement of hydrogen and carbon atoms in the molecule, the kind of bonds existing in the molecule, the conformation, and configuration (Feng et al. 2020).

4.11 CONCLUSION

Since ancient times, wild edible plants have been used for nutritional and medicinal purposes. The climate, environment, and origin of the plant all influence variations in biologically active chemical constituents. Due primarily to their antioxidant properties, wild edible plants have multiple applications in the food, drug, and cosmetic industries. In human physiological systems, cell protection and regulation may be just as dependent on biological functions that go beyond their antioxidant activity. Wild edible plants, such as spices and herbs, will undoubtedly continue to be extensively utilized in the food industry as they comprise abundant sources of bioactive phytonutrients, aromatics, flavoring, and coloring compounds.

REFERENCES

Abubakar, Abdullahi R., and Mainul Haque. 2020. 'Preparation of Medicinal Plants: Basic Extraction and Fractionation Procedures for Experimental Purposes'. *Journal of Pharmacy and Bioallied Sciences* 12 (1): 1. doi:10.4103/jpbs.JPBS_175_19

Actis-Goretta, Lucas, Antoine Lévèques, Maarit Rein, Alexander Teml, Christian Schäfer, Ute Hofmann, Hequn Li, Matthias Schwab, Michel Eichelbaum, and Gary Williamson. 2013. 'Intestinal Absorption, Metabolism, and Excretion of (–)-Epicatechin in Healthy Humans Assessed by Using an Intestinal Perfusion Technique'. *The American Journal of Clinical Nutrition* 98 (4): 924–33. doi:10.3945/ajcn.113.065789

Alaca, Kevser, Emine Okumuş, Emre Bakkalbaşi, and Issa Javidipour. 2022. 'Phytochemicals and Antioxidant Activities of Twelve Edible Wild Plants from Eastern Anatolia, Turkey'. *Food Science and Technology (Brazil)* 42: 1–7. doi:10.1590/fst.18021

Alam, Mohammad Khairul, Ziaul Hasan Rana, Sheikh Nazrul Islam, and Mohammad Akhtaruzzaman. 2020. 'Comparative Assessment of Nutritional Composition, Polyphenol Profile, Antidiabetic and Antioxidative Properties of Selected Edible Wild Plant Species of Bangladesh'. *Food Chemistry* 320 (August): 126646. doi:10.1016/J.FOODCHEM.2020.126646

Alamri, Eman S., Haddad A. El Rabey, Othman R. Alzahrani, Fahad M. Almutairi, Eman S. Attia, Hala M. Bayomy, Renad A. Albalwi, and Samar M. Rezk. 2022. 'Enhancement of the Protective Activity of Vanillic Acid against Tetrachloro-Carbon (CCl4) Hepatotoxicity in Male Rats by the Synthesis of Silver Nanoparticles (AgNPs)'. *Molecules* 27 (23). doi:10.3390/molecules27238308

Alharbi, Mudi H., Daniel J. Lamport, Georgina F. Dodd, Caroline Saunders, Laura Harkness, Laurie T. Butler, and Jeremy P. E. Spencer. 2016. 'Flavonoid-Rich Orange Juice Is Associated with Acute Improvements in Cognitive Function in Healthy Middle-Aged Males'. *European Journal of Nutrition* 55 (6): 2021–29. doi:10.1007/s00394-015-1016-9

Altemimi, Ammar, Naoufal Lakhssassi, Azam Baharlouei, Dennis Watson, and David Lightfoot. 2017. 'Phytochemicals: Extraction, Isolation, and Identification of Bioactive Compounds from Plant Extracts'. *Plants* 6 (4): 42. doi:10.3390/plants6040042

Alu'datt, Muhammad H., Taha Rababah, Mohammad N. Alhamad, Sana Gammoh, Majdi A. Al-Mahasneh, Carole C. Tranchant, and Mervat Rawshdeh. 2018. *Pharmaceutical, Nutraceutical and Therapeutic Properties of Selected Wild Medicinal Plants: Thyme, Spearmint, and Rosemary. Therapeutic, Probiotic, and Unconventional Foods.* Elsevier Inc. doi:10.1016/B978-0-12-814625-5.00014-5

Aune, Dagfinn. 2019. 'Plant Foods, Antioxidant Biomarkers, and the Risk of Cardiovascular Disease, Cancer, and Mortality: A Review of the Evidence'. *Advances in Nutrition* 10 (November): S404–21. doi:10.1093/advances/nmz042

Baek, H. I., K. C. Ha, H. M. Kim, E. K. Choi, E. O. Park, B. H. Park, H. J. Yang, M. J. Kim, H. J. Kang, and S. W. Chae. 2016. 'Randomized, Double-Blind, Placebo-Controlled Trial of Ficus Carica Paste for the Management of Functional Constipation'. *Asia Pacific Journal of Clinical Nutrition* 25 (3): 487–96.

Baenas, Nieves, Francisco J. Salar, Raúl Domínguez-Perles, and Cristina García-Viguera. 2019. 'New UHPLC-QqQ-MS/MS Method for the Rapid and Sensitive Analysis of Ascorbic and Dehydroascorbic Acids in Plant Foods'. *Molecules* 24 (8): 1632. doi:10.3390/molecules24081632

Barsalani, R., E. Riesco, J-M. Lavoie, and I. J. Dionne. 2012. 'Effect of Exercise Training and Isoflavones on Hepatic Steatosis in Overweight Postmenopausal Women'. *Climacteric* 16 (1): 88–95. doi:10.3109/13697137.2012.662251

Basumatary, Sanjay, and Hwiyang Narzary. 2017. 'Nutritional Value, Phytochemicals and Antioxidant Property of Six Wild Edible Plants Consumed by the Bodos of North-East India'. *Mediterranean Journal of Nutrition and Metabolism* 10 (3): 259–71. doi:10.3233/mnm-17168

Bot, Arjen. 2019. 'Phytosterols'. *Encyclopedia of Food Chemistry*, January. Academic Press, 225–28. doi:10.1016/B978-0-08-100596-5.21626-0

Bunea, Andrea, Dumiţriţa Rugină, Zoriţa Sconţa, Raluca M. Pop, Adela Pintea, Carmen Socaciu, Flaviu Tăbăran, Charlotte Grootaert, Karin Struijs, and John VanCamp. 2013. 'Anthocyanin Determination in Blueberry Extracts from Various Cultivars and Their Antiproliferative and Apoptotic Properties in B16-F10 Metastatic Murine Melanoma Cells'. *Phytochemistry* 95 (November): 436–44. doi:10.1016/J.PHYTOCHEM.2013.06.018

Burca-Busaga, Cristina Gabriela, Noelia Betoret, Lucía Seguí, Jorge García-Hernández, Manuel Hernández, and Cristina Barrera. 2021. 'Antioxidants Bioaccessibility and Lactobacillus Salivarius (CECT 4063) Survival Following the In Vitro Digestion of Vacuum Impregnated Apple Slices: Effect of the Drying Technique, the Addition of Trehalose, and High-Pressure Homogenization'. *Foods* 10 (9): 2155. doi:10.3390/foods10092155

Cámara, Montaña, Virginia Fernández-Ruiz, and Brígida María Ruiz-Rodríguez. 2016. 'Wild Edible Plants as Sources of Carotenoids, Fibre, Phenolics and Other Non-Nutrient Bioactive Compounds'. In *Mediterranean Wild Edible Plants: Ethnobotany and Food Composition Tables*, edited by María de Cortes Sánchez-Mata and Javier Tardío, 187–205. New York, NY: Springer New York. doi:10.1007/978-1-4939-3329-7_9

Chandrasekaran, Akshaya, Maria del Pilar Sosa Idelchik, and J. Andrés Melendez. 2017. 'Redox Control of Senescence and Age-Related Disease'. *Redox Biology* 11 (April): 91–102. doi:10.1016/J.REDOX.2016.11.005

Chen, Yipeng, Yueming Jiang, Lingrong Wen, and Bao Yang. 2022. 'Structure, Stability and Bioaccessibility of Icaritin-Loaded Pectin Nanoparticle'. *Food Hydrocolloids* 129 (August): 107663. doi:10.1016/j.foodhyd.2022.107663

Cheraghpour, Makan, Hossein Imani, Shahrzad Ommi, Seyed Moayed Alavian, Elahe Karimi-Shahrbabak, Mehdi Hedayati, Zahra Yari, and Azita Hekmatdoost. 2019. 'Hesperidin Improves Hepatic Steatosis, Hepatic Enzymes, and Metabolic and Inflammatory Parameters in Patients with Nonalcoholic Fatty Liver Disease: A Randomized, Placebo-controlled, Double-blind Clinical Trial'. *Phytotherapy Research* 33 (8): 2118–25. doi:10.1002/ptr.6406

Chikara, Shireen, Lokesh Dalasanur Nagaprashantha, Jyotsana Singhal, David Horne, Sanjay Awasthi, and Sharad S. Singhal. 2018. 'Oxidative Stress and Dietary Phytochemicals: Role in Cancer Chemoprevention and Treatment'. *Cancer Letters* 413 (January): 122–34. doi:10.1016/J.CANLET.2017.11.002

Christophoridou, Stella, Photis Dais, Li-Hong Tseng, and Manfred Spraul. 2005. 'Separation and Identification of Phenolic Compounds in Olive Oil by Coupling High-Performance Liquid Chromatography with Postcolumn Solid-Phase Extraction to Nuclear Magnetic Resonance Spectroscopy (LC-SPE-NMR)'. *Journal of Agricultural and Food Chemistry* 53 (12): 4667–79. doi:10.1021/jf040466r

Chua, Lee Suan, Norliza Abd Latiff, and Muna Mohamad. 2016. 'Reflux Extraction and Cleanup Process by Column Chromatography for High Yield of Andrographolide Enriched Extract'. *Journal of Applied Research on Medicinal and Aromatic Plants* 3 (2): 64–70. doi:10.1016/j.jarmap.2016.01.004

Connelly, A. Erin, Amy J. Tucker, Hilary Tulk, Marisa Catapang, Lindsey Chapman, Natasha Sheikh, Svitlana Yurchenko, et al. 2014. 'High-Rosmarinic Acid Spearmint Tea in the Management of Knee Osteoarthritis Symptoms'. *Journal of Medicinal Food* 17 (12): 1361–67. doi:10.1089/jmf.2013.0189

Cox-Georgian, Destinney, Niveditha Ramadoss, Chathu Dona, and Chhandak Basu. 2019. 'Therapeutic and Medicinal Uses of Terpenes'. In *Medicinal Plants: From Farm to Pharmacy*, 333–59. Springer International Publishing. doi:10.1007/978-3-030-31269-5_15

D'Amelia, Vincenzo, Riccardo Aversano, Pasquale Chiaiese, and Domenico Carputo. 2018. 'The Antioxidant Properties of Plant Flavonoids: Their Exploitation by Molecular Plant Breeding'. *Phytochemistry Reviews* 17 (3): 611–25. doi:10.1007/s11101-018-9568-y

Dasgupta, Amitava, and Kimberly Klein. 2014. 'Antioxidant Vitamins and Minerals'. *Antioxidants in Food, Vitamins and Supplements*, 277–94. doi:10.1016/b978-0-12-405872-9.00015-x

Datta, Sudeshna, B. K. Sinha, Soumen Bhattacharjee, and Tapan Seal. 2019. 'Nutritional Composition, Mineral Content, Antioxidant Activity and Quantitative Estimation of Water Soluble Vitamins and Phenolics by RP-HPLC in Some Lesser Used Wild Edible Plants'. *Heliyon* 5 (3): e01431. doi:10.1016/J.HELIYON.2019.E01431

Desai, Sapna D., Dhruv G. Desai, and Harmeet Kaur. 2009. 'Saponins and Their Biological Activities'. *Pharma Times* 41 (3): 13–16.

Dias, Maria Celeste, Diana C. G. A. Pinto, and Artur M. S. Silva. 2021. 'Plant Flavonoids: Chemical Characteristics and Biological Activity'. *Molecules* 26 (17): 1–16. doi:10.3390/molecules26175377

Domínguez-Fernández, Maite, Iziar A. Ludwig, María-Paz De Peña, and Concepción Cid. 2021. 'Bioaccessibility of Tudela Artichoke (*Cynara Scolymus* Cv. Blanca de Tudela) (Poly)Phenols: The Effects of Heat Treatment, Simulated Gastrointestinal Digestion and Human Colonic Microbiota'. *Food & Function* 12 (5): 1996–2011. doi:10.1039/D0FO03119D

Doughari, James Hamuel. 2012. 'Phytochemicals: Extraction Methods, Basic Structures and Mode of Action as Potential Chemotherapeutic Agents'. *Phytochemicals - A Global Perspective of Their Role in Nutrition and Health*. doi:10.5772/26052

Dryden, Gerald W., Allan Lam, Karen Beatty, Hassan H. Qazzaz, and Craig J. McClain. 2013. 'A Pilot Study to Evaluate the Safety and Efficacy of an Oral Dose of (−)-Epigallocatechin-3-Gallate–Rich Polyphenon E in Patients With Mild to Moderate Ulcerative Colitis'. *Inflammatory Bowel Diseases*. doi:10.1097/MIB.0b013e31828f5198

Ebert, Andreas. 2014. 'Potential of Underutilized Traditional Vegetables and Legume Crops to Contribute to Food and Nutritional Security, Income and More Sustainable Production Systems'. *Sustainability* 6 (1): 319–35. doi:10.3390/su6010319

Falcone, Paul H., Aaron C. Tribby, Roxanne M. Vogel, Jordan M. Joy, Jordan R. Moon, Chantelle A. Slayton, Micah M. Henigman, et al. 2018. 'Efficacy of a Nootropic Spearmint Extract on Reactive Agility: A Randomized, Double-Blind, Placebo-Controlled, Parallel Trial'. *Journal of the International Society of Sports Nutrition* 15 (1). doi:10.1186/s12970-018-0264-5

Favaro-Trindade, Carmen S., Fernando E. de Matos Junior, Paula K. Okuro, João Dias-Ferreira, Amanda Cano, Patricia Severino, Aleksandra Zielińska, and Eliana B. Souto. 2021. 'Encapsulation of Active Pharmaceutical Ingredients in Lipid Micro/Nanoparticles for Oral Administration by Spray-Cooling'. *Pharmaceutics* 13 (8): 1186. doi:10.3390/pharmaceutics13081186

Feng, Weisheng, Meng Li, Zhiyou Hao, and Jingke Zhang. 2020. 'Analytical Methods of Isolation and Identification'. In *Phytochemicals in Human Health*. IntechOpen. doi:10.5772/intechopen.88122

Fernández-García, Elisabet, Irene Carvajal-Lérida, Manuel Jarén-Galán, Juan Garrido-Fernández, Antonio Pérez-Gálvez, and Dámaso Hornero-Méndez. 2012. 'Carotenoids Bioavailability from Foods: From Plant Pigments to Efficient Biological Activities'. *Food Research International* 46 (2): 438–50. doi:10.1016/j.foodres.2011.06.007

Ferreira, Isabel C. F. R., Patricia Morales, and Lillian Barros, eds. 2016. *Wild Plants, Mushrooms and Nuts: Functional Food Properties and Applications. Wild Plants, Mushrooms and Nuts: Functional Food Properties and Applications.* Chichester, West Sussex, United Kingdom: Wiley. doi:10.1002/9781118944653

Forni, Cinzia, Francesco Facchiano, Manuela Bartoli, Stefano Pieretti, Antonio Facchiano, Daniela D'Arcangelo, Sandro Norelli, et al. 2019. 'Beneficial Role of Phytochemicals on Oxidative Stress and Age-Related Diseases'. *BioMed Research International* 2019 (Figure 1). doi:10.1155/2019/8748253

Gambacciani, M., M. Ciaponi, B. Cappagli, L. Piaggesi, and A. R. Genazzani. 1997. 'Effects of Combined Low Dose of the Isoflavone Derivative Ipriflavone and Estrogen Replacement on Bone Mineral Density and Metabolism in Postmenopausal Women'. *Maturitas* 28 (1): 75–81. doi:10.1016/S0378-5122(97)00059-5

Gidley, Michael J., and Gleb E. Yakubov. 2019. 'Functional Categorisation of Dietary Fibre in Foods: Beyond "Soluble" vs "Insoluble"'. *Trends in Food Science & Technology* 86 (April): 563–68. doi:10.1016/J.TIFS.2018.12.006

Gómez-Maqueo, Andrea, Dora Steurer, Jorge Welti-Chanes, and M. Pilar Cano. 2021. 'Bioaccessibility of Antioxidants in Prickly Pear Fruits Treated with High Hydrostatic Pressure: An Application for Healthier Foods'. *Molecules* 26 (17): 5252. doi:10.3390/molecules26175252

Gratton, Gabriele, Samuel R. Weaver, Claire V. Burley, Kathy A. Low, Edward L. Maclin, Paul W. Johns, Quang S. Pham, Samuel J. E. Lucas, Monica Fabiani, and Catarina Rendeiro. 2020. 'Dietary Flavanols Improve Cerebral Cortical Oxygenation and Cognition in Healthy Adults'. *Scientific Reports* 10 (1): 19409. doi:10.1038/s41598-020-76160-9

Grosso, Giuseppe, Roberto Bei, Antonio Mistretta, Stefano Marventano, Giorgio Calabrese, Laura Masuelli, Maria Gabriella Giganti, Andrea Modesti, Fabio Galvano, and Diego Gazzolo. 2013. 'Effects of Vitamin C on Health: A Review of Evidence'. *Front Biosci (Landmark Ed)* 18 (3): 1017–29.

Guan, Ruirui, Quyet Van Le, Han Yang, Dangquan Zhang, Haiping Gu, Yafeng Yang, Christian Sonne, et al. 2021. 'A Review of Dietary Phytochemicals and Their Relation to Oxidative Stress and Human Diseases'. *Chemosphere* 271 (May): 129499. doi:10.1016/J.CHEMOSPHERE.2020.129499

Hachul, Helena, Letícia Campos Brandão, Vânia D'Almeida, Lia Rita Azeredo Bittencourt, Edmund Chada Baracat, and Sergio Tufik. 2011. 'Isoflavones Decrease Insomnia in Postmenopause'. *Menopause* 18 (2): 178–84. doi:10.1097/gme.0b013e3181ecf9b9

Hao, Jia, Jianzhong Xu, Wenguan Zhang, Xiaoyu Li, Dandan Liang, Duoxia Xu, Yanping Cao, and Baoguo Sun. 2022. 'The Improvement of the Physicochemical Properties and Bioaccessibility of Lutein Microparticles by Electrostatic Complexation'. *Food Hydrocolloids* 125 (April): 107381. doi:10.1016/j.foodhyd.2021.107381

Havaux, Michel. 2014. 'Carotenoid Oxidation Products as Stress Signals in Plants'. *The Plant Journal* 79 (4): 597–606. doi:10.1111/tpj.12386

Hayes, Micaela, Sydney Corbin, Candace Nunn, Marti Pottorff, Colin D. Kay, Mary Ann Lila, Massimo Iorrizo, and Mario G. Ferruzzi. 2021. 'Influence of Simulated Food and Oral Processing on Carotenoid and Chlorophyll *in Vitro* Bioaccessibility among Six Spinach Genotypes'. *Food & Function* 12 (15): 7001–16. doi:10.1039/D1FO00600B

He, Jing-Ru, Jing-Jing Zhu, Shou-Wei Yin, and Xiao-Quan Yang. 2022. 'Bioaccessibility and Intracellular Antioxidant Activity of Phloretin Embodied by Gliadin/Sodium Carboxymethyl Cellulose Nanoparticles'. *Food Hydrocolloids* 122 (January): 107076. doi:10.1016/j.foodhyd.2021.107076

Heywood, Vernon Hilton. 1999. *Use and Potential of Wild Plants in Farm Households.* Food & Agriculture Org.

Hu, Yao, Qianzhu Lin, Hui Zhao, Xiaojing Li, Shangyuan Sang, David Julian McClements, Jie Long, Zhengyu Jin, Jinpeng Wang, and Chao Qiu. 2023. 'Bioaccessibility and Bioavailability of Phytochemicals: Influencing Factors, Improvements, and Evaluations'. *Food Hydrocolloids* 135 (February): 108165. doi:10.1016/j.foodhyd.2022.108165

Hu, Yao, Chao Qiu, Zhengyu Jin, Yang Qin, Chen Zhan, Xueming Xu, and Jinpeng Wang. 2020. 'Pickering Emulsions with Enhanced Storage Stabilities by Using Hybrid β-Cyclodextrin/Short Linear Glucan Nanoparticles as Stabilizers'. *Carbohydrate Polymers* 229 (February): 115418. doi:10.1016/j.carbpol.2019.115418

Huang, Yancui, Di Xiao, Britt M. Burton-Freeman, and Indika Edirisinghe. 2016. 'Chemical Changes of Bioactive Phytochemicals during Thermal Processing'. *Reference Module in Food Science.* doi:10.1016/B978-0-08-100596-5.03055-9

Javadi, Fatemeh, Arman Ahmadzadeh, Shahryar Eghtesadi, Naheed Aryaeian, Mozhdeh Zabihiyeganeh, Abbas Rahimi Foroushani, and Shima Jazayeri. 2017. 'The Effect of Quercetin on Inflammatory Factors and Clinical Symptoms in Women with Rheumatoid Arthritis: A Double-Blind, Randomized Controlled Trial'. *Journal of the American College of Nutrition* 36 (1): 9–15. doi:10.1080/07315724.2016.1140093

Jo, Myeongsu, Choongjin Ban, Kelvin K. T. Goh, and Young Jin Choi. 2021. 'Enhancement of the Gut-Retention Time of Resveratrol Using Waxy Maize Starch Nanocrystal-Stabilized and Chitosan-Coated Pickering Emulsions'. *Food Hydrocolloids* 112 (March): 106291. doi:10.1016/j.foodhyd.2020.106291

Jones, Dean P. 2008. 'Radical-Free Biology of Oxidative Stress'. *American Journal of Physiology - Cell Physiology*. doi:10.1152/ajpcell.00283.2008

Joshi, Apurva, Satyendra K. Prasad, Vinod Kumar Joshi, and Siva Hemalatha. 2016. 'Phytochemical Standardization, Antioxidant, and Antibacterial Evaluations of Leea Macrophylla: A Wild Edible Plant'. *Journal of Food and Drug Analysis* 24 (2): 324–31. doi:10.1016/j.jfda.2015.10.010

Kamiloglu, Senem, Merve Tomas, Tugba Ozdal, and Esra Capanoglu. 2021. 'Effect of Food Matrix on the Content and Bioavailability of Flavonoids'. *Trends in Food Science & Technology* 117 (November): 15–33. doi:10.1016/j.tifs.2020.10.030

Kan, Juntao, Feng Wu, Feijie Wang, Jianheng Zheng, Junrui Cheng, Yuan Li, Yuexin Yang, and Jun Du. 2022. 'Phytonutrients: Sources, Bioavailability, Interaction with Gut Microbiota, and Their Impacts on Human Health'. *Frontiers in Nutrition* 9. doi:10.3389/fnut.2022.960309

Kapoor, Mahendra P., Masamitsu Moriwaki, Kamiya Uguri, Derek Timm, and Yuichi Kuroiwa. 2021. 'Bioavailability of Dietary Isoquercitrin-γ-Cyclodextrin Molecular Inclusion Complex in Sprague–Dawley Rats and Healthy Humans'. *Journal of Functional Foods* 85 (October): 104663. doi:10.1016/j.jff.2021.104663

Karthikeyan, Ramadoss, Chapala Devadasu, and Puttagunta Srinivasa Babu. 2015. 'Isolation, Characterization, and RP-HPLC Estimation of P-Coumaric Acid from Methanolic Extract of Durva Grass (Cynodon Dactylon Linn.) (Pers.)'. *International Journal of Analytical Chemistry* 2015. doi:10.1155/2015/201386

Koche, Deepak, Rupali Shirsat, and Mahesh Kawale. 2016. 'An Overerview of Major Classes of Phytochemicals: Their Types and Role in Disease Prevention'. *Hislopia Journal* 9 (1/2): 1–11.

Kris-Etherton, Penny M., Kari D. Hecker, Andrea Bonanome, Stacie M. Coval, Amy E. Binkoski, Kirsten F. Hilpert, Amy E. Griel, and Terry D. Etherton. 2002. 'Bioactive Compounds in Foods: Their Role in the Prevention of Cardiovascular Disease and Cancer'. *The American Journal of Medicine* 113 (9): 71–88. doi:10.1016/S0002-9343(01)00995-0

Krüger, Karsten, Frank C Mooren, Klaus Eder, and Robert Ringseis. 2014. 'Immune and Inflammatory Signaling Pathways in Exercise and Obesity'. *American Journal of Lifestyle Medicine* 10 (4): 268–79. doi:10.1177/1559827614552986

Kumar, Ashwani, P. Nirmal, Mukul Kumar, Anina Jose, Vidisha Tomer, Emel Oz, Charalampos Proestos, et al. 2023. 'Major Phytochemicals: Recent Advances in Health Benefits and Extraction Method.' *Molecules (Basel, Switzerland)* 28 (2): 1–41. doi:10.3390/molecules28020887

Kumar, Kshitiz, Shivmurti Srivastav, and Vijay Singh Sharanagat. 2021. 'Ultrasound Assisted Extraction (UAE) of Bioactive Compounds from Fruit and Vegetable Processing by-Products: A Review'. *Ultrasonics Sonochemistry* 70 (January): 105325. doi:10.1016/j.ultsonch.2020.105325

Lamport, Daniel J, Clare L Lawton, Natasha Merat, Hamish Jamson, Kyriaki Myrissa, Denise Hofman, Helen K Chadwick, Frits Quadt, JoLynne D Wightman, and Louise Dye. 2016. 'Concord Grape Juice, Cognitive Function, and Driving Performance: A 12-Wk, Placebo-Controlled, Randomized Crossover Trial in Mothers of Preteen Children'. *The American Journal of Clinical Nutrition* 103 (3): 775–83. doi:10.3945/ajcn.115.114553

Law, Yat-Yin, Hui-Fang Chiu, Hui-Hsin Lee, You-Cheng Shen, Kamesh Venkatakrishnan, and Chin-Kun Wang. 2016. 'Consumption of Onion Juice Modulates Oxidative Stress and Attenuates the Risk of Bone Disorders in Middle-Aged and Post-Menopausal Healthy Subjects'. *Food & Function* 7 (2): 902–12. doi:10.1039/C5FO01251A

Lee, Mi-Kyung. 2020. 'Liposomes for Enhanced Bioavailability of Water-Insoluble Drugs: In Vivo Evidence and Recent Approaches'. *Pharmaceutics* 12 (3): 264. doi:10.3390/pharmaceutics12030264

Leitzmann, Claus. 2016. 'Characteristics and Health Benefits of Phytochemicals'. *Forschende Komplementarmedizin*. doi:10.1159/000444063

Leyane, Thobekile S., Sandy W. Jere, and Nicolette N. Houreld. 2022. 'Oxidative Stress in Ageing and Chronic Degenerative Pathologies: Molecular Mechanisms Involved in Counteracting Oxidative Stress and Chronic Inflammation'. *International Journal of Molecular Sciences* 23 (13): 7273. doi:10.3390/ijms23137273

Li, Chunyang, Dian Liu, Meigui Huang, Wuyang Huang, Ying Li, and Jin Feng. 2022. 'Interfacial Engineering Strategy to Improve the Stabilizing Effect of Curcumin-Loaded Nanostructured Lipid Carriers'. *Food Hydrocolloids* 127 (June): 107552. doi:10.1016/j.foodhyd.2022.107552

Li, Xiao-Min, Xuehong Li, Zhengzong Wu, Ying Wang, Jie-Shun Cheng, Ting Wang, and Bao Zhang. 2020. 'Chitosan Hydrochloride/Carboxymethyl Starch Complex Nanogels Stabilized Pickering Emulsions for

Oral Delivery of β-Carotene: Protection Effect and in Vitro Digestion Study'. *Food Chemistry* 315 (June): 126288. doi:10.1016/j.foodchem.2020.126288

Lin, Qianzhu, Shengju Ge, David Julian McClements, Xiaojing Li, Zhengyu Jin, Aiquan Jiao, Jinpeng Wang, Jie Long, Xueming Xu, and Chao Qiu. 2021. 'Advances in Preparation, Interaction and Stimulus Responsiveness of Protein-Based Nanodelivery Systems'. *Critical Reviews in Food Science and Nutrition*, November, 1–14. doi:10.1080/10408398.2021.1997908

Liu, Rui Hai. 2013. 'Health-Promoting Components of Fruits and Vegetables in the Diet'. *Advances in Nutrition* 4 (3): 384S–392S. doi:10.3945/an.112.003517

Liu, Ying, Yury Tikunov, Rob E. Schouten, Leo F. M. Marcelis, Richard G. F. Visser, and Arnaud Bovy. 2018. 'Anthocyanin Biosynthesis and Degradation Mechanisms in Solanaceous Vegetables: A Review'. *Frontiers in Chemistry* 6 (MAR). doi:10.3389/fchem.2018.00052

Lu, Linlin, Qian Feng, Tao Su, Yuanyuan Cheng, Zhiying Huang, Qiuju Huang, and Zhongqiu Liu. 2019. 'Pharmacoepigenetics of Chinese Herbal Components in Cancer'. *Pharmacoepigenetics*, January. Academic Press, 859–69. doi:10.1016/B978-0-12-813939-4.00035-8

Łuczaj, Łukasz, Andrea Pieroni, Javier Tardío, Manuel Pardo-De-Santayana, Renata Sõukand, Ingvar Svanberg, and Raivo Kalle. 2012. 'Wild Food Plant Use in 21st Century Europe: The Disappearance of Old Traditions and the Search for New Cuisines Involving Wild Edibles'. *Acta Societatis Botanicorum Poloniae* 81 (4): 359–70. doi:10.5586/asbp.2012.031

Majnooni, Mohammad Bagher, Sajad Fakhri, Syed Mustafa Ghanadian, Gholamreza Bahrami, Kamran Mansouri, Amin Iranpanah, Mohammad Hosein Farzaei, and Mahdi Mojarrab. 2023. 'Inhibiting Angiogenesis by Anti-Cancer Saponins: From Phytochemistry to Cellular Signaling Pathways'. *Metabolites* 13 (3): 323. doi:10.3390/metabo13030323

Mandal, Mamun, Manisha Sarkar, Azmi Khan, Moumita Biswas, Antonio Masi, Randeep Rakwal, Ganesh Kumar Agrawal, Amrita Srivastava, and Abhijit Sarkar. 2022. 'Reactive Oxygen Species (ROS) and Reactive Nitrogen Species (RNS) in Plants– Maintenance of Structural Individuality and Functional Blend'. *Advances in Redox Research* 5 (July): 100039. doi:10.1016/J.ARRES.2022.100039

Manousi, Natalia, Ioannis Sarakatsianos, and Victoria Samanidou. 2019. 'Extraction Techniques of Phenolic Compounds and Other Bioactive Compounds From Medicinal and Aromatic Plants'. In Alexandru Mihai Grumezescu and Alina Maria Holban (eds). *Engineering Tools in the Beverage Industry*, 283–314. Elsevier. doi:10.1016/B978-0-12-815258-4.00010-X

Manuja, Rohini, Shikha Sachdeva, Akash Jain, and Jasmine Chaudhary. 2013. 'A Comprehensive Review on Biological Activities of P-Hydroxy Benzoic Acid and Its Derivatives'. *International Journal of Pharmaceutical Sciences Review and Research* 22 (2): 109–15.

Marrocco, Ilaria, Fabio Altieri, and Ilaria Peluso. 2017. 'Measurement and Clinical Significance of Biomarkers of Oxidative Stress in Humans'. *Oxidative Medicine and Cellular Longevity* 2017. doi:10.1155/2017/6501046

Mastroiacovo, Daniela, Catherine Kwik-Uribe, Davide Grassi, Stefano Necozione, Angelo Raffaele, Luana Pistacchio, Roberta Righetti, et al. 2015. 'Cocoa Flavanol Consumption Improves Cognitive Function, Blood Pressure Control, and Metabolic Profile in Elderly Subjects: The Cocoa, Cognition, and Aging (CoCoA) Study—a Randomized Controlled Trial1–4'. *The American Journal of Clinical Nutrition* 101 (3): 538–48. doi:10.3945/ajcn.114.092189

Matos, M., A. Marefati, P. Barrero, M. Rayner, and G. Gutiérrez. 2021. 'Resveratrol Loaded Pickering Emulsions Stabilized by OSA Modified Rice Starch Granules'. *Food Research International* 139 (January): 109837. doi:10.1016/j.foodres.2020.109837

Mattila, Pirjo, and Jarkko Hellström. 2007. 'Phenolic Acids in Potatoes, Vegetables, and Some of Their Products'. *Journal of Food Composition and Analysis* 20 (3–4): 152–60. doi:10.1016/j.jfca.2006.05.007

McClements, David Julian. 2012. 'Nanoemulsions versus Microemulsions: Terminology, Differences, and Similarities'. *Soft Matter* 8 (6): 1719–29. doi:10.1039/C2SM06903B

Miao, Jinyu, Na Xu, Ce Cheng, Liqiang Zou, Jun Chen, Yi Wang, Ruihong Liang, David Julian McClements, and Wei Liu. 2021. 'Fabrication of Polysaccharide-Based High Internal Phase Emulsion Gels: Enhancement of Curcumin Stability and Bioaccessibility'. *Food Hydrocolloids* 117 (August): 106679. doi:10.1016/j.foodhyd.2021.106679

Monari, Stefania, Maura Ferri, Beatrice Montecchi, Mirko Salinitro, and Annalisa Tassoni. 2021. 'Phytochemical Characterization of Raw and Cooked Traditionally Consumed Alimurgic Plants'. *PLoS ONE* 16 (8 August): 1–17. doi:10.1371/journal.pone.0256703

Monjotin, Nicolas, Marie Josèphe Amiot, Jacques Fleurentin, Jean Michel Morel, and Sylvie Raynal. 2022a. 'Clinical Evidence of the Benefits of Phytonutrients in Human Healthcare'. *Nutrients* 14 (9): 1–54. doi:10.3390/nu14091712

Monjotin, Nicolas, Marie Josèphe Amiot, Jacques Fleurentin, Jean Michel Morel, and Sylvie Raynal. 2022b. 'Clinical Evidence of the Benefits of Phytonutrients in Human Healthcare'. *Nutrients* 14 (9): 1712. doi:10.3390/nu14091712

Mzoughi, Zeineb, Hassiba Chahdoura, Yasmine Chakroun, Montaña Cámara, Virginia Fernández-Ruiz, Patricia Morales, Habib Mosbah, Guido Flamini, Mejdi Snoussi, and Hatem Majdoub. 2019. 'Wild Edible Swiss Chard Leaves (Beta Vulgaris L. Var. Cicla): Nutritional, Phytochemical Composition and Biological Activities'. *Food Research International* 119 (April 2018): 612–21. doi:10.1016/j.foodres.2018.10.039

Narzary, Hwiyang, Anuck Islary, and Sanjay Basumatary. 2016. 'Phytochemicals and Antioxidant Properties of Eleven Wild Edible Plants from Assam, India'. *Mediterranean Journal of Nutrition and Metabolism* 9 (3): 191–201. doi:10.3233/MNM-16116

Nowak, Emilia, Yoav D. Livney, Zhigao Niu, and Harjinder Singh. 2019. 'Delivery of Bioactives in Food for Optimal Efficacy: What Inspirations and Insights Can Be Gained from Pharmaceutics?' *Trends in Food Science & Technology* 91 (September): 557–73. doi:10.1016/j.tifs.2019.07.029

Nuutila, A. M., K Kammiovirta, and K.-M. Oksman-Caldentey. 2002. 'Comparison of Methods for the Hydrolysis of Flavonoids and Phenolic Acids from Onion and Spinach for HPLC Analysis'. *Food Chemistry* 76 (4): 519–25. doi:10.1016/S0308-8146(01)00305-3

Nwafor, Ebuka Olisaemeka, Peng Lu, Ying Zhang, Rui Liu, Hui Peng, Bin Xing, Yiting Liu, et al. 2022. 'Chlorogenic Acid: Potential Source of Natural Drugs for the Therapeutics of Fibrosis and Cancer'. *Translational Oncology* 15 (1): 101294. doi:10.1016/J.TRANON.2021.101294

Pan, Yi, Xiao-Min Li, Ran Meng, Bao-Cai Xu, and Bao Zhang. 2021. 'Investigation of the Formation Mechanism and Curcumin Bioaccessibility of Emulsion Gels Based on Sugar Beet Pectin and Laccase Catalysis'. *Journal of Agricultural and Food Chemistry* 69 (8): 2557–63. doi:10.1021/acs.jafc.0c07288

Patel, Manisha. 2016. 'Targeting Oxidative Stress in Central Nervous System Disorders'. *Trends in Pharmacological Sciences* 37 (9): 768–78. doi:10.1016/J.TIPS.2016.06.007

Periyannan, Velu, and Vinothkumar Veerasamy. 2018. 'Syringic Acid May Attenuate the Oral Mucosal Carcinogenesis via Improving Cell Surface Glycoconjugation and Modifying Cytokeratin Expression'. *Toxicology Reports* 5 (January): 1098–1106. doi:10.1016/J.TOXREP.2018.10.015

Piironen, Vieno, and Anna Maija Lampi. 2014. 'Rye as a Source of Phytosterols, Tocopherols, and Tocotrienols'. *Rye and Health*, January, 131–58. doi:10.1016/B978-1-891127-81-6.50009-2

Pinela, José, Ana Maria Carvalho, and Isabel C. F. R. Ferreira. 2017. 'Wild Edible Plants: Nutritional and Toxicological Characteristics, Retrieval Strategies and Importance for Today's Society'. *Food and Chemical Toxicology* 110 (December): 165–88. doi:10.1016/J.FCT.2017.10.020

Pohl, Pawel. 2007. 'What Do Metals Tell Us about Wine?' *TrAC Trends in Analytical Chemistry* 26 (9): 941–49. doi:10.1016/j.trac.2007.07.005

Poiroux-Gonord, Florine, Luc P. R. Bidel, Anne-Laure Fanciullino, Hélène Gautier, Félicie Lauri-Lopez, and Laurent Urban. 2010. 'Health Benefits of Vitamins and Secondary Metabolites of Fruits and Vegetables and Prospects To Increase Their Concentrations by Agronomic Approaches'. *Journal of Agricultural and Food Chemistry* 58 (23): 12065–82. doi:10.1021/jf1037745

Prado, Juliana M., Renata Vardanega, Isabel C. N. Debien, Maria Angela de Almeida Meireles, Lia Noemi Gerschenson, Halagur Bogegowda Sowbhagya, and Smain Chemat. 2015. 'Conventional Extraction'. In *Food Waste Recovery*, 127–48. Elsevier. doi:10.1016/B978-0-12-800351-0.00006-7

Proestos, Charalampos, Konstantina Lytoudi, Olga Mavromelanidou, Panagiotis Zoumpoulakis, and Vassileia Sinanoglou. 2013. 'Antioxidant Capacity of Selected Plant Extracts and Their Essential Oils'. *Antioxidants* 2 (1): 11–22. doi:10.3390/antiox2010011

Ramírez-Melo, Lisette Monsibaez, Nelly del Socorro Cruz-Cansino, Luis Delgado-Olivares, Esther Ramírez-Moreno, Quinatzin Yadira Zafra-Rojas, José Luis Hernández-Traspeña, and Ángela Suárez-Jacobo. 2022. 'Optimization of Antioxidant Activity Properties of a Thermosonicated Beetroot (Beta Vulgaris L.) Juice and Further in Vitro Bioaccessibility Comparison with Thermal Treatments'. *LWT* 154 (January): 112780. doi:10.1016/j.lwt.2021.112780

Rathaur, Pooja, and S. R. Johar Kaid. 2019. 'Metabolism and Pharmacokinetics of Phytochemicals in the Human Body'. *Current Drug Metabolism*. doi:10.2174/1389200221666200103090757

Ribas-Agustí, Albert, Olga Martín-Belloso, Robert Soliva-Fortuny, and Pedro Elez-Martínez. 2018. 'Food Processing Strategies to Enhance Phenolic Compounds Bioaccessibility and Bioavailability in Plant-Based Foods'. *Critical Reviews in Food Science and Nutrition* 58 (15): 2531–48. doi:10.1080/1040839 8.2017.1331200

Rocha, M. I., M. J. Rodrigues, C. Pereira, H. Pereira, M. M. da Silva, N. Rosa da Neng, J. M. F. Nogueira, J. Varela, L. Barreira, and L. Custódio. 2017. 'Biochemical Profile and in Vitro Neuroprotective Properties of Carpobrotus Edulis L., a Medicinal and Edible Halophyte Native to the Coast of South Africa'. *South African Journal of Botany* 111: 222–31. doi:10.1016/j.sajb.2017.03.036

Rodrigues, Daniele Bobrowski, Marcella Camargo Marques, Adriele Hacke, Paulo Sérgio Loubet Filho, Cinthia Baú Betim Cazarin, and Lilian Regina Barros Mariutti. 2022. 'Trust Your Gut: Bioavailability and Bioaccessibility of Dietary Compounds'. *Current Research in Food Science* 5: 228–33. doi:10.1016/j. crfs.2022.01.002

Rohilla, Shubham, Hemanta Chutia, Vegonia Marboh, and Charu Lata Mahanta. 2022. 'Ultrasound and Supercritical Fluid Extraction of Phytochemicals from Purple Tamarillo: Optimization, Comparison, Kinetics, and Thermodynamics Studies'. *Applied Food Research* 2 (2): 100210. doi:10.1016/j. afres.2022.100210

Rohmah, Miftakhur, Anton Rahmadi, and Sri Raharjo. 2022. 'Bioaccessibility and Antioxidant Activity of β-Carotene Loaded Nanostructured Lipid Carrier (NLC) from Binary Mixtures of Palm Stearin and Palm Olein'. *Heliyon* 8 (2): e08913. doi:10.1016/j.heliyon.2022.e08913

Saitou, Katsuyoshi, Ryuji Ochiai, Kazuya Kozuma, Hirotaka Sato, Takashi Koikeda, Noriko Osaki, and Yoshihisa Katsuragi. 2018. 'Effect of Chlorogenic Acids on Cognitive Function: A Randomized, Double-Blind, Placebo-Controlled Trial'. *Nutrients* 10 (10): 1337. doi:10.3390/nu10101337

Saleem, Hammad, Gokhan Zengin, Irshad Ahmad, Thet Thet Htar, Rakesh Naidu, Mohamad Fawzi Mahomoodally, and Nafees Ahemad. 2020. 'Therapeutic Propensities, Phytochemical Composition, and Toxicological Evaluation of Anagallis Arvensis (L.): A Wild Edible Medicinal Food Plant'. *Food Research International* 137 (November): 109651. doi:10.1016/J.FOODRES.2020.109651

Šamec, Dunja, Erna Karalija, Ivana Šola, Valerija Vujčić Bok, and Branka Salopek-Sondi. 2021. 'The Role of Polyphenols in Abiotic Stress Response: The Influence of Molecular Structure'. *Plants*. doi:10.3390/plants10010118

Satter, Mohammed Abdus Miah, Mohammed Murtaza Reza Linkon Khan, Syeda Absha Jabin, Nusrat Abedin, Mohammed Faridul Islam, and Badhan Shaha. 2016. 'Nutritional Quality and Safety Aspects of Wild Vegetables Consume in Bangladesh'. *Asian Pacific Journal of Tropical Biomedicine* 6 (2): 125–31. doi:10.1016/J.APJTB.2015.11.004

Sauer, Lydia, Binxing Li, and Paul S Bernstein. 2019. 'Ocular Carotenoid Status in Health and Disease'. *Annual Review of Nutrition* 39 (1): 95–120. doi:10.1146/annurev-nutr-082018-124555

Scholey, Andrew, Luke A. Downey, Joseph Ciorciari, Andrew Pipingas, Karen Nolidin, Melissa Finn, Melissa Wines, et al. 2012. 'Acute Neurocognitive Effects of Epigallocatechin Gallate (EGCG)'. *Appetite* 58 (2): 767–70. doi:10.1016/j.appet.2011.11.016

Seal, Tapan. 2016. 'Quantitative HPLC Analysis of Phenolic Acids, Flavonoids and Ascorbic Acid in Four Different Solvent Extracts of Two Wild Edible Leaves, Sonchus Arvensis and Oenanthe Linearis of North-Eastern Region in India'. *Journal of Applied Pharmaceutical Science* 6 (2): 157–66. doi:10.7324/JAPS.2016.60225

Seal, Tapan, Kausik Chaudhuri, and Basundhara Pillai. 2017. 'Nutraceutical and Antioxidant Properties of Cucumis Hardwickii Royle: A Potent Wild Edible Fruit Collected from Uttarakhand, India'. *Journal of Pharmacognosy and Phytochemistry* 6 (6): 1837–47.

Seal, Tapan, Basundhara Pillai, and Kausik Chaudhuri. 2016. 'Identification and Quantification of Phenolic Acids by HPLC, in Three Wild Edible Plants Viz. Viburnum Foetidum, Houttuynia Cordata and Perilla Ocimoides Collected from North-Eastern Region in India'. *International Journal of Current Pharmaceutical Review and Research* 7 (5): 267–74.

Sęczyk, Łukasz, Fethi Ahmet Ozdemir, and Barbara Kołodziej. 2022. 'In Vitro Bioaccessibility and Activity of Basil (Ocimum Basilicum L.) Phytochemicals as Affected by Cultivar and Postharvest Preservation Method – Convection Drying, Freezing, and Freeze-Drying'. *Food Chemistry* 382 (July): 132363. doi:10.1016/j.foodchem.2022.132363

Shahbazi, Yasser, and Nassim Shavisi. 2021. 'Limonene'. *A Centum of Valuable Plant Bioactives*, January. Academic Press, 77–91. doi:10.1016/B978-0-12-822923-1.00016-9

Shaheen, Shabnum, Mushtaq Ahmad, and Nidaa Haroon. 2017. 'Edible Wild Plants: An Alternative Approach to Food Security'. *Edible Wild Plants: An Alternative Approach to Food Security*, 1–183. doi:10.1007/978-3-319-63037-3

Shiny, C. T., Anuj Saxena, and Sharad Prakash Gupta. 2013. 'Phytochemical and Hypoglycaemic Activity Investigation of Costus Pictus Plants from Kerala and Tamilnadu'. *International Journal of Pharmaceutical Science Invention ISSN* 2 (January 2013): 11–18.

Si, Hongwei, and Dongmin Liu. 2014. 'Dietary Antiaging Phytochemicals and Mechanisms Associated with Prolonged Survival'. *The Journal of Nutritional Biochemistry* 25 (6): 581–91. doi:10.1016/J. JNUTBIO.2014.02.001

Sies, Helmut. 2020. 'Oxidative Stress: Concept and Some Practical Aspects'. *Antioxidants* 9 (9): 852. doi:10.3390/antiox9090852.

Silva, Marluci Palazzolli, Eduarda Grecco Farsoni, Cricia Fernanda Gobato, Marcelo Thomazini, and Carmen S. Favaro-Trindade. 2022. 'Simultaneous Encapsulation of Probiotic and Guaraná Peel Extract for Development of Functional Peanut Butter'. *Food Control* 138 (August): 109050. doi:10.1016/j. foodcont.2022.109050

Singh, Mukesh Kumar, and Annika Singh. 2022. 'Nuclear Magnetic Resonance Spectroscopy'. In *Characterization of Polymers and Fibres*, 321–39. Elsevier. doi:10.1016/B978-0-12-823986-5.00011-7

Stompor-Goracy, Monika, and Maciej Machaczka. 2021. 'Recent Advances in Biological Activity, New Formulations and Prodrugs of Ferulic Acid'. *International Journal of Molecular Sciences* 22 (23): 12889. doi:10.3390/ijms222312889

Suksathan, Ratchuporn, Apinya Rachkeeree, Ratchadawan Puangpradab, Kuttiga Kantadoung, and Sarana Rose Sommano. 2021. 'Phytochemical and Nutritional Compositions and Antioxidants Properties of Wild Edible Flowers as Sources of New Tea Formulations'. *NFS Journal* 24 (August): 15–25. doi:10.1016/J. NFS.2021.06.001

Sun, Tianhu, Sombir Rao, Xuesong Zhou, and Li Li. 2022. 'Plant Carotenoids: Recent Advances and Future Perspectives'. *Molecular Horticulture* 2 (1): 3. doi:10.1186/s43897-022-00023-2

Sun, Tianhu, Hui Yuan, Hongbo Cao, Mohammad Yazdani, Yaakov Tadmor, and Li Li. 2018. 'Carotenoid Metabolism in Plants: The Role of Plastids'. *Molecular Plant* 11 (1): 58–74. doi:10.1016/J. MOLP.2017.09.010

Surendran, Shelini, Fatimah Qassadi, Geyan Surendran, Dash Lilley, and Michael Heinrich. 2021. 'Myrcene— What Are the Potential Health Benefits of This Flavouring and Aroma Agent?' *Frontiers in Nutrition* 8 (July): 1–14. doi:10.3389/fnut.2021.699666

Swargiary, Ananta, Abhijita Daimari, Manita Daimari, Noymi Basumatary, and Ezekiel Narzary. 2016. 'Phytochemicals, Antioxidant, and Anthelmintic Activity of Selected Traditional Wild Edible Plants of Lower Assam.' *Indian Journal of Pharmacology* 48 (4): 418–23. doi:10.4103/0253-7613.186212

Tahara, Satoshi. 2007. 'A Journey of Twenty-Five Years through the Ecological Biochemistry of Flavonoids'. *Bioscience, Biotechnology, and Biochemistry* 71 (6): 1387–1404. doi:10.1271/bbb.70028

Teng, Hui, and Lei Chen. 2019. 'Polyphenols and Bioavailability: An Update'. *Critical Reviews in Food Science and Nutrition* 59 (13): 2040–51. doi:10.1080/10408398.2018.1437023

Thakur, Arti, Somvir Singh, Kanika Dulta, Nitesh Singh, Baber Ali, Aqsa Hafeez, Dan C. Vodnar, and Romina Alina Marc. 2022. 'Nutritional Evaluation, Phytochemical Makeup, Antibacterial and Antioxidant Properties of Wild Plants Utilized as Food by the Gaddis-a Tribal Tribe in the Western Himalayas'. *Frontiers in Agronomy* 4 (December): 1–12. doi:10.3389/fagro.2022.1010309

Thakur, Nitasha, Pinky Raigond, Yeshwant Singh, Tanuja Mishra, Brajesh Singh, Milan Kumar Lal, and Som Dutt. 2020. 'Recent Updates on Bioaccessibility of Phytonutrients'. *Trends in Food Science & Technology* 97 (March): 366–80. doi:10.1016/j.tifs.2020.01.019

Tian, Li. 2015. 'Recent Advances in Understanding Carotenoid-Derived Signaling Molecules in Regulating Plant Growth and Development'. *Frontiers in Plant Science* 6 (September): 1–4. doi:10.3389/fpls.2015.00790

Tomé-Sánchez, Irene, Ana Belén Martín-Diana, Elena Peñas, Juana Frias, Daniel Rico, Iván Jiménez-Pulido, and Cristina Martínez-Villaluenga. 2021. 'Bioprocessed Wheat Ingredients: Characterization, Bioaccessibility of Phenolic Compounds, and Bioactivity During in Vitro Digestion'. *Frontiers in Plant Science* 12 (December). doi:10.3389/fpls.2021.790898

Torija-Isasa, María Esperanza, and María Cruz Matallana-González. 2016. 'A Historical Perspective of Wild Plant Foods in the Mediterranean Area'. In *Mediterranean Wild Edible Plants*, edited by Maria de Cortes Sanchez-Mata and Javier Tardio, 3–15. New York: Springer. doi:10.1007/978-1-4939-3329-7

Tsao, Rong, and Hongyan Li. 2013. 'Analytical Techniques for Phytochemicals'. In *Handbook of Plant Food Phytochemicals*, 434–51. Oxford: John Wiley & Sons Ltd. doi:10.1002/9781118464717.ch19

Tsikrika, Konstantina, Boon-Seang Chu, David H. Bremner, and Adilia Lemos. 2022. 'Effect of Ultrasonic Treatment on Enzyme Activity and Bioactives of Strawberry Puree'. *International Journal of Food Science & Technology* 57 (3): 1739–47. doi:10.1111/ijfs.15550

Tsikrika, Konstantina, Adília Lemos, Boon-Seang Chu, David H. Bremner, and Graham Hungerford. 2022. 'Effect of Ultrasound on the Activity of Mushroom (Agaricus Bisporous) Polyphenol Oxidase and Observation of Structural Changes Using Time-Resolved Fluorescence'. *Food and Bioprocess Technology* 15 (3): 656–68. doi:10.1007/s11947-022-02777-5

Tsikrika, Konstantina, Nora O'Brien, and Dilip K. Rai. 2019. 'The Effect of High Pressure Processing on Polyphenol Oxidase Activity, Phytochemicals and Proximate Composition of Irish Potato Cultivars.' *Foods* 8 (10): 517. doi:10.3390/foods8100517

Valcarcel, Jesus, Kim Reilly, Michael Gaffney, and Nora O'Brien. 2015. 'Total Carotenoids and L-Ascorbic Acid Content in 60 Varieties of Potato (Solanum Tuberosum L.) Grown in Ireland'. *Potato Research* 58 (1): 29–41. doi:10.1007/s11540-014-9270-4

Veronese, Nicola, Marco Solmi, Maria Gabriella Caruso, Gianluigi Giannelli, Alberto R. Osella, Evangelos Evangelou, Stefania Maggi, Luigi Fontana, Brendon Stubbs, and Ioanna Tzoulaki. 2018. 'Dietary Fiber and Health Outcomes: An Umbrella Review of Systematic Reviews and Meta-Analyses'. *The American Journal of Clinical Nutrition* 107 (3): 436–44. doi:10.1093/AJCN/NQX082

Wang, Chenxi, David Julian McClements, Aiquan Jiao, Jinpeng Wang, Zhengyu Jin, and Chao Qiu. 2022. 'Resistant Starch and Its Nanoparticles: Recent Advances in Their Green Synthesis and Application as Functional Food Ingredients and Bioactive Delivery Systems'. *Trends in Food Science & Technology* 119 (January): 90–100. doi:10.1016/j.tifs.2021.11.025

Wang, Ling-Na, Wei Wang, Masao Hattori, Mohsen Daneshtalab, and Chao-Mei Ma. 2016. 'Synthesis, Anti-HCV, Antioxidant and Reduction of Intracellular Reactive Oxygen Species Generation of a Chlorogenic Acid Analogue with an Amide Bond Replacing the Ester Bond'. *Molecules* 21 (6): 737. doi:10.3390/molecules21060737

Wang, W.-Y., H.-B. Qu, and X.-C. Gong. 2020. 'Research Progress on Percolation Extraction Process of Traditional Chinese Medicines'. *China Journal of Chinese Materia Medica* 45 (5): 1039–46.

Wang, Xuanbin, Yan Ma, Qihe Xu, Alexander N. Shikov, Olga N. Pozharitskaya, Elena V. Flisyuk, Meifeng Liu, Hongliang Li, Liliana Vargas-Murga, and Pierre Duez. 2023. 'Flavonoids and Saponins: What Have We Got or Missed?' *Phytomedicine* 109 (January): 154580. doi:10.1016/J.PHYMED.2022.154580

Wannasaksri, Werawat, Piya Temviriyanukul, Amornrat Aursalung, Yuraporn Sahasakul, Sirinapa Thangsiri, Woorawee Inthachat, Nattira On-Nom, et al. 2021. 'Influence of Plant Origins and Seasonal Variations on Nutritive Values, Phenolics and Antioxidant Activities of Adenia Viridiflora Craib., an Endangered Species from Thailand'. *Foods* 10 (11). doi:10.3390/foods10112799

Wattanathorn, Jintanaporn, Woraluk Somboonporn, Sudarat Sungkamanee, Wipawee Thukummee, and Supaporn Muchimapura. 2018. 'A Double-Blind Placebo-Controlled Randomized Trial Evaluating the Effect of Polyphenol-Rich Herbal Congee on Bone Turnover Markers of the Perimenopausal and Menopausal Women'. *Oxidative Medicine and Cellular Longevity* 2018 (November): 1–11. doi:10.1155/2018/2091872

Wei, Junchi, Yun Yang, Ye Peng, Shaoying Wang, Jing Zhang, Xiaobo Liu, Jianjun Liu, Beibei Wen, and Meifeng Li. 2023. 'Biosynthesis and the Transcriptional Regulation of Terpenoids in Tea Plants (Camellia Sinensis)'. *International Journal of Molecular Sciences* 24 (8): 6937. doi:10.3390/ijms24086937

Wellala, Chandi Kanchana Deepali, Jinfeng Bi, Xuan Liu, Xinye Wu, Jian Lyu, Jianing Liu, Dazhi Liu, and Chongting Guo. 2022. 'Effect of High Pressure Homogenization on Water-Soluble Pectin Characteristics and Bioaccessibility of Carotenoids in Mixed Juice'. *Food Chemistry* 371 (March): 131073. doi:10.1016/j.foodchem.2021.131073

Yan, Yongwang, Xu Zhou, Kangxiao Guo, Feng Zhou, and Hongqi Yang. 2020. 'Use of Chlorogenic Acid against Diabetes Mellitus and Its Complications'. *Journal of Immunology Research* 2020. doi:10.1155/2020/9680508

Yang, Li, Kui Shan Wen, Xiao Ruan, Ying Xian Zhao, Feng Wei, and Qiang Wang. 2018. 'Response of Plant Secondary Metabolites to Environmental Factors'. *Molecules* 23 (4): 1–26. doi:10.3390/molecules23040762

Yang, Lu, Xing Qiao, Jiayu Gu, Xuemin Li, Yunrui Cao, Jie Xu, and Changhu Xue. 2021. 'Influence of Molecular Structure of Astaxanthin Esters on Their Stability and Bioavailability'. *Food Chemistry* 343 (May): 128497. doi:10.1016/j.foodchem.2020.128497

Yao, Kangfei, David Julian McClements, Chang Yan, Jie Xiao, Han Liu, Zhiqing Chen, Xiaoning Hou, Yong Cao, Hang Xiao, and Xiaojuan Liu. 2021. 'In Vitro and in Vivo Study of the Enhancement of Carotenoid Bioavailability in Vegetables Using Excipient Nanoemulsions: Impact of Lipid Content'. *Food Research International* 141 (March): 110162. doi:10.1016/j.foodres.2021.110162

Yegin, Sirma, Aneta Kopec, David D. Kitts, and Jerzy Zawistowski. 2020. 'Dietary Fiber: A Functional Food Ingredient with Physiological Benefits'. *Dietary Sugar, Salt and Fat in Human Health*, January. Academic Press, 531–55. doi:10.1016/B978-0-12-816918-6.00024-X

Zhang, Man, Siyue Zhu, Wenjian Yang, Qingrong Huang, and Chi-Tang Ho. 2021. 'The Biological Fate and Bioefficacy of Citrus Flavonoids: Bioavailability, Biotransformation, and Delivery Systems'. *Food & Function* 12 (8): 3307–23. doi:10.1039/D0FO03403G

Zhou, Qingxin, Jie Xu, Lu Yang, Caixia Gu, and Changhu Xue. 2019. 'Thermal Stability and Oral Absorbability of Astaxanthin Esters from *Haematococcus Pluvialis* in Balb/c Mice'. *Journal of the Science of Food and Agriculture* 99 (7): 3662–71. doi:10.1002/jsfa.9588

Zou, Chao, Ling Huang, Donghui Li, Yu Ma, Yixiang Liu, Yanbo Wang, Min-Jie Cao, Guang-Ming Liu, and Lechang Sun. 2021. 'Assembling Cyanidin-3-O-Glucoside by Using Low-Viscosity Alginate to Improve Its in Vitro Bioaccessibility and in Vivo Bioavailability'. *Food Chemistry* 355 (September): 129681. doi:10.1016/j.foodchem.2021.129681

5 Wild Edible Plants and the Importance of Vitamins as a Constituent

Srijit Das and S. Srinivasa Rao

5.1 INTRODUCTION

Wild edible plants (WEPs) refer to plant species that can grow naturally without cultivation or domestication. Wild edible plants are species which are neither cultivated nor harvested but grow wild in different geographical regions and are edible (Beluhan and Ranogajec, 2011). Interestingly, some of the wild edible plants possessed more nutritious values than the cultivated ones (Burlingame, 2000). They are easily available in nature and serve as a natural source of food (de Cortes Sánchez-Mata & Tardío, 2016). WEPs are usually found in forests, cultivable fields, roadsides, and wastelands. They are widely distributed over different geographical regions all over the world. WEPs play a significant role in providing the subsistence and livelihood of various ethnic communities and local populations in developing countries (Schippmann et al., 2002). To date, there have been more than 7000 WEPs that were used by humans for their livelihood (Grivetti and Ogle, 2000). They are the richest sources of vitamins, micronutrients, minerals, dietary fibers, and fatty acids. Some of them also possess various medicinal properties (Dansi et al., 2008). In many parts of the world, they form the global food basket (Chakravarty et al., 2016). In addition to nutritional supplementation, they provide employment to many people through sales of raw materials, juice, and local drinks (Ruffo et al., 2002). However, their importance has been overlooked in the rural population (Teketay et al., 2010).

Besides supplementing basic nutrition, WEPs contain physiologically active compounds, including vitamins, which provide various health benefits (Pinela et al., 2016). Vitamins are micronutrients, which are considered essential nutrients. The human body acquires them through diet. They possess various diverse biological functions and are involved in the regulation of cell and tissue growth (Pinela et al., 2017). Interestingly, various WEPs are important sources of vitamins, in particular hydrophilic vitamins. However, they possess low amounts of lipophilic vitamins, as these plants are fat-poor foods (de Cortes Sánchez-Mata & Tardío, 2016). Various researchers have performed analyses of phytochemical and/or chemical compounds in WEPs from various geographical regions. In two different books written on the WEPs of Ireland (Curtis and Whelan, 2019) and California (Nyerges, 2020), the presence of vitamins, mainly A and C, has been documented. In a recent review, it was mentioned that WEPs are part of the ecosystem, and they possess almost 52 compounds and ions. Furthermore, the same review highlighted that all green edible plants share various compounds, promote human health, and increase the longevity (Åhlberg, 2021). Unfortunately, there is paucity of reported literature on the identification, composition, or nutritional value, use, management, and consumers' preferences (Frison et al., 2006; Vinceti et al., 2012). Hence, it is necessary to document the nutritional benefits of wild edible plants.

5.2 VITAMINS AND THEIR ROLE

Vitamins are essential for various functions and metabolic processes in the body. While vitamin B3 (niacin) and D are synthesized by the human body, others such as vitamins A, B1 (thiamine), B2

DOI: 10.1201/9781003395935-5

(riboflavin), B5 (pantothenic acid), B6 (pyridoxine), B7 (biotin), B9 (folate), B12 (cobalamin), E, and K cannot be synthesized by human body, and have to be obtained from external sources (Drouin et al., 2011). Vitamin C (ascorbic acid) is synthesized by many vertebrate and invertebrate species (Chatterjee et al., 1961; Birney et al., 1980). Vitamin C is produced by the kidneys of fishes, amphibians, reptiles, and older bird orders, while it is produced by the liver of few birds and mammals (Gupta Dutta et al., 1973). Among the vitamins present, only A, D, E, and K are fat soluble. Fruits and vegetables are the main source of vitamins.

Each vitamin may be present in different forms, known as vitamers, that have similar biological activity upon ingestion (Seal et al., 2018). Vitamins may be present in food at relatively low levels, and they are susceptible to degradation because of exposure to light, air, heat, and high pH (Seal et al., 2018).

5.2.1 Vitamin A

The precursors of vitamin A are carotenoids, and they possess potent antioxidant properties, which have beneficial effects on the body. β-carotene, a colored pigment present in many vegetables and fruits, is converted into vitamin A which is an important antioxidant.

Vitamin A (retinol, retinoic acid) is an important nutrient that is needed for healthy vision, growth, cell division, growth, development, and better immunity (Gilbert, 2013). According to published reports, consumption of 6 to 14 mg of lutein/day was associated with more than 50% reduction in risk for development of age-related macular degeneration and cataract of the eye (Alves-Rodrigues and Shao, 2004).

It cannot be forgotten that foods possess their distinct color due to their carotenoid content (Fratianni et al., 2005). The main carotenoids present in wild edible plants include lutein, β-carotene, neoxanthin, and violaxanthin. These active compounds are present in plants such as *Asparagus acutifolius, Humulus lupulus, Bryonia dioica*, and *Tamus communis*(García-Herrera et al., 2014). The leaves and stems of food crop *Amaranthus caudatus* L. found to be rich in vitamin A and vitamin C (Joshi and Verma, 2020).

Carotenoids have been reported in a few wild edible fruits such as sea buckthorn and mulberry, adding to the attractiveness of the fruit and acting as chemo-protective agents and antioxidants (Young & Lowe, 2001; Eccleston et al., 2002; McGhie & Ainge, 2002; Ercisli, 2007). Wild edible fruits like berries are known to contain β-carotene (Beekwilder et al., 2005; Abu Bakar et al., 2009).

An earlier research study showed the mean total carotenoid content in the plant Crepis vesicara and S. asper was reported to be 10.0 mg/100g and 15.0/100 g, respectively (Fratianni et al., 2005). The main carotenoid was lutein (3–4 mg/100 g), and the other carotenoid was β-carotene (2–3 mg/100) (Fratianni et al., 2005). Lutein (35%) and β-carotene (20%) were the main carotenoids present in *Crepis vesicaria*, while lutein (25%), violaxanthin (27%), and β-carotene (20%) constituted the total carotenoids present (Fratianni et al., 2005). The concentration of vitamin A in wild blueberries was reported to be 0.46 µg/g (Calderwood & Brogan, 2020).

Interestingly, the qualitative distribution of various carotenoids varied in *Sonchus* spp according to the harvest year and the variability could be explained due to different conditions in the environment such as humidity and temperature (Fratianni et al., 2005). Hence, it could be concluded that the carotenoid content may vary according to the season during which it grows.

According to the Mayo Clinic, United States, the recommended dose of vitamin A is 900 micrograms (mcg) for adult males and 700 mcg for adult females (Mayo Clinic, 2023). According to reports, it has been observed that vitamin A deficiency-related ocular symptoms begin to appear at vitamin A levels lower than 10 micrograms/dL (Miller et al., 2002). A schematic diagram showing beneficial effects of vitamin A is shown in Figure 5.1.

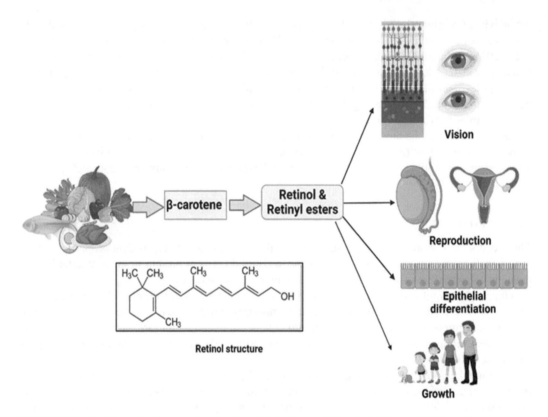

FIGURE 5.1 A schematic diagram showing beneficial effects of vitamin A. Created in Biorender.

5.2.2 Vitamin C

Vitamin C, or ascorbic acid, functions as an important redox buffer and as a cofactor for different enzymes that are involved in regulation of photosynthesis and biosynthesis of hormones (Gallie, 2013). In plants, vitamin C has various functions on the chloroplasts and an important role in photosynthesis. Vitamin C acts as an antioxidant, which acts on the free radicals and checks the oxidative damage to the cells. Vitamin C is a powerful oxidant, and it plays a big role in the body's immune system. Vitamin C (i) acts on the immune system of the body, (ii) helps in collagen production needed for wound healing, (iii) plays an important role in absorption of iron, and (iv) has an action on the skin and musculoskeletal structures. Thus, antioxidants like vitamin C may be helpful in reducing the oxidative damage. It cannot be forgotten that the majority of the diseases including diabetes; atherosclerosis; Alzheimer's; Parkinson's; cardiovascular, neurological, and metabolic diseases; and cancer arise due to oxidative stress. In oxidative stress, there is accumulation of free radicals, which causes much damage to the cellular structures and various organs in the body. An earlier study conducted on ten wild edible fruits from the Himalayan region reported the ascorbic acid (mg CGE/g fresh weight) content in *Berberis asiatica* (2.70 ± 0.13), *Celtis australis* (2.02 ± 0.03), *Ficus palmata* (7.27 ± 0.12), *Fragaria indica* (5.93 ± 0.28), *Morus alba* (29.53 ± 0.87), *Myrica esculenta* (3.20 ± 0.37), *Phyllanthus emblica* (33.15 ± 0.53), *Prunus armeniaca* (3.59 ± 0.72), *Pyracantha crenulata* (3.30 ± 0.34), *Terminalia chebula* (6.26 ± 0.02) (Bhatt et al., 2017). The presence of a comparable amount of ascorbic acid in these fruits can be promoted as a source of natural ascorbic acid (Rymbai et al., 2023).

In patients with diabetes, hypertension, hyperlipidemias, renal failure, or chronic heart disease and smokers, the elderly, and individuals at risk for infection, further studies of vitamin C dose-concentration correlations and functional outcomes are warranted (Levine et al., 2001).

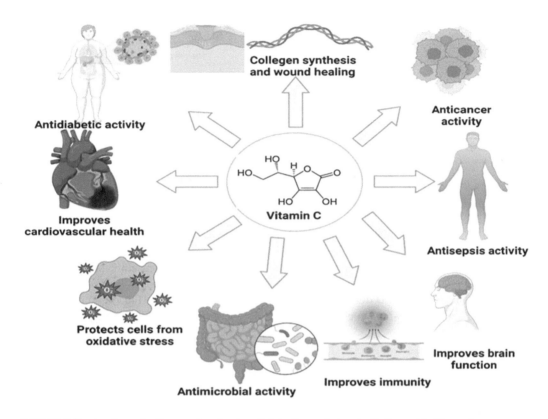

FIGURE 5.2 A schematic diagram showing beneficial effects of vitamin C. Created in Biorender.

During the coronavirus disease 2019 (COVID-19) pandemic, all cases were supplemented with vitamin C. Ascorbic acid is considered as a powerful antioxidant and its deficiency results in scurvy, gingivitis, and disturbance in cellular functions related to the immune system. Vitamin C deficiency may be prevented in individuals by a recommended daily intake of 75 mg (females) and 90 mg (males) (Levine et al., 2001). A schematic diagram showing beneficial effects of vitamin C is shown in Figure 5.2.

5.2.3 VITAMIN D

Vitamin D is required for bone metabolism, phosphate metabolism, and calcium absorption and utilization. Bone mineralization influences bone mineral density. Vitamin D is also responsible for promoting osteoclastic mediated bone resorption by increasing the osteoclastic number and activity (Suda et al., 1992). Vitamin D also has anticancer and anti-inflammatory properties, possesses a positive effect on the muscles and immune system, and improves brain function. It also has action on parathormone secretion. Deficiency of vitamin D results in osteomalacia and rickets. Deficiency of vitamin D was reported to be associated with the activation of the pro-inflammatory mechanism resulting in atherogenesis (Nitsa et al., 2018). It has been found that vitamin D controls the maturation and infiltration of macrophages into the vasculature and expression of pro-inflammatory cytokines and adhesion molecules, both of which are necessary for the development of atherosclerosis (Nitsa et al., 2018).

Many chronic diseases, including cancer, autoimmune diseases, metabolic disorders, infectious diseases, and oxidative stress response were found to be closely correlated with the level of vitamin D in the body (Ahmed et al., 2015). A decrease in the plasma level of vitamin D is known to be

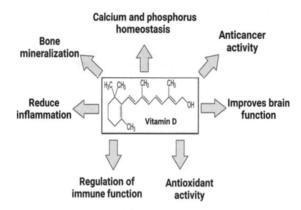

FIGURE 5.3   A schematic diagram showing beneficial effects of vitamin D.

associated with severe respiratory tract infections (Hadizadeh, 2021). Hence, the supplementation of vitamin D to COVID-19 patients was justified. Depending on age, body weight, disease conditions, and ethnicity, vitamin D doses range between 400 and 2000 IU/day (Pludowski et al., 2018). A schematic diagram showing beneficial effects of vitamin D is shown in Figure 5.3.

5.2.4   VITAMIN E

Vitamin E is found to be naturally occurring in the diet. It is an antioxidant, which acts as a scavenging agent on free radicals. Vitamin E is a naturally occurring compound that has tocotrienols and four tocopherols, namely α, β, γ, and δ. Although vitamin E serves as an antioxidant, humans only require dietary tocopherol since the other forms of vitamin E are not well-recognized by the hepatic tocopherol transfer protein and are not transformed into α tocopherol in the body (Mustacich et al., 2007). Vitamin E as an antioxidant is important in the prevention of cardiovascular diseases as it checks lipoprotein oxidation and the inhibition of platelet aggregation (Clarke et al., 2008). Vitamin E protects skin, reduces cholesterol, improves cardiovascular health, prevents cancer, and improves cognitive health. The Mayo Clinic recommended daily dose is 15mg/day for adults (Mayo Clinic, 2023). A schematic diagram showing beneficial effects of vitamin E is shown in Figure 5.4.

5.3   EXAMPLES OF A FEW WILD EDIBLE PLANTS THAT POSSESS VITAMINS

5.3.1   DIPLAZIUM ESCULENTUM

The genus *Diplazium* (family: Athyriaceae) consists of approximately 350 different species of pteridophytes (Semwal et al., 2021). *Diplazium esculentum* (Retz.) Sw. is commonly found in different parts of Asia and Oceania. According to published literature, *D. esculentum* was traditionally used for the prophylaxis and treatment of many diseases including diabetes, diarrhea, rheumatism, dysentery, headache, fever, wounds, pain, measles, hypertension, constipation, and oligospermia (Semwal et al., 2021). In Indonesia, *Diplazium esculentum*, locally known as 'bajei', has been reported to be traditionally used by the Dayak individuals in the Central Kalimantan region of Indonesia as a traditional medicine to treat tumors, asthma, and acne (Zannah et al., 2017). This plant grows up to an average height of 0.5–2.5 meters and is locally consumed as 'ulam' or green edible leaves, along with hot sauce, which is known as 'krawoo' in Malaysia (Rahmat et al., 2004; Alderwerelt, 1989).

Diplazium esculentum contains phenols, flavonoids, and antioxidants like vitamin C. Among antioxidants, vitamin C (mg/100 g), vitamin A (mg/100 g), and vitamin D (mg/100 g) were also reported (Gupta et al., 2020). Boiling the leaves in water may also result in diminished levels of antioxidants (Alderwerelt, 1989).

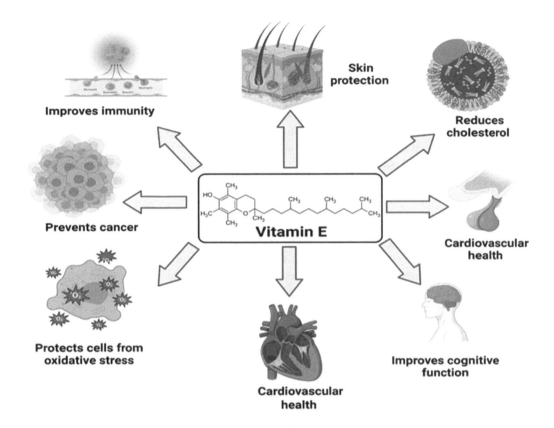

FIGURE 5.4 A schematic diagram showing beneficial effects of vitamin E. Created in Biorender.

5.3.2 Fumaria indica

Fumaria indica (Hausskn.) Pugsley (Fumariaceae), commonly known as "Fumitory", is an annual herb, which grows as a common weed in different plains of India and Pakistan (Gupta et al., 2012). The plant has been traditionally used as an anthelmintic, diuretic, diaphoretic, laxative, cholagogue, stomachic, smooth muscle relaxant, spasmogenic and spasmolytic, analgesic, anti-inflammatory, and neuropharmacological, antibacterial, and sedative agent (Gupta et al., 2012). Vitamin A and C are present as constituents in this plant extract. According to an earlier report, the nutritional constituent of *Fumaria indica* included 916.52–1666.05 IU/ of vitamin A and 0.03–0.065% of vitamin C (Ravikanth et al., 2014).

5.3.3 Taraxacum campylodes

Taraxacum campylodes is a perennial herb found in many parts of Alabama. *Taraxacum campylodes* is consumed as a vegetable in many parts of India. The leaves of the plant have been used in traditional medicine and used as salads, while the flowers were used to prepare dandelion wine (Alabama Plant Atlas, 2023). The plant was reported to be a rich source of vitamin C (Alabama Plant Atlas, 2023). There are reports of the plant being used in traditional medicine for its action on the liver, urinary tract, and skeletal system (Trapani, 2019). One of its active compounds, taraxacin, acts on the gallbladder to contract to increase bile flow, thereby helping in purification; it also acts on the secretions of various glands of the gastrointestinal system, the muscles of the digestive tract, and has secondary laxative action (Trapani, 2019).

It was reported that the root of the plant has potential healing properties and contains vitamins A, B1, B2 B3, C, E, and K (Trapani, 2019). The plant is rich in proteins, calcium, iron, carbohydrate, phosphorus, dietary fiber, and vitamin A (Escudero et al., 2003; Rutto et al., 2013).

5.3.4 URTICA DIOICA L.

Urtica dioica L. is found in different temperate climate areas of Europe, parts of Africa, and temperate Asia and America; it contains nutritional ingredients, mineral salts, vitamins, and antioxidants (Wolska et al., 2016). It is also known as stinging nettle and it is rich in vitamins A, C, and D; iron; proteins; manganese; and potassium (Wolska et al., 2016). Nettle leaves have been reported to contain vitamin C (270 mg %) and vitamins B, K, (200 mg %) and vitamin E (Wolska et al., 2016).

5.3.5 PHYLLANTHUS EMBLICA

Phyllanthus emblica (Emblica officinalis) is a deciduous herb that is found in India, China, Pakistan, Indonesia, and Sri Lanka (Kumar et al., 2016). The plant has been used traditionally to treat digestive problems, normalize blood pressure, and enhance hair growth, and for heart and hepatic reinforcement (Kumar et al., 2016). Among other active compounds present in *Phyllanthus emblica*, vitamin C is an important constituent.

5.3.6 CORDIA DICHOTOMA

Cordia dichotoma Forst, found in the subtropics, is a small or moderate size plant that belongs to the Boraginaceae family and is known as bhokar, lasura, gonda, Indian cherry, and shlesmataka (Jamkhande et al., 2013). The plant grows in the sub-Himalayan region and may attain a height up to 1500 metres (Kirtikar and Basu, 1935). The fruit, bark, seed, and leaves are reported to possess antidiabetic, anti-inflammatory, analgesic, antiulcer and immunomodulatory actions (Jamkhande et al., 2013). The fruit pulp extract acts as an antioxidant as it was reported to contain vitamin C (Ibrahim et al., 2019). An earlier study showed that *C. dichotoma* exhibited antioxidant activity comparable with standard ascorbic acid at varying concentrations of 10, 15, 25, 50, 60 µg/mL (Nariya et al., 2013). The same study found that there was an increase in the antioxidant activity (percentage) in a dose-dependent manner for different concentrations that were tested (Nariya et al., 2013). An earlier study showed that the reducing power of *C. dichotoma* extract was higher compared to that of ascorbic acid (El-Newary et al., 2018).

5.3.7 ALOCASIA

This plant is found in many tropical and subtropical regions including parts of India and Sri Lanka. The three major types of Alocasia are *Alocasia cuculata* (L.), *Alocasia indica Schott*, and *Alocasia macrorrhiza* (L.) (Jana, 2015). A research report from Philippines showed that *Alocasia macrorrhiza* contain flavonoids, cyanogenetic glycosides, and ascorbic and gallic acid besides being a rich source of vitamin C (Mollejon & Tibe, 2019). *Alocasia indica* Schott is an indigenous herb and has been used traditionally as a digestive, laxative, diuretic, astringent, and a remedy for rheumatic arthritis (Nadkarni, 1996). There are various active compounds present in the plant extract; a few to mention are flavonoids, cyanogenetic glycosides, ascorbic acid, gallic acid, malic acid (Prajapati, 2003). The leaves of Alocasia are rich in vitamin B6, vitamin C, and niacin (Jana, 2015). The antioxidant ascorbic acid in *Alocasia indica* Schott was found to be 76.6 ± 4.03 mg/100 g dry weight and 76.65 mg/100 g in two different experimental studies (Basu et al., 2014; Nakhuru et al., 2021).

5.3.8 AMARANTHUS

Amaranthus viridis (*Tete abalaye* in Yoruba) is found in many parts of Nigeria. Malabar spinach (*Basella alba* L.) contains different amounts of carotenoids and vitamins A and C (Oliveira et al., 2013). Seeds were reported to possess potent antioxidant action and were reported to be beneficial for treating hypertension, cardiovascular diseases, and cholesterol (Samota, 2017). The leaves of mg/100 g (DW) were beta-carotene (3.29), thiamine (2.75), riboflavin (4.24), niacin (1.54), pyridoxine (2.33), ascorbic acids (25.40), and alpha-tocopherol (0.50) (Akubugwo et al., 2007).

The plant contains vitamin A, folic acid, and vitamin C. One cup of amaranth leaves subjected to cooking, boiling, and draining was reported to contain 90% vitamin C and 73% vitamin A (Samota, 2017). One of the species, *A. dubius*, contained the highest amount of vitamin C while another variety, *A. viridus*, showed the highest amount of protein and vitamin A (Samota, 2017).

5.3.9 POLYGONACEAE

Rumex vesicarius L. (Polygonaceae) is an annual green herb commonly found in desert and semi-desert regions of Northern parts of Africa, South Asia, and Australia (Rao et al., 2012). Polygonaceae species such as *R. ulmifolius, R. induratus, R. acetosella,* and *R. papillaris* have been reported to contain a high amount of ascorbic acid (Periera et al., 2011, 2013; Morales et al., 2014). Higher levels of ascorbic acid in the leaves of *R.vesicarius* (44.95 ± 2.57 mg 100 g^{-1} fresh weight) were reported (Manoj et al., 2019).

5.4 CONCLUSION

Inadequate levels of vitamins in the body lead to a deficiency state that may be treated with adequate supplementation. Vitamins are essential for metabolism, growth, and development of the human body. Vitamins may also be helpful in combating malnutrition problems in remote villages. Vitamins also act as potent antioxidants to check oxidative stress related diseases. Besides fruits and vegetables, wild edible plants may be an additional source of vitamins. While external agents and compounds are added to increase the nutraceutical activity of food products, the presence of vitamins in natural form in WEP may be better for consumption. There is a need to conduct experimental and toxicity studies on the active components of various WEP. Based on the results of the experiments, WEP may also be cultivated in the future.

REFERENCES

Åhlberg, Mauri Kalervo. "A profound explanation of why eating green (wild) edible plants promote health and longevity." *Food Frontiers* 2, no. 3 (2021): 240–267.

Ahmed, Syed Zaryab, Anila Jaleel, Kamran Hameed, Salman Qazi, and Ahsan Suleman. "Does vitamin D deficiency contribute to the severity of asthma in children and adults?" *Journal of Ayub Medical College Abbottabad* 27, no. 2 (2015): 458–463.

Akubugwo, I.E., Obasi, N.A., Chinyere, G.C., and Ugbogu, A.E. Nutritional and chemical value of Amaranthus hybridus L. leaves from Afikpo, Nigeria. *African Journal of Biotechnology* 6, no. 24 (2007): 2833–2839.

Alabama Plant Atlas. 2023. Accessed from website- http://floraofalabama.org/Plant.aspx?id=1031 Accessed on 26 March 2023.

Alderwerelt, V. *Malaysian ferns*. Amsterdam: Asher and Co (1989).

Alves-Rodrigues, Alexandra, and Andrew Shao. "The science behind lutein." *Toxicology Letters* 150, no. 1 (2004): 57–83.

Bakar, Mohd Fadzelly Abu, Maryati Mohamed, Asmah Rahmat, and Jeffrey Fry. "Phytochemicals and antioxidant activity of different parts of bambangan (Mangifera pajang) and tarap (Artocarpus odoratissimus)." *Food Chemistry* 113, no. 2 (2009): 479–483.

Basu, Subhashree, Moumita Das, Anurupa Sen, Utpal Roy Choudhury, and Gouriprosad Datta. "Analysis of complete nutritional profile and identification of bioactive components present in Alocasia indica tuber cultivated in Howrah District of West Bengal, India." *Asian Pacific Journal of Tropical Medicine* 7 (2014): S527–S533.

Beekwilder, Jules, Robert D. Hall, and C. H. De Vos. "Identification and dietary relevance of antioxidants from raspberry." *Biofactors* 23, no. 4 (2005): 197–205.

Beluhan, S., and A. Ranogajec. "Chemical composition and non-volatile components of Croatian wild edible mushrooms." *Food Chemistry* 124, no. 3 (2011): 1076–1082.

Bhatt, Indra D., Sandeep Rawat, Amit Badhani, and Ranbeer Rawal. "Nutraceutical potential of selected wild edible fruits of the Indian Himalayan region." *Food Chemistry* 215 (2017): 84–91. https://doi.org/10.1016/j.foodchem.2016.07.143

Birney, Elmer C., Robert Jenness, and Ian D. Hume. "Evolution of an enzyme system: ascorbic acid biosynthesis in monotremes and marsupials." *Evolution* (1980): 230–239.

Burlingame, Barbara. "Wild nutrition." *Journal of Food Composition and Analysis* 2, no. 13 (2000): 99–100.

Calderwood, Lily and Tooley Brogan. *Wild Blueberry Concentrations: Antioxidants, Vitamins and Minerals.* Cooperative Extension: Maine Wild Blueberries. 2020. Accessed from https://extension.umaine.edu/blueberries/factsheets/quality/wild-blueberry-concentrations-antioxidants-vitamins-and-minerals/ on 25.5.2024

Chakravarty, Sumit, Karma D. Bhutia, C. P. Suresh, Gopal Shukla, and Nazir A. Pala. "A review on diversity, conservation and nutrition of wild edible fruits." *Journal of Applied and Natural Science* 8, no. 4 (2016): 2346–2353.

Chatterjee, I. B., N. C. Kar, N. C. Ghosh, and B. C. Guha. "Aspects of ascorbic acid biosynthesis in animals." *Annals of the New York Academy of Sciences* 92, no. 1 (1961): 36–56.

Clarke, Michael W., John R. Burnett, and Kevin D. Croft. "Vitamin E in human health and disease." *Critical Reviews in Clinical Laboratory Sciences* 45, no. 5 (2008): 417–450.

Curtis, T. G. F., and Paul Whelan. *The Wild Food Plants of Ireland: The Complete Guide to Their Recognition, Foraging, Cooking, History and Conservation.* Orla Kelly Publishing, 2019. Accessed from https://www.summerfieldbooks.com/product/the-wild-food-plants-of-ireland-the-complete-guide-to-their-recognition-foraging-cooking-history-and-conservation/. Accessed on 30 March 2023.

Dansi, A., A. Adjatin, H. Adoukonou-Sagbadja, V. Faladé, H. Yedomonhan, D. Odou, and B. Dossou. "Traditional leafy vegetables and their use in the Benin Republic." *Genetic Resources and Crop Evolution* 55 (2008): 1239–1256.

de Cortes Sánchez-Mata, Maria, and Tardío, J. (Eds.). *Mediterranean wild edible plants: ethnobotany and food composition tables.* Springer, 2016. Accessed from https://link.springer.com/book/10.1007/978-1-4939-3329-7. Accessed on 30 March 2023.

Drouin, Guy, Jean-Rémi Godin, and Benoît Pagé. "The genetics of vitamin C loss in vertebrates." *Current Genomics* 12, no. 5 (2011): 371–378.

Eccleston, Clair, Yang Baoru, Raija Tahvonen, Heikki Kallio, Gerald H. Rimbach, and Anne M. Minihane. "Effects of an antioxidant rich juice (sea buckthorn) on risk factors for coronary heart disease in humans." *The Journal of Nutritional Biochemistry* 13, no. 6 (2002): 346–354.

El-Newary, Samah A., Abeer Y. Ibrahim, Samir M. Osman, and Michael Wink. "Evaluation of possible mechanisms of Cordia dichotoma fruits for hyperlipidemia controlling in Wistar albino rats." *Asian Pacific Journal of Tropical Biomedicine* 8, no. 6 (2018): 302–312.

Ercisli, Sezai. "Chemical composition of fruits in some rose (Rosa spp.) species." *Food Chemistry* 104, no. 4 (2007): 1379–1384.

Escudero, Nora Lilian, M. L. De Arellano, S. Fernández, G. Albarracín, and S. Mucciarelli. "Taraxacum officinale as a food source." *Plant Foods for Human Nutrition* 58 (2003): 1–10. https://doi.org/10.1023/B:QUAL.0000040365.90180.b3

Fratianni, Alessandra, Mario Irano, Gianfranco Panfili, and Rita Acquistucci. "Estimation of color of durum wheat. Comparison of WSB, HPLC, and reflectance colorimeter measurements." *Journal of Agricultural and Food Chemistry* 53, no. 7 (2005): 2373–2378.

Frison, Emile A., Ifeyironwa Francisca Smith, Timothy Johns, Jeremy Cherfas, and Pablo B. Eyzaguirre. "Agricultural biodiversity, nutrition, and health: making a difference to hunger and nutrition in the developing world." *Food and Nutrition Bulletin* 27, no. 2 (2006): 167–179.

Gallie, Daniel R. "L-ascorbic acid: a multifunctional molecule supporting plant growth and development." *Scientifica* 2013 (2013): 795964.

García-Herrera, P., M. C. Sánchez-Mata, M. Cámara, V. Fernández-Ruiz, C. Díez-Marqués, M. Molina, and J. Tardío. "Nutrient composition of six wild edible Mediterranean Asteraceae plants of dietary interest." *Journal of Food Composition and Analysis* 34, no. 2 (2014): 163–170.

Gilbert, C. "What is vitamin A and why do we need it?" *Community Eye Health*, 26, no. 84 (2013): 65.

Grivetti, Louis E., and Britta M. Ogle. "Value of traditional foods in meeting macro-and micronutrient needs: the wild plant connection." *Nutrition Research Reviews* 13, no. 1 (2000): 31–46.

Gupta, Prakash Chandra, Nisha Sharma, and Ch V. Rao. "A review on ethnobotany, phytochemistry and pharmacology of Fumaria indica (Fumitory)." *Asian Pacific Journal of Tropical Biomedicine* 2, no. 8 (2012): 665–669.

Gupta Dutta, S., P. K. Choudhury, and I. B. Chatterjee. "Synthesis of l-ascorbic acid from d-glucurono-1, 4-lactone conjugates by different species of animals." *International Journal of Biochemistry* 4, no. 21 (1973): 309–314.

Gupta, Sanjay Mohan, Basant Ballabh, Pradeep Kumar Yadav, Ankur Agarwal, and Madhu Bala. "Nutrients Analysis of Diplazium esculentum: Underutilized wild wetland Pteridophytes ensure food and nutritional security." *Acta Scientific Nutritional Health* 4, no. 11 (2020): 46–49.

Hadizadeh, Fatemeh. "Supplementation with vitamin D in the COVID-19 pandemic?" *Nutrition Reviews* 79, no. 2 (2021): 200–208.

Ibrahim, Abeer Y., Samah A. El-Newary, and Gamil E. Ibrahim. "Antioxidant, cytotoxicity and anti-tumor activity of Cordia dichotoma fruits accompanied with its volatile and sugar composition." *Annals of Agricultural Sciences* 64, no. 1 (2019): 29–37.

Jamkhande, Prasad G., Sonal R. Barde, Shailesh L. Patwekar, and Priti S. Tidke. "Plant profile, phytochemistry and pharmacology of Cordia dichotoma (Indian cherry): A review." *Asian Pacific Journal of Tropical Biomedicine* 3, no. 12 (2013): 1009–1012.

Jana, B. R. "Alocasia: a promising food crop for next generation." In *III International Symposium on Underutilized Plant Species* 1241 (2015): 31–36.

Joshi, N. and Verma, K.C. "A review on nutrition value of Amaranth (Amaranthus caudatus L.): The crop of future." *Journal of Pharmacognosy and Phytochemistry* 9, no. 4 (2020): 1111–1113.

Kirtikar, Kanhoba Ranchoddas, and Baman Das Basu. *Indian Medicinal Plants*, 11th Edition, Vol 3, Orient Enterprises, 1935, pp. 1029–1030.

Kumar, Avneesh, Sunil Kumar, Savita Bains, Vanya Vaidya, Baljinder Singh, Ravneet Kaur, Jagdeep Kaur, and Kashmir Singh. "De novo transcriptome analysis revealed genes involved in flavonoid and vitamin C biosynthesis in Phyllanthus emblica (L.)." *Frontiers in Plant Science* 7 (2016): 1610.

Levine, Mark, Yaohui Wang, Sebastian J. Padayatty, and Jason Morrow. "A new recommended dietary allowance of vitamin C for healthy young women." *Proceedings of the National Academy of Sciences* 98, no. 17 (2001): 9842–9846.

McGhie, T.K., and Ainge, G.D. "Color in fruit of the genus actinidia: carotenoid and chlorophyll compositions." *Journal of Agricultural and Food Chemistry* 50, no. 1 (2002): 117–21. doi: 10.1021/jf0106771

Manoj, Prabhakaran, Sandopu Sravan Kumar, and Parvatam Giridhar. "In vitro shoot multiplication of Rumex vesicarius L., and quantification of ascorbic acid and major phenolics from its leaf derived callus." *3 Biotech* 9 (2019): 1–9.

Mayo Clinic. Accessed from https://www.mayoclinic.org/drugs-supplements-vitamin-a/art-20365945, 2023. Accessed on 26 March 2023.

Miller, Melissa, Jean Humphrey, Elizabeth Johnson, Edmore Marinda, Ron Brookmeyer, and Joanne Katz. "Why do children become vitamin A deficient?" *The Journal of Nutrition* 132, no. 9 (2002): 2867S–2880S.

Mollejon, C. V., and Tibe, J. E. Nutritional and nutraceutical content of alocasia macrorrhizos (l.) G. Don (Talyan). GSJ, 7, no. 3 (2019): 584–652.

Morales, Patricia, Isabel CFR Ferreira, Ana Maria Carvalho, Mª Cortes Sánchez-Mata, Montaña Cámara, Virginia Fernández-Ruiz, Manuel Pardo-de-Santayana, and Javier Tardío. "Mediterranean non-cultivated vegetables as dietary sources of compounds with antioxidant and biological activity." *LWT-Food Science and Technology* 55, no. 1 (2014): 389–396.

Mustacich, Debbie J., Bruno, Richard S., and Traber, Maret G. (2007). Vitamin E. *Vitamins and Hormones*, 76, 1–21.

Nadkarni, K. M. *Indian Materia Medica*. Mumbai: Popular Prakashan, 1996, pp. 72–3.

Nakhuru, Khonamai Sewa, Adani Lokho, Mridusmita Barman, Jayshree Das, and Sanjai Kumar Dwivedi. "Evaluation of vitamin C of ethno-wild edible plants in Northeast India." *Plant Science Today* 8, no. 3 (2021): 473–481.

Nariya, Pankaj B., Nayan R. Bhalodia, Vinay J. Shukla, Rabinarayan Acharya, and Mukesh B. Nariya. "In vitro evaluation of antioxidant activity of Cordia dichotoma (Forst f.) bark." *Ayu* 34, no. 1 (2013): 124.

Nitsa, Alkippi, Marina Toutouza, Nikolaos Machairas, Anargyros Mariolis, Anastassios Philippou, and Michael Koutsilieris. "Vitamin D in cardiovascular disease." *In Vivo* 32, no. 5 (2018): 977–981.

Nyerges, Christopher. *Foraging Arizona: Finding, Identifying, and Preparing Edible Wild Foods in Arizona*. Rowman & Littlefield, 2020.

Oliveira, Danyela de Cássia da S., Carmen Wobeto, Marcio R. Zanuzo, and Cristiane Severgnini. "Mineral composition and ascorbic acid content in four non-conventional leafy vegetables species." *Horticultura Brasileira* 31 (2013): 472–475.

Pereira, Carla, Lillian Barros, Ana Maria Carvalho, and Isabel C.F.R. Ferreira. "Nutritional composition and bioactive properties of commonly consumed wild greens: Potential sources for new trends in modern diets." *Food Research International* 44, no. 9 (2011): 2634–2640.

Pereira, Carla, Lillian Barros, Ana Maria Carvalho, and Isabel CFR Ferreira. "Use of UFLC-PDA for the analysis of organic acids in thirty-five species of food and medicinal plants." *Food Analytical Methods* 6 (2013): 1337–1344.

Pinela, José, Ana Maria Carvalho, and Isabel CFR Ferreira. "Wild edible plants: Nutritional and toxicological characteristics, retrieval strategies and importance for today's society." *Food and Chemical Toxicology* 110 (2017): 165–188.

Pinela, Jose, Márcio Carocho, Maria Ines Dias, Cristina Caleja, Lillian Barros, and Isabel CFR Ferreira. "Wild plant-based functional foods, drugs, and nutraceuticals." *Wild Plants, Mushrooms and Nuts: Functional Food Properties and Applications* (2016): 315–351.

Pludowski, Pawel, Michael F. Holick, William B. Grant, Jerzy Konstantynowicz, Mario R. Mascarenhas, Afrozul Haq, Vladyslav Povoroznyuk et al. "Vitamin D supplementation guidelines." *The Journal of Steroid Biochemistry and Molecular Biology* 175 (2018): 125–135.

Prajapati, Narayan Das. *A Handbook of Medicinal plants*. India: Agrobois Publication (2003): 32.

Rahmat, Asmah, Vijay Kumar, Loo Mei Fong, Susi Endrini, and Huzaimah Abdullah Sani. "Determination of total antioxidant activity in three types of local vegetables shoots and the cytotoxic effect of their ethanolic extracts against different cancer cell lines." *Asia Pacific Journal of Clinical Nutrition* 13, no. 3 (2004): 308–311d.

Rao, K. N., Sunitha Ch, S. Sandhya, and T. Rajeshwar. "Anthelminthic activity of different extracts on aerial parts of Rumex vesicarius Linn." *International Journal of Pharmaceutical Sciences Review and Research* 12 (2012): 64–66.

Ravikanth, Kotagiri, Anil Kanaujia, Deepak Thakur, Anirudh Sharma, and Bhupesh Gautam. "Nutritional Constituents of the Plants Fumaria indica and Caesalpinia bonducella." *International Journal of Advances Pharmacy, Biology and Chemistry* 3, no. 3 (2014): 698–702.

Ruffo, Christopher K., Ann Birnie, and Bo Tengnäs. "Edible wild plants of Tanzania." 2002. Accessed from https://apps.worldagroforestry.org/downloads/Publications/PDFS/B11913.pdf. Accessed on 30 March 2023.

Rutto, Laban K., Yixiang Xu, Elizabeth Ramirez, and Michael Brandt. "Mineral properties and dietary value of raw and processed stinging nettle (Urtica dioica L.)." *International Journal of Food Science* 2013 (2013): 857120.

Rymbai, Heiplanmi, Veerendra Kumar Verma, Hammylliende Talang, S. Ruth Assumi, M. Bilashini Devi, Rumki Heloise C.H. Sangma, Kamni Paia Biam et al. "Biochemical and antioxidant activity of wild edible fruits of the eastern Himalaya, India." *Frontiers in Nutrition* 10 (2023): 1039965.

Samota, M. K. "Nutritional value of Amaranthus. Biotech articles." 2017. Accessed from https://biotecharticles.com/Agriculture-Article/Nutritional-Value-of-Amaranthus-3958.html. Accessed on 26 March 2023.

Schippmann, Uwe, Danna J. Leaman, and Anthony B. Cunningham. "Impact of cultivation and gathering of medicinal plants on biodiversity: global trends and issues." *Biodiversity and the ecosystem approach in agriculture, forestry and fisheries* (2002). Accessed from https://www.cabdirect.org/cabdirect/abstract/20033097198. Accessed on 30 March 2023.

Seal, Tapan, Kausik Chaudhuri, and Basundhara Pillai. "A rapid high-performance liquid chromatography method for the simultaneous estimation of water-soluble vitamin in ten wild edible plants consumed by the tribal people of North-eastern Region in India." *Pharmacognosy Magazine* 14, no. Suppl 1 (2018): S72–S77.

Semwal, Prabhakar, Sakshi Painuli, Kartik M. Painuli, Gizem Antika, Tugba Boyunegmez Tumer, Ashish Thapliyal, William N. Setzer et al. "Diplazium esculentum (Retz.) Sw.: ethnomedicinal, phytochemical, and pharmacological overview of the Himalayan ferns." *Oxidative Medicine and Cellular Longevity* 2021 (2021): 1917890.

Suda, Tatsuo, Naoyuki Takahashi, and Etsuko Abe. "Role of vitamin D in bone resorption." *Journal of Cellular Biochemistry* 49, no. 1 (1992): 53–58.

Teketay, Demel, Mulugeta Lemenih, Tesfaye Bekele, Yonas Yemshaw, Sisay Feleke, Wubalem Tadesse, Yitebitu Moges, Tesfaye Hunde, and Demeke Nigussie. "Forest resources and challenges of sustainable forest management and conservation in Ethiopia." In *Degraded Forests in Eastern Africa*, pp. 19–63. Routledge, 2010.

Trapani, G. "Dandelion: benefits, active ingredients, and contraindications of dandelion". 2019. Accessed from https://blog.yamamotonutrition.com/en/dandelion-benefits-active-ingredients-and-contraindications-of-dandelion-a786. Accessed on 26 March 2023.

Vinceti, Barbara, Pablo Eyzaguirre, and Timothy Johns. "The nutritional role of forest plant foods for rural communities." In Carol J. Pierce Colfer (ed.) *Human Health and Forests*, pp. 63–96. Routledge, 2012.

Wolska, Jolanta, Michał Czop, Karolina Jakubczyk, and Katarzyna Janda. "Influence of temperature and brewing time of nettle (Urtica dioica L.) infusions on vitamin C content." *Roczniki Państwowego Zakładu Higieny* 67, no. 4 (2016): 367–371.

Young, Andrew J., and Gordon M. Lowe. "Antioxidant and prooxidant properties of carotenoids." *Archives of Biochemistry and Biophysics* 385, no. 1 (2001): 20–27.

Zannah, Fathul, Mohammad Amin, Hadi Suwono, and Betty Lukiati. "Phytochemical screening of Diplazium esculentum as medicinal plant from Central Kalimantan, Indonesia." In *AIP Conference Proceedings*, vol. 1844, no. 1. AIP Publishing, 2017.

6 Solanum and Asparagus Oils Are Good Sources of Omega Fatty Acids

Zhihong Xu

6.1 INTRODUCTION TO OMEGA FATTY ACIDS

Natural products play an important role as sources of energy and nutritional supplements and provide various health benefits to patients. For example, galactolipids, the most abundant lipids in thylakoid membranes, are widely found in edible plants such as beans, kale, chili, asparagus, broccoli, and pumpkin, consisting mainly of the mono- and digalactosyl diacylglycerols, have shown in vitro and/or in vivo anti-inflammation and antitumor activities (Christensen 2009). Fatty acids (FA) are naturally occurring monocarboxylic acids with an unbranched carbon chain. There are two general types – saturated and unsaturated fatty acids (SFAs and UFAs). UFAs can be further classified by the double bond (DB) number and the position of the first DB away from the terminal methyl (omega, ω) location. An ω-3 fatty acid (ω-3 FA) has the terminal DB to be the third bond from its methyl end, and an ω-6 fatty acid (ω-6 FA) has the terminal DB to be the sixth bond from its methyl end, and so on. Among the UFAs, omega fatty acids (ω-FAs) play important roles in biological activities and are recognized as major biological regulators in health promotion and disease risk reduction. For example, the central nervous system (CNS) is highly enriched by long-chain polyunsaturated ω-3 and ω-6 FAs which share similar elongation and desaturation enzymes but have certain contrasting effects on metabolic functions (Martins et al. 2008; Dhull and Punia 2021; Saini and Keum 2018). For example, dietary linoleic acid (LA, ω-6) favors low density lipoprotein oxidation and increases platelet response to aggregation; while α-linolenic acid (ALA, ω-3) inhibits platelet clotting activity and arachidonic acid (ARA) metabolism and lowers blood pressure and decreases the risk of coronary heart disease in men (Simopoulos 2004a). Marine origin ω-3 FAs may change the trajectories of human diseases associated with aging and influence the exercise, skeletal muscle metabolism, and nutritional response of skeletal muscles (Jeromson et al. 2015). High level long-chain ω-3 FAs from deep-sea fish oils exert protective effects against cancers, especially breast, colon, and prostate cancers, via inhibition of angiogenicity, neoplastic transformation or cell growth, or enhancement of apoptosis (Rose and Connolly 1999).

A few ω-FAs from natural and agricultural resources have shown multiple functional properties. For example, *Acer truncatum* Bunge seed oil is enriched with ω-6 (30.7%) and ω-9 (53.93%) FAs, and characterized by nervonic acid (NA, C24:1, ω-9, 3–7%) which participates in the processes of cognitive improvement and promotes nerve regeneration and damaged nerve cell/tissue repair (Yang et al. 2018; Song et al. 2022). LA is a 5-α-reductase inhibitor and has exhibited antiarthritic, antieczemic, antianaphylactic, anticoronary, antiatherosclerotic, antifibrinolytic, and anticancer activities (Chakraborty et al. 2016). Numerous modifications of ω-3 FAs have produced various plant hormones (Cholewski et al. 2018). Well-known biochemically important monounsaturated and polyunsaturated ω-FAs (MUFAs and PUFAs) in our body include palmitoleic acid (POA, 16:1, Δ^9, ω-7), oleic acid (OA, 18:1, Δ^9, ω-9), linoleic acid (LA, 18:2, $\Delta^{9,12}$, ω-6), α-linolenic acid (ALA, 18:3, $\Delta^{9,12,15}$, ω-3), stearidonic acid (SDA, 18:4, $\Delta^{6,9,12,15}$, ω-3), arachidonic acid (ARA, 20:4, $\Delta^{5,8,11,14}$, ω-6), eicosapentaenoic acid (EPA, 20:5, $\Delta^{5,8,11,14,17}$, ω-3), docosapentaenoic acid (DPA; 22:5, $\Delta^{7,10,13,16,19}$, ω-3),

DOI: 10.1201/9781003395935-6

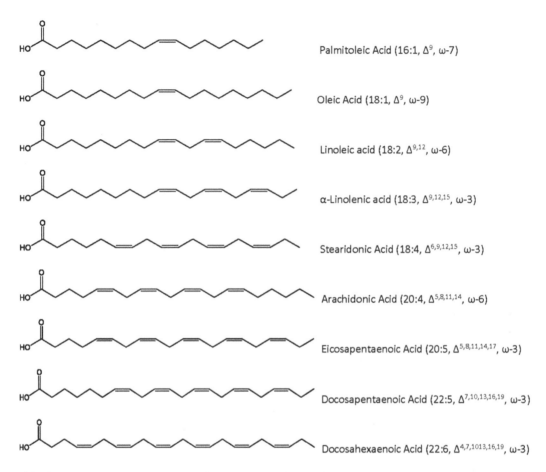

Palmitoleic Acid (16:1, Δ^9, ω-7)

Oleic Acid (18:1, Δ^9, ω-9)

Linoleic acid (18:2, $\Delta^{9,12}$, ω-6)

α-Linolenic acid (18:3, $\Delta^{9,12,15}$, ω-3)

Stearidonic Acid (18:4, $\Delta^{6,9,12,15}$, ω-3)

Arachidonic Acid (20:4, $\Delta^{5,8,11,14}$, ω-6)

Eicosapentaenoic Acid (20:5, $\Delta^{5,8,11,14,17}$, ω-3)

Docosapentaenoic Acid (22:5, $\Delta^{7,10,13,16,19}$, ω-3)

Docosahexaenoic Acid (22:6, $\Delta^{4,7,1013,16,19}$, ω-3)

FIGURE 6.1 Structures and sources of representative bioactive important unsaturated fatty acids (UFAs).

and docosahexaenoic acid (DHA, 22:6, $\Delta^{4,7,1013,16,19}$, ω-3). Their structures are shown in Figure 6.1. OA contributes to the metabolism of essential fatty acids (EFAs) (Dhopeshwarkar and Mead 1961; Lowry and Tinsley 1966), ARA is a structural moiety of membrane phospholipids involved in signal transduction and important for eicosanoid synthesis, and DHA is needed for membrane structure and optimal visual keenness/neural development (Spector 1999). Among all the FAs, the human body can totally synthesize SFAs and certain UFAs, except the ω-6 & ω-3 FAs which are considered as EFAs. The EFAs must be obtained from diet or supplementation. ARA can be synthesized from LA, and ALA is the precursor of EPA, DPA, and DHA (ω-3 PUFAs). These ω-3 PUFAs can be synthesized via chain elongation and desaturation (metabolic pathway: ALA→SDA→eicosatetraenoic acid→EPA→DPA→tetracosapentaenoic acid→tetracosahexaenoic acid→DHA); however, this conversion rate is less than 4% at best in humans (Shahidi and Ambigaipalan 2018). DHA is the major component of brain gray matter (McNamara and Carlson 2006), and both DHA and EPA are found to be critical for fetal development (Swanson et al. 2012). DHA and EPA also have an impact on eicosanoid generation, gene expression, cell signaling, the physical nature of cell membranes, and related protein-mediated responses (Calder and Yaqoob 2009a), and influence cardiovascular function in coronary events, peripheral artery disease, anticoagulation, and during inflammation (Swanson et al. 2012). In addition, bioactive DHA and EPA metabolites, such as resolvins and protectins, are more potent anti-inflammatory mediators (Weylandt et al. 2012). ARA is rich in human immune cells, and eicosanoid hormones produced from ARA and other ω-FAs have roles in the central nervous system, heart, blood pressure/clotting conditions (Dhull and Punia 2021), inflammation, and B and T lymphocyte function regulations (Calder 2007). Generally, ARA increases the amount of pro-inflammatory eicosanoids, while DHA and EPA increase the number of anti-inflammatory eicosanoids (Saini and Keum 2018).

The most important EFAs are linoleic and linolenic acids, as they are essential for cell membrane structure and biosynthesis of longer-chain ω-6 & ω-3 FAs. Deficiencies of linoleic and linolenic acids may cause skin redness, infections, dehydration, and liver abnormalities. Since we can only convert EFAs LA and ALA to more bioactive longer chain ω-6 and ω-3 FAs or PUFAs like ARA, EPA, and DHA with a very low efficiency (Domenichiello et al. 2015; Calder and Yaqoob 2009b), we also need to intake certain amounts of these PUFAs.

Both ω-3 and ω-6 PUFAs are entirely obtained from the diet. Western diets were recommended to contain increasingly larger amounts of ω-6 LA for its cholesterol-lowering effect (Harris et al. 2009), but diets rich in LA and poor in ω-3 might result in autoimmune disorders (Simopoulos 2002a), cancer, hypertension, chronic inflammation, and cardiovascular problems (Simopoulos 2004b; Dhull and Punia 2021). Recent evidence has also supported that the ω-6 PUFAs have a role in the origins of metabolic diseases and may lead to obesity (Muhlhausler and Ailhaud 2013). A high intake of ω-6 FAs can lead to a prothrombotic and pro-aggregatory biological state, while ω-3 FAs exhibit antithrombotic, anti-inflammatory, anti-arrhythmic, vasodilatory, and hypolipidemic activities (Vidrih et al. 2009). Evidence shows that a high ω-6 FA intake also inhibits ω-3 FA inflammation-resolving and anti-inflammatory functions (Innes 2008; Innes and Calder 2018). Sufficient ω-3 PUFA diet is essential for neural development and neuroprotective potential, and optimal visual functions (Dyall and Michael-Titus 2008). The ω-3 FAs are famous for their effects on reducing cardiovascular morbidity/mortality and ω-3 PUFAs are found to be able to revert the microbiota composition in inflammatory bowel diseases effectively, and at the same time, the production of anti-inflammatory short-chain FAs are increased. In addition, ω-3 PUFAs have an impact on the gut–brain axis, and the interplay of gut microbiota and immunity helps interact with host immune cells and maintain the intestinal wall integrity (Costantini et al. 2017). The ω-3 PUFAs are also found to exert major alterations on both the innate and the adaptive immune cell activations (Gutierrez et al. 2019). The marine ω-3 FAs EPA and DHA have shown good cardio protection (Saravanan et al. 2010), and their effects on cardiovascular inflammations and cancer have led people to believe that these compounds are more beneficial compared to other dietary supplements (Gogus and Smith 2010). Increasing evidence has shown that ω-3 FAs might be important to mental health, as reduced levels of ω-3 FAs in the membranes of red blood cells have been reported in both schizophrenic and depressive patients (Peet and Stokes 2005), and ω-3 FAs were reported to regulate locomotor and exploratory activities, cognitive functions, and multiple emotions in rodents (Fedorova and Salem 2006).

A balanced ω-6/ω-3 ratio in foods is very important (Simopoulos 2009), and the ideal ω-6 FA to ω-3 FA ratio can be 4–10 g to 1 g, and the optimal ratio to assist patient treatments might vary with typical diseases. According to WHO/FAO 1994 (Dhull and Punia 2021), a ω-6/ω-3 FA ratio of 5:1 or less for most people is desirable in the reduction of many chronic diseases. For example, the Greenland Inuit and Danish population, whose diets are rich in marine mammals and fish ω-FAs, have a low incidence of heart diseases (Shahidi and Miraliakbari 2006). In contrast, conventional Western diets have the ω-6 to ω-3 FA ratio of 15:1 to 16.7:1, which might have contributed to high heart disease rates (Yashodhara et al. 2009). In addition to cardiovascular diseases (Simopoulos 2002b), a very high ω-6 to ω-3 ratio or excessive ω-6 PUFAs in today's Western diets have caused several other health problems, including cancer, inflammation, obesity, and autoimmune diseases. Dietary amounts of LA and the LA/ALA ratio appeared to be important for the synthesis of longer-chain ω-3 PUFAs from ALA (Simopoulos 2004a). Inadequate ω-3 FA intake will decrease the amount of DHA, and a DHA decrease in a developing brain may lead to deficits in neurotransmitter metabolism and neurogenesis, altered learning, and a notable impact on visual functions in animals (Innis 2008). As an alternative to the conventional vegetables available in stores, a variety of wild edible plants generally contain a good balance of ω-6 and ω-3 FAs (Morales et al. 2012).

Fatty acids are important structural components of various esters like triacylglycerides, and other lipids like phospholipids in all species. A triacylglyceride contains three fatty acyl groups on one glycerol molecule. In humans, the 'long-term' storage for energy is concentrated within adipocytes where triacylglycerols are stored. Plant oil is a triacylglycerol mixture that is a liquid at room temperature (rt), and an animal fat is a triacylglycerol mixture that is a solid or semi-solid at rt. People with a high saturated fat diet tend to be associated with higher rates of heart diseases and/or typical cancers. Therefore, a healthy diet should consist of a daily saturated fat intake smaller than 10% of total calories. Generally, cis-monounsaturated FAs, ω-3 and ω-6 UFAs, especially many PUFAs are considered 'good', while saturated fats and trans fats are considered 'bad'. For healthier foods, more people are now paying attention to the ω-6/ω-3 FA intake ratio and thinking about having more ω-3 FAs. Scientists are paying more attention to the availability of the ω-3 PUFAs because of their multiple protective impacts (Yashodhara et al. 2009; Almeida et al. 2014; Shahidi and Miraliakbari 2005; Shahidi and Ambigaipalan 2018; Saini and Keum 2018) on atherosclerosis and other cardiovascular disorders (Balk et al. 2006; Balk and Lichtenstein 2017; von Schacky 2007, 2015; Mozaffarian and Wu 2011; Harper and Jacobson 2001; Harris et al. 2006; Harris 2007a; Harris et al. 2008a, 2008b; Harris et al. 2013; Mori 2014, 2017), vascular function, blood pressure (Cabo et al. 2012; Mori 2010), antiplatelet effect (Mori 2006), stroke (especially ischemic stroke), inflammation, asthma, and allergy (Mori and Beilin 2004; Kang and Weylandt 2008; Miyata and Arita 2015; Teitelbaum and Allan Walker 2001), viral infection (Chou et al. 2012), and cancer (Reddy 2004; Wendel and Heller 2009; Gerber 2012; Freitas and Campos 2019; Fabian et al. 2015; Gleissman et al. 2010) (unlikely to prevent cancer though, MacLean et al. 2006). Further, ω-3 PUFAs have shown benefits in neurological (neurological/neuropsychiatric) and visual symptoms such as eye diseases including age-related macular degeneration (Merle et al. 2013; Surangi and Fiaz 2016), cognitive decline and dementia incidences including Alzheimer's disease (Sydenham et al. 2012; Bousquet et al. 2011; Burckhardt et al. 2016; Jicha and Markesbery 2010), brain development/function (McNamara and Carlson 2006), anxiety (Ross 2009), attention-deficit (Bloch and Qawasmi 2011), psychiatric and mental illness including depression (McNamara and Strawn 2013; Logan 2003; Ross 2006; Bloch and Hannestad 2012), gene expression (Price et al. 2000; Mozaffarian and Wu 2011; Cabo et al. 2012), fetus development, and maternal and child health (Yashodhara et al. 2009; Larque et al. 2012; Shahidi and Ambigaipalan 2018). In addition, ω-3 PUFAs are effective in the treatment of metabolic disorders and obesity, fatty liver disease (Scorletti and Byrne 2013), abnormal lipid profile/dyslipidemia, diabetes (Iwasaki et al. 2012; Wu et al. 2012a; Hartweg et al. 2008), gastrointestinal disorders, and rheumatological and other medical conditions.

Omega FA supplements are generally beneficial in assisting certain patients with treatment and disease prevention. However, there are also safety issues associated with taking related food supplements. In addition to the mild dyspepsia and belching associated with the intake of dietary fish oils (Yashodhara et al. 2009) and potential antithrombotic/platelet inhibition effect to increase the risk for bleeding (Harris 2007b, Bays 2006), toxicity from the susceptibility of ω-3 FA oxidation (related to patient intolerance), environmental toxins like mercury and other contaminants (Bays 2006), and possible undesirable effects of fish oils on heart, inflammatory bowel disease, and increased risks for typical cancers (Dhull and Punia 2021; MacLean et al. 2006) are major concerns.

6.2 SOURCES OF OMEGA FATTY ACIDS

Large quantities of ω-FAs can be found in common foods such as various meats from fish, poultry, beef and lamb/sheep; agriculture plants including soybean, cottonseed, peanut, canola, sunflower, palm, coconut, olive, and so on; and they are also obtained from nature, such as wild plants and animals, marine sources, bacteria, fungi, microalgae, and alternative sources like genetically engineered plants and microorganisms, and fortified foods (Ghafoorunissa 1993; Dhull and Punia 2021), which contents can be determined by numerous factors, including changes in soil microbial community (Hamel et al. 2005). The majority (~98%) of plant- and animal-based long-chain PUFAs are found in triacylglycerol form, followed by phospholipids (higher uptake by brain) and diacylglycer-ols, as well as in cholesterol and vitamin esters (Saini and Keum 2018). Edible wild greens are found to be valuable sources of EFAs (Vardavas et al. 2006; Liu et al. 2002), and cultivated sunflower, rapeseed, almond, walnut, safflower, corn, olive, soybean, and sesame contain more than 80% total UFAs (Kaur et al. 2014).

6.2.1 MAIN SOURCES OF OMEGA-3 FATTY ACIDS

Various ω-3 FAs can be mainly obtained from plants, animals, poultry, marine sources, and microbes. Some ω-3 MUFAs are pheromone precursors in insects. Long-chain PUFAs are commonly found in the CNS, mammalian testes, and in sponge organisms. Edible wild plants have been found to produce ALA and higher amounts of vitamins E and C than cultivated plants (Simopoulos 2004a). The primary source of the parent ω-3 FA ALA is plants, although plants are unable to synthesize long-chain multiple DB containing ω-3 FAs (Haslam et al. 2013). ALA is concentrated mainly in the seeds, nuts, and vegetable oils (Shahidi and Ambigaipalan 2018), and can be obtained primarily from flax, perilla, chia, echium, and canola and their seeds (Dhull and Punia 2021). Flaxseed oil has the highest ALA content compared to other nuts and plant seeds, and it contains up to 50% ALA; echium seed oil has been found to contain up to 33% ALA and 12.5% SDA (Dhull and Punia 2021). Seed oils from the Boraginaceae family plants, such as borage, *Echium vulgare*, *Buglossoides arvensis*, and hemp, and fish oils are great sources of SDA which increases the EPA level much more effectively than ALA does (Guil-Guerrero 2007).

Freshwater ecosystems are relatively rich in EPA (Kang 2011; Tocher 2009), and seafood is the most readily available and edible source of major ω-3 PUFAs. Omega-3 PUFAs are mainly found in marine organisms including several algae, red/white meats, and egg yolk. Wild fish contain more ω-3 PUFAs than cultivated or farmed fishes, and cold-water fishes produce higher proportions of long chain ω-3 PUFAs (Saini and Keum 2018). The major ω-3 PUFAs from marine sources are EPA and DHA, and low DPA level is present in most fish oils (Shahidi and Ambigaipalan 2018). The highest levels of DHA and EPA are found in the oily fishes from *Scombridae*, *Clupeidae*, and *Salmonidae* families (Dhull and Punia 2021), while cod, clams, shrimps, crabs, and oysters were reported to have lower DHA content (Arbex et al. 2015). Among all fishes, skipjack tuna, halibut, and cod flesh contain the highest percentages of DHA (30% of total FAs), while flounder species, cod flesh, and haddock contain the highest percentages of EPA (15–19% of total FAs) (Shahidi and Miraliakbari 2006).

While marine protists and dinoflagellates are DHA rich, microalgae are good sources of EPA. Microalgae were found to be the most abundant EPA and DHA producers in the initial aquatic food chain (Zahringer et al. 2005). Many marine microalgae species have 10–50% (w/w) oil content and produce up to 30–70% of total lipids (on dry weight basis) (Ward and Singh 2005). However, limiting factors such as cost and process methods have restricted their potential for use by humans on larger scales (Adarme-Vega et al. 2012). Omega-3 PUFAs synthesized by wild phytoplankton and algae are transferred and deposited into fish and marine mammal lipids in the food chain (Alasalvar et al. 2002), and algae *Crypthecodinium cohnii* and *Schizochytrium* spp. are major sources of DHA with 55% and 40% of the total FAs (Senanayake and Fichtali 2006). DHA and EPA are also found in animals, fungi, many microorganisms, and transgenic plants (Adarme-Vega et al. 2012). Several marine gamma-proteobacteria and deep-sea bacteria produce EPA and DHA (Allemann et al. 2019), and bacteria *Shewanella putrefaciens*, *Pneumatophorus japonicas*, and *Alteromonas putrefaciens* were found to be good sources of EPA (Yazawa 1996; Yazawa et al. 1988). Species of *Mortierella* (lower fungi) produce high amounts of ARA, and under certain conditions, EPA and DHA can also be produced (Ward and Singh 2005; Dhull and Punia 2021). Fungus *Thraustochytrium* sp. can synthesize a high amount of ω-3 PUFAs with 69–73% of DHA and 21–15% of DPA if cultured (Dhull and Punia 2021). Mushrooms have a good balance of unsaturated to saturated FAs, and mainly consist of MUFA and PUFA healthy lipids (Abugri et al. 2016). It is encouraging that fungus

Pythium acanthium can accumulate high amounts of DHA when cultured with linseed oil in the media (Salunke et al. 2014).

6.2.2 Main Sources of Omega-6 and Omega-9 Fatty Acids

Most crop seeds and vegetable oils are major sources of ω-6 FA (LA), but with relatively low proportions of ω-3 FAs (ALA) except a few green vegetables (60–70% ALA out of total FAs) (Saini and Keum 2018). LA is found in many plant species, and is especially rich in safflower, sunflower, corn, soybean, canola, peanut, walnut, cottonseed, and other vegetable oils (Shahidi and Miraliakbari 2004; Yashodhara et al. 2009). The ω-6 FAs can also be obtained from different whole grains, wheat germ, dried fruits, and nuts (Meyer et al. 2003; Dhull and Punia 2021). In addition, meat, poultry, eggs, and all types of fish are rich sources of ω-6 FAs, especially ARA (Meyer et al. 2003). Dietary γ-linolenic acid (GLA, C18:3, ω-6) supplementation with sources such as borage oil (richest source, 17–25%), black currant oil (15–20%), and evening primrose oil (7–10%) is needed when the conditionally EFA GLA produced in the body from LA is not efficient and in the cases of diabetes, high cholesterol level and viral infected patients, high consumption of alcohol and/or saturated fat, and old age (Benatti et al. 2004).

Omega-9 FAs, commonly found in safflower, sunflower, macadamia nut, hazelnut, olive oil, cashew nut oil, almond oil, soybean oil, avocado oil, canola oil, and other seeds/nuts and vegetables (Eddey 2008), exhibit diverse bioactivities including modulating inflammatory, lipid, cardiovascular, and cancer diseases. Olive oil and macadamia oil are rich dietary sources of OA (ω-9), whereas NA (C24:l, ω-9) is mainly found in king salmon, yellow mustard seed, and flaxseed (Johnson and Bradford 2014).

6.2.3 New Sources of Omega-Fatty Acids

Many people take flaxseed or fish oils (rich in ω-3 PUFAs) as food supplements to decrease cholesterol in the blood vessels. Commercial fish oils are sold as food supplements or as ethyl esters or acylglycerols in a concentrated ω-3 PUFA form, and particularly, krill oil contains EPA and DHA in both triacylglycerol and phospholipid forms. Fungal and single-cell oils are also available in the market currently (Shahidi and Ambigaipalan 2018). However, to reach the intake amount recommended for EPA and DHA, the cold-water fish supply will not be sufficient. Recent research has been directed toward expansion based on other sustainable and alternative sources.

Alternative approaches such as the genetic modifications of the higher plants and microorganisms have been attempted (Dhull and Punia 2021). As a result, a few transgenic plants capable of producing DHA and EPA have been developed (Haslam et al. 2013; Napier and Graham 2010; Rogalski and Carrer 2011; Sayanova and Napier 2011). For example, the engineered species of *Camelina saliva* was found to successfully convert OA, LA, and ALA mainly into EPA and DHA (Petrie et al. 2014). Microorganisms and their fermentation production of dietary ω-3 fatty acid techniques, genetic engineering of plant lipids, and microalgal biotechnology and production platforms have been available for a while (Yongmanitchai and Ward 1989; Ursin 2003; Adarme-Vega et al. 2012; Adarme-Vega et al. 2014). Biotechnology can provide solutions for animals and plants to efficiently convert highly abundant ω-6 PUFA to ω-3 PUFAs (Kang 2011). Fortified foods are the foods enriched with ω-3 PUFAs by (1) modification of the animal diet using ω-3 PUFA-rich products; (2) oils obtained from transgenic microbial and plant sources; and (3) fortification of compounding foods, such as infant formula, and so on, with plant or microalgal oils (Arbex et al. 2015). For example, soybean oil can be enriched by SDA to ensure efficient conversion to EPA, which provides a biological/cost effective approach to obtain ω-3 FA from a sustainable plant source (Deckelbaum and Torrejon 2012), and genetically engineered cyanobacterium *Synechococcus elongatus* PCC 7942 with improved ω-3 PUFA producing pathways is very productive (Santos-Merino et al. 2018). Yeast *Yarrowia lipolytica* engineered by E. I. DuPont Company has been used for commercial production of EPA (Xie et al. 2015), and gene clusters from *Shewanella baltica* MACl expressed in *Lactococcus lactis* subsp. *Cremoris* MG1363 have yielded high percentages of EPA and DHA in the final products (Amiri-Jami et al. 2014).

However, several issues regarding the acceptance of implementing these genetically modified crops for FA production have been raised (Zarate et al. 2016). In the meanwhile, there is an increasing concern of over-exploitation of wild sources; therefore, scientists are now paying more attention to locating valuable sources from conventional cultural foods that can provide large quantity of useful ω-FAs. Aquaculture has been considered as the most reliable way to supply global consumptions of ω-3 PUFAs, and one major advantage of farmed fish is better controlled contaminants. However, there are some environmental impacts of aquaculture, as aquaculture is still heavily dependent on marine-derived foods (Glencross 2009), and fishmeal and fish oil are the two basic ingredients for aquafeeds (Almeida et al. 2014). To both meet the current demand for ω-3 FAs and reduce our dependance on wild fishes, the aquaculture industry has been increasingly choosing vegetable resources as alternative foods for farmed fishes (Almeida et al. 2014). There are a limited number of plant oilseeds that are conventionally recognized as rich sources of PUFAs, so it's urgent that more cost-effective, productive, and safe PUFA sources are to be identified to satisfy the demands of ω-FA supply.

Solanum and asparagus oils have been reported to be rich sources of ω-FAs but have not yet been used as part of the main materials for industrial ω-FA production. The wild and cultivated plants from the genus *Solanum*, such as potatoes, tomatoes, and eggplants, are among the most consumed vegetables globally, and they have been found to contain various FAs. In 1972, Brieskorn et al. reported the isolation of 12 saturated, 6 unsaturated (1 DB) and 1 unsaturated (2 DBs) mono-carboxylic acids, 8 saturated and 3 unsaturated (1 DB) dicarboxylic acids, 11 hydroxycarboxylic acids, and 3 unsaturated ω-monohydroxy acids from the suberin-rich cork of potato peels (Brieskorn and Binnemann 1972). Later, both *Asparagus adscendens* (5.9% oil content) and *Solanum indicum* (7.1% oil content) were reported to have high percentages of unsaturated acids (83.7% and 80.3%, respectively) in the respective total oils (Ahmad et al. 1978). In addition, ω-3 PUFA DHA could be produced successfully from cull potatoes using an algae culture process (Chi et al. 2007). Despite the many publications about the impact of dietary ω-FAs on human health, the solanum and asparagus oils as good sources of ω-FAs are less recognized, especially the possible applications of agro-industrial byproducts from these plants. In this chapter, published information (up to May 2023) of ω-FAs from solanum and asparagus oils will be introduced and summarized.

6.3 OMEGA FATTY ACIDS FROM SOLANUM AND ASPARAGUS OILS

6.3.1 Overview

6.3.1.1 Solanum Fatty Acids

The genus *Solanum* contains 1,500–2,000 species (Jarret et al. 2016). *Solanum macranthum* Dunal (Solanaceae) oil has been used in traditional medicine for multiple purposes, and its leaves and fruits contain 49.1% and 11.9% FAs, respectively (Essien et al. 2012). Like ω-3 PUFA, tomatoes reduce platelet aggregation (McEwen 2014). In addition to the anti-oxidative effect of lycopene in tomatoes (Shukla et al. 2010), evidence has shown that tomato-related products can reduce the risk of cardiovascular diseases. One of the cardiovascular risk factors is the low level of high-density lipoprotein cholesterol (HDL-C), and raw tomato intake has significantly increased HDL-C levels in overweight women (Cuevas-Ramos et al. 2013). The seeds from *Solanum* genus plants have been considered as a rich source of FAs. For example, tomato seeds generally yield 20–23% oils (Giuffre et al. 2015), eggplant *Solanum melongena* seeds have 23.7% (SD 2.1) oil content (Jarret et al. 2016), and main FA constituents were found to be myristic, palmitic (PA), stearic and arachidic acids, OA, LA, and ALA (Kashimoto and Noda 1958). Overall, *Solanaceae* pulp contains more polar lipids and less neutral lipid than do lipid extracts from their seeds (Sanchez and Cattaneo 1987). Some wild

Solanum species contain higher FAs. For example, *Solanum anguivi* Lam. has the highest FA content among several African *Solanum* species studied by Halinski et al. (2019). Dhellot et al. (2006) and Nzikou et al. (2007) reported that *Solanum nigrum* L. (synonym-*S. americanum* Mill.) seed oil yields were between 34.5–37.5% and 37.1–38.8%, respectively. Dhellot et al. stated that the green oil extracted by Soxhlet, Bligh and Dyer, and Folch methods contained linoleic (67.9%), and palmitic, stearic and oleic FAs (18: 2, ω-6 > 18: 1, ω-9 > 16: 0 > 18: 0).

The FAs in the triglyceride form primarily include capric, linoleic, lauric, oleic, palmitic, and stearic acids. For example, *Solanum melongena* variety Baladi has the following triglyceride composition: OLL 18, LLL 12.8, PLL 12.2, PLO 8.6, and OLO 6.3%; while variety Romy contains PLL 5.1, OLL 5.1, LLL 5.9, POL 9.5, OLL 9.1 and LOL 10.5% (Osman et al. 1968), where L, O, P, and S stand for linoleoyl, oleoyl, palmitoyl, and stearoyl fatty acyl structures. Representative FA composition% of selected plants/fruits/seeds are compared in Table 6.1. It's obvious that total PUFAs in the popular *Solanum* plants, that is, tomato seed oil, eggplant fruits, and potato tubers, do not show significant difference to those in other family ω-FA rich plants, such as walnut, flaxseed, corn, grapeseed, safflower and pumpkin seed oils, but instead, have even higher PUFA contents compared to almond oil.

6.3.1.1.1 Tomato fatty acids

Elevated ALA level in tomato extra-plastidial membranes was reported to improve plant tolerance to two insects, *Heliothis peltigera* and *Spodoptera littoralis* (Zhang et al. 2019). The building blocks of the tomato cutin biopolymer mainly include ω-hydroxy FAs (Hauff et al. 2010). In addition to high contents of lycopene and β-carotene in tomato products (Li and Wei 2003), tomato seed oil contains PA (C16:0), OA (C18:1), LA (C18:2) and ALA (C18:3) (a total of 21.79% MUFAs and 53.70% PUFAs) (Yang et al. 2018). Besides other antimicrobial and antioxidant constituents of tomato processing byproducts in agro-industry, seeds are the major tomato waste component, from which oil content (13.3–19.3 %) can be isolated for nutritive or other industrial purposes (Szabo et al. 2019; Botineştean et al. 2015).

When the ω-3 FA desaturases (FAD3 and FAD7 that catalyze the reaction from LA to ALA) were overexpressed, along with more tolerance to chilling, the tomato *Solanum lycopersicum* volatile profile was changed with an increase to the 18:3/18:2 ratio in leaves/fruits (Dominguez et al. 2010). Yu et al. stated that overexpression of endoplasmic reticulum ω-3 FA desaturase gene LeFAD3 could lead to increased level of 18:3 FA and improve chilling tolerance in tomatoes (Yu et al. 2009). According to Kolotilin et al., expressing yeast SAMdc gene confers broad changes in tomato gene expressions, including FA composition, and notably increased ω-3 FAs at the expense of other lipids (Kolotilin et al. 2011). Tomato chloroplast ω-3 FA desaturase gene (LeFAD7, induced by chilling stress) silence has reduced the amount of ALA and enhanced the tomato high-temperature tolerance (Liu et al. 2010). In case of drought, the LeFAD7 gene protects tomato plants – ALA decreases less under drought stress (Huang 2010).

6.3.1.1.2 Eggplant fatty acids

Eggplants were reported to contain many SFAs and UFAs (Hanifah et al. 2018). Tropical green leafy vegetable *Solanum macrocarpon* L. (gboma eggplant) was found to contain rich ALA, which accounted for 43–45% of the total FAs, along with plentiful essential amino acids (2–3% of the dry leaf weight) (Halinski et al. 2015). The nutrition of *S. melongena* var. *insanum* has been highlighted by its PUFA (120 ± 1.36 mg OA per 100 g fresh fruit) and MUFA (181 ± 2.77 mg LA and 57.2 ± 1.21 mg ALA per 100 g fresh fruit) contents (Nadeeshani et al. 2021). The most abundant FA in eggplants (*Solanum melongena* L., etc.) was found to be LA (Ayaz et al. 2015; Deineka and Deineka 2004). In addition to C18 PUFA, leaf cuticular waxes of the gboma eggplant (*S. macrocarpon* L.) cultivars also contain a lot of sterols (19, 32% of total waxes), but the hydrocarbon content was 6–13 times lower than those alkanes in *S. melongena* L. (Halinski et al. 2009; Halinski et al. 2012, 2015).

TABLE 6.1
Representative Fatty Acid Composition% of Selected Plants

Fatty Acid	C16:0	C18:0	C18:1	C18:2	C18:3	C20:0	Total SFA	Total MUFA	Total PUFA	Ref.
Tomato seed oil	17.14	5.21	21.79	53.7		2.13	24.48	21.79	53.7	Xiao et al. 2000
Eggplant fruits	16.60–23.40	11.39–14.40	3.04–11.71	39.14–53.81	7.55–14.59	1.47–2.55	31.44–36.92	3.04–11.71	53.66–62.71	Ayaz et al. 2015
Potato tubers 2 sp.	19/19.6	5.3/3.9	1.4/1.7	47.1/47.5	21.1/21.5	2.2/1.8	29.8/28.8	1.9/2.1	68.3/69.1	Dobson et al. 2004
Asparagus roots (wild)	22.82	8.21	19.3	26.6	5.3	3.7	41.1	26.3	32.7	Adouni et al. 2022
Asparagus rhizomes (wild)	24.1	8.57	20.3	28	5.5	3.99	42.6	23.3	34.4	Adouni et al. 2022
Flaxseed oil	7.31	5.04	23.29	13.86	49.17	0.13	12.71	23.54	63.2	Li and Ding 2005
Walnut oil	5.96	2.42	18.24	58.59	10.3		8.38	18.24	68.89	Xu et al. 2016
Corn oil	13.38	2.01	31.34	51.64	1.18		15.39	31.34	52.82	Yang et al. 2018
Grape seed oil	7.27	3.42	16.69	71.48	0.32	0.16	11.07	17.07	71.8	Yang et al. 2018
Almond oil	4.59	1.18	66.97	26.31	0.27		5.77	67.65	26.58	Yang et al. 2018
Safflower oil	6.52	2.02	11.4	77.93	0.19	0.43	9.79	11.65	78.12	Yang et al. 2018
Pumpkin seed oil	15.53	3.2	28.25	52.92		0.37	18.84	28.25	52.92	Yang et al. 2018

Zhao et al. reported that eggplant calyx contained higher levels of cytotoxic FA ketodiene 9-oxo-10(E),12(E)-octadecadienoic acid than the edible part of the eggplants, and this FA induced apoptosis in human ovarian cancer cells via mitochondrial regulations (Zhao et al. 2014; Zhao et al. 2015).

Innovative food technologies have been applied to eggplants: Yeast Δ-9 desaturase gene expression in eggplants can increase 16:1, 18:1, and 16:3 FA levels, and cis-Δ9 16:1 was found to directly enhance the plant resistance to a fungal plant pathogen *Verticillium dahliae* (Xing and Chin 2000).

6.3.1.1.3 Potato fatty acids

In addition to the natural antioxidants in potato peel polar compound extracts (Franco et al. 2016), a few FAs have been identified as bioactive components in potatoes. Potato leaf FA compositions may influence the plant growth, tuber taste, and even its defense mechanisms against insects (Clements et al. 2022). Functional ω-FAs are involved in establishing suberin (a cell wall lipophilic macro polyester in the periderm cork cells) structure and functions including resistance to greening in *Solanum tuberosum* tuber periderm (Tanios et al. 2020; Serra et al. 2009a). Periderm lipids have been intensively investigated for their prevention of dehydration and responses to various stresses. Molecular adaptation to life stresses is a complex process in plants, mainly based on the gene transcriptional modifications and protein-protein interaction alterations, including adaptations to low temperatures to promote high levels of constitutive expression of chloroplast ω-3 FA desaturase FAD7 (Nosenko et al. 2016). Wound-induced suberin formation is characterized by the synthesis of long chain FAs, 1-alkanols, ω-hydroxy FAs, and α,ω-dioic acids (Woolfson et al. 2018). Even under storage, FAs, α,ω-diacids and some aromatic compounds were found to be more abundant in red-skinned potato *Solanum tuberosum* L. cv. Asterix suberin (Jarvinen et al. 2011). The abundant long-chain FAs in lipophilic extracts of *S. tuberosum* L. (the smooth-skinned Yukon Gold potato) suberin-enriched wound-healing tissues suggest accelerated suberization, and 2-hydroxysebacic acid (10:0), tetradecanoic acid (14:0), heneicosanoic acid (21:0), tetracosanoic acid (24:0), and pentacosanic acid (25:0) are considered as important potential markers of wound healing (Dastmalchi et al. 2015). Kolattukudy et al. reported that potato tuber skin (suberin) contained good amounts of C20-C28 FAs, ω-hydroxy acids and fatty alcohols (Kolattukudy and Agrawal 1974). Newly formed FAs in potato (*S. tuberosum*) tubers are subjected to one of the two main metabolic pathways during wound-induced suberization: (1) desaturation and then oxidation to form the 18:1 ω-hydroxy and dioic acids; (2) elongation to form very long chain FAs (C20 to C28), associated with reduction to form 1-alkanols, decarboxylation to form n-alkanes, or small portion of hydroxylation (Yang and Bernards 2006).

In *S. tuberosum* L., more than 65% aliphatics are ω-hydroxy FAs and α,ω-dioic acids (Bjelica et al. 2016). In different species and various genotypes of the same species potatoes, FA contents and compositions vary (Dobson et al. 2004, 2010). For example, the total FA content in *Solanum phureja* (data averaged across all storage dates and genotypes) was 37%, which was higher than that in *S. tuberosum*, no matter whether data were expressed on a dry weight (4073 vs 3115 μg/g) or fresh (885 vs 648 μg/g) plant basis (Dobson et al. 2004).

Agriculture technology has influenced the total amount of the FAs in potatoes. For example, compared to the wild-type *S. tuberosum* L. cv. Desiree (49% LA, 10% PA and 14% ALA), an increase in lipid content by 69% more total fat in the transgenic potato tubers modified by the 14-3-3 gene (derived from *Cucurbita pepo*) overexpression was reported (Prescha et al. 2001).

6.3.1.2 Asparagus Fatty Acids

The genus *Asparagus* belongs to the family Asparagaceae, has around 300 species, and arose from Siberia to southern Africa. *Asparagus* has been identified as an alternate source of oil, ω-FAs, dietary fiber, and so on (Peters 1902; Scora et al. 1986; Neue Ölquellen 1916; Adouni et al. 2022), and the asparagus seed oil was reported to be like corn oil in composition and has at least 2% of the

C20 acid series (Hopkins and Chisholm 1957). According to Table 6.1, wild asparagus roots and rhizomes contain higher PUFAs than almond oil does. *Asparagus scandens* oil contains unusually high quantities of LA (63.8%, Jamal et al. 1986). The seed oil (15.3%) of *Asparagus officinalis* Linn. contains stearic (3.59%), palmitic (11.52%), arachidic (43.47%), and behenic (5.78%) acids, OA (22.16%), LA (11.34%), and ALA (2.14%) (Prasad and Nigam 1982). In comparison to that (0.33 g/100 g) in the cultivated *A. officinalis* L., the total lipid content (0.90 g/100 g) was found to be higher in the wild *Asparagus acutifolius* L., while the two essential LA (ω-6) and ALA (ω-3) and the saturated PA accounted for ~90% of the total FAs in both species analyzed (Ferrara et al. 2011). Morales et al. reported that the wild asparagus *A. acutifolius* contained 42.29% LA (Morales et al. 2012). LA and ALA metabolism has been reported to be quite active during green asparagus growth (Liu et al. 2022).

Asparagus racemosus L. root extract has antibacterial activity (Beldal et al. 2017), and it has been used widely in traditional medicine. Its extract contains some bioactive metabolites, including FAs, isoflavones, sapogenins, racemosols, mucilage, shatavarins, asparosides, glycosides, fructo-oligosaccharides, and polysaccharides. It promotes fertility, helps modulate stresses and hormones, and has been used in treating stomach ulcers, kidney, and Alzheimer's diseases (Kohli et al. 2023). Its seed oil (16.7% yield) contains ~ 67% PUFAs (Parveen et al. 2018). Unusual cyclopropenoid FAs, 7-(2-octacyclopropen-1-yl)heptanoic acid (3.2%) and 8-(2-octacyclopropen-1-yl)octanoic acid (2.8%) were identified in *A. racemosus* seed oil (Katagi and Hosamani 2013).

6.3.2 ISOLATION, STRUCTURE IDENTIFICATION, AND ANALYSIS OF REPRESENTATIVE OMEGA FATTY ACIDS

6.3.2.1 General Information

Solvent (Hanifah et al. 2018; Brieskorn and Binnemann 1972) or supercritical CO_2 (Yuan and Ge 2009) extraction can be applied to extract the total oils out of crude materials. The isolation of various FAs could be achieved using column chromatography (Brieskorn and Binnemann 1972).

FA samples can be analyzed by thin-layer chromatography (TLC), followed by quantification using an Iatroscan TLC–flame ionization detection system (reported by Rao et al. 1984), and high-high-performance liquid chromatography (HPLC) and gas chromatography (GC) are more frequently used recently. The chemical/biochemical methods used to investigate the FA composition in plants include chemolysis, pyrolysis, and enzymatic degradation. In chemolysis, the lipids or waxes are extracted by organic solvents before depolymerizing. Pyrolysis-GC can be performed in presence of tetramethylammonium hydroxide (for hydrolysis), followed by (trans)methylation of hydroxyl/carboxyl groups (Santos Bento et al. 2001). Cutins/cuticles/suberin can be chemolysed first using base, followed by derivatization (Topolewska et al. 2015), and often used reagents include KOH, tetramethylammonium hydroxide (TMAH, Del Rio and Hatcher 1998), BH_3, and so on. For example, after alkaline hydrolysis, the following acid salts could be detected in potato peel suberin: monobasic acids (C10, C12-C31), α,ω-dibasic acids (C15-C29), ω-hydroxymonobasic acids (C16-C28, C30), 10,16-dihydroxyhexadecanoic acid, 8,9-dihydroxyheptadecane-1,17-dioic acid; the major compounds were found to be 9-octadecen-1,18-dioic acid (39%) and 18-hydroxy-9-octadecenoic acid (21%) (Briekorn and Binemann 1974). The most common FA derivatives used for GC analysis are their methylesters synthesized by boron trifluoride in methanol (BF_3/MeOH, Lough 1964), diazomethane, trimethylsilyldiazomethane (TMSD, safer) or transesterification. N,O-bis(trimethylsilyl) trifluoroacetamide (BSTFA) and N-(tert-butyldimethylsilyl)-N-methyltrifluoroacetamide (MTBSTFA, allowing more specific mass fragmentation) are common silylation reagents to treat suberin and wax FAs (Li-Beisson et al. 2013). HCl or H_2SO_4 in methanol solution, or acetyl chloride (less frequently used) can esterify and transesterify FAs in common lipids (Topolewska et al. 2015). Topolewska et al. discussed on the FA derivatization yields and compositions using four different methods (differing in reaction mechanisms, concentrations of reagents, optimal reaction time, and possible interferences formed during derivatizations) and concluded that the most useful method in plant FA analysis by GC was based on derivatization using TMSD. Jarvinen et al. compared the cutinase CcCut1 hydrolysis and NaOMe methanolysis of potato *Solanum tuberosum* var. Nikola peel suberin and found that methanolysis procedure released more $CHCl_3$-soluble compounds, while enzyme hydrolysis generated higher percentages of aliphatic monomers. Monomers produced from both reactions were mainly α,ω-dioic acids and ω-hydroxy acids, at 40.0 and 32.7% for methanolysis, and 64.6 and 8.2% for cutinase hydrolysis. Obviously, cutinase CcCut1 has shown higher activity toward α,ω-dioic acid ester bond (Jarvinen et al. 2009).

FA chemical structures can be analyzed and identified by chromatographic and spectral methods, using GC, HPLC, gas chromatography-mass spectrometry (GC-MS), HPLC-DAD-ESI/MS (Adouni et al. 2022), UV, MS, IR, NMR, and so on, often assisted by chemical derivatization (Topolewska et al. 2015) and/or standard comparison. With available standards, known FAs could be identified readily by TLC, IR, HPLC, GC, and combined liquid chromatography-mass spectrometry (LC-MS) and GC-MS techniques.

6.3.2.2 Research Examples

Example 1

Analysis of fatty acid composition in tomato seed oil by GC-MS (Botineştean et al. 2012; Botineştean et al. 2015).

Solanum lycopersicum syn. *Lycopersicon lycopersicum* is a common vegetable available in conventional grocery stores and processed by manufactories to obtain food products such as puree, sauce, paste, ketchup, juices, and tomato powder. The resulting wastes contain tomato skin and fibrous solids, seeds, and cull tomatoes, which have been considered an ecological/technical/environmental problem. Therefore, it is very important to evaluate the major components (e.g., ~2.1% seeds) in tomato wastes for beneficial use in other areas. It was reported that tomato seeds contain ~20% vegetable oils, and it's very important to find out the degrees of unsaturation of the FAs in the mixtures. Botineştean et al. successfully analyzed the composition of FAs in *Lycopersicon lycopersicum* seeds using GC-MS and found that the major compound in this tomato seed oil was LA (48.22%), followed by PA (17.18%) and OA (9.20%), and so on.

Seed oil extraction and fatty acid methylation: In the work by Botineştean et al., *Solanum lycopersicum* seed oil was obtained via cold-pressed extraction or solvent extraction. To bring FAs into a vaporous phase for GC test, the oil samples were esterified to produce fatty acid methyl esters (FAMEs): tomato seed oil 100 mg or its 5 mL hexane extraction solution was treated with 5 mL BF_3/MeOH and refluxed for 2 minutes on a water bath before the addition of 5 mL hexane. After refluxing for another 1 minute, 15 mL saturated $NaCl/H_2O$ solution was added in, under vigorous stirring. Then the organic layer was separated and dried by anhydrous $CaCl_2$.

Composition Analysis of FAMEs: A Hewlett Packard HP 6890 Series gas chromatography coupled with a Hewlett Packard 5973 mass spectroscopy detector (GC-MS) system was used for the FAME analysis, connected with a 30-meter–long HP-5 capillary GC column (0.25 mm i.d., 0.25 μm film thickness). Helium was used as the carrier gas and the temperature increase was set at 4°C/min from 50°C to 250°C. Both the injector and detector temperatures were 280°C. EI 70 eV ionization energy was used for the MS detector, with a source temperature at 150°C, and a scan range of 50–300 amu at 1 s^{-1}. The compounds including FAs identified in the order of retention time (R_t) are listed in Table 6.2. Obviously, *S. lycopersicum* seed oil (13.3–19.3 % content) is a valuable source of EFAs LA (ω-6) and OA (ω-9). GC-MS has shown an oil composition profile with a high percentage of LA (20.8–39.9 mg/mL), followed by PA (6.3–19.3 mg/mL), OA (2.5–14.2 mg/mL), ALA (0.7–4.9 mg/mL), stearic acid (0.1–0.8 mg/mL), POA (0.03–0.5 mg/mL), arachidic (0.08–0.4 mg/mL), myristic (0.05–0.2 mg/mL), and margaric (0.02–0.11 mg/mL) acids.

TABLE 6.2

GC-MS Identification of Compounds from Derivatized Tomato Seed Oil*

Peak R_t (min)	Peak Area%	Compound Identified
6.30	0.15	Decane
6.91	0.15	Hexanal dimethyl acetal
12.98	0.42	Tetralin
23.27	0.13	Myristic acid, methyl ester
26.69	13.52	Palmitic acid, methyl ester
29.71	18.85	Oleic acid, methyl ester
29.87	54.91	Linoleic acid, methyl ester
30.18	2.94	Linolenic acid, methyl ester
30.58	1.08	Oleic acid, ethyl ester
30.73	3.54	9,12-Octadecadienoic acid, ethyl ester
32.1	1.36	9,12-Octadecadienoic acid, ethyl ester

* Modified from the originally data reported by Botineştean et al. (2012).

Example 2

Analysis of fatty acids in eggplant *Solanum melongena* L. and selected *Solanum* species seed oils (Jarret et al. 2016).

Eggplants arose from Africa and later dispersed globally. *Solanum melongena* L. is a popular non-starchy vegetable rich in dietary fiber. For future food and medical applications, it is very important to examine the FA composition in eggplant seed oil which is rich in the related agro-industry wastes. In the research by Jarret et al., all seeds used were obtained from the USDA/ARS Plant Germplasm Collection in Griffin, GA, and a total of 386 gene bank accessions were investigated. Oil standards were prepared from *S. melongena* cv. Black Beauty (Eden Bros., Asheville, NC) (Jarret et al. 2013).

Seed oil content determination by NMR: The seed oil contents of 305 gene bank accessions of eggplant *S. melongena*, five related phytochemical/medicinal active species (*S. aethiopicum* L., *S. incanum* L., *S. anguivi* Lam., *S. linnaeanum* Hepper and P.M.L. Jaeger, and *S. macrocarpon* L., 22.3–25.6% oil), and 27 additional *Solanum* species were analyzed by TD-NMR on intact seeds using the method reported by Krygsman and Barrett (2004) and Jarret et al. (2013). Jarret et al. studied the seeds drawn from single inventories without replacement, but collected the data in triplicate, and the results were averaged. Among the 305 samples, *S. melongena* seed oil contents varied from 17.2%–28.0%. Seed oil contents of other *Solanum* species varied from 11.8% (*S. capsicoides*) to 44.9% (*S. aviculare*), and the oil contents of six *S. amercanum* seed samples in this research averaged 34%.

Isolation and methylation of fatty acids: Seeds (100 mg) were ground to fine powder using a mortar and pestle in liquid nitrogen. About half the amount of each was transferred into a 16 × 100 mm test tube, and 5.0 mL n-heptane was added in and mixed well. Then the solution was treated with 500 µL 0.5 M NaOMe/MeOH for 2 hours. After the FAMEs were formed, 7 mL DI water was added into the reaction mixture, and after work-up for 45 min, an aliquot of the organic layer (~1.5 mL, containing the methyl esters) was collected for analysis.

Analysis of fatty acids by GC-MS: FAME stock solution was diluted 100-fold with heptane containing methyl nonadecanoate (C19:0, 25 µg/mL, used as internal standard), and then analyzed by a Thermo Quest Finnigan DSQII GC-MS system (Thermo Fisher, San Jose, CA, USA). FAMEs were separated by a 30-meter DB5® column (0.25 mm i.d., 0.25 µ film, helium flow rate at 1.5 mL/min). The injection was set in the splitless mode, with port temperature at 220°C. The oven was set at 60°C for 1 minute and increased to 250°C at 8°C/min, and then held for 5 minutes.

TABLE 6.3

Fatty Acid Composition (%) in Seeds of Selected Solanum Species*

Solanum Species	No. of Accessions	16:00	18:00	18:1, ω-9	18:2, ω-6	18:3, ω-6
S. aethiopicum	10	9.70 ± 0.80	4.10 ± 0.40	16.50 ± 3.60	67.80 ± 4.10	1.00 ± 0.03
S. anguivi	4	9.80 ± 0.60	3.80 ± 0.30	14.40 ± 3.10	70.50 ± 3.10	0.90 ± 0.06
S. incanum	4	9.60 ± 0.07	3.80 ± 0.20	15.20 ± 0.90	69.80 ± 1.70	0.85 ± 0.05
S. linnaeanum	3	9.40 ± 0.60	3.70 ± 0.40	15.60 ± 1.80	70.10 ± 2.30	0.98 ± 0.70
S. macrocarpon	2	8.60 ± 0.30	4.00 ± 0.40	16.20 ± 2.30	69.50 ± 3.10	1.00 ± 0.03
S. melongena	55	9.68 ± 0.60	3.90 ± 0.40	16.70 ± 2.80	68.30 ± 2.80	0.98 ± 0.10

* Modified from the original data reported by Jarret et al. (2016).

The mass spectrometer was set in the electron impact mode, scanning from m/z = 50 to 400 for data acquisition. Peak analysis was based on a FAME mixture (GLC-10) from Matreya LLC (Pleasant Gap, PA, USA). Principal FAs in seeds of *S. melongena* accessions and its five related species were found to be C18:2n-6 > C18:1n-9 > C16:0 > C18:0 > C18:3n-6. Trace amount (< 0.25%) C14:0, C16:1n-7, C20:0, C22:0, C24:0 and C26:0 FAs were also detected. It's not surprising that the FA profiles and composition% in seed oils of *S. melongena* and its five related species are quite similar, as shown in Table 6.3.

FAs were also determined by HPLC for samples made from *S. melongena* (55), *S. aethiopicum* (10), *S. anguivi* (4), *S. incanum* (4) and *S. macrocarpon* (2). In all samples studied, the major FAs were found to be LA (18:2, 57–74.5%), OA (18: 1, 11.3–25.2%), and PA (16: 0, 8.4–11.2%).

Example 3

Identification/quantification of the compositional monomers from lipid polyesters in potato root and tuber periderm suberin (Company-Arumí et al. 2016).

Suberin, cutin, and associated waxes contain fatty acyl-derived lipophilic polyesters that protect plants. Suberin, composed of both aliphatic and aromatic domains, is biosynthesized naturally and in case of wounding. The aliphatic polyester is mainly made from long-chain 1-alkanols, ω-hydroxyacids, α,ω-alkanedioc and alkanoic acids. Suberin-associated waxes contain fatty acyl derivatives (Pollard et al. 2008), and typical fatty acyl chains (>18 carbons) in potato periderm include alkanes, 1-alkanols, alkanoic acids, and alkyl ferulates (Serra et al. 2009b).

Extraction of waxes and suberin: After being stored for 21 days at rt to allow post-harvest maturation, sheets of suberized layers of *Solanum tuberosum* cv. Desirée periderms and suberized roots were obtained after a cellulase/pectinase treatment (Schreiber et al. 2005). Suberin and wax monomers from roots and periderms were extracted with chloroform-methanol (1:1 v/v) followed by methanolysis. The suberin was depolymerized via transesterification by immersing the wax-free residues in BF_3/MeOH at 70°C for 16–18 hours (Franke et al. 2005). After the reaction, 10 µg dotriacontane was added in as a surrogate and the methanolysate was transferred to 2 mL saturated $NaHCO_3$ solution. The aqueous phase was then extracted twice with chloroform, then the organic layers were combined and rinsed with water, and then dried by anhydrous Na_2SO_4 powder before filtration and solvent evaporation.

Derivatization and data analysis: MTBSTFA and isooctane (20 µL) were added to the dry residues obtained from the crude standard solution, wax, and depolymerized suberin extracts, and after this derivatization step was conducted at 120°C for 60 minutes, more isooctane was added to reach the desired concentrations for GC analysis.

Fatty acyl tert-butyldimethylsilyl (tBDMS) derivatives were detected, identified, and quantified by gas chromatography coupled to ion trap-mass spectrometry (GC-IT-MS) based on the specific peak at M-57 and representative ions in the mass spectra. Quantitative data were obtained using external calibration curves constructed by plotting peak areas versus concentrations of the standards derived from MTBSTFA. A 30-meter BPX-5 capillary column (0.25 mm i.d., 0.25 μm film thickness) was used to separate the derivatives with helium as a carrier gas at 1 mL/min. The injector temperature was set at 280°C. For wax tBDMS derivatives, the oven was initially set to 120°C for 2 minutes, then 25°C/min increase up to 270°C, holding for 2 minutes, and then 2°C/min increase up to 340°C, holding for 26 minutes. For suberin tBDMS derivatives, the initial oven temperature was set to 100°C for 2 minutes, then 25°C/min increase up to 200°C, holding for 1 minute, followed by 3°C/min increase up to 235°C and then by 2°C/min up to 330°C, holding for 30 minutes. MS analysis was conducted with electron impact ionization at 70 eV and the ion source temperature at 225°C. The transfer line was set to 290°C and acquisition was conducted in full-scan mode (50–700 amu). Xcalibur 1.4 software was used to analyze all chromatographic data. The mass data were compared with reference spectra (NIST MS Search 2.0) and derivatized standards.

Composition results: The quantity of each monomer in the extracts was calculated based on the concentration of the specific monomer and normalized by the sample dry weight in μg/mg. The total amount of suberin in potato periderm (119.4 ± 1.4 μg/mg) was calculated from the sum of the monomers, containing 42.3 ± 3.1 μg/mg alkanoic acid, 49.5 ± 3.4 μg/mg α,ω-alkanedioc acid, 17.9 ± 2.4 μg/mg ω-hydroxyacid, and so on. Obviously, total suberin in roots was much less (12.5 ± 0.4 μg/mg). Suberin found in either potato root or periderm had similar percentages of ω-hydroxyacids (18.0 ± 2.3% and 14.9 ± 2.0%, respectively), but alkanoic acids in roots were much less (15.2 ± 2.4%) than that in periderm (35.4 ± 2.3%), whereas α,ω-diacid content (56.7 ± 3.5%) in roots was higher than that in periderm (41.5 ± 3.1%). The data also showed that potato root suberin contained shorter fatty acyl chains and α,ω-alkanedioic acids were more predominant than those in the periderm.

Example 4

Isolation, identification, and quantification of unsaturated fatty acids from potato peel (Wu et al. 2012b).

Isolation: Small piece peels from potatoes (2 kg) produced by Wulanchabumeng, Inner Mongolia of China, were extracted three times (2, 1, and 0.5 hr) with methanol under reflux. After solvent evaporation, the methanol extract was partitioned between ethyl acetate and water (EtOAc–H$_2$O). The EtOAc extract was then subjected to a silica gel column chromatography, and the pure EtOAc eluting fractions were collected. After solvent evaporation, the sample was run on an ODS (C-18) column using a gradient mobile phase (0–100% MeOH in H$_2$O). FAs **1** (43 mg) and **2** (98 mg) were obtained from 70% and 75% MeOH fractions, respectively.

Structure Identification: Compound structures were identified as 9,10,11-trihydroxy-12(Z), 15(Z)-octadecadienoic acid (**1**) and 9,10,11-trihydroxy-12(Z)-octadecenoic acid (**2**) by NMR/MS spectra and compared with reported spectral data.

Quantification by UPLC-DAD–MS: Freeze-dried potato peel/flesh powder (1 g) was extracted with 20 mL methanol for 20 min, assisted by ultrasonication. The filtrates were collected, and solvent was evaporated under reduced pressure. Then the extract was dissolved in 20 mL methanol, passed through a Sep-Pak cartridge (Vac C18 3 cc, WAT020805, Waters) and filtered through a 0.22 μm microfilter.

Samples were analyzed by a UPLC-DAD–QQQMS system using an Agilent ZORBAX SB-C18 RRHT column (50 × 2.1 mm, 1.8 μm) at 30°C. The eluting program for mobile phase (solvent A: 0.1% formic acid in H$_2$O; solvent B: 0.1% formic acid in acetonitrile) at 0.4 mL/min was arranged as: 0 min, 0% B; 0.4 min, 13% B; 1 min, 13% B; 2 min, 21% B; 3 min, 31% B; 3.5 min, 40% B; 4 min, 48% B; 4.01–5.5 min, 100% B. Post run time: 1.5 min. The injection volume was 2 μL. MS2 scan mode (negative) was chosen for qualitative analysis. Selective ion monitoring (SIM) mode was used with negative polarity for both **1** and **2** quantifications. The ion source parameters were set as following: capillary, 4 kV; gas temperature, 350°C; gas flow, 11 L/min; and nebulizer, 45 psi. FA **1** content in dry peer powder was found to be 0.01–0.02 mg/g, and FA **2** content was in the range of 0.02–0.03 mg/g.

Example 5

Fatty acids from *Asparagus sprengeri* (Hassan et al. 2014).

The herb *Asparagus sprengeri*, which is normally cultivated as an ornamental plant, was collected from the garden of Egypt National Research Centre. Based on HPTLC results, the plant contains L-α-phosphatidyl-DL-glycerol, L-α-phosphatidyl ethanol amine and neutral lipids.

Extraction and saponification of lipids: Around 500 g dried plant powder was extracted with petroleum ether using a Soxhlet apparatus. Then the extract was passed through fuller's earth to remove pigments and dried over anhydrous sodium sulfate before solvent evaporation in vacuo at 40°C to yield 7.8 g pale yellow extract. Around 3 g extract was saponified by 100 mL 0.5 N KOH in alcohol under refluxing. After the reaction, the solution was concentrated to ~25 mL and diluted with 100 mL cold DI H_2O, and then the unsaponifiable matters were extracted with ether (3×100 mL).

Extraction of total FAs and preparation of FAMEs: The pH value of the saponification solution was adjusted with 5% H_2SO_4 to reach pH 2, and then the total FAs were extracted with ether several times. The combined ether extract was washed with DI water and then dried over anhydrous sodium sulfate. After solvent evaporation in vacuo at ~40°C, 1.6 g dry total FAs were obtained. Around 1.5 g FAs were dissolved in 7.5 mL dry methanol, then 7.5 mL of 12% BF_3 solution was added in and refluxed on a boiling water bath for 5 minutes. After solvent evaporation, the residue was mixed with 10 mL DI water and the FAMEs were extracted with ether (3×10 mL). The combined ether solution was washed with DI H_2O, dried over anhydrous Na_2SO_4, and then filtered. Aliquots of 2 μL were injected into Hewlett Packard HP 6890 series GC system equipped with HP-70 capillary column (60 m, 320 μm diameter and 0.25 μm film thickness). Oven: 70°C for 2 min, then 6°C/min to 220°C; inlet temperature: 270°C (split 15:1); FID detector temperature: 300°C; Carrier gas: H_2 flow at 40 mL/min, N_2 flow at 40 mL/min, air flow at 45 mL/min. FAME GC data showed the existence of nine FAs in *A. sprengeri*; five are unsaturated FAs (~ 43.79%), four are saturated FAs (~56.2%), and myristic (C14: 0, 42.21%) and linoleic (C18: 2, 10.93%) acids are the main constituents, as shown in Table 6.4.

TABLE 6.4
Fatty Acids in *Asparagus sprengeri*

Fatty Acid	Me Ester R_t (min)	Composition%
Myristic acid (C14: 0)	12.21	42.21
Myristoleic acid (C14:1)	12.72	11.72
Palmitic acid (C16: 0)	16.02	9.59
Palmitoleic acid(C16:1)	16.87	9.83
Heptadecanoic acid (C17: 0)	17.98	4.4
Oleic acid (C18: 1)	19.62	1.77
Linoleic acid (C18: 2)	20.51	10.93
Linolenic acid (C18: 3)	22.34	1.01
Arachidonic acid (C20: 4)	24.32	8.53

Source: Reported by Hassan et al. 2024.

Example 6

Standardization and quantification of linoleic acid in *Solanum nigrum* Berries by HPTLC (Chakraborty et al. 2016).

The medicinal plant *Solanum nigrum* (Solanaceae) has been extensively used to treat several disorders including anasarca and heart disease. Its berries have been used as an antidiarrheal and for ophthalmopathy or hydrophobia, and seeds can treat dipsia and giddiness. As LA is the most abundant UFA in *Solanum nigrum* oil (Saleem et al. 2009), its content was selected by Chakraborty et al. to standardize and quantify this plant berry quality by high performance thin layer chromatography (HPTLC). The CAMAG HPTLC system uses WINCATS software, LINOMAT V automatic sample applicator, automatic development chamber, scanning densitometer CAMAG scanner 3, and photo documentation apparatus CAMAG reprostar-3. HPTLC is a simple and cost-effective method to standardize and identify the chemical ingredients that have light-absorptions and/or can be stained, and typical ingredients are expected to be present in a sample, although it has a limited developing distance, and lower efficiency compared to HPLC and GC.

Preparation of standard solution: To a 1 mL Eppendorf tube, ~1 mg LA standard and 1.0 mL methanol were added in and mixed well before filtration through a 0.45 μ syringe filter.

Preparation of linoleic acid calibration curve: The HPTLC plate was developed with the mobile phase n-hexane–ethyl acetate (5:4 v/v) in a twin-trough glass chamber at 25°C. The standard solution made from commercially available LA was loaded 2, 4, 6, 8, 10 μL in a band wise fashion. After the development and solvent evaporation, the plate was sprayed by sulfuric acid–anisaldehyde solution. Then the plate was kept at 110°C for 5 minutes in a hot air oven and evaluated at 366 and 540 nm. The calibration curve was obtained by plotting peak areas versus LA concentrations. The data fit well in the equation of Y=3141.508 × X+1366.840 (correlation coefficient = 0.9954), where X is the amount of LA applied, and Y is the area under the curve. The R_f value of standard LA was found to be 0.65. Instrumental precision (1.97–2.08, RSD = 1.65%) was checked by repeated scanning (n = 5) of the standard spots (2 μg LA per spot) and expressed as relative standard deviation (RSD). The repeatability (1.98–2.08, RSD = 1.95%) of the method was evaluated by analyzing 2 μg/spot of LA in 5 TLC plates and expressed as RSD. The accuracy of the analysis was assessed via recovery study at three different levels (50%, 100%, and 125% LA addition), and the percentage recoveries were calculated. The average recovery percentage was found to be 100.63%.

Quantification: In each Eppendorf tube, ~5 mg of *S. nigrum* berry methanol extract was dissolved in 1 mL methanol, assisted with ultrasonication. After filtration through a 0.45 μm syringe filter, 4, 8, and 12 μL sample solutions were applied consequently onto the HPTLC plate, and then the plate was subjected to the same chromatography described earlier. Based on the areas under curve and the R_f value, LA% in *S. nigrum* berry methanol extract was found to be 9.32% w/w. The unique image of HPTLC coupled with the digital scanning profile was found to be useful in constructing herbal chromatographic fingerprint, as shown in Figure 6.2 (modified from the figures reported in Chakraborty et al. 2016).

Example 7

Solanum americanum Miller oil extraction (Dzondo-Gadet et al. 2006).

Solanum americanum is a traditional medicine used for the treatment of type-2 diabetes. It is one of the most widespread and morphologically variable species in the *Solanum* genus. The green fruit and leaves are poisonous mainly because of the high levels of the toxic glycoalkaloids, solanine and solamargine. Other toxins include chaconine, solasonine, solanigrine, gitogenin, saponins, and the tropane alkaloids. The toxicity seems to diminish with ripening, though (Nehring et al. 1979). The seminal report by Dzondo-Gadet et al. has shown that this plant is a rich source for ω-FAs, especially the total PUFAs.

Extraction: Brazzaville (Congo) dried ripe *S. Americanum* berry seeds were ground by a coffee grinder and separated in two groups. One group was heated at 105°C for 2 hours to inactivate endogenous enzymes for pectinase (0.2%, v/v) treatment (3v/w water added to make a slurry, pH adjusted from 6.21 to 5.5 using 1N HCl, enzyme digestion at 40°C in an incubator overnight, stirred) and extraction. Pre-extraction treatment by pectinase can reduce viscosity of the slurry

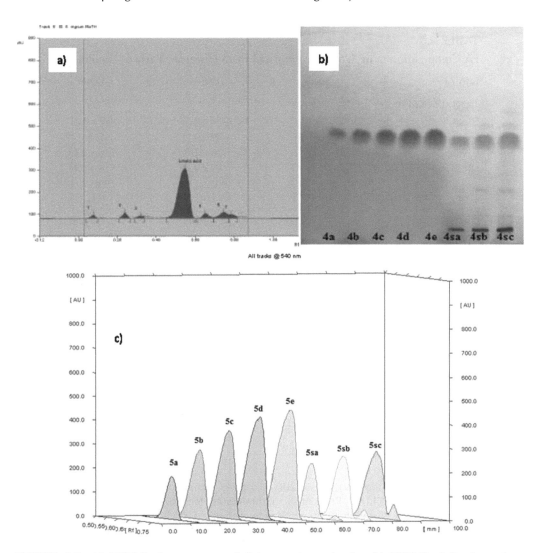

FIGURE 6.2 (a) HPTLC chromatogram of *Solanum nigrum* berries (b) HPTLC of *S. nigrum* berry methanol extract at 540 nm. Track 4a, 4b, 4c, 4d and 4e: LA standard sample loading of 2, 4, 6, 8, and 10 μL, respectively. Track 4sa, 4sb and 4sc: *S. nigrum* sample loading of 4, 8, and 12 μL, respectively (c) HPTLC 3D chromatogram of (b).

and hydrolyze the cell membrane. The supernatant oil was obtained after the mixture was centrifuged at 10,000 xg under nitrogen. Oil extraction from dry powdered seeds was obtained by a Soxhlet extractor over 5 hours using petroleum ether. The Nancy (France) leaves and roots were cleaned with DI water and crushed, and then the oil was extracted by methanol-chloroform (2:1, v/v) using the Bligh and Dyer method (Bligh and Dyer 1959).

Fatty acid composition analysis: FAMEs were made through treating the crude oil aliquot (50 mg) with 1 mL BF$_3$/MeOH (8% w/v) for 10 minutes in a shaking water bath at 90°C before the GC (PerichromTM 2000, Saulxles-Chartreux, France) analysis. The GC system was equipped with a flame ionization detector and a fused silica capillary column (25 m × 0.25 mm, 0.5 μm, BPX70 SGE Australia Pty. Ltd.). Column temperature program: 145°C for 20 minutes, 5°C/min from 145 to 210°C, and held for 15 minutes. The injection port end and the detector were set to 230°C and 260°C, respectively. The FAs were identified by comparison to standards. Extracted by solvents, mansa (*S. americanum* Miller) seeds were found to contain up to 35% weight/weight of oil, but only 2.7% in roots and 1.3% in leaves. However, the ALA, EPA, and DHA levels in leaves and roots were found to be much higher than those in the seeds, as shown in Table 6.5 (modified from

TABLE 6.5

FA Composition% in *S. americanum* Oil from Different Parts of the Plant

Fatty Acids	Seeds	Leaves	Roots
Total Oil Yielded	35	1.3	2.7
C14:0	0.15 ± 0.01	0.96 ± 0.23	0.9 ± 0.02
C16:0 (PA)	12.6 ± 0.81	6.99 ± 0.71	12.43 ± 0.79
C17:0		1.18 ± 0.8	1.05 ± 0.13
C18:0	4.4 ± 0.06	2.63 ± 0.12	5.09 ± 0.44
C20:0		1.02 ± 0.09	1.53 ± 0.56
Total SFA	17.15	12.78	21
C16:1 ω-7	0.2 ± 0.03	0.2 ± 0.07	0.59 ± 0.01
C17:1 ω-8		1.2 ± 0.1	1.1 ± 0.1
C18:1, ω-9 (OA)	14.6 ± 1.01	18.44 ± 1.16	19.74 ± 1.51
C20:1, ω-9	0.3 ± 0.01	1.4 ± 0.15	0.4 ± 0.01
C22:1, ω-11		1.21 ± 0.17	1 ± 0.22
Total MUFA	15.1	22.45	22.83
C18:2, ω-6 (LA)	66.5 ± 3.01	55.15 ± 2.21	41.7 ± 1.56
C18:3, ω-3	0.91 ± 0.18	1.41 ± 0.28	4.67 ± 0.85
C20:5, ω-3 (EPA)	0.17 ± 0.01	2.44 ± 0.24	1.42 ± 0.91
C22:6, ω-3 (DHA)	0.17 ± 0.02	3.91 ± 0.13	1.39 ± 0.41
Total PUFA	67.75	62.91	49.18

Note: Data: mean ± SEM, $n = 3$.

the data reported by Dzondo-Gadet et al. 2006). The FAs identified in *S. americanum* seed oil include LA (66.5%), OA (14.6%), PA (12.6%), and minor contents of DHA and EPA (Table 6.5). The detected PUFA/SFA of the two groups of crude seed powder oils were 3.95/4.86. It's clear that the extraction process did affect the composition of FAs in oil, as extraction after enzyme pectinase pretreatment only yielded 18% weight/weight of oil. Although the minor DHA and EPA contents were maintained in the enzyme-pretreated extraction method, they were susceptible to heat accelerated oxidation.

6.3.3 Applications

Information about USDA national collections of animal, microbial, and plant genetic resources important for food and agricultural production can be found at the www.ars-grin.gov website. Vegetable-based nutritional products are important for people's daily life globally, and a highly balanced ω-6 to ω-3 ratio in diets is essential for health. Solanum and asparagus oils can be functional food compositions (Seo 2014; Banavar et al. 2021) and new sources of ω-FAs. For example, *Solanum nigrum* seed oil has been proposed as a diversified source of diet lipids (Chakraborty et al. 2016).

Many fruit and vegetable wastes from agro-industry have been used as substrates to produce organic acids, essential oils, exogenous enzymes and single-cell proteins, bioethanol or methanol, biosorbents, biodegradable plastics, biofertilizers, biopesticides, biopreservatives, and edible mushrooms (Wadhwa et al. 2015). Since fatty acid–amino acid conjugates can function as herbivore-associated molecular patterns, the alarm signals for insects, a new generation of pest-control agents might be made (Grissett et al. 2020) from *Solanum* and asparagus oils. Other applications may include body-care products. Hair and skin protecting products can be made from asparagus (Wen and Zhou 2021; Lv et al. 2021; He 2016) and eggplants (Zhang 2018). For example, a nonirritating moisturizing skin care product containing *Asparagus schoberioides* UFAs was reported to give good moisturization and have a mild effect on skin (Lv et al. 2021).

6.3.4 PERSPECTIVES

The concentrations of phenolics and other secondary metabolites like ω-FAs in plant leaves, pulp/pomace, peels, and seeds of many fruits/vegetables such as tomatoes are often higher than those contained in their edible tissues, and these phytochemicals help treat various diseases such as cardiovascular and capillary fragility; inhibit platelet aggregation; prevent thrombosis, osteoporosis and diabetes; and relieve oxidative stress in vertebrates (Wadhwa et al. 2015). Even FA amount in leaf cuticular waxes from selected potato varieties is in the range of 3.6–285.7 ng/cm^2 (Szafranek and Synak 2006), which might have beneficial function against cancer and microbial and pathogen infections, and exert immune-modulatory effects.

Large amounts of fruit and vegetable wastes produced by the agro-industry can cause environmental contamination. Valuable components like fatty acids, oils, and other lipids; fibers; isoprenoids; proteins; saponins; antioxidants; and phyto-estrogens in the wastes have possible applications in pharmaceutical excipients, food compositions, pharmaceutical formulations, or food matrixes to generate nutritional and/or functional products (Jimenez-Moreno et al. 2020). Vegetable-processing origin wastes might be excellent raw material for oil production, and for other nutritive or industrial purposes. Industrial peel wastes from potato chips and fries can be a valuable source of beneficial fatty acids (Wu et al. 2012b). Trimmed asparagus byproducts may be further processed to produce dietary supplements for health improvement (Nopparatmaitree et al. 2022), and eggplant wastes can be used for next-step industrial applications (Silva et al. 2021).

Some plant species have been cultivated on a large scale for mass food production in the agro-industry where various seeds are left as food wastes. New possible applications of vegetable seed oils like biodiesel production have prompted investigations on the yields, FA compositions, and physicochemical characteristics of oils from tomato and eggplant seeds (Giuffre et al. 2015; Jarret et al. 2016). In addition to valuable food additives for more nutritional and functional properties, agro-industrial asparagus and potato wastes can also be used for biofuel production (Siciliano et al. 2019). Production of biodiesel from leaves of certain plants such as eggplant *Solanum melongena* L. is a possibility (Karthika 2022). According to EN 14214 European standard, vegetables with more than 30% fixed oil content in seeds are most suitable as alternative sources for biodiesel production (Sotiroudis et al. 2010). Certain *Solanum* seeds have met this requirement (Dhellot et al. 2006; Dzondo-Gadet et al. 2006; Nzikou et al. 2007; Jarret et al. 2016), and, hopefully, more wild asparagus species with higher oil content could be discovered in the future. Further, since the ability for the human body to produce ω-3 PUFAs from their precursors is quite limited, people should take in more ω-3 PUFAs from all possible food resources; however, the known sources for this group of fatty acids are very limited. Therefore, more research is needed to search for ω-fatty acids from less-recognized species or sources.

REFERENCES

Abugri, D.; McElhenney, W.H.; Willian, K.R. Fatty acid profiling in selected cultivated edible and wild medicinal mushrooms in the Southern United States. *J. Exp. Food Chem.* 2016, *2*, 1–7.

Adarme-Vega, C.T.; Lim, D.K.Y.; Timmins, M.; Vernen, F.; Li, Y.; Schenk, P.M. Microalgal biofactories: a promising approach towards sustainable omega-3 fatty acid production. *Microb. Cell Factories* 2012, *11*, 96.

Adarme-Vega, T.C.; Thomas-Hall, S.R.; Schenk, P.M. Towards sustainable sources for omega-3 fatty acids production. *Curr. Opin. Biotechnol.* 2014, *26*, 14–18.

Adouni, K.; Julio, A.; Santos-Buelga, C.; Gonzalez-Paramas, A.M.; Filipe, P.; Rijo, P.; Costa Lima, S.A.; Reis, S.; Fernandes, A.; Ferreira, I.C.F.R.; et al. Roots and rhizomes of wild Asparagus: Nutritional composition, bioactivity and nanoencapsulation of the most potent extract. *Food Biosci.* 2022, *45*, 101334.

Ahmad, M.S.; Ahmad, M.U.; Ansari, A.A.; Osman, S.M. Studies on herbaceous seed oils. V. *Fette, Seifen, Anstrichmittel* 1978, *80*, 353–354.

Alasalvar, C.; Shahidi, F.; Quantick, P. Food and health applications of marine nutraceuticals: a review; Alasalvar, C.; Taylor T., Eds.; *Seafoods: Quality, Technology and Nutraceutical Applications*; Springer: Heidelberg, Ger., 2002, pp. 175–204.

Allemann, M.N.; Shulse, C.; Allen, E.E. Linkage of marine bacterial polyunsaturated fatty acid and long-chain hydrocarbon biosynthesis. *Front. Microbiol.* 2019, *10*, 702.

Almeida, I.; Fernandes, T.J.R.; Guimarães, B.M.R.; Oliveira, M.B.P.P. Omega-3 dietary intake: Review on supplementation, health benefits and resources sustainability; Khan, W., Ed.; Omega-3 Fatty Acids: Chemistry, Dietary Sources and Health Effects; Nova Science Publishers, Inc.: Canada, United Kingdom, Germany, India. 2014, pp. 191–211.

Amiri-Jami, M.; LaPointe, G.; Griffiths, M.W. Engineering of EPA/DHA omega-3 fatty acid production by *Lactococcus lactis* subsp. *cremoris* MG1363. *Appl. Microbiol. Biotechnol.* 2014, *98*, 3071–3080.

Arbex, A.K.; Bizarro, V.R.; Santos, J.C.S.; Araujo, L.M.M.; de Jesus, A.L.C.; Fernandes, M.S.A.; Salles, M.M.; Rocha, D.R.T.W.; Marcadenti, A. The impact of the essential fatty acids (EFA) in human health. *Open J. Endocr. Metab. Dis.* 2015, *5*, 98–104.

Ayaz, F.A.; Colak, N.; Topuz, M.; Tarkowski, P.; Jaworek, P.; Seiler, G.; Inceer, H. Comparison of nutrient content in fruit of commercial cultivars of eggplant (*Solanum melongena* L.). *Pol. J. Food Nutr. Sci.* 2015, *65*, 251–260.

Balk, E.M.; Lichtenstein, A.H. Omega-3 fatty acids and cardiovascular disease: summary of the 2016 agency of healthcare research and quality evidence review. *Nutrients* 2017, *9*, 865/1–865/13.

Balk, E.M.; Lichtenstein, A.H.; Chung, M.; Kupelnick, B.; Chew, P.; Lau, J. Effects of omega-3 fatty acids on serum markers of cardiovascular disease risk: A systematic review. *Atherosclerosis* (Amsterdam, Netherlands) 2006, *189*, 19–30.

Banavar, G.S.; Messier, H.; Basseda, R.; Heald, D.; Perlina, A. Methods for and compositions for determining food item recommendations. United States, US20210398449 A1 2021-12-23.

Bays, H. Clinical overview of Omacor: a concentrated formulation of omega-3 polyunsaturated fatty acids. *Am. J. Cardiol.* 2006, *98*, 71i–76i.

Beldal, B.S.; Meenakshi, S.C.; Londonkar, R.L. Preliminary phytochemical screening, antibacterial activity and GC-MS analysis of *Asparagus racemosus* root extract. *Int. Res. J. Pharm.* 2017, *8*, 91–94.

Benatti, P.; Peluso, G.; Nicolai, R.; Calvani, M. Polyunsaturated fatty acids: Biochemical, nutritional and epigenetic properties. *J. Am. Coll. Nutr.* 2004, *23*, 281–302.

Botineştean, C.; Gruia, A.T.; Jianu, I. Utilization of seeds from tomato processing wastes as raw material for oil production. *J. Mater. Cycles Waste Manag.* 2015, *17*, 118–124.

Botineştean, C.; Hădărugă, N.G.; Hădărugă, D.I.; Jianu, I. Fatty acids composition by gas chromatographymass spectrometry (GC-MS) and most important physicalchemicals parameters of tomato seed oil. *J. Agroaliment. Processes Technol.* 2012, *18*, 89–94.

Bjelica, A.; Haggitt, M.L.; Woolfson, K.N.; Lee, D.P.N.; Makhzoum, A.B.; Bernards, M.A. Fatty acid ω-hydroxylases from Solanum tuberosum. *Plant Cell Rep.* 2016, *35*, 2435–2448.

Bligh, E.G.; Dyer, W.J. A rapid method of total lipid extraction and purification. *Can. J. Biochem. Physiol.* 1959, *37*, 911–917.

Bloch, M.H.; Hannestad, J. Omega-3 fatty acids for the treatment of depression: systematic review and metaanalysis. *Mol. Psychiatry* 2012, *17*, 1272–1282.

Bloch, M.H.; Qawasmi, A. Omega-3 fatty acid supplementation for the treatment of children with attentiondeficit/hyperactivity disorder symptomatology: systematic review and meta-analysis. *J. Am. Acad. Child Adolesc. Psychiatry* 2011, *50*, 991–1000.

Bousquet, M.; Calon, F.; Cicchetti, F. Impact of omega-3 fatty acids in Parkinson's disease. *Ageing Res. Rev.* 2011, *10*, 453–463.

Brieskorn, C.H.; Binnemann, P.H. Chemical composition of potato cork. *Tetrahedron Lett.* 1972, *12*, 1127–1130.

Briekorn, C.H.; Binemann, P.H. Chemical composition of potato peel suberin. *Zeitschrift fuer Lebensmittel-Untersuchung und -Forschung* 1974, *154*, 213–222.

Burckhardt, M.; Herke, M.; Wustmann, T.; Watzke, S.; Langer, G.; Fink, A. Omega-3 fatty acids for the treatment of dementia. *Cochrane Database Syst. Rev.* 2016, *4*, CD009002.

Cabo, J.; Alonso, R.; Mata, P. Omega-3 fatty acids and blood pressure. *Br. J. Nutr.* 2012, *107*, S195–S200.

Calder, P.C. Immunomodulation by omega-3 fatty acids. *Prostaglandins Leukot. Essent. Fatty Acids* 2007, *77*, 327–335.

Calder, P.C.; Yaqoob, P. Omega-3 polyunsaturated fatty acids and human health outcomes. *Biofactors* 2009a, *35*, 266–272.

Calder, P.C.; Yaqoob, P. Understanding omega-3 polyunsaturated fatty acids. *Postgrad. Med.* 2009b, *121*, 148–157.

Chakraborty, A.; Bhattacharjee, A.; Dasgupta, P.; Manna, D.; Oh, W.C.; Mukhopadhyay, G. Simple method for standardization and quantification of linoleic acid in *Solanum nigrum* berries by HPTLC. *J. Chromatogr. Sep. Tech.* 2016, *7*, 342/1–342/4.

Chi, Z.; Hu, B.; Liu, Y.; Frear, C.; Wen, Z.; Chen, S. Production of ω–3 polyunsaturated fatty acids from cull potato using an algae culture process. *Appl. Biochem. Biotechnol.* 2007, *137–140*, 805–815.

Cholewski, M.; Tomczykowa, M.; Tomczyk, M. A comprehensive review of chemistry, sources and bioavailability of omega-3 fatty acids. *Nutrients* 2018, *10*, 1662/1–1662/33.

Chou, S.-C.; Huang, T.-J.; Lin, E.-H.; Huang, C.-H.; and Chou, C.-H. Antihepatitis B virus constituents of *Solanum erianthum*. *Nat. Prod. Commun.* 2012, *7*, 153–156.

Christensen, L.P. Galactolipids as potential health promoting compounds in vegetable foods. *Recent Pat. Food Nutr. Agric.* 2009, *1*, 50–58.

Clements, J.; Bradford, B.Z.; Lipke, M.; Jansky, S.; Olson, J.; Groves, R.L. Difference in foliar fatty acid composition in potato cultivars over a growing season may influence the host location preference of *Leptinotarsa decemlineata*. *Am. J. Potato Res.* 2022, *99*, 40–47.

Company-Arumí, D.; Figueras, M.; Salvadó, V.; Molinas, M.; Serra, O.; Anticó, E. The identification and quantification of suberin monomers of root and tuber periderm from potato (*Solanum tuberosum*) as fatty acyl tert-butyldimethylsilyl derivatives. *Phytochem. Anal.* 2016, *27*, 326–335.

Costantini, L.; Molinari, R.; Farinon, B.; Merendino, N. Impact of omega-3 fatty acids on the gut microbiota. *Int. J. Mol. Sci.* 2017, *18*, 2645/1–2645/18.

Cuevas-Ramos, D.; Almeda-Valdes, P.; Chavez-Manzanera, E.; Meza-Arana, C.E.; Brito-Cordova, G.; Mehta, R.; Perez-Mendez, O.; Gomez-Perez, F.J. Effect of tomato consumption on high-density lipoprotein cholesterol level: a randomized, singleblinded, controlled clinical trial. *Diabetes Metab. Syndr. Obes.: Targets Ther.* 2013, *6*, 263–273.

Dastmalchi, K.; Kallash, L.R.; Phan, V.C.; Huang, W.; Serra, O.; Stark, R. Investigating the potato's defensive shield: Metabolites profiling and solid-state NMR compositional analysis of suberin-enriched wound-healing tissues. Abstracts of Papers, *250th ACS National Meeting & Exposition*, Boston, MA, United States, August 16–20, 2015, AGFD–321.

Deckelbaum, R.J.; Torrejon, C. The omega-3 fatty acid nutritional landscape: health benefits and sources. *J. Nutr.* 2012, *142*, 587S–591S.

Deineka, V.I.; Deineka, L.A. Triglyceride types of seed oils. I. Certain cultivated plants of the Solanaceae family. *Khimiya Prirodnykh Soedinenii* 2004, *40*, 184–185.

Del Rio, J.C.; Hatcher, P.G. Analysis of aliphatic biopolymers using thermochemolysis with tetramethylammonium hydroxide (TMAH) and gas chromatography-mass spectrometry. *Org. Geochem.* 1998, *29*, 1441–1451.

Dhellot, J.R.; Matouba, E.; Maloumbi, M.G.; Nzikou, J.M.; Dzondo, M.G.; et al. Extraction and nutritional properties of *Solanum nigrum* L. seed oil. *Afric. J. Biotech.* 2006, *5*, 987–991.

Dhopeshwarkar, G.A.; Mead, J.R. Role of oleic acid in the metabolism of essential fatty acids. *JAOCS* 1961, *38*, 297–301.

Dhull, S.B.; Punia, S. Sources: plants, animals and microbial; Dhull, S.B.; Punia, S.; Sandhu, K.S., Eds.; *Essential Fatty Acids: Sources, Processing Effects, and Health Benefits*; CRC: Boca Raton, London, New York, 2021; pp. 19–55.

Dobson, G.; Griffiths, D.W.; Davies, H.V.; Mcnicol, J.W. Comparison of fatty acid and polar lipid contents of tubers from two potato species, *Solanum tuberosum* and *Solanum phureja*. *J. Agric. Food Chem.* 2004, *52*, 6306–6314.

Dobson, G.; Shepherd, T.; Verrall, S.R.; Griffiths, W.D.; Ramsay, G.; Mcnicol, J.W.; Davies, H.V.; Stewart, D. A metabolomics study of cultivated potato (*Solanum tuberosum*) groups Andigena, Phureja, Stenotomum, and Tuberosum using gas chromatography-mass spectrometry. *J. Agric. Food Chem.* 2010, *58*, 1214–1223.

Domenichiello, A.R.; Kitson, A.R.; Bazinet, R.P. Is docosahexaenoic acid synthesis from a-linolenic acid sufficient to supply the adult brain? *Prog. Lipid Res.* 2015, *59*: 54–66.

Dominguez, T.; Hernandez, M.L.; Pennycooke, J.C.; Jimenez, P.; Martinez-Rivas, J.M.; Sanz, C.; Stockinger, E.J.; Sanchez-Serrano, J.J.; Sanmartin, M. Increasing ω–3 desaturase expression in tomato results in altered aroma profile and enhanced resistance to cold stress. *Plant Physiol.* 2010, *153*, 655–665.

Dyall, S.C.; Michael-Titus, A.T. Neurological benefits of omega-3 fatty acids. *Neuromolecular Med.* 2008, *10*, 219–235.

Dzondo-Gadet, M.; Dellhot, J.; Silva, P.H.A., Desobry, S. Yield and chemical characterization of Congolese mansa (*Solanum americanum* Miller) oil extracted from plant by solvents and enzymes. *J. Food Technol.* 2006, *4*, 259–263.

Eddey, S. Omega-6 and 9 fatty acids. *J. Complement Med.* 2008, *7*, 34.

Essien, E.E.; Ogunwande, I.A.; Setzer, W.N.; Ekundayo, O. Chemical composition, antimicrobial, and cytotoxicity studies on *S. erianthum* and *S. macranthum* essential oils. *Pharm. Biol.* 2012, *50*, 474–480.

Fabian, C.J.; Kimler, B.F.; Hursting, S.D. Omega-3 fatty acids for breast cancer prevention and survivorship. *Breast Cancer Res.* 2015, *17*, 1–11.

Fedorova, I.; Salem, N. Omega-3 fatty acids and rodent behavior. *Prostaglandins Leukot. Essent. Fatty Acids* 2006, *75*, 271–289.

Ferrara, L.; Dosi, R.; Di Maro, A.; Guida, V.; Cefarelli, G.; Pacifico, S.; Mastellone, C.; Fiorentino, A.; Rosati, A.; Parente, A. Nutritional values, metabolic profile and radical scavenging capacities of wild asparagus (*A. acutifolius* L.). *J. Food Compost Anal.* 2011, *24*, 326–333.

Franco, D.; Pateiro, M.; Rodriguez Amado, I.; Lopez Pedrouso, M.; Zapata, C.; Vazquez, J.A.; Lorenzo, J.M. Antioxidant ability of potato (*Solanum tuberosum*) peel extracts to inhibit soybean oil oxidation. *Eur. J. Lipid Sci. Technol.* 2016, *118*, 1891–1902.

Franke, R.; Briesen, I.; Wojciechowski, T.; Faust, A.; Yephremov, A.; Nawrath, C.; Schreiber, L. Apoplastic polyesters in Arabidopsis surface tissues – a typical suberin and a particular cutin. *Phytochem.* 2005, *66*, 2643–2658.

Freitas, R.D.S.; Campos, M.M. Protective effects of omega-3 fatty acids in cancer-related complications. *Nutrients* 2019, *11*, 945.

Gerber, M. Omega-3 fatty acids and cancers: a systematic update review of epidemiological studies. *Br. J. Nutr.* 2012, *107*, S228–S239.

Ghafoorunissa, J.P. Vegetables as sources of α-linolenic acid in Indian diets. *Food Chem.* 1993, *47*, 121–124.

Giuffre, A.M.; Sicari, V.; Capocasale, M.; Zappia, C.; Pellicano, T.M.; et al. Physico-chemical properties of tomato seed oil (*Solanum lycopersicum* L.) for biodiesel production. *Acta Horticult.* 2015, *1081*, 237–242.

Gleissman, H.; Johnsen, J.I.; Kogner, P. Omega-3 fatty acids in cancer, the protectors of good and the killers of evil? *Exp. Cell Res.* 2010, *316*, 1365–1373.

Glencross, B.D. Exploring the nutritional demand for essential fatty acids by aquaculture species. *Rev. Aquac.* 2009, *1*, 71–124.

Gogus, U.; Smith, C. n–3 Omega fatty acids: A review of current knowledge. *Int. J. Food Sci. Technol.* 2010, *45*, 417–436.

Grissett, L.; Ali, A.; Coble, A.-M.; Logan, K.; Washington, B.; Mateson, A.; McGee, K.; Nkrumah, Y.; Jacobus, L.; Abraham, E.; et al. Survey of sensitivity to fatty acid-amino acid conjugates in the Solanaceae. *J. Chem. Ecol.* 2020, *46*, 330–343.

Guil-Guerrero, J.L. Stearidonic acid (18:4n–3): metabolism, nutritional importance, medical uses and natural sources. *Eur. J. Lipid Sci. Technol.* 2007, *109*, 1226–1236.

Gutierrez, S.; Svahn, S.L.; Johansson, M.E. Effects of omega 3 fatty acids on immune cells. *Int. J. Mol. Sci.* 2019, *20*, 5028.

Halinski, L.P.; Paszkiewicz, M.; Golebiowski, M.; Stepnowski, P. The chemical composition of cuticular waxes from leaves of the gboma eggplant (*Solanum macrocarpon* L.). *J. Food Compost Anal.* 2012, *25*, 74–78.

Halinski, L.P.; Puckowski, A.; Stepnowski, P. Glycoalkaloid, phytosterol and fatty acid content of raw and blanched leaves of the gboma eggplant (*Solanum macrocarpon* L.). *J. Food Nutr. Res.* 2015, *54*, 9–20.

Halinski, L.P.; Szafranek, J.; Szafranek, B.M.; Golebiowski, M.; Stepnowski, P. Chromatographic fractionation and analysis of the main components of eggplant (*Solanum melongena* L.) leaf cuticular waxes. *Acta Chromatogr.* 2009, *21*, 127–137.

Halinski, L.P.; Topolewska, A.; Rynkowska, A.; Mika, A.; Urasinska, M.; Czerski, M.; Stepnowski, P. Impact of plant domestication on selected nutrient and anti-nutrient compounds in Solanaceae with edible leaves (Solanum spp.). *Genet. Resour. Crop Evol.* 2019, *66*, 89–103.

Hamel, C.; Vujanovic, V.; Jeannotte, R.; Nakano-Hylander, A.; St-Arnaud, M. Negative feedback on a perennial crop: Fusarium crown and root rot of asparagus is related to changes in soil microbial community structure. *Plant Soil.* 2005, *268*, 75–87.

Hanifah, A.; Maharijaya, A.; Putri, S.P.; Laviña, W.A.; Sobir. Untargeted metabolomics analysis of eggplant (*Solanum melongena* L.) fruit and its correlation to fruit morphologies. *Metabolites* 2018, *8*, 49.

Harper, C.R.; Jacobson, T.A. The fats of life: the role of omega-3 fatty acids in the prevention of coronary heart disease. *Arch. Intern. Med.* 2001, *161*, 2185–2192.

Harris, W.S. Omega-3 fatty acids and cardiovascular disease: A case for omega-3 index as a new risk factor. *Pharmacol. Res.* 2007a, *55*, 217–223.

Harris, W.S. Expert opinion: Omega-3 fatty acids and bleeding-cause for concern? *Am. J. Card.* 2007b, *99*, 44C–46C.

Harris, W.S.; Assaad, B.; Poston, W.C. Tissue omega-6/omega-3 fatty acid ratio and risk for coronary artery disease. *Am. J. Card.* 2006, *98*, 19i–26i.

Harris, W.S.; Dayspring, T.D.; Moran, T.J. Omega-3 fatty acids and cardiovascular disease: new developments and applications. *Postgrad. Med.* 2013, *125*, 100–113.

Harris, W.S.; Kris-Etherton, P.M.; Harris, K.A. Intakes of long-chain omega-3 fatty acid associated with reduced risk for death from coronary heart disease in healthy adults. *Curr. Atheroscler. Rep.* 2008a, *10*, 503–509.

Harris, W.S.; Miller, M.; Tighe, A.P.; Davidson, M.H.; Schaefer, E.J. Omega-3 fatty acids and coronary heart disease risk: Clinical and mechanistic perspectives. *Atherosclerosis* (Amsterdam, Netherlands) 2008b, *197*, 12–24.

Harris, W.S.; Mozaffarian, D.; Rimm, E.; Kris-Etherton, P.; Rudel, L.L.; Appel, L.J.; Engler, M.M.; Engler, M.B.; Sacks, F. Omega-6 fatty acids and risk for cardiovascular disease. *Circulation* 2009, *119*, 902–907.

Hartweg, J.; Perera, R.; Montori, V.; Dinneen, S.; Neil, H.A.W.; Farmer, A. Omega-3 polyunsaturated fatty acids (PUFA) for type 2 diabetes mellitus. *Cochrane Database Syst. Rev.* 2008, *1*, CD003205.

Haslam, R.R.; Ruiz-Lopez, N.; Eastmond, R.; Moloney, M.; Sayanova, O.; Napier, J.A. The modification of plant oil composition via metabolic engineering-better nutrition by design. *Plant Biotechnol. J.* 2013, *11*, 157–168.

Hassan, R.A.; Tawfeek, W.A.; Habeeb, A.A.; Mohamed, M.S.; Abdelshafeek, K.A. Investigation of some chemical constituents and antioxidant activity of *Asparagus sprengeri*. *Int. J. Pharm. Pharm. Sci.* 2014, *6*, 46–51.

Hauff, S.; Chefetz, B.; Shechter, M.; Vetter, W. Determination of hydroxylated fatty acids from the biopolymer of tomato cutin and their fate during incubation in soil. *Phytochem. Anal.* 2010, *21*, 582–589.

He, T. A whitening traditional Chinese medicine mask. China, CN105520893 A 2016–04–27.

Hopkins, C.Y.; Chisholm, M.J. Fatty acids of asparagus-seed oil. *J. Am. Oil Chem. Soc.* 1957, *34*, 477–479.

Huang, X. Effect of LeFAD7 gene on drought resistance of tomato plants. *Beifang Yuanyi* 2010, *21*, 145–149.

Innes, J.K.; Calder, P.C. Omega-6 fatty acids and inflammation. *Prostaglandins Leukot. Essent. Fatty Acids* 2018, *132*, 41–48.

Innis, S.M. Dietary omega 3 fatty acids and the developing brain. *Brain Res.* 2008, *1237*, 35–43.

Iwasaki, M.; Hoshian, F.; Tsuji, T.; Hirose, N.; Matsumoto, T.; Kitatani, N.; Sugawara, K.; Usui, R.; Kuwata, H.; Sugizaki, K.; et al. Predicting efficacy of dipeptidyl peptidase–4 inhibitors in patients with type 2 diabetes: association of glycated hemoglobin reduction with serum eicosapentaenoic acid and docosahexaenoic acid levels. *J. Diabetes Investig.* 2012, *3*, 464–467.

Jamal, S.; Ahmad, M.; Osman, S.M.; Ahmad, I. Studies on minor seed oils. XIII. *JOTAI* 1986, *18*, 81–83.

Jarret, R.; Levy, I.; Potter, T.; Cermak, S. Oil and fatty acids in seed of eggplant (*Solanum melongena* L.) and some related and unrelated Solanum species. *Am. J. Agric. Biol. Sci.* 2016, *11*, 76–81.

Jarret, R.L.; Levy, I.J.; Potter, T.L.; Cermak, S.C. Seed oil and fatty acid composition in *Capsicum* spp. *J. Food Comp. Anal.* 2013, *30*, 102–108.

Jarvinen, R.; Rauhala, H.; Holopainen, U.; Kallio, H. Differences in suberin content and composition between two varieties of potatoes (*Solanum tuberosum*) and effect of post-harvest storage to the composition. *LWT - Food Sci. Technol.* 2011, *44*, 1355–1361.

Jarvinen, R.; Silvestre, A.J.D.; Holopainen, U.; Kaimainen, M.; Nyyssola, A.; Gil, A.M.; Neto, C.P.; Lehtinen, P.; Buchert, J.; Kallio, H. Suberin of potato (*Solanum tuberosum* Var. Nikola): Comparison of the effect of cutinase CcCut1 with chemical depolymerization. *J. Agric. Food Chem.* 2009, *57*, 9016–9027.

Jeromson, S.; Gallagher, I.J.; Galloway, S.D.R.; Hamilton, D.L. Omega-3 fatty acids and skeletal muscle health. *Mar. Drugs* 2015, *13*, 6977–7004.

Jicha, G.A.; Markesbery, W.R. Omega-3 fatty acids: potential role in the management of early Alzheimer's disease. *Clin. Interv. Aging* 2010, *5*, 45–61.

Jimenez-Moreno, N.; Esparza, I.; Bimbela, F.; Gandia, L.M.; Ancin-Azpilicueta, C. Valorization of selected fruit and vegetable wastes as bioactive compounds: opportunities and challenges. *Crit. Rev. Environ. Sci. Technol.* 2020, *50*, 2061–2108.

Johnson, M.; Bradford, C. Omega-3, omega-6 and omega-9 fatty acids: implications for cardiovascular and other diseases. *J. glycom. lipidom.* 2014, *4*, 2153–0637.

Kang, J.X. Omega-3: a link between global climate change and human health. *Biotechnol Adv.* 2011, *29*, 388–390.

Kang, J.X.; Weylandt, K.H. Modulation of inflammatory cytokines by omega-3 fatty acids. *Sub-Cell. Biochem.* 2008, *49*, 133–143.

Karthika, P. Production of biodiesel from leaves of *Solanum melongena* L. and product. India, IN202241029720 A 2022–06–17.

Kashimoto, T.; Noda, K. Vegetable oils and fats. VI. *Nippon Kagaku Zasshi* 1958, *79*, 873–876.

Katagi, K.S.; Hosamani, K.M. Unique occurrence of cyclopropenoid fatty acids in *Asparagus racemosus* seed oil: a rich source of oil and its possible industrial application. *J. Appl. Chem.* 2013, *2*, 561–566.

Kaur, N.; Chugh, V.; Gupta, A.K. Essential fatty acids as functional components of foods: a review. *J. Food Sci. Technol.* 2014, *51*, 2289–2303.

Kohli, D.; Champawat, P.S.; Mudgal, V.D. Asparagus (*Asparagus racemosus* L.) roots: nutritional profile, medicinal profile, preservation, and value addition. *J. Sci. Food Agric.* 2023, *103*, 2239–2250.

Kolattukudy, P.E.; Agrawal, V.P. Structure and composition of aliphatic constituents of potato tuber skin (suberin). *Lipids* 1974, *9*, 682–691.

Kolotilin, I.; Koltai, H.; Bar-Or, C.; Chen, L.; Nahon, S.; Shlomo, H.; Levin, I.; Reuveni, M. Expressing yeast SAMdc gene confers broad changes in gene expression and alters fatty acid composition in tomato fruit. *Physiol. Plant.* 2011, *142*, 211–223.

Krygsman, P.H.; Barrett, A.E. Simple methods for measuring total oil content by bench-top NMR; Luthria, D.L., Ed.; *Oil Extraction and Analysis: Critical Issues and Competitive Studies*; The American Oil Chemists' Society: Urbana, Illinois, 2004; pp. 152–165.

Larque, E.; Gil-Sanchez, A.; Prieto-Sanchez, M.T.; Koletzko, B. Omega 3 fatty acids, gestation and pregnancy outcomes. *Br. J. Nutr.* 2012, *107*, S77–S84.

Li, G.Y.; Ding, X.L. Analysis of fatty acids of flaxseed oil with GC-MS. *Food & Machinery* 2005, *21*, 30–32.

Li, H.L.; Wei, S.P. Determination of β-carotene in tomato seed oil by HPLC. *Food Sci.* 2003, *24*, 122–123.

Li-Beisson, Y.; Shorrosh, B.; Beisson, F.; Andersson, M.X.; Arondel, V.; Bates, P.D.; Baud, S.; Bird, D.; Debono, A.; Durrett, T.P.; et al. Acyl-lipid metabolism. *Arabidopsis Book*. 2013, *11*, e0161.

Liu, P.; Gao, R.; Gao, L.; Bi, J.; Jiang, Y.; Zhang, X.; Wang, Y. Distinct quality changes of asparagus during growth by widely targeted metabolomics analysis. *J. Agric. Food Chem.* 2022, *70*, 15999–16009.

Liu, L.; Howe, P.; Zhou, Y.-F.; Hocart, C.; Zhang, R. Fatty acid profiles of leaves of nine edible wild plants: An Australian study. *J. Food Lipids* 2002, *9*, 65–71.

Liu, X.; Yang, J.-H.; Li, B.; Yang, X.-M.; Meng, Q.-W. Antisense expression of tomato chloroplast omega-3 fatty acid desaturase gene (LeFAD7) enhances the tomato high-temperature tolerance through reductions of trienoic fatty acids and alterations of physiological parameters. *Photosynthetica* 2010, *48*, 59–66.

Logan, A.C. Neurobehavioral aspects of omega-3 fatty acids: possible mechanisms and therapeutic value in major depression. *Altern. Med. Rev.* 2003, *8*, 410–425.

Lough, A.K. The production of methoxy-substituted fatty acids as artifacts during the esterification of unsaturated fatty acids with methanol containing borontrifluoride. *Biochem. J.* 1964, *90*, 4c–5c.

Lowry, R.R.; Tinsley, I.J. Oleic and linoleic acid interaction in polyunsaturated fatty acid metabolism in the rat. *J. Nutr.* 1966, *88*, 26–32.

Lv, Y.; Li, L.; Luo, H. Non-irritating moisturizing skin care product containing *Asparagus schoberioides* unsaturated fatty acids and its preparation method. China, CN113679633 A 2021–11–23.

MacLean, C.H.; Newberry, S.J.; Mojica, W.A.; Khanna, P.; Issa, A.M.; Suttorp, M.J.; Lim, Y.-W.; Traina, S.B.; Hilton, L.; Garland, R.; et al. Effects of omega-3 fatty acids on cancer risk. A systematic review. *J. Am. Med. Assoc. JAMA*. 2006, *295*, 403–415.

Martins, M.B.; Suaiden, A.S.; Piotto, R.R.; Barbosa, M. Properties of omega-3 polyunsaturated fatty acids obtained of fish oil and flaxseed oil. *Revista do Instituto de Ciencias da Saude* 2008, *26*, 153–156.

McEwen, B.J. The influence of diet and nutrients on platelet function. *Semin. Thromb. Hemost.* 2014, *40*, 214–226.

McNamara, R.K.; Carlson, S.E. Role of omega-3 fatty acids in brain development and function: Potential implications for the pathogenesis and prevention of psychopathology. *Prostaglandins Leukot. Essent. Fatty Acids* 2006, *75*, 329–349.

McNamara, R.K.; Strawn, J.R. Role of long-chain omega-3 fatty acids in psychiatric practice. *PharmaNutrition* 2013, *1*, 41–49.

Merle B.M.; Delyfer, M.N.; Korobelnik, J.F.; Rougier, M.B., Malet, F.; Féart, C.; Le Goff, M.; Peuchant, E.; Letenneur, L.; Dartigues, J.F.; Colin, J.; Barberger-Gateau, P.; Delcourt, C. High concentrations of plasma n3 fatty acids are associated with decreased risk for late age-related macular degeneration. *J. Nutr.* 2013, *143*, 505–511.

Meyer, B.; Mann, N.J.; Lewis, J.L.; Milligan, G.C.; Sinclair, A.J.; Howe, P.R. Dietary intakes and food sources of omega-6 and omega-3 polyunsaturated fatty acids. *Lipids* 2003, *38*, 391–398.

Miyata, J.; Arita, M. Role of omega-3 fatty acids and their metabolites in asthma and allergic diseases. *Allergol. Int.* 2015, *64*, 27–34.

Morales, P.; Ferreira, I.C.F.R.; Carvalho, A.M.; Sánchez-Mata, M.C.; Cámara, M.; Tardío, J. Fatty acids profiles of some Spanish wild vegetables. *Food Sci. Technol. Int.* 2012, *18*, 281–290.

Mori, T.A. Marine omega-3 fatty acids in the prevention of cardiovascular disease. *Fitoterapia* 2017, *123*, 51–58.

Mori, T.A. omega-3 fatty acids and blood pressure. *Cell. Mol. Biol.* (Sarreguemines, France) 2010, *56*, 83–92.

Mori, T.A. omega-3 fatty acids and cardiovascular disease: epidemiology and effects on cardiometabolic risk factors. *Food Funct.* 2014, *5*, 2004–2019.

Mori, T.A. omega-3 fatty acids and hypertension in humans. *Clin. Exp. Pharmacol.* 2006, *33*, 842–846.

Mori, T.A.; Beilin, L.J. omega-3 fatty acids and inflammation. *Curr. Atheroscler. Rep.* 2004, *6*, 461–467.

Mozaffarian, D.; Wu, J.H.Y. omega-3 fatty acids and cardiovascular disease: effects on risk factors, molecular pathways, and clinical events. *J. Am. Coll. Cardiol.* 2011, *58*, 2047–2067.

Muhlhausler, B.S.; Ailhaud, G.P. omega-6 polyunsaturated fatty acids and the early origins of obesity. *Curr. Opin. Endocrinol. Diabetes Obes.* 2013, *20*, 56–61.

Nadeeshani, H.; Samarasinghe, G.; Wimalasiri, S.; Silva, R.; Hunter, D.; Madhujith, T. Comparative analysis of the nutritional profiles of selected Solanum species grown in Sri Lanka. *J. Food Compos.* 2021, *99*, 103847.

Napier, J.A.; Graham, I.A. Tailoring plant lipid composition: designer oilseeds come of age. *Curr. Opin. Plant Biol.* 2010, *13*, 329–336.

Nehring, C.; Rayfield, S.; Olmstead, N.C.; Niering, W.A. *National Audubon Society Field Guide to North American Wildflowers (Eastern Region)*; Knopf: New York, 1979; p.804.

Neue Ölquellen, N.H. New sources of oil. *Angew. Chem.* 1916, *29*, 337–338.

Nopparatmaitree, M.; Nava, M.; Chumsangchotisakun, V.; Saenphoom, P.; Chotnipat, S.; Kitpipit, W. Effect of trimmed asparagus by-products supplementation in broiler diets on performance, nutrients digestibility, gut ecology, and functional meat production. *Vet. World* 2022, *15*, 147–161.

Nosenko, T.; Boendel, K.B.; Kumpfmueller, G.; Stephan, W. Adaptation to low temperatures in the wild tomato species *Solanum chilense. Mol. Ecol.* 2016, *25*, 2853–2869.

Nzikou, J.M.; Mvoula-Tsieri, M.; Matos, L.; Matouba, E.; Ngakegni-Limbili, A.C.; et al. *Solanum nigrum* L. seeds as an alternative source of edible lipids and nutriment in Congo Brazzaville. *J. Appl. Sci.* 2007, *7*, 1107–1115.

Osman, F.; Subbaram, M.R.; Achaya, K.T. Glyceride structure of some Egyptian vegetable oils. *Fette, Seifen, Anstrichmittel* 1968, *70*, 69–73.

Parveen, S.; Taufeeque, M.; Malik, A. Characterisation of seed oils of two plant species from arid and semi arid region of Rajasthan. *Int. J. Pharm. Biol.* 2018, *8*, 989–993.

Peet, M.; Stokes, C. omega-3 fatty acids in the treatment of psychiatric disorders. *Drugs* 2005, *65*, 1051–1059.

Peters, W. Investigation of the asparagus seeds. *Archiv der Pharmazie* (Germany) 1902, *240*, 53–56.

Petrie, J.R.; Shrestha, R.; Belide, S.; Kennedy, Y.; Lester, G.; Liu, Q.; Divi, U.K.; Mulder, R.J.; Mansour, M.P.; Nichols, P.D.; Singh, S.P. Metabolic engineering *Camelina sativa* with fish oil-like levels of DHA. *PLoS One* 2014, *9*, e85061.

Pollard, M.; Beisson, F.; Li, Y.; Ohlrogge, J.B. Building lipid barriers: biosynthesis of cutin and suberin. *Trends Plant Sci.* 2008, *13*, 236–246.

Prasad, Y.R.; Nigam, S.S. Chemical examination of the seeds of *Asparagus officinalis* Linn. I. Investigation of the seed oil. *P. Natl. A. Sci. India A.* 1982, *52*, 396–398.

Prescha, A.; Swiedrych, A.; Biernat, J.; Szopa, J. Increase in lipid content in potato tubers modified by 14–3–3 gene overexpression. *J. Agric. Food Chem.* 2001, *49*, 3638–3643.

Price, P.T.; Nelson, C.M.; Clarke, S.D. omega-3 polyunsaturated fatty acid regulation of gene expression. *Curr. Opin. Lipidol.* 2000, *11*, 3–7.

Rao, G.S.R.L.; Willison, J.H.M.; Ratnayake, W.M.N. Suberin production by isolated tomato fruit protoplasts. *Plant Physiol.* 1984, *75*, 716–719.

Reddy, B.S. omega-3 fatty acids in colorectal cancer prevention. *Int. J. Cancer* 2004, *112*, 1–7.

Rogalski, M.; Carrer, H. Engineering plastid fatty acid biosynthesis to improve food quality and biofuel production in higher plants. *Plant Biotechnol. J.* 2011, *9*, 554–564.

Rose, D.P.; Connolly, J.M. omega-3 fatty acids as cancer chemopreventive agents. *Pharmacol. Ther.* 1999, *83*, 217–244.

Ross, B.M. ω–3 Fatty acid deficiency in major depressive disorder is caused by the interaction between diet and a genetically determined abnormality in phospholipid metabolism. *Med. Hypotheses* 2006, *68*, 515–524.

Ross, B.M. omega-3 polyunsaturated fatty acids and anxiety disorders. *Prostaglandins Leukot. Essent. Fatty Acids* 2009, *81*, 309–312.

Saini, R.K.; Keum, Y.-S. omega-3 and omega-6 polyunsaturated fatty acids: dietary sources, metabolism, and significance - A review. *Life Sci.* 2018, *203*, 255–267.

Saleem, T.M.S.; Chetty, C.M.; Ramkanth, S.; Alagusundaram, M.; Gnanaprakash, K.; et al. *Solanum nigrum* Linn.- A Review. *Pharmacogn. Rev.* 2009, *3*, 342–345.

Salunke, D.; Mangalekar, R.; Kuvalekar, A. Bioconversion of alpha-linolenic acid into long chain polyunsaturated fatty acids by oleaginous fungi. *Int. J. Pharm. Sci.* 2014, *5*, 27–35.

Sanchez, M.A.; Cattaneo, P. Contents and fatty acid composition values of total lipids (Folch) from pulps of non-oily edible fruits. *An. des la Asoc.* 1987, *75*, 531–549.

Santos Bento, M.F.; Pereira, H.; Cunha, M.Á.; Moutinho, A.M.M.C.; van den Berg, K.J.; Boon, J.J. Study of variability of suberin composition in cork from *Quercus suber* L. using thermally assisted transmethylation GC-MS. *J. Anal. Appl. Pyrolysis* 2001, *57*, 45–55.

Santos-Merino, M.; Garcillan-Barcia, M.R.; de la Cruz, F. Engineering the fatty acid synthesis pathway in *Synechococcus elongatus* PCC 7942 improves omega-3 fatty acid production. *Biotechnol. Biofuels* 2018, *11*, 239.

Saravanan, P.; Davidson, N.C.; Schmidt, E.B.; Calder, P.C. Cardiovascular effects of marine omega-3 fatty acids. *Lancet* 2010, *376*, 540–550.

Sayanova, O.; Napier, J.A. Transgenic oilseed crops as an alternative to fish oils. *Prostaglandins Leukot. Essent. Fatty Acids* 2011, *85*, 253–260.

Schreiber, L.; Franke, R.; Hartmann, K. Wax and suberin development of native and wound periderm of potato (*Solanum tuberosum* L.) and its relation to peridermal transpiration. *Planta* 2005, *220*, 520–530.

Scora, R.W.; Mueller, E.; Guelz, P.G. Wax components of *Asparagus officinalis* L. (Liliaceae). *J. Agric. Food Chem.* 1986, *34*, 1024–1026.

Scorletti, E.; Byrne, C.D. omega-3 fatty acids, hepatic lipid metabolism, and nonalcoholic fatty liver disease. *Annu. Rev. Nutr.* 2013, *33*, 231–248.

Senanayake, SPJN; Fichtali, J. Marine oils: single cell oil as sources of nutraceuticals and specialty lipids: processing technologies and application; Shahidi, F., ed.; *Nutraceutical and Speciality Lipids and their Co-Products*; CRC: Boca Raton, London, New York, 2006; pp. 251–280.

Seo, D.H. Composition comprising eggplant extract for treating or preventing disease related to sebaceous gland. Korea, Republic of, KR2014090794 A 2014–07–18.

Serra, O.; Soler, M.; Hohn, C.; Sauveplane, V.; Pinot, F.; Franke, R.; Schreiber, L.; Prat, S.; Molinas, M.; Figueras, M. CYP86A33-targeted gene silencing in potato tuber alters suberin composition, distorts suberin lamellae, and impairs the periderm's water barrier function. *Plant Physiol.* 2009a, *149*, 1050–1060.

Serra, O.; Soler, M.; Hohn, C.; Franke, R.; Schreiber, L.; Prat, S.; Molinas, M.; Figueras, M. Silencing of StKCS6 in potato periderm leads to reduced chain lengths of suberin and wax compounds and increased peridermal transpiration. *J. Exp. Bot.* 2009b, *60*, 697–707.

Shahidi, F; Ambigaipalan, P. omega-3 polyunsaturated fatty acids and their health benefits. *Annu. Rev. Food Sci. Technol.* 2018, *9*, 345–381.

Shahidi F, Miraliakbari H. Marine oils: compositional characteristics and health effects; Shahidi, F., Ed.; *Nutraceutical and Specialty Lipids and their Co-Products*; CRC: Boca Raton, FL, 2006; pp. 227–250.

Shahidi, F.; Miraliakbari, H. Omega-3 (n–3) fatty acids in health and disease: part 1—cardiovascular disease and cancer. *J. Med. Food* 2004, 7, 387–401.

Shahidi, F.; Miraliakbari, H. Omega-3 fatty acids in health and disease: Part 2-health effects of omega-3 fatty acids in autoimmune diseases, mental health, and gene expression. *J. Med. Food* 2005, *8*, 133–148.

Shukla, S.K.; Gupta, S.; Ojha, S.K.; Sharma, S.B. Cardiovascular friendly natural products: a promising approach in the management of CVD. *Nat. Prod. Res.* 2010, *24*, 873–898.

Siciliano, A.; Limonti, C.; Mehariya, S.; Molino, A.; Calabro, V. Biofuel production and phosphorus recovery through an integrated treatment of agro-industrial waste. *Sustainability* 2019, *11*, 52.

Silva, G.F.P.; Pereira, E.; Melgar, B.; Stojkovic, D.; Sokovic, M.; Calhelha, R.C.; Pereira, C.; Abreu, R.M.V.; Ferreira, I.C.F.R.; Barros, L. Eggplant fruit (*Solanum melongena* L.) and bio-residues as a source of nutrients, bioactive compounds, and food colorants, using innovative food technologies. *Appl. Sci.* 2021, *11*, 151.

Simopoulos, A.P. Evolutionary aspects of the dietary omega-6:omega-3 fatty acid ratio: medical implications. *World Rev. Nutr. Diet* 2009, *100*, 1–21.

Simopoulos, A.P. Omega-3 fatty acids in inflammation and autoimmune diseases. *J. Am. Coll. Nutr.* 2002a, *21*, 495–505.

Simopoulos, A.P. The importance of the ratio of omega-6/omega-3 essential fatty acids. *Biomed. Pharmacother.* 2002b, *56*, 365–379.

Simopoulos, A.P. Omega-3 fatty acids and antioxidants in edible wild plants. *Biol. Res.* 2004a, *37*, 263–277.

Simopoulos A.P. Omega-6/omega-3 essential fatty acid ratio and chronic diseases. *Food Rev. Int.* 2004b, *20*, 77–90.

Song, W.; Zhang, K.; Xue, T.; Han, J.; Peng, F.; Ding, C.; Lin, F.; Li, J.; Sze, F.T.A.; Gan, J.; Chen, X. Cognitive improvement effect of nervonic acid and essential fatty acids on rats ingesting *Acer truncatum* Bunge seed oil revealed by lipidomics approach. *Food Funct.* 2022, *13*, 2475–2490.

Sotiroudis, V.T.; Sotiroudis, T.G.; Kolisis, F.N. The potential of biodiesel production from fatty acid methyl esters of some European/Mediterranean and cosmopolitan halophyte seed oils. *J. ASTM Int.* 2010, *7*, 421–433.

Spector, A.A. Essentiality of fatty acids. *Lipids* 1999, *34*, S1–S3.

Surangi, K.G.; Fiaz, S. Ayurveda and modern perspective on eye care and nutrition. *J. Biol. Sci. Opin.* 2016, *4*, 55–58.

Swanson, D.; Block, R.; Mousa, S.A. Omega-3 fatty acids EPA and DHA: health benefits throughout life. *Adv. Nutr.* 2012, *3*, 1–7.

Sydenham, E.; Dangour, A.D.; Lim, W.-S. Omega 3 fatty acid for the prevention of cognitive decline and dementia. *Cochrane Database Syst. Rev.* 2012, *6*, CD005379.

Szabo, K.; Dulf, F.V.; Diaconeasa, Z.; Vodnar, D.C. Antimicrobial and antioxidant properties of tomato processing byproducts and their correlation with the biochemical composition. *LWT - Food Sci. Technol.* 2019, *116*, 108558.

Szafranek, M.; Synak, E.E. Cuticular waxes from potato (*Solanum tuberosum*) leaves. *Phytochem.* 2006, *67*, 80–90.

Tanios, S.; Thangavel, T.; Eyles, A.; Tegg, R.S.; Nichols, D.S.; Corkrey, R.; Wilson, C.R. Suberin deposition in potato periderm: a novel resistance mechanism against tuber greening. *New Phytol.* 2020, *225*, 1273–1284.

Teitelbaum, J.E.; Allan Walker, W. Review: the role of omega 3 fatty acids in intestinal inflammation. *J. Nutr. Biochem.* 2001, *12*, 21–32.

Tocher, D.R. Issues surrounding fish as a source of omega-3 long-chain polyunsaturated fatty acids. *Lipid Technol.* 2009, *21*, 13–16.

Topolewska, A.; Czarnowska, K.; Halin'ski, Ł.P.; Stepnowski, P. Evaluation of four derivatization methods for the analysis of fatty acids from green leafy vegetables by gas chromatography. *J. Chromatogr. B* 2015, *990*, 150–157.

Ursin, V.M. Modification of plant lipids for human health: Development of functional land-based omega-3 fatty acids. *J. Nutr.* 2003, *133*, 4271–4274.

Vardavas, C.I.; Majchrzak, D.; Wagner, K.H.; Elmadfa, I.; Kafatos, A. Lipid concentrations of wild edible greens in Crete. *Food Chem.* 2006, *99*, 822–834.

Vidrih, R.; Filip, S.; Hribar, J. Content of higher fatty acids in green vegetables. *Czech J. Food Sci.* 2009, *27*, S125–S129.

von Schacky, C. Omega-3 fatty acids and cardiovascular disease. *Curr. Opin. Clin. Nutr. Metab. Care* 2007, *10*, 129–135.

von Schacky, C. Omega-3 fatty Acids in cardiovascular disease - An uphill battle. *Prostaglandins Leukot. Essent. Fat. Acids* 2015, *92*, 41–47.

Wadhwa, M.; Bakshi, M.P.S.; Makkar, H.P.S. Wastes to worth: value added products from fruit and vegetable wastes. *CAB Rev.* 2015, *10*, 43/1–43/25.

Ward, O.R.; Singh, A. Omega-3/6 fatty acids: alternative sources of production. *Process Biochem.* 2005, *40*, 3627–3652.

Wen, Y.; Zhou, R. *Asparagus cochinchinensis* (lour.) merr. shampoo with beautifying and caring hair effects. China, CN113679650 A 2021–11–23.

Wendel, M.; Heller, A.R. Anticancer actions of omega-3 fatty acids - current state and future perspectives. *Anti-Cancer Agents Med. Chem.* 2009, *9*, 457–470.

Weylandt, K.H.; Chiu, C.-Y.; Gomolka, B.; Waechter, S.F.; Wiedenmann, B. Omega-3 fatty acids and their lipid mediators: towards an understanding of resolvin and protectin formation. *Prostaglandins Other Lipid Mediat.* 2012, *97*, 73–82.

Woolfson, K.N.; Haggitt, M.L.; Zhang, Y.; Kachura, A.; Bjelica, A.; Rey Rincon, M.A.; Kaberi, K.M.; Bernards, M.A. Differential induction of polar and non-polar metabolism during wound-induced suberization in potato (*Solanum tuberosum* L.) tubers. *Plant J.* 2018, *93*, 931–942.

Wu, J.H.Y.; Micha, R.; Imamura, F.; Pan, A.; Biggs, M.L.; Ajaz, O.; Djousse, L.; Hu, F.B.; Mozaffarian, D. Omega-3 fatty acids and incident type 2 diabetes: a systematic review and meta-analysis. *Br. J. Nutr.* 2012a, *107*, S214–S227.

Wu, Z.-G.; Xu, H.-Y.; Ma, Q.; Cao, Y.; Ma, J.-N.; Ma, C.-M. Isolation, identification and quantification of unsaturated fatty acids, amides, phenolic compounds and glycoalkaloids from potato peel. *Food Chem.* 2012b, *135*, 2425–2429.

Xiao, G.; Sun, Q.J.; Yang, L.Q. Composition of fatty acids and glycerides in tomato seed oil. *Journal of Wuxi University of Light Industry*, 2000, *19*, 177–180.

Xie, D.; Jackson, E.N.; Zhu, Q. Sustainable source of omega-3 eicosapentaenoic acid from metabolically engineered *Yarrowia lipolytica*: from fundamental research to commercial production. *Appl. Microbiol. Biotechnol.* 2015, *99*, 1599–1610.

Xing, J.; Chin, C.-K. Modification of fatty acids in eggplant affects its resistance to *Verticillium dahlia*. *Physiol. Mol. Plant Pathol.* 2000, *56*, 217–225.

Xu, F.; Shi, A.M.; Liu, H.Z.; Liu, L.; Wang, Q. The content components of fatty acids and endogenous antioxidant of walnut oil and their correlation with oxidative stability index. *J. Chin. Cereals Oils Assoc.* 2016, *31*, 53–58.

Yang, W.-L.; Bernards, M.A. Wound-induced metabolism in potato (*Solanum tuberosum*) tubers: Biosynthesis of aliphatic domain monomers. *Plant Signal. Behav.* 2006, *1*, 59–66.

Yang, R.; Zhang, L.; Lia, P.; Yua, L.; Mao, J.; Wang, X.; Zhang, Q. A review of chemical composition and nutritional properties of minor vegetable oils in China. *Trends Food Sci. Technol.* 2018, *74*, 26–32.

Yashodhara, B.M.; Umakanth, S.; Pappachan, J.M.; Bhat, S.K.; Kamath, R.; Choo, B.H. Omega-3 fatty acids: a comprehensive review of their role in health and disease. *Postgrad. Med. J.* 2009, *85*, 84–90.

Yazawa, K. Production of eicosapentaenoic acid from marine bacteria. *Lipids* 1996, *31*, 297–300.

Yazawa, K.; Araki, K.; Okazaki, N.; Watanabe, K.; Ishikawa, C.; Inoue, A.; Numao, N.; Kondo, K. Production of eicosapentaenoic acid by marine bacteria. *J. Biochem.* 1988, *103*, 5–7.

Yongmanitchai, W.; Ward, O.P. Omega-3 fatty acids: alternative sources of production. *Process Biochem.* (UK) 1989, *24*, 117–125.

Yu, C.; Wang, H.-S.; Yang, S.; Tang, X.-F.; Duan, M.; Meng, Q.-W. Overexpression of endoplasmic reticulum omega-3 fatty acid desaturase gene improves chilling tolerance in tomato. *Plant Physiol. Biochem.* 2009, *47*, 1102–1112.

Yuan, Y.; Ge, F. Extraction of *Asparagus cochinchinensis* seed oil by supercritical carbon dioxide extraction. *Zhongyaocai* 2009, *32*, 804–807.

Zahringer, U.; Domergue, F.; Abbadi, A.; Moreau, H.X.E.; Heinz, E. In vivo characterization of the first acyl-CoA Delta6-desaturase from a member of the plant kingdom, the microalga *Ostreococcus tauri*. *Biochem. J.* 2005, *389*, 483–490.

Zarate, R.; el Jaber-Vazdekis, N.; Ramirez-Moreno, R. Importance of polyunsaturated fatty acids from marine algae; Hegde, M., Zanwar, A., and Adekar, S., Eds.; *Omega-3 Fatty Acids*; Springer: Cham, 2016; pp. 101–126.

Zhang, L. Camellia oil with nourishing heart and soothe the nerve function and processing method thereof. China, CN108935749 A 2018–12–07.

Zhang, M.; Demeshko, Y.; Dumbur, R.; Iven, T.; Feussner, I.; Lebedov, G.; Ganim, M.; Barg, R.; Ben-Hayyim, G. Elevated α-linolenic acid content in extra-plastidial membranes of tomato accelerates wound-induced jasmonate generation and improves tolerance to the herbivorous insects *Heliothis peltigera* and *Spodoptera littoralis*. *J. Plant Growth Regul.* 2019, *38*, 723–738.

Zhao, B.; Sakurai, Y.; Shibata, K.; Kikkawa, F.; Tomoda, Y.; Mizukami, H. Cytotoxic fatty acid ketodienes from eggplants. *Nippon Shokuhin Kagaku Gakkaishi* 2014, *21*, 42–47.

Zhao, B.; Tomoda, Y.; Mizukami, H.; Makino, T. 9-Oxo-(10E,12E)-octadecadienoic acid, a cytotoxic fatty acid ketodiene isolated from eggplant calyx, induces apoptosis in human ovarian cancer (HRA) cells. *J. Nat. Med.* 2015, *69*, 296–302.

7 Wild Edible Plants as Alternate Food Sources and Agents of Food Security

Muhammad Majeed, Hakim Ali Sahito,
Wali Muhammad Mangrio, Murad Muhammad,
Mumtaz Hussain, Robina Aziz, Allah Nawaz Khan and
Allah Bakhsh Gulshan

7.1 INTRODUCTION

The exploitation of native flora can be a buffer against periodical famines which are becoming prevalent in tropical areas. Collection of wild edible plants (WEPs) is mostly done by poor and illiterate people, where such activities have perhaps been normalized as survival strategies during the dry periods of the year when there are insufficient resources available for human survival. Within context of evolutionary connections among humans and the environment, WEPs refer to the whole spectrum of non-domesticated species of plants that are already utilized by humans. WEPs seem to be plants that have edible components that may be grown locally in their natural habitats, such as woods, farms, and uncultivated land. The many components of these plants are used to produce a wide variety of foods, including fresh produce, cooked meals, condiments, snacks, beverages, and even minerals and carbs (seeds, leaves and fruits, nuts, roots, barks, and tubers).

Many traditional indigenous agricultural and hunting cultures rely heavily on WEPs. If cultivated food output is low or fails, they provide a viable alternative for underprivileged communities. Using this material to make famine meals helps to reduce hunger and improves nutrition to a little extent. WEPs' value as backup plans, or even as means of survival, cannot be overstated in drought- and famine-prone developing countries. Regarding the tribal belt's food supply, it's important to remember that individuals are impoverished and often rely on WFPs, particularly in the region under scrutiny. The majority of their meals are consumed at home, and they seldom stay in hotels or dine at restaurants. This further solidifies the connections between humans and their wild plant food sources. Cultural traditions of wild food species have been preserved and passed down verbally from generation to generation among indigenous peoples. Since ancient times, people and plants have worked together as part of a system of checks and balances meant to protect the planet's biodiversity. The impact of wild fruits on human health against different diseases and disorders has attracted the world's attention. The wild fruits reveal different therapeutic applications depending on the chemical composition and nutritional content. Large portions of these communities must find household food security to meet their needs. Indigenous societies in this region have long relied on wild edible plants for their nutritional, food security, and economic needs, as shown by medicinal and aromatic plants. Edible plants as well as mushrooms, even with their ubiquity and diverse possible positive values, haven't yet received as much focus as domesticated food products, even though becoming progressively identified mostly by researchers like an essential alternative or additional supply source to cope with necessities in rural areas. Most of these investigations have been on useful plants for medical purposes, whereas wild edibles have received less attention. The traditional food system has noticed that the usage of conventional healers and medicinal herbs in most developing nations

DOI: 10.1201/9781003395935-7

are a normative foundation for the preservation of good health. More than 90% the plant species employed in the herbal industry are harvested from the wild, and over 70% of herbal medicines in Indian Himalaya are susceptible to disruptive harvests; the bulk of these plants originates in sub-alpine and alpine parts of the Himalaya.

When seasonal supplies of farm-raised produce are low, consumers increasingly turn to wild-harvested options. The current study was compiled with the goals of (1) documenting the diversi-fication of WFPs accumulated and imbibed by tribal societies; (2) assessing traditional knowledge about WFPs; and (3) comparing the Hindu Kush valleys with some other regions on the basis of WFPs, based on the significance of WFPs in the primitive tribal food system.

7.2   CONSERVATION OF WILD FOOD PLANTS AND THEIR POTENTIAL FOR COMBATTING FOOD INSECURITY

Plant-derived foods such as fruits and nuts provide many nutritional benefits to the body. Fruits can improve the nutrition of poor people who may suffer from deficiencies in vitamins, minerals, and other macronutrients (Ehrhardt et al., 2006). Many fruits are also important sources of vitamins A and C which may be lacking in the diet. For example, vitamin C, which is found in significant quan-tities in many fruits, is essential for protecting body cells and improves the adsorption of nonheme iron from plant-based foods. As a result of low intake of vitamin A, an estimated 50 million children in Africa are at risk of its deficiency, making it the third greatest health problem on the continent, preceded by malaria and HIV/AIDS. Traditionally, children used to eat wild foods such as fruits and nuts during herding, which served them with nutritional benefits. However, foods from wild plants may not at the present time form a major part of the diet of the local communities (Richardson et al., 2023).

Furthermore, some traditional local vegetable species such as leaf amaranth are sold in the local markets in Kenya. The local people have the knowledge on preparation and production of traditional foods, which require minimal additional inputs that are affordable to many, including poor people (Moyo et al., 2022). Some wild foods also have medicinal properties to the human body and can be processed through various methods such as boiling, fermentation, and sun drying by the local people (Majeed, Bhatti, Pieroni, et al., 2021a). However, many wild indigenous fruits are sold locally in Kenyan markets such as fruits of *Adansonia digitata*, which is also processed by coloring the seed pulp to make a snack (Onomu, 2023). Its products are also in global demand for novel foods, phar-maceuticals, and cosmetics where the European Union, United States, Japan, and South Africa are reportedly potential markets.

Since the dawn of human history, WEP have been an integral part of our food supply. When people talk about "wild food plants," they're referring to the varieties of vegetation that aren't farmed yet are nevertheless used in human diets. Many foodstuffs consumed in tropical Africa are derived from wild plants. Those plants are utilized in different ways such as fruits, vegetables, cere-als, roots, and tubers. About 60% of the Kenyan population faces starvation due to lack of physical and economic access to adequate calories. Kenya is endowed with diverse plant species, which are estimated to comprise about 6293 indigenous vascular plants. These include an estimated 800 food plants some of which are underutilized food plants such as *Amaranthus* spp. (leaf amaranth), *Solanum americanum* (African nightshade), *Cleome gynandra* (spider plant), *Cucumis dipsaceus* (Hedgehog cucumber), *Commelina forskaolei* (Rat's ear), and *Cucur bita* spp. (pumpkin leaves), which can all be utilized as green leafy vegetables. Such plant species were relied upon in the past as sources of vitamins, minerals, and proteins by rural societies. Despite their importance, ethnobotanical knowledge of traditional wild foods is declining in Kenya (Alva-Alvarado et al., 2023). In northern Kenya for example, collection of gums is mostly done by married women in an effort to provide an additional income for their households, perhaps adopting the role of single parenthood, especially the widowed. In addition, some wild food plants considered to be of minor significance are gathered by little children and are at times used as diet supplements and emergency foods. During collection of some wild edible plants in Kenya such as *Ficus* fruits (figs), *Vangueria* fruits, *Craibia laurentii* nuts, and *Maerua kirkii* nuts, children accompany their mothers to help in gathering while collection of some species such as tubers of *Cyphia glandulosa* is reportedly done by children as they go on with their duties. For example, among the hunter-gatherer communities in Kenya, wild foods may comprise the main diet of the day at certain times such as during famines (Mutie et al., 2023).

Introduction of exotic vegetables has diverted the focus on indigenous vegetables. Recognizing the value of wild food plants can be useful in conservation of germplasm for future generations as well as buffering against famine in changing climatic conditions. The need for the recognition of the nutritional value of traditional foods has resulted in campaigns for them to be incorporated into both rural and urban diets. It has focused on documentation of wild food plants at local levels. In Kitui county, studies have mostly focused on documentation of medicinal plants with little attention given to wild edible plants. Local utilization and acceptance of underutilized vegetable species have been reported in Kitui county where cowpeas are the most popular vegetable species (Sarfo et al., 2023). The overall aim of this study is to highlight the potential of wild edible plants in Kitui county as resources that can be utilized in combatting food insecurity and famine by the rural dry land communities.

7.2.1 Use of Edible Plants

About 690 million people (8.90% of the world population) worldwide suffer from hunger, with an increase of approximately 10 million people every year and a total increase of nearly 60 million people in the last 5 years. As such, the challenge of achieving food security has been listed as one of the 17 Sustainable Development Goals (SDGs) aimed to be accomplished by all the United Nations (UN) members in 2030. Achieving food security, eradicating hunger, improving human nutrition, and promoting sustainable agriculture are the goals of the UN SDG 2. Furthermore, the UN highlighted the potential benefits of indigenous knowledge in attaining food security through sustainable and proper utilization of biocultural resources. Failed agriculture, hunger, and undernutrition were problems faced by many developing countries in the 1972–1973 and 2007–2008 world food crises triggered by the global scarcity and unaffordability of rice. A number of countries were impacted together with the Philippines, which is part of the current global food system that primarily depends on three staples: wheat, maize, and rice.

According to the FAO, food and nutrition security can be met if everyone has physical, economic, and social access to enough food with adequate quantity and quality, that is, to suit their food preferences and nutritional requirements in terms of food diversity, nutrient quality, and safety aspects. Unfortunately, local varieties and breeds of domesticated plants and animals are disappearing worldwide along with the associated indigenous and local knowledge due to a shift toward intensification of agriculture with a small number of improved crop species and varieties (Intergovernmental Science-Policy Platform on Biodiversity and Ecosystem Services [IPBES], 2018). The decline in traditional agro-biodiversity and genetic diversity poses a serious risk to global food security by undermining the resilience of many local communities (Constas, 2023). Apart from rice, several species of edible plants have been used by various local and ethnic cultures, in particular by indigenous peoples. Many ethnic groups and indigenous communities have extensive knowledge of edible plants, many of which are neglected and underutilized species. The traditional food system of the indigenous peoples plays an important role in local food security by making them independent from the market system. The food system of indigenous peoples can be defined as foods that are free and locally accessible to the indigenous peoples. Moreover, indigenous peoples possess indigenous knowledge about edible plants in their local natural environment, either through farming or wild harvesting. This is in contrast to "market foods" that can be purchased from various commercial producers (i.e., cooking oil, canned foods, sugar, instant noodles, etc.), although in some circumstances, indigenous peoples may purchase some of their foods (i.e., coffee beans, poultry products, wild meat, rice, and wild vegetables) from others who own land and/or have time to collect or harvest in their community.

However, based on the IPBES, the consumption of traditional foods in the Asia-Pacific region is decreasing due to urbanization, rapid economic development coupled with growth in international trade, rural out-migration, and changes in lifestyles and dietary habits. As such, the rich indigenous knowledge about the use of neglected and underutilized edible plant species offers a largely unexploited resource to support food security, combat malnutrition, and achieve sustainable agriculture, that is, the key goals of SDG 2 (World Health Organization, 2023). An ethnobotanical approach is particularly promising in a biodiversity-rich yet food-insecure country.

Guinand and Lemessa classified WEPs into four broad categories, depending on plant parts consumed, consumer's perception (ordinary vs. starving), and trade associations (adults, children, men, women). There are four types of plants that fall into this category: (a) the typical 'mass starvation' plants, (b) the 'wild-food' plants to 'mass starvation' components, (c) the 'wild-food' plants that entice enhanced consumer groups throughout shortage-of-food durations, and (d) on-food plants with 'mass starvation' components.

All around the globe, wild edible plants play a critical role in maintaining the supply of food available to humans. An estimated one billion people throughout the globe rely on wild foods (mainly plants) on a regular basis. Most rural residents (pastoralists, shifting farmers, constant croppers, or gatherers) rely on them to provide food, income, and subsistence since they are cheap, easy to get, and useful in a variety of ways. Women and children, who are disproportionately represented among the poor and the excluded, rely heavily on such plants to survive periods of serious drought. WEPs have the potential to enhance dietary intake; cushion the effects of hunger, drought, and shocks; provide additional revenue; and supply genetic material for use in experiments, medicine, and the preservation of cultural and spiritual traditions in rural areas. Minerals and vitamins may be found, for example, in foods such as vegetables, fruit, and seeds (Khan et al., 2017). Domesticating edible wild fruits includes assessing species viability, selecting species, developing improved germplasm, propagating species, developing production systems, and commercialization products.

Ethiopia is a worldwide biodiversity hotspot and the birthplace of several edible plants and extensive ancient traditions of their use in WEPs. As a result, WEP consumption is a fundamental part of local dietary patterns. In all, Ethiopia used up about 413 WEPs. Their contribution toward the family's food supply might be supplementary, periodic, or emergency in nature. Consumption, though, is higher in regions with less access to nutritious food. Recurrent weather shocks, like those experienced in Konso, Derashe, and Burji of southern Ethiopia, have reduced agriculture productivity and contributed to food shortages. Weed-like edible plants such *Adansonia digitata*, *Balanites aegyptiaca*, *Carissa spinarum*, *Cordia africana*, *Tamarindus indica*, *Ximenia americana*, and *Ziziphus spina-christi* are also widely consumed in Northern Ethiopia. Many young children like WEPs. Children eat a wide variety of plant-based foods, including fruit of varieties including *Ficus*, *Carissa*, and *Rosa abyssinica*. However, in Ethiopia, the intake of wild plants, their relevance for food intake of remote regions, and their social and historical features are still poorly assessed and need adequate attention (Duguma, 2020).

7.2.2 WILD FRUITS AS A STRONG SOURCE OF FOOD SECURITY

Food and nutrition security are crucial for a healthy and productive life and are vital components of socio-economic development. People in the Himalayas suffer from significant food and nutrition security issues. Therefore, sustainable management and utilization of the Himalayan Mountains' natural resources such as their forests, pastures, and water supplies provide numerous possibilities for improving their socioeconomic position and food security. Wild plants and their fruits play a significant role in the lives of the Himalayan people/communities and are considered a major source of food, nutrition, and livelihood. These wild fruits are the native species that grow naturally without being cultivated and are known to possess a wide variety of secondary metabolites due to critical/harsh climatic conditions (Aziz et al., 2023b; Bashir et al., 2023; Arshad et al., 2022). According to our research, the nutritional content of Himalayan wild foods was far more diverse than that of commercially grown foods. One in six people on the planet relies on these wild fruits for their sustenance and they are an open-source supply of food and nutrition (Sharma, 2022).

Research and development activities focused on wild edible fruits could help in filling up the gap between population expansion and food scarcity, a growing problem in developing countries. Assessment of species potential, prioritization, enhanced germplasm development, propagation, production techniques, and product commercialization are all part of domesticating wild edible fruits. This selection of prospective wild species (Muhammad et al., 2023; Majeed et al., 2023a, 2023b) necessitates a thorough understanding of botanic traits, yield potential, producer and customer preferences, and needs. Research data on existing production and consumption patterns, cultivar development, nutritional content analyses, postharvest storage methods, and product processing are all needed to expand the cultivation of indigenous fruits.

These phytochemicals are responsible for different therapeutic activities such as anticancer, anti-inflammatory, neuroprotective, anti-diabetes, cardioprotective, antimicrobial, and antioxidant activities. The present review will give detailed information on different applications of wild fruits from the Himalaya region. Our research indicates that Himalayan wild fruits, including their nutraceuticals and chemical makeup as well as their potential health benefits, are the first of their kind to have been thoroughly examined (Abdullah et al., 2021).

7.2.3 CONTRIBUTION OF INDIGENOUS FOOD PREPARATION AND PRESERVATION TECHNIQUES TO ATTAINMENT OF FOOD SECURITY

Despite the fact that there is an undernutrition issue in the nation, postharvest food wastage is substantial. This is mostly due to inadequate food storage and processing facilities. The typical postharvest losses in the nation are estimated at between 30 and 80% for vegetables and fruits and between 20 and 40% for dairy and its substitutes by various researchers. Recognizing, promoting, and using indigenous knowledge, abilities, and traditions in food handling, preparation, preservation, and storage is one strategy to reduce massive postharvesting losses and boost food production. Indigenous knowledge is what has been passed from generation to generation within a community and has been shown to be adaptable in the face of change. The World Bank has acknowledged the value of indigenous knowledge as novel and distinctive among local producers, which may aid in the battle over malnutrition and poverty. This is due to people have relied on indigenous knowledge to make judgments about food safety at the neighborhood level.

However, documenting and disseminating traditional knowledge about food preparation, conservation, and storage remains a major difficulty in the country. Due to the breadth of this information's potential use in situations of crisis or adaptation, it is crucial that this knowledge and its associated behaviors be well recorded and widely disseminated. This might serve as a portal for academics and developmental mediators to validate and bolster indigenous knowledge with cutting-edge technology, allowing communities to better weather the challenges of a changing climate and advance their own role in achieving food and nutrition security. Due to the aforementioned, this research investigates the literature upon that importance of indigenous knowledge for ensuring food security (Pawlak and Kołodziejczak, 2020).

Vegetables, fruits, tubers, and nuts come from these areas, making them essential to maintaining food supplies and improving the quality of the food available. Some of the most important nutrients for a child's healthy growth and cognitive development come from the wild green vegetables that grow abundantly, as in the tropics. It is also crucial to many people, particularly the impoverished and the disadvantaged, in rural areas as well as metropolitan areas.

Since the Philippines is frequently struck by natural calamities like storms and droughts, if not by protracted dry and wet seasons, that generate seasonal or sometimes catastrophic food deficiencies, wild food plant collections have become commonplace. The widespread poverty of indigenous peoples makes them particularly vulnerable to food shortages. When food is scarce, indigenous families may turn to foraging, hunting, or even occasional menial work in metropolitan areas to supplement their food supplies. Reducing demand, consuming low-quality diets, missing one or even more meals each day, and even begging are some emergency methods. The indigenous Negrito people have been particularly affected by reports of food shortages. The estimated 33,000 indigenous Filipinos may be broken down into more than 30 linguistic and cultural groupings (Ong and Kim, 2017). Once nomadic gangs, people toward the close of the twentieth century had totally settled down in one place (Headland and Blood, 2002). However, despite the increasing tendencies in occasional transitioning to agricultural, wage labor, and currency economies, certain tribes, such as the Ati Negrito (Ati here forward) in the central Philippines, still exist by gathering and hunting of forest goods.

7.2.3.1 Role of Wild Edible Plants in Household Food Security

More than 80% of the people in Asia rely on medicinal plants as a main source of treatment since they are easily available, inexpensive, and culturally suitable. Particularly reliant on this traditionally known, scientifically simple, financially accessible, and often household food security are marginalized populations. Conventional medical practices in some countries of South and East Asia rely on the usage of hundreds of plant species in household food security. Protecting and promoting the cultural and spiritual qualities of household food security has widespread and consistent popular support in the area. In addition, it has been suggested that indigenous knowledge systems be tried to intervene with any other interference to achieve the objective of "food for all" in such a cost-effective manner as 80% of the people in developing countries rely on conventional foods for their own primary health care because it is natural, secure, non-narcotic, possesses no side effects, and is both preventive and curative (Iqbal et al., 2021).

"Finished labelled food goods that include substances from aerial or subterranean sections of plants or other plant matter or combinations, whether in the crude condition or in plant preparations," is how the World Health Organizations (WHO) defines herbal medicines. At least 25% of the medications in today's foods are derived from plants, and many of the synthetic counterparts are modelled by molecules first identified from these sources.

Pakistan, India, and China, the world's two most populous nations, are the primary source countries since they contain 40 percent of the world's biodiversity and are home to many rare species. These regions are famous for being the source of raw materials for foods. According to the United Nations, 33% of foods in the advanced industrialized nations come from plants.

7.2.3.2 Wild Food Plants and Food System of Tribal Cultures

Pashtunwali (sometimes spelled Pukhtunwali) is the culture that has been practiced and enjoyed by the local population for ages. Along the Pakistani and Afghan border in the Hindu Kush Mountains, Pashtunwali is a shared cultural legacy among the tribal groups of Pathans. It's the rules by which they live their lives, whether in the form of law or social norms. It states that tribal cultures' most fundamental institution is a tradition dating back two to three millennia. Pashtunwali is seen as unique culture in the present day, with a major impact on how people act, think, and feel. The three principles of Pashtunwali are still a code of honor (Nang), hospitality (Melmastia), and retribution (Badal). Pashtun culture relies heavily on and is distinguished by a strict adherence to an honor code that emphasizes the maintenance of one's own dignity and honor. In contrast to Badal, which is like a debt that is paid off via vengeance, Melmastia is a collection of norms and an instantaneous recompense from local communities with respect to local cultural values. The collective expectation of a tribe's members and of visitors may be summed up in these three cultural norms. In keeping with WFPs, Pathans embrace the giving and receiving of food among one another and the protection of their natural resources. For this reason, the researchers focused on the Bajaur region, which is the most populous, climatically diverse, and anthropologically rich of the Hindu Kush mountain range's tribal areas. This region is representative of the culture and traditional knowledge shared by the two countries along the entire Pakistan-Afghanistan border. The lowlands of the Hindu Kush Mountains are a generally arid environment. Barang in the south; Nawagai and Chamrkand in the southwest; and Salarzai, Mamund, and Utmankhel tehsils in the north and northeast all have significantly diverse climates, making it one of the key factors impacting all aspects of life, especially wild food and food systems. People in the area forage for food in the mountains, valleys, and farmland. Based on their extensive experience and understanding of the natural world, they are aware of the optimal times to harvest various wild vegetables. The nutritional content of different WEPs varies greatly depending on when and how they are collected. Tribal women traditionally prepare them in a certain way (Khan et al., 2021).

7.2.3.3 Himalayan Wild Fruits as a Strong Source of Food Security

Having reliable access to nutritious food is an important factor in maintaining good health and fostering economic growth. The people of the Himalayas have serious challenges in ensuring their access to adequate food and nourishment. Therefore, there are various opportunities to enhance their socioeconomic standing and nutritional security via ecological sustainability and exploitation of natural resources, such as pasture, forest, and water resources, as found in the Himalayan mountain range. The Himalayan inhabitants and towns rely heavily on wild plants and the fruits that provide for food, nutrition, and economic survival (Sinha et al., 2023; Bashir et al., 2023; Bhattacharya et al., 2023). Due to critical/harsh environmental circumstances, such wild fruits are indeed the native species that naturally occur without being farmed and thus are reported to exhibit a large diversity of secondary metabolites. Our investigation revealed that the nutritional profiles of the Himalayan forest produce differed greatly from those of mass-produced meals. Wild fruits are a freely available source of food and nourishment for one of every six people worldwide. Many studies have shown that biologically active compounds, antioxidants, minerals, and vitamins are found in Himalayan wild fruits such as *Berberis asiatica*, *Celtis australis*, *Ficus palmata*, *Fragaria indica*, *Morus alba*, *Myrica esculenta*, *Phyllanthus emblica*, *Prunus armeniaca*, *Pyracantha crenulata*, and *Terminalia chebula*. However, the range of Himalayan wild fruits lacks in-depth research on well-being properties, chemical makeup, and nutraceutical profile.

In order to make an informed decision on which wild species to pursue, you must have an in-depth familiarity with botanic characteristics, yield potential, producer preferences, and consumer demands. Expansion of farming this indigenous fruit requires data on current usage and production trends, cultivar improvement, nutrient content assessments, postharvest storage strategies, and approach to process (Haq et al., 2022a).

7.2.3.4 Wild Edible Plants as a Food Security

Wild edible plants and mushrooms (WEPMs) are "plants and mushroom which develop spontaneously form self-maintaining communities in wild or semi-natural ecosystems and may persist independently of direct human intervention," as defined by the Food and Agriculture Organization. (FAO). They have historically played a significant role in the diets of the vast majority of human communities, and both the species and the method of usage have adapted to suit varying environments and cultural norms. Wild edible plant assets have been studied by experts throughout the globe (Jamil et al., 2022) for centuries because of their potential to improve global food security by serving as a dietary supplement during food shortages and by offering certain uncommon nutrients. Numerous studies conducted all across Africa have shown that wild fruits may serve as a healthy addition to the average diet and as a suitable replacement for more expensive exotic fruits. As a result, they are considered a crucial tool for helping rural African families adapt to climate change (Akinola et al., 2020).

Major socioeconomic developments and the rural migration driving the concentrations of people in big cities have contributed to the loss of most of the traditional education and experience of gathering spontaneous plants of culinary importance in the Bamenda Highlands during the last century.

Many rural populations are becoming more impoverished and food insecure as a result of rising habitat destruction brought on by agricultural and pastoral growth. However, the "rediscovery" of wild food and medicinal species is happening all over the globe right now. Communities in rural areas are more likely to have preserved ethnobotanical expertise of wild edible plants. Traditional uses of edible wild plants and mushrooms in the research area were researched by interviewing locals about recent developments along with the historical accounts of the preceding few decades (Fongnzossie et al., 2020).

7.2.3.5 Wild Edible Plants Traditionally Collected and Used

In addition to hunting, gathering wild plants was a primary source of nutrition for early humans. This activity persisted even after the development of agriculture and livestock husbandry. As a result, the field known as "ethnobotany" investigates how people interact with plants. Its purpose is to catalogue the many wild plants (Majeed, Tariq, et al., 2022a) used by indigenous communities for medicinal, dietary, and commercial purposes (Cheek, et al., 2021).

The native people of Yemen still retain a close connection to WEPs, despite the availability of modern agricultural technologies. Economic downturns, endless war, and continuing political and cultural disputes are just a few of the factors that have prompted locals to look to wild plants as a significant food source. A lack of precipitation over a period of years is another defining feature of the climate in the remote southern parts of Yemen; as a result, agriculture failed to provide enough food. Until the early nineteenth century, the local economy relied solely on the sale of agricultural products grown within the region.

However, contemporary scientific research also shows that these plants are significant sources of resistance to disease. Al-Razi (865–910 C.E.), a physician practicing ancient Arab-Islamic medicine, made the first connection between diet and health. The phrase "your medicine is in your food" comes from his best-selling book, which is devoted to the "benefits of foods and prevent their negative impact." Because of this, we may say that Al-Razi was the pioneer of the field now known as nutraceutical sciences or heath nutrition (Al-Fatimi, 2021).

Studies of the phytochemical composition of several WEPs have so far concentrated on their nutritive and antioxidants components and how they could impact human health and nutrition. Primary metabolites, which are required for an organism's survival, include carbohydrates, lipids, protein, and mineral components; secondary metabolites, which are primarily antioxidant compounds like polyphenols and micronutrients, preserve plants and are found in lesser amounts in WEPs. When it comes to inorganic metabolites (minerals) necessary for animal and human health balances, wild food plants are superior to their domesticated counterparts. Differences among wild and farmed food plants are mostly due to the former's higher mineral composition.

Many people throughout the globe struggle with hunger and poor nutrition. Micronutrient inadequacies are believed to affect around two billion people worldwide, or one in ten individuals. These deficiencies increase people's susceptibility to illness and may be a major drag on economic development. Due to emerging environmental damage, dependence on conventional farming, and limited access to technology, the predicted crop production declines caused by climate change were also predicted to be particularly extreme in largely resourced regions of the world of sub-Saharan Africa, for which price fluctuations in 2006–2008 impacted an approximated 12 million people.

Resilience, defined as "a mix of adaptability in the face of disruption and ability to adapt to change," is increasingly being highlighted in treatments and policies aimed at lessening the negative effects of climate change for food production. Current climate change evaluations have found that low levels of adaptability in Africa are a major contributor to the continent's vulnerability. This is due to the continent's high rates of extreme poverty, weather extremes (especially water stress), low rates of output technology, and overall insufficient infrastructure. To our understanding, few studies have examined the role of current techniques within native welfare systems, despite the fact that many have attempted to address the significance of adaptability in farming systems, especially when viewed through nationwide socioeconomic and administration signifiers and in response to disasters (Djagba et al., 2019).

7.2.3.6   Wild Edible Plants as a Strategy to Increase Social-Ecological Resilience

In order to reduce their susceptibility to environmental changes, rural communities may use a number of different methods. During times of food scarcity, these "portfolio managers of choices" might range from simple, short-term activities like limiting food consumption or selling animals for money, to much more complicated, long-term acts like migration. Household members in rural Africa have found that eating WEPs, which the FAO defines as "plants that grow instantaneously in self-maintaining communities in pure or semi-natural eco-systems and also can remain independent of direct human action," is an effective way to mitigate the effects of climate change and adapt to less favorable conditions. Plants collected from agricultural regions, uncultivated regions, or forest areas are all included in the definition of "wild" for this research since they are not farmed.

The socioeconomic systems that enable the collection of these natural resources must be adequately safeguarded, managed, and valued to prevent overexploitation and degradation, especially given the significance of WEPs to family food security. In order to guide agricultural production, management of natural resources, and nutrition security protocols that might facilitate extra self-sustaining use of these assets and even boost their beneficial influence on community adaptability, a greater understanding of both ethnobotanical and WEP users is required. Poorly designed limitations may exclude people that depend on WEPs as a primary nutrition supply (Falconer, 1990) or force them to buy alternative crops in the market using limited economic resources, while unchecked unrestricted access can lead to unsustainable harvesting levels and deterioration (Stewart, 2003). The identification and safeguarding of localized indigenous knowledge systems to guide the collecting and use of WEPs is urgently needed to address this pressing problem (Majeed, Lu, et al., 2022b).

Diarrhea and other ailments have been linked to households lacking consistent access to nutritious meals. When diets missing sufficient nutrients diminish the body's capacity to fight illness, increasing vulnerability to infections in polluted water or food, the synergistic link among food insecurity and bad health becomes evident. Unfortunately, many communities waste opportunities to address food poverty and related health issues by neglecting to use wild edible plants (Majeed et al., 2022a), even though doing so might have significant health advantages. Food insecure communities in underdeveloped nations may benefit from eating wild edible plants.

People in sub-Saharan Africa often turn to uncultivated plants when they have insufficient access to modern food, healthcare, or medication because they have an intimate cultural and historical understanding of these plants and their uses and advantages. Particularly in locations where infectious illnesses are widespread and modern healthcare is limited, wild plants give traditional and culturally acceptable solutions for managing health requirements. Because of their antioxidant and antibacterial characteristics, several edible plants are used to supplement diet in poor nations, particularly during times of food crisis, when they are seen as having health advantages. Wild edible plants are an important part of the diet of many rural populations in sub-Saharan Africa, and their use is not limited to times of food scarcity (Okello et al., 2021).

There is a pressing need to quadruple protein output and expand worldwide food production through an estimated 60 percent from present levels to meet the demands of a growing human population (anticipated to exceed 9.6 billion by 2050). About 70 million new humans join the planet's population every year. By 2050, the world's population is projected to reach over 9 billion, more than doubling China's present population. Roughly 40 percent of the biomass produced by land and coastal oceans is eaten by humans. Livestock production accounts for around 30% of all land on Earth and 70% of agricultural land. While the demand for food is predicted to rise by 50% by 2030, food stocks are at a 50-year low. Attempts to feed the world's rapidly expanding population only via agricultural practices like animal farming seem hopeless. Insects, as a cheap and nutritious alternative to conventional food crops and animals, are being investigated as a means of coping with this predicament and improving the bleak outlook for global food security (Majeed et al., 2022a).

Just as with animals, the first usage of therapeutic herbs was based purely on instinct. Everything has been based on experience since there was not enough knowledge available at the time to determine the causes of the diseases or study the specific plant and how it might be used as a remedy.

Houses, clothes, food, flavorings, scents, and last but not least, medicines, have all been offered to man by plants. Intricate traditional medicine (TM) systems, some of which date back thousands of years, have relied on plants as a source of healing for humankind. Ancient societies were renowned for their extensive and precisely outlined herbal pharmacopoeias, which were the result of a methodical gathering of data on plants. Medicinal plant treatment is founded on the empirical discoveries of hundreds and maybe thousands of years, despite the fact that many of the therapeutic characteristics ascribed to plants have shown to be erroneous.

Plants that are used to treat, prevent, or cure illness, or to modify physiological and pathological processes, fall within OPS's definition of medicinal plants, as are plants that are utilized as a source of pharmaceuticals or their precursors. Any synthetic medication derived only from plants (or their components, juices, resins, and oils) in their raw or processed forms is considered phytopharmaceutical preparations or herbal medicine. Researchers focusing on ethnobotany are discovering more and more applications in fields like medicine and environmental protection.

There has been an incredible surge in interest and use of "green" medications. Many useful plant-derived drugs were initially discovered through ethnobotany and ethnopharmacology; however, in order to be successful, future efforts at mass bioprospecting will need to take into account the scientific expertise (in all relevant disciplines) of a team, as well as the expertise of those skilled in a broad range of human undertakings, including diplomacy (Van Andel et al., 2014).

The findings of ethnobotanical research have been invaluable in shaping global healthcare and conservation initiatives. Green medications are gaining a lot of attention and are becoming more mainstream. Many effective pharmaceuticals derived from plants were first discovered with the help of ethno-botany and ethno-pharmacology. Soejarto et al. (2005) argued that any future massive bio prospecting effort should take into account not only the scientific expertise of the team (in all relevant disciplines), but also the expertise of team members in a wide variety of human endeavors, such as diplomacy, international regulations and legal interpretations, social sciences, politics, anthropology, and good common sense. A thorough study of the abundance and distribution of target species is essential for the long-term usage of wildlife populations of medicinal herbs. Self-sustaining cultivation of wild populations is "one of the most misinterpreted and mishandled concepts in today's sustainability domain", and sustainable use seems to have no direct connection to the broader concept of "environmental conservation (Hussain et al., 2023; Hussain et al., 2023)", but in profession, resource managers often have few other options, particularly because of the long history of reliance of rural areas on cultivation from nature.

Human modifications of ecosystems have a profound effect on the planet's biodiversity, but the converse is also true: biodiversity, in the broadest sense, influences the characteristics of ecosystems and, by extension, the advantages that people get from them. Poor farmers, rural poor, and ancient culture, all of which depend heavily on a variety of ecosystem services (Haq et al., 2022b, 2022c; Hassan et al., 2022b), are among those who are most seriously threatened by biodiversity loss, and this is also true of medicinal plants. This research aims to assess how medicinal plants have been used and how they are being managed to ensure the sustainability of local communities.

Healing plants impact the environment (Majeed et al., 2022d). Traditional remedies used in India have been gathered from the country's woods for centuries. In India, there are 17,000 described varieties of higher plants, 7500 of which have been scientifically documented as having therapeutic effects. The Charak Samhita, a literature on herbal treatment compiled about 300 BC, details 340 herbal remedies and their indigenous applications. More and more people are turning to alternative medicine due to its low price and rising popularity of herbal remedies (Verma et al., 2023). Although there are many disorders that may be cured by allopathic medication, the exorbitant costs and unpleasant side effects are driving many people back to safer, more natural remedies like herbal medicines (Kala, 2004). The basic needs and income regeneration of rural families in the Mahipal area of semi-arid South-West Madagascar is mostly dependent on the natural resource exploitation. The overuse of resounding with much more than 90% of its species of plants and animals being unique, Madagascar is one of the significant biodiversity hotspots on Earth. However, ecological degradation poses a serious danger to these precious resources.

About a quarter of allopathic medications now are based on chemicals discovered in plants, and many more are synthetic counterparts developed from molecules first identified from these sources. WHO estimates that 80 percent of the global population relies on traditional medicine for their primary health care? The Himalayan mountains in northern India is home to a wide variety of useful plants for healing. About 1748 plant species within Indian Himalaya are used for medicinal purposes, out of a total of about 8000 (Singh et al., 2017). They include 8000 angiosperms, 44 gymnosperms, and 600 pteridophytes. Uttarakhand, located in the Himalayan region's northwest, has a comparatively verdant landscape (at 65%).

Uttarakh is home to the greatest diversity of known medicinal plant species. In contrast, the trans-Himalaya only manages to support around 337 kinds of medicinal plants owing to its unique topography and ecologically marginal circumstances (Singh et al., 2005). Since the demand for herbal remedies and medications derived from plants has increased dramatically in recent years, several species of medicinal plants have been overharvested. More than 150 species of plants have indeed gone extinct in the wild due to habitat degradation, inappropriate harvest, and overexploitation to suit the needs of the mostly illicit trade in medicinal herbs (Cunningham, 2000).

7.2.4   INDIGENOUS KNOWLEDGE AND WILD EDIBLE PLANTS

The tribe belt's ecosystem is very varied, including plains, woodlands, and pastures, which together provide a home for many different kinds of wild veggies and fruits. These WFPs are significant not just for their nutritious worth but also for their therapeutic significance; several WFPs have been examined for their pharmaceutical qualities in many different regions of the globe. For their health benefits, certain WFPs have now been labelled "functional foods" in recent years. They make for a nutritious meal and may help stave off disease.

Several reports have highlighted their value in terms of economic growth, poverty alleviation, dietary diversity, food safety, and agricultural diversification. Traditional societies often lack access to diverse and stable food sources, therefore WFPs play a crucial role in ensuring their members' nutritional well-being. Current instances include the COVID-19 pandemic catastrophe and the geopolitics and tribal instability throughout the globe, both of which highlight the critical role that wild food plants play in human survival throughout times of hunger and shortages (Gbashi et al., 2021).

Pakistan, with the fifth-highest population in the world, is classified as a lower-middle-income nation. Despite having all four seasons and a plenty of environmental assets, notably plant life, it is the eleventh least undernourished nation in the world. Sixty percent of the population is at risk of hunger. Disputes, isolation from urban centers, and a generally arid environment all contribute to higher rates of food insecurity in the country's tribal belt. Since ancient times, the geographical location of the Hindu Kush Mountainous region along the Pakistani–Afghan boundary has been the source of conflict. Other key causes of poverty and food insecurity within tribal areas are natural disasters, the tremendous expansion of the human species, restricted availability of food, and indigenous livelihood methods. Proper management and use of wild food plants may be a huge assist in reducing malnutrition and hunger. Making them an essential resource for low-income populations. Little is known about the indigenous vegetable and fruit varieties used inside the food production of Pakistan and even the Pakistani-Afghan surrounding regions as in the Hindu Kush mountainous region (Abdullah et al., 2021).

7.2.5   ECOLOGICAL STATUS OF WILD FOOD PLANTS

There is always a vested interest in exploring plant life for potential new food and medical sources. In order to combat human health issues and food insecurity, it is crucial to have access to the fruit and leaves of many tree species (both natural and farmed), particularly in the developing countries. A person's risk of developing diabetes, obesity, cancer, and cardiovascular disease is decreased when they eat fruit from trees. Most people in underdeveloped countries rely on medicinal plants as the foundation of their conventional healthcare systems. It is estimated that 500 million individuals in south Asian nations look to plants for protection against illness (Flood, 2010). With rising populations and often insufficient access to modern treatment, demand has been rising. Medicinal plants (Hassan et al., 2022a; Khoja et al., 2022; Tassadduq et al., 2022) are in high demand in Ethiopia for a number of reasons, including cultural linkages, the religious communities have had in herbal medicine, and the inexpensive cost of treatment. For as far as history can trace back, humans have hunted and collected to meet their basic necessities. Plant and animal products utilized for culinary, medical, cosmetic, or cultural purposes are only a few examples. For both cultural and economic reasons, people in affluent nations continue to forage for high-value items like mushrooms (matsutake, morels, and truffle) and medicinal herbs (ginseng, black cohosh, and goldenseal).

7.2.5.1   Wild Food Plants and Human Health

WEPs are edible wild plants. WEPs have been eaten for millennia. WEPs protect food sovereignty, security, and vulnerable households during food crop shortages. WEPs can enhance local market actors and reduce consumer-producer distance to reduce globalized value chains. The global food system can feed everyone, but many go hungry or lack good nutrition. Studies show processed food is unhealthy. A healthier food system is needed to combat climate change and extreme malnutrition. Hence, WEPs may improve certain people's diets. Traditional phytotherapy uses many species to treat many ailments. Functional foods have more vitamins, phenols, flavonoids, antioxidants, microelements, and fiber than cultivated crops and are healthier. Many prefer wild plants to pesticide-contaminated farmed ones. Hence, wild species may produce new colors, flavors, pharmaceuticals, and nutritional supplements. Decades of study relate nutrition and dietary variables to disease. Numerous studies show the Mediterranean Diet (MD) is healthy. In 2013, UNESCO included the MD to its Intangible Cultural Heritage of Humanity list. MD foods are tasty, nutritious, and heart-healthy and the source of daily life nutrients. Olive oil reduces MD diabetes and cancer. Wild and greenhouse-grown produce are in the MD. Rural diets always used the former. Finally, eating wild plants benefits the local economy and ecosystem. Eating and gathering them benefits cultural ecosystems.

Preparing, preserving, and nutritionally assessing wild foods is as vital as listing species. Researchers have studied rural communities' plant and product uses, particularly WEPs, in recent decades. European, American, African, and Asian floristic inventories have found widespread application. Italian WEPs lists 1103 alimurgic taxa. This study illuminates Italy's edible vascular plants and cultural benefits. European rural home garden food plants are less investigated than tropical ones. Ethnobotanical and ecological knowledge differs in regions with distinct linguistic populations, past political borders, geographic remoteness, or cultural influences (Chaughule and Barve, 2023).

7.2.5.2   Socio-Economic Benefits of Wild Food Plants

Gross national income per capita in Madagascar is $828, placing the country at position 151 out of 187 just on Human Development Index. (HDI). Seventy-four percent of the population is located in rural regions; of this group, 78 percent are classified as low-income and rely heavily on the direct extraction of natural resources (such as farmland, water, and forestland) to make ends meet. Some of the greatest regions of MAP diversity are located in the two global 'hotspots' of the eastern Himalayas and the western part of the Indian subcontinent, hence the region's natural rich biodiversity is intrinsically related to its medicinal plant variety (Majeed et al., 2022e). The region's biodiversity is under danger from a number of different reasons, including environmental, socioeconomic, and institutional issues. Local people in the Khasi area of Meghalaya reportedly earn roughly $0.75 million yearly from the sale of 2,800 tons of *Cinamomom tamala*. However, traditional and indigenous practices pertaining to these plants are dwindling and, in many instances, disappearing from the public sphere. Over half of India's higher blooming plant species are medicinal plants. This amounts to about 8,000 species (Bargali et al., 2022). Numerous rural homes, maybe in the millions, rely on medicinal plants for treatment. For preventative, primary, and secondary care, about 1.5 million Indian Systems of Medicine professionals employ medicinal plants. It is believed that India is home to more than 7800 factories.

The demand for herbal products has increased dramatically in recent years, leading to a dramatic increase in the number of organic materials exchanged both domestically and internationally. The worldwide market for commerce connected to medicinal plants is estimated by the EXIM Bank to be US$ 60 billion annually, rising at a pace of barely 7% (Lenin, 2010). While India has a wide variety of plant and animal life, the country's increasing population is placing a significant pressure on the country's limited supply. Some medicinal plants are becoming more endangered in their native environment (Aziz et al., 2023a, 2023b), despite the rising demand for them. The future demands can only be met if medicinal plant cultivation is promoted. In an all-India ethno-biological study undertaken by the Ministry of Environmental & Forestry, India's government found that the people of India make use of more than 8,000 different plant species (Kamble & Nitave, 2022). Nearly every person in Africa has used traditional herbal medicine at some point in their lives, and the yearly market for these goods is approaching US$60 billion. Research into traditional herbal therapy is believed by many to be of vital importance to the health of people everywhere. Traditional herbal medicine research has received significant funding from India, China and Nigeria, the United States, and the World Health Organization (WHO). The pharmaceutical and chemical industries have both spent millions of dollars on the search for new and effective chemical compounds and therapeutic plants. When compared to the rest of the pharmaceutical business, this is still a pretty small expenditure, but it does bring up some important ethical problems that aren't often encountered in the course of developing new medications.

7.2.5.3 Nutritional Benefits of Wild Food Plants

People have been eating wild food plants for hundreds of years, not just because of the one-of-a-kind flavors they impart, but also because of the myriad nutritional benefits they bring. Certain plants, such blackberries, elderberries, and blueberries, often contain a high concentration of antioxidants; these antioxidants protect our cells from the damage that is created by free radicals. In addition, wild plants have a high amount of fiber, which not only contributes to healthy digestion but also reduces the risk of a variety of diseases, including coronary heart disease, obesity, and diabetes. These plants are frequently an extremely rich source of a wide range of vitamin and mineral compounds. Such examples include stinging nettle, which is rich in both iron and calcium, and dandelion greens, which are strong in vitamin K. Dandelion greens are also an example (Baldi et al., 2022). Another good example is stinging nettle, which is rich in the mineral potassium. Wild plants, in general, have a low-calorie count, which makes them a great choice of food for those who wish to lower their weight while maintaining their nutrition levels at a healthy level. This is the most essential aspect of why eating wild plants is such a good idea. Wild food plants, in general, offer a wide variety of health benefits, which is one of the reasons why they continue to play such an important role in the human diet even in the modern day. Another reason is that wild food plants can be found in a wider number of habitats.

Roughly 80% of the world's marginal populations rely on medicinal plants as their major source of healthcare. In addition to containing pharmacologically significant photochemical, each pharmaceutical species of plants has its own unique nutritional content. These vitamins and minerals are crucial to human health since the body needs them to operate properly. Carbohydrates, lipids, and proteins, among others, are essential biochemicals and micronutrients for human survival. More than 300 kinds of annual or perennial plants and shrubs, *Hibiscus* (family Malvaceae) are also known as Roselle or crimson sorrel. One of them is the multipurpose *Hibiscus sabdariffa* L. Originally from West Africa, this crop is now cultivated extensively in Northeastern India's Brahmaputra Valley (Besharati et al., 2022).

A lot of people in India, Africa, and Mexico utilize this plant for medicinal purposes. Previously, infusion of leaves or calyces have been employed for its diuretic, choleretic, mixing up, and hypotensive effects; these benefits include a reduction in blood viscosity and an increase in intestinal peristalsis. It's also helpful for treating things like cancer and liver damage as well as heart and nerve disorders. Tea is used as a major component in industrially manufactured teas and drinks all over the globe, in addition to its value as a foodstuff or conventional healers in the nations of its geographic origin. The tender, immature leaves and stems may be eaten fresh in salads or cooked to use as a spice in dishes like curry. The fresh calyx (the flower's outer whorl) may be eaten raw in salads or prepared and utilized as a flavor in cakes, jellies, sauces, soups, pickles, puddings, and so on.

The human body's innate enzymatic and nonenzymatic antioxidants defense are very complex and effective at scavenging free radicals as well as other oxidants. Consuming foods rich in antioxidants is a simple way to increase your immune system of the body against free radicals. The phenol component concentration of plant materials has been shown to correlate positively with their antioxidant potential. The human diet relies heavily on phenol chemicals, notably phenol acids and flavonoids, which may be found in a wide variety of plant foods such as vegetables, fruit, seeds, tea, wines, and juices. They have been essential in the development of the medical field. A diet rich in minerals and antioxidants may help strengthen the body's immune system. Including foods high in antioxidants in our diets may help strengthen our immune systems while also being more cost-effective and less risky than commercially supplied antioxidants. Antioxidants may be found in abundance in hibiscus (Nguyen et al., 2022).

7.2.6 Conservation and Management of Wild Food Plants

Protecting medicinal plants and making efficient use of resources are crucial to the pharmaceutical industry's long-term viability. Medicinal herbs have long been utilized as a foundation for natural remedies for a variety of health issues. Millions of people still use these herbs every year, putting their faith in their purported healing properties. This highlights the critical need of preserving these ecosystems (Khan et al., 2022).

The increasing demand for natural remedies is directly responsible for the fast depletion of these natural resources. Overharvesting, deforestation, and urbanization are only a few examples of unsustainable practices that contribute to the dwindling supply of medicinal plants (Majeed, Bhatti, & Amjad, 2021b). This underscores the need for developing and implementing a sustainable strategy for managing medicinal plant supply.

Planting rare and endangered species and employing efficient harvesting techniques are two examples of sustainable resource management strategies that help conserve medicinal plants (Majeed et al., 2021b). The need to preserve genetic diversity in the fight against genetic deterioration should also be emphasized. Taking these steps will ensure that future generations may enjoy our natural resources. Protecting and making ethical use of the world's medicinal plant resources is crucial to the pharmaceutical industry's future. Quick action is needed from stakeholders to implement strategies that promote the preservation and long-term utilization of these assets. That's the only way to guarantee that future generations will have access to these natural riches (Zehra et al., 2023).

So over the last three decades, the worldwide environmental crisis has prompted the long-overdue recognition that humans are an integral part of the natural world, ushering in a new paradigm for the field of biology and ecology (Majeed, Khan, et al., 2022a), which had previously tended to view natural objects as completely divorced from the realms of society and government (Haq et al., 2022d). Much emphasis has been paid in resource management to the value of ethno knowledge of biology (here we regard ethnobotany or ethnoecology as separate areas under the general field known as ethnobiology).

The World Wildlife Foundation and UNESCO's joint Peoples and Plant project has also encouraged studies of indigenous people's botanical expertise and the use of such insights into local managing resources (Cunningham, 2000). Understanding the interplay between society and the environment processes has been facilitated by the incorporation of local-use trends and the cultural and institutional context that directs human interactions with the natural world into ecologic and biological investigations (Majeed et al., 2023b; Tariq et al., 2023). In a dialectical interaction, local practices and ethno-biological knowledge change the environment and its plant populations. Local communities' knowledge, beliefs, and values linked with the many components of the environment are challenged with global perspectives on biodiversity protection and scientific knowledge of ecological systems within the framework of community-based programmers.

Accurate management techniques that include both science and local information need an examination of local policies and expertise. It has been shown in recent research that indigenous practices, like complex adaptive systems, can cope with uncertainty and adjust to environmental changes (Spannring & Hawke, 2022). New hypotheses for management-related research studies may be produced by combining local knowledge and practices into the scientific research process. There hasn't been a lot of focus on the fact that knowledge differs between and within cultural groups, nor has there been a clear connection drawn between this information's bearers' intentions and actions.

7.2.7 CHALLENGES OF WILD FOOD PLANTS

Despite the widespread availability of Western medicine, significant populations in developing nations, particularly in rural regions where 80% or more of the population resides, continue to depend on traditionally religious medicine (TRM) for primary care. The vast majority (95%) of TRM is made up of medicinal herbs (Kindie & Mengistu, 2022). The lack of adequate healthcare infrastructure, trained medical professionals, pharmaceuticals, or other emergency aid is a key driving factor in the reliance on TRM and its adherents in these nations. Beneficial herbal products are often gathered from their natural habitats (Muhammad et al., 2023; Majeed et al., 2021a). TRM and its adherents are deeply entwined with the values and norms of the local community. Sacred woods, which were culturally forbidden and were regions where herbal medicines were located, have been encroached upon as a consequence of the current pattern of population expansion. Matambiko rituals also take place in sacred woodlands (traditional sacrifice). Not only that, but many underdeveloped nations are introducing an increase in the number of enterprises engaged in mines, road and railroad building, and big or estate farming. In addition to these issues, a great deal of biodiversity is being lost in Africa due to bush fires, logging, and bio prospectors looking for medicinal and fragrant species. A recent FAO research, for instance, found that global tropical forest cover was cut by such an averaged approximately 15.4 million hectare per year (8% per year) between 1980 and 1990. There are more trees being cut down in Africa than are being planted each year. A yearly depletion rate of 300,000–400,000 hectares is predicted for Tanzania (MoussaKourouma et al., 2022).

7.3 CONCLUSION

This study looked at indigenous tribes' dietary practices throughout times of transition to see how they deal with shifts in both social and natural contexts, specifically in terms of food availability and security. Focusing on indigenous populations in transit, including those with a gathering and hunting past who now reside in the peri-urban area, this study sought to demonstrate that preferences in kinds of WEPs might transmit information about their circumstances. In our opinion, this research has the potential to provide something new and interesting to the body of knowledge on wild plants and food safety. Since WEP productivity, consumption, and variety have all fallen due to a lack of attention in scientific study, development, and policy, it is probable that this has had a negative impact on rural family nutrition. The consumption of wild plants and animals continues to provide a vital source of nutrition and money for a large number of people, primarily in developing nations. The primary source of TRM's resources, medicinal plants, are affected by these human socioeconomic activities. Wild plants in underdeveloped nations have been the subject of several studies documenting their botanical, nutritional, and/or ethnomedicinal properties. However, it would seem that no research has looked at the links between food insecurity, health consequences, and the use of edible wild plants. The goals of this research were to (1) explain how communities cope with food scarcity by using wild edible plants, and (2) determine whether or not self-reported diarrhea is associated with ingestion of natural, edible plant material across various categories of domestic food security.

REFERENCES

Abdullah, A., S. M. Khan, A. Pieroni, A. Haq, Z. U. Haq, Z. Ahmad, S. Sakhi, A. Hashem, A. B. F. Al-Arjani, A. A. Alqarawi, and E. F. Abd-Allah. 2021. "A Comprehensive Appraisal of the Wild Food Plants and Food System of Tribal Cultures in the Hindu Kush Mountain Range; A Way Forward for Balancing Human Nutrition and Food Security." *Sustainability* 13, no. 9: 5258.

Akinola, R., L. M. Pereira, T. Mabhaudhi, F. M. De Bruin, and L. Rusch. 2020. "A Review of Indigenous Food Crops in Africa and the Implications for More Sustainable and Healthy Food Systems." *Sustainability* 12, no. 8: 3493.

Al-Fatimi, M. 2021. "Wild Edible Plants Traditionally Collected and Used in Southern Yemen." *Journal of Ethnobiology and Ethnomedicine* 17, no. 1: 1–21.

Ali, H., Z. Muhammad, M. Majeed, R. Aziz, A. Khan, W. M. Mangrio, ... A. A. Al Dughairi. 2023. "Vegetation Diversity Pattern during Spring Season in Relation to Topographic and Edaphic Variables in Sub-tropical Zone." *Botanical Studies* 64, no. 1: 25.

Arif, U., K. H. Bhatti, M. Ajaib, N. A. Wagay, M. Majeed, J. Zeb, A. Hameed, and J. Kiani. 2021. "Ethnobotanical Indigenous Knowledge of Tehsil Charhoi, District Kotli, Azad Jammu and Kashmir, Pakistan." *Ethnobotany Research and Applications* 22. https://doi.org/10.32859/ERA.22.50.1-24

Aziz, A., M. M. Anwar, M. Majeed, J. A. Albanai, H. Almohamad, and H. G. Abdo. 2023a. "Quantifying the Impacts of Urbanization on Urban Green, Evidences from Maga City, Lahore Pakistan." *Discover Sustainability* 4, no. 1: 1–11.

Aziz, A., M. M. Anwar, M. Majeed, S. Fatima, S. S. Mehdi, W. M. Mangrio, A. Elbouzidi, M. Abdullah, S. Shaukat, N. Zahid, E. A. Mahmoud, R. Casini, K. Yessoufou, and H. O. Elansary. 2023b. "Quantifying Landscape and Social Amenities as Ecosystem Services in Rapidly Changing Peri-Urban Landscape." *Land* 12, no. 2. https://doi.org/10.3390/land12020477

Baldi, A., P. Bruschi, S. Campeggi, T. Egea, D. Rivera, C. Obón, and A. Lenzi. 2022. "The Renaissance of Wild Food Plants: Insights from Tuscany (Italy)." *Foods* 11, no. 3: 300.

Bargali, H., A. Kumar, and P. Singh. 2022. "Plant Studies in Uttarakhand, Western Himalaya: A Comprehensive Review." *Trees, Forests and People*, 8: 100203.

Bashir, S. M., M. Altaf, T. Hussain, M. Umair, M. Majeed, W. M. Mangrio, A. M. Khan, A. B. Gulshan, M. H. Hamed, S. Ashraf, M. S. Amjad, R. W. Bussmann, A. M. Abbasi, R. Casini, A. Alataway, A. Z. Dewidar, M. Al-Yafrsi, M. Amin, and H. O. Elansary. 2023. "Vernacular Taxonomy, Cultural and Ethnopharmacological Applications of Avian and Mammalian Species in the Vicinity of Ayubia National Park, Himalayan Region." *Biology* 12, no. 4. https://doi.org/10.3390/biology12040609

Besharati, J., M. Shirmardi, H. Meftahizadeh, M.D. Ardakani, and M. Ghorbanpour. 2022. "Changes in Growth and Quality Performance of Roselle (*Hibiscus sabdariffa* L.) in Response to Soil Amendments with Hydrogel and Compost under Drought Stress." *South African Journal of Botany* 145: 334–347.

Bhattacharya, R., R. Bhattacharya, M. Majeed, S. Bhandari, R. Aziz, D. Sinha, ... S. S. Niyogi. 2023. "Environmental Rehabilitation of Industrial Waste Dumping Site." In *Biohydrometallurgical Processes*, 44–89. CRC Press.

Chaughule, R.S., and R.S. Barve. 2023. "Role of Herbal Medicines in the Treatment of Infectious Diseases." *Vegetos*, 37: 1–11.

Cheek, M., S. Hatt, and J.M. Onana. 2021. "The Endemic Plant Species of Bali Ngemba Forest Reserve, Bamenda Highlands Cameroon, with a New Endangered Cloud-Forest Tree Species *Vepris onanae* (Rutaceae)." *BioRxiv*, 2021–10.

Constas, M.A. 2023. "Food Security and Resilience: The Potential for Coherence and the Reality of Fragmented Applications in Policy and Research." In *Resilience and Food Security in a Food Systems Context*, 147–184. Cham: Springer International Publishing.

Djagba, J.F., S.J. Zwart, C.S. Houssou, B.H. Tenté, and P. Kiepe. 2019. "Ecological Sustainability and Environmental Risks of Agricultural Intensification in Inland Valleys in Benin." *Environment, Development and Sustainability* 21 (2019): 1869–1890

Duguma, H.T. 2020. "Wild Edible Plant Nutritional Contribution and Consumer Perception in Ethiopia." *International Journal of Food Science* (2020), 2958623.

Ehrhardt, S., G.D. Burchard, C. Mantel, J.P. Cramer, S. Kaiser, M. Kubo, R.N. Otchwemah, U. Bienzle, and F.P. Mockenhaupt. 2006. "Malaria, Anemia, and Malnutrition in African Children—Defining Intervention Priorities." *The Journal of Infectious Diseases* 194, no. 1: 108–114.

Flood, J. 2010. "The Importance of Plant Health to Food Security." *Food Security* 2, no. 3: 215–231.

Fongnzossie, E.F., C.F.B. Nyangono, A.B. Biwole, P.N.B. Ebai, N.B. Ndifongwa, J. Motove, and S.D. Dibong. 2020. "Wild Edible Plants and Mushrooms of the Bamenda Highlands in Cameroon: Ethnobotanical Assessment and Potentials for Enhancing Food Security." *Journal of Ethnobiology and Ethnomedicine* 16 (2020): 1–10.

Gbashi, S., O. Adebo, J.A. Adebiyi, S. Targuma, S. Tebele, O.M. Areo, B. Olopade, J.O. Odukoya, and P. Njobeh. 2021. "Food Safety, Food Security, and Genetically Modified Organisms in Africa: A Current Perspective." *Biotechnology and Genetic Engineering Reviews* 37, no. 1: 30–63

Haq, S. M., A. Tariq, Q. Li, U. Yaqoob, M. Majeed, M. Hassan, S. Fatima, M. Kumar, R. W. Bussmann, M. F. U. Moazzam, and M. Aslam. 2022d. "Influence of Edaphic Properties in Determining Forest Community Patterns of the Zabarwan Mountain Range in the Kashmir Himalayas." *Forests* 13, no. 8. https://doi.org/10.3390/f13081214.

Haq, S. M., U. Yaqoob, M. Majeed, M. S. Amjad, M. Hassan, R. Ahmad, and A. Morales-de la Nuez. 2022b. "Quantitative Ethnoveterinary Study on Plant Resource Utilization by Indigenous Communities in High-Altitude Regions." *Frontiers in Veterinary Science* 9: 944046.

Haq, S.M., M. Hassan, H.A. Jan, A.A. Al-Ghamdi, K. Ahmad, and A.M. Abbasi. 2022c. "Traditions for Future Cross-National Food Security—Food and Foraging Practices among Different Native Communities in the Western Himalayas." *Biology* 11, no. 3: 455.

Haq, S.M., M. Hassan, R.W. Bussmann, E.S. Calixto, I.U. Rahman, S. Sakhi, F. Ijaz, A. Hashem, A.B.F. Al-Arjani, K.F. Almutairi, and E.F. Abd-Allah. 2022a. "A Cross-Cultural Analysis of Plant Resources Among Five Ethnic Groups in the Western Himalayan Region of Jammu and Kashmir." *Biology* 11, no. 4: 491.

Hassan, M., S. M. Haq, M. Majeed, M. Umair, H. A. Sahito, M. Shirani, M. Waheed, R. Aziz, R. Ahmad, R. W. Bussmann, A. Alataway, A. Z. Dewidar, T. K. Z. El-Abedin, M. Al-Yafrsi, H. O. Elansary, and K. Yessoufou. 2022a. "Traditional Food and Medicine: Ethno-Traditional Usage of Fish Fauna across the Valley of Kashmir: A Western Himalayan Region." *Diversity* 14, no. 6: 455. https://doi.org/10.3390/d14060455

Hassan, M., S. M. Haq, R. Ahmad, M. Majeed, H. A. Sahito, M. Shirani, I. Mubeen, M. A. Aziz, A. Pieroni, R. W. Bussmann, A. Alataway, A. Z. Dewidar, M. Al-Yafrsi, H. O. Elansary, and K. Yessoufou. 2022b. "Traditional Use of Wild and Domestic Fauna among Different Ethnic Groups in the Western Himalayas—A Cross Cultural Analysis." *Animals* 12, no. 17: 2276. https://doi.org/10.3390/ani12172276

Hussain, S., A. Raza, H. G. Abdo, M. Mubeen, A. Tariq, W. Nasim, M. Majeed, H. Almohamad, and A. A. Al Dughairi. 2023. "Relation of land surface temperature with different vegetation indices using multi-temporal remote sensing data in Sahiwal region, Pakistan." *Geoscience Letters* 10, no. 1: 33. https://doi.org/10.1186/s40562-023-00287-6

Iqbal, M.S., K.S. Ahmad, M.A. Ali, M. Akbar, A. Mehmood, F. Nawaz, S.A. Hussain, N. Arshad, S. Munir, H. Arshad, and K. Shahbaz. 2021. "An Ethnobotanical Study of Wetland Flora of Head Maralla Punjab Pakistan." *PLOS ONE* 16, no. 10: e0258167.

Jamil, M. D., M. Waheed, S. Akhtar, N. Bangash, S. K. Chaudhari, M. Majeed, M. Hussain, K. Ali, and D. A. Jones. 2022. "Invasive Plants Diversity, Ecological Status, and Distribution Pattern in Relation to Edaphic Factors in Different Habitat Types of District Mandi Bahauddin, Punjab, Pakistan." *Sustainability (Switzerland)* 14, no. 20. https://doi.org/10.3390/su142013312

Khan, A. M., Q. Li, Z. Saqib, N. Khan, T. Habib, N. Khalid, M. Majeed, and A. Tariq. 2022. "MaxEnt Modelling and Impact of Climate Change on Habitat Suitability Variations of Economically Important Chilgoza Pine (*Pinus gerardiana Wall.*) in South Asia." *Forests* 13, no. 5. https://doi.org/10.3390/f13050715

Khan, F.A., S.A. Bhat, and S. Narayan. 2017. *Wild Edible Plants as a Food Resource: Traditional Knowledge.* University of Agricultural Science and Technology, Research Gate, March.

Khan, S., W. Hussain, S. Shah, H. Hussain, A.E. Altyar, M.L. Ashour, and A. Pieroni. 2021. "Overcoming Tribal Boundaries: The Biocultural Heritage of Foraging and Cooking Wild Vegetables among Four Pathan Groups in the Gadoon Valley, NW Pakistan." *Biology* 10, no. 6: 537.

Khoja, A. A., S. M. Haq, M. Majeed, M. Hassan, M. Waheed, U. Yaqoob, R. W. Bussmann, A. Alataway, A. Z. Dewidar, M. Al-Yafrsi, H. O. Elansary, K. Yessoufou, and W. Zaman. 2022. "Diversity, Ecological and Traditional Knowledge of Pteridophytes in the Western Himalayas." *Diversity* 14(8). https://doi.org/10.3390/d14080628

Majeed, M., A. M. Khan, T. Habib, M. M. Anwar, H. A. Sahito, N. Khan, and K. Ali. 2022a. "Vegetation Analysis and Environmental Indicators of an Arid Tropical Forest Ecosystem of Pakistan." *Ecological Indicators* 142. https://doi.org/10.1016/j.ecolind.2022.109291

Majeed, M., A. Tariq, S. M. Haq, M. Waheed, M. M. Anwar, Q. Li, … A. Jamil. 2022d. "A Detailed Ecological Exploration of the Distribution Patterns of Wild Poaceae from the Jhelum District (Punjab), Pakistan." *Sustainability* 14, no. 7: 3786.

Majeed, M., K. H. Bhatti, A. Pieroni, R. Sõukand, R. W. Bussmann, A. M. Khan, and M. S. Amjad. 2021a. "Gathered Wild Food Plants among Diverse Religious Groups in Jhelum District, Punjab, Pakistan." *Foods* 10, no. 3: 594.

Majeed, M., K. H. Bhatti, and M. S. Amjad. 2021b. "Impact of Climatic Variations on the Flowering Phenology of Plant Species in Jhelum District, Punjab, Pakistan." *Applied Ecology and Environmental Research* 19, no. 5: 3343–3376. https://doi.org/10.15666/aeer/1905_33433376

Majeed, M., L. Lu, S. M. Haq, M. Waheed, H. A. Sahito, S. Fatima, … M. Aslam. 2022b. "Spatiotemporal Distribution Patterns of Climbers along an Abiotic Gradient in Jhelum District, Punjab, Pakistan." *Forests* 13, no. 8: 1244.

Majeed, M., Lu, L., Anwar, M. M., Tariq, A., Qin, S., El-Hefnawy, M. E., El-Sharnouby, M., Li, Q., & Alasmari, A. 2023c. "Prediction of Flash Flood Susceptibility Using Integrating Analytic Hierarchy Process (AHP) and Frequency Ratio (FR) Algorithms." *Frontiers in Environmental Science* 10. https://doi.org/10.3389/fenvs.2022.1037547

Majeed, M., M. Muhammad, S. Nawaz, T. Naz, M. M. Iqbal, N. Zahid, … M. Mehboob. 2023b. "Nanopriming for Crop Management for Sustainable Agriculture: An Overview." In *Nanopriming Approach to Sustainable Agriculture*. 110–141.

Majeed, M., M. Muhammad, T. Hussain, M. Ali, T. Naz, S. Nawaz, … G. Abbas. 2023a. "Nanopriming-Based Management of Biotic Stresses for Sustainable Agriculture." In *Nanopriming Approach to Sustainable Agriculture*. 263–289. IGI Global.

Majeed, M., Tariq, A., Haq, S. M., Waheed, M., Anwar, M. M., Li, Q., Aslam, M., Abbasi, S., Mousa, B. G., & Jamil, A. 2022c. "A Detailed Ecological Exploration of the Distribution Patterns of Wild Poaceae from the Jhelum District (Punjab), Pakistan." *Sustainability (Switzerland)* 14, no. 7. https://doi.org/10.3390/su14073786

MoussaKourouma, J., D. Phiri, A.T. Hudak, and S. Syampungani. 2022. "Land Use/Cover Spatiotemporal Dynamics, and Implications on Environmental and Bioclimatic Factors in Chingola District, Zambia." *Geomatics, Natural Hazards and Risk* 13, no. 1: 1898–1942.

Moyo, M. P., S. Tatsvarei, and T. Rukasha. 2022. "Commercialization of Indigenous Vegetable Value Chains: A Review of Selected African Countries." *International Journal of Development and Sustainability* 11: 184–201.

Muhammad, M., A. Basit, M. Majeed, A. A. Shah, I. Ullah, H. I. Mohamed, … A. M. Ghanaim. n.d. "Bacterial Pigments and Their Applications." In *Bacterial Secondary Metabolites*, 277–298. Elsevier.

Muhammad, M., A. Waheed, A. Wahab, M. Majeed, M. Nazim, Y. H. Liu, … W. J. Li. 2023. "Soil Salinity and Drought Tolerance: An Evaluation of Plant Growth, Productivity, Microbial Diversity, and Amelioration Strategies." *Plant Stress* 100319.

Mutie, F.M., Y.M. Mbuni, P.C. Rono, E.M. Mkala, J.M. Nzei, M. Phumthum, G.W. Hu, and Q.F. Wang. 2023. "Important Medicinal and Food Taxa (Orders and Families) in Kenya, Based on Three Quantitative Approaches." *Plants* 12, no. 5: 1145.

Nguyen, Q.D., T.T. Dang, T.V.L. Nguyen, T.T.D. Nguyen, and N.N. Nguyen. 2022. "Microencapsulation of Roselle (*Hibiscus sabdariffa* L.) Anthocyanins: Effects of Different Carriers on Selected Physicochemical Properties and Antioxidant Activities of Spray-Dried and Freeze-Dried Powder." *International Journal of Food Properties* 25, no. 1: 359–374.

Nsevolo, M.P., N. Kiatoko, M.B. Kambashi, F. Francis, and R.C. Megido. 2023. "Reviewing Entomophagy in the Democratic Republic of Congo: Species and Host Plant Diversity, Seasonality, Patterns of Consumption and Challenges of the Edible Insect Sector." *Journal of Insects as Food and Feed* 9, no. 2: 225–244.

Okello, M., J. Lamo, M. Ochwo-Ssemakula, and F. Onyilo. 2021. "Challenges and Innovations in Achieving Zero Hunger and Environmental Sustainability through the Lens of Sub-Saharan Africa." *Outlook on Agriculture* 50, no. 2: 141–147.

Ong, H.G. and Y.D. Kim. 2017. "The Role of Wild Edible Plants in Household Food Security Among Transitioning Hunter-Gatherers: Evidence from the Philippines." *Food Security* 9: 11–24.

Onomu, A.R. 2023. "Pitfalls and Potential Pathways to Commercialization of Indigenous Food Crops, Fruits, and Vegetables in Africa." *Asian Journal of Agriculture and Rural Development* 13, no. 1: 25–38.

Pawlak, K. and M. Kołodziejczak. 2020. "The Role of Agriculture in Ensuring Food Security in Developing Countries: Considerations in the Context of the Problem of Sustainable Food Production." *Sustainability* 12, no. 13: 5488.

Richardson, K., R. Calow, L. Mayhew, G. Jobbins, G. Daoust, A. Waterson, ... L. Burgin. 2023. *Climate Risk Report for the Southern Africa Region.*

Sabourjanati, M., & Ghalandarian, I. 2022. Empowering Urban Ecotourism: A New approach to Development Local Communities Empowerment. *Quarterly Journals of Urban and Regional Development Planning* 7, no. 23: 87–120.

Sarfo, J., E. Pawelzik, and G.B. Keding. 2023. "Fruit and Vegetable Processing and Consumption: Knowledge, Attitude, and Practices among Rural Women in East Africa." *Food Security* 2023: 1–19.

Singh, A., M. C. Nautiyal, R. M. Kunwar, and R. W. Bussmann. 2017. "Ethnomedicinal Plants Used by Local Inhabitants of Jakholi Block, Rudraprayag District, Western Himalaya, India." *Journal of Ethnobiology and Ethnomedicine* 13: 1–29.

Sinha, D., A. K. Maurya, G. Abdi, M. Majeed, R. Agarwal, R. Mukherjee, ... J. T. Chen. 2023. "Integrated Genomic Selection for Accelerating Breeding Programs of Climate-Smart Cereals." *Genes* 14, no. 7: 1484.

Spannring, R., and S. Hawke. 2022. "Anthropocene Challenges for Youth Research: Understanding Agency and Change through Complex, Adaptive Systems." *Journal of Youth Studies* 25, no. 7: 977–993.

Tariq, A., Mumtaz, F., Majeed, M., & Zeng, X. 2023. "Spatio-temporal assessment of land use land cover based on trajectories and cellular automata Markov modelling and its impact on land surface temperature of Lahore district Pakistan." *Environmental Monitoring and Assessment* 195, no. 1. https://doi.org/10.1007/s10661-022-10738-w

Tassadduq, S. S., Akhtar, S., Waheed, M., Bangash, N., Nayab, D. E., Majeed, M., Abbasi, S., Muhammad, M., Alataway, A., Dewidar, A. Z., Elansary, H. O., & Yessoufou, K. 2022. "Ecological Distribution Patterns of Wild Grasses and Abiotic Factors." *Sustainability (Switzerland)* 14, no. 18. https://doi.org/10.3390/su141811117

Ullah, I., Aslam, B., Shah, S. H. I. A., Tariq, A., Qin, S., Majeed, M., & Havenith, H. B. 2022. "An Integrated Approach of Machine Learning, Remote Sensing, and GIS Data for the Landslide Susceptibility Mapping." *Land* 11, no. 8. https://doi.org/10.3390/land11081265

Van Andel, T., H.J. de Boer, J. Barnes, and I. Vandebroek. 2014. "Medicinal Plants Used for Menstrual Disorders in Latin America, the Caribbean, Sub-Saharan Africa, South and Southeast Asia and Their Uterine Properties: A Review." *Journal of Ethnopharmacology* 155, no. 2: 992–1000.

Verma, V. K., A. Kumar, H. Rymbai, H. Talang, P. Chaudhuri, M. B. Devi, ... V. K. Mishra. 2023. "Assessment of Ethnobotanical Uses, Household, and Regional Genetic Diversity of Aroid Species Grown in Northeastern India." *Frontiers in Nutrition* 10: 1065745.

World Health Organization. 2023. Five Keys for Safer Traditional Food Markets: Risk Mitigation in Traditional Food Markets in the Asia-Pacific Region, WHO Regional Office for the Western Pacific.

Zehra, A., M. Meena, D.M. Jadhav, P. Swapnil, and H. Mangesh. 2023. "Regulatory Mechanisms for the Conservation of Endangered Plant Species, *Chlorophytum tuberosum* Potential Medicinal Plant Species." *Sustainability* 15, no. 8: 6406.

8 Cancer Drug Development from the Wild Edible Plants
Needs and Current Status

Noopur Khare, Pragati Khare and Sachidanand Singh

8.1 INTRODUCTION

Cancer, a powerful enemy of human health, continues to take a terrible toll on people and society all over the world. Cancer still poses a tremendous challenge since many patients are left with few therapy options and frequently disabling side effects, despite great progress in our knowledge of its basic causes and the development of creative treatments. A search for novel and unconventional sources of potential medicines has been sparked by the continuous fight against cancer (Gullett et al., 2010). Wild edible plants are one such area of investigation, a vast and underutilized store of natural chemicals that have long supported a variety of societies. Originally valued for their nutritional value, these plants are now gaining more recognition for their potential use in the creation of cancer treatments (Antonelli et al., 2020). The knowledge of indigenous and traditional healing techniques is at the foundation of the use of wild edible plants as a source of medicinal substances, which dates back millennia. These plants have developed special biochemical defenses and produce a wide variety of secondary metabolites as a result of frequently existing in biodiverse environments and adapting to various climatic circumstances (Yao et al., 2018). These secondary metabolites play a variety of roles in plant life, including defense against herbivores, diseases, and environmental stresses. They range from alkaloids and flavonoids to terpenoids and polyphenols. A significant number of these bioactive substances also have powerful anticancer effects, which is of great interest to modern medicine (Ahmed et al., 2017).

The need to investigate new sources of therapeutic molecules is underscored by the growing demand for alternative cancer treatments. Despite considerable advancements in traditional cancer treatments like surgery, chemotherapy, radiation therapy, and targeted therapies, cancer continues to be the world's top cause of death. There are numerous drawbacks to current therapies, such as the emergence of drug resistance, serious side effects, and exorbitant costs (Dasari et al., 2022). Additionally, the complexity and variety of cancer necessitate a multidimensional strategy that draws from a wide range of therapeutic choices. Wild edible plants offer a rich and mostly untapped reservoir of bioactive chemicals with the ability to supplement current treatments or serve as the basis for whole new therapeutic techniques, making them an exciting candidate for extending this arsenal. Although the investigation of wild edible plants for the development of cancer drugs offers enormous promise, it is not without difficulties and restrictions (Savoia, 2012). Sustainability issues are major since indiscriminate harvesting of these plants can result in habitat damage and overexploitation, endangering not just the plants but also the ecosystems they depend on. The need to sustainably use their medical potential must be balanced with conservation measures. The use of natural goods comes with the possibility of side effects and interactions with conventional treatments, which raises legitimate safety and toxicity concerns as well. To assure both safety and effectiveness, rigorous preclinical and clinical testing is crucial (Chen et al., 2016).

DOI: 10.1201/9781003395935-8

Furthermore, there is a lack of comprehensive knowledge in the area of wild edible plants and cancer research. Research initiatives frequently lack systematic coordination and collaboration among many disciplines, such as botany, pharmacology, oncology, and ethnobotany. This results in isolated research activities. To close these gaps and realize these plants' full therapeutic potential, a thorough study of the potential of these plants in cancer therapy is required (Peduruhewa et al., 2021). The goal of this chapter is to give a thorough investigation of the relationship between the development of cancer medications and wild edible plant use. It starts by looking at the bioactive substances present in these plants and how they work to combat cancer cells. It also explores the difficulties and restrictions, such as the necessity for methodical study, safety concerns, and sustainability considerations. We will highlight prominent wild food plants with anticancer potential from diverse parts of the world through a review of recent research and case stories. To give a complete picture of the development achieved in using these plants for cancer therapy, laboratory investigations, preclinical research, and clinical trials will all be included.

8.2 WILD EDIBLE PLANTS AS A SOURCE OF ANTICANCER COMPOUNDS

Wild edible plants have a tremendous potential for the creation of anticancer medications due to their capacity to create a wide variety of secondary metabolites, many of which have substantial anticancer activities as seen in Figure 8.1. These plants have developed a remarkable array of biochemical defenses against herbivores, pests, and environmental stressors. They have been prized for generations for their nutritional content and traditional medicinal purposes. Scientists have found a wealth of bioactive molecules with the ability to fight cancer in a variety of inventive ways within these chemical arsenals (Padam et al., 2014). The enormous diversity of wild edible plants, both in terms of species and geographic distribution, distinguishes them. Their secondary metabolites, which include alkaloids, flavonoids, terpenoids, polyphenols, and a variety of other substances, also

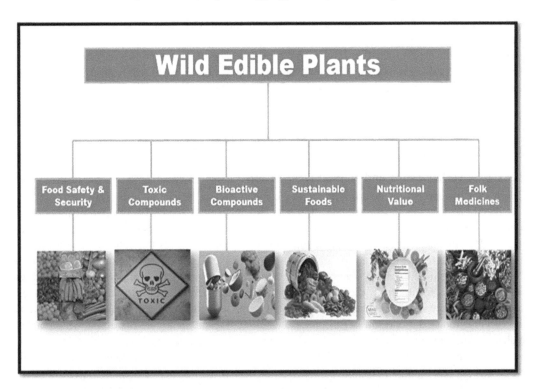

FIGURE 8.1 Uses of wild edible plants.

exhibit this diversity (Huntley et al., 2006). Numerous of these secondary metabolites are known to have anticancer properties that target different phases of cancer genesis and progression. By interfering with cell division and reducing tumor growth, alkaloids like vincristine and vinblastine, which are extracted from the Madagascar periwinkle (*Catharanthus roseus*), have proven to be extremely helpful in treating several types of cancer (Das and Sharangi, 2017).

These secondary metabolites have the ability to fight cancer in a variety of ways that selectively target cancer cells, not just one mechanism of action. Inducing apoptosis, a process of planned cell death that stops the unchecked proliferation of cancer cells, is one typical approach. For instance, it has been demonstrated that the polyphenol resveratrol, which is found in berries and wild grapes, causes cancer cells to undergo apoptosis (Seca and Pinto, 2018). Another mechanism is the suppression of angiogenesis, the growth of new blood vessels that feed tumors. The turmeric plant contains chemicals like curcumin, which have been shown to have the capacity to prevent angiogenesis, hence limiting tumor growth and metastasis. Additionally, some secondary metabolites disrupt the cell cycle, inhibiting the division and spread of cancer cells (Abd El-Hack et al., 2021). Cancer cells can develop multidrug resistance (MDR), which poses a substantial obstacle to the treatment of cancer. A possible answer to this issue is provided by the wild edible plants. By blocking the efflux pumps that remove medications from cancer cells, substances like saponins, which are present in a variety of wild plants, have demonstrated the ability to reverse MDR. This can make cancer cells that had previously been resistant to conventional therapies vulnerable, enhancing the efficacy of chemotherapy regimens (Jiang et al., 2018).

The possibility for wild edible plants to work in conjunction with currently used cancer therapies is arguably one of the most interesting parts of this approach (Piawah and Venook, 2019). Chemotherapy or radiation therapy can be combined with plant-derived chemicals to improve therapeutic results while minimizing the adverse side effects that are frequently connected to these therapies. In addition to increasing the overall effectiveness of cancer therapy, this synergistic strategy also improves the quality of life for patients receiving treatment (Hussain et al., 2021). A valuable and varied source of anticancer agents, wild edible plants provide a plentiful supply of secondary metabolites with complex modes of action. Their potential for developing new cancer treatments goes beyond conventional medicine and into the cutting edge of contemporary oncology. But as we explore these floral gems more deeply, it is crucial to address issues with sustainability, security, and the necessity of methodical research. (Sugumaran et al., 2022).

8.3   CHALLENGES AND LIMITATIONS

8.3.1   SUSTAINABILITY CONCERNS

8.3.1.1   Overharvesting and Habitat Destruction

Overharvesting and habitat destruction are two critical challenges when considering the utilization of wild edible plants in cancer drug development (Karjalainen et al., 2010). Overharvesting occurs when the demand for these plants exceeds their natural replenishment rates, endangering their populations. This threatens not only the plants themselves but also disrupts the delicate ecological balance of their habitats. It can lead to diminished genetic diversity, local extinctions, and the loss of cultural and traditional practices among indigenous communities (Liu et al., 2019). On the other side, habitat loss results from the transformation of natural landscapes into human-altered ones, such as agriculture, urban development, or industry. The area that wild edible plants can grow and flourish in is reduced as a result of this process. Their ecological niches are disrupted, which makes it challenging for them to thrive and procreate. Decreased biodiversity caused by habitat degradation has a ripple effect on the ecosystem as a whole, with potentially serious ecological and economic repercussions (Zhang et al., 2017).

For the sustained development of cancer drugs from wild edible plants, it is essential to address these issues. To find a balance between using these plants' potential and protecting their natural habitats and cultural importance, effective conservation methods, moral sourcing, and legal frameworks are required (Jasovský et al., 2016).

8.3.1.2   Conservation Efforts

When evaluating wild edible plants for cancer medication development, conservation initiatives are crucial in resolving the problems of overharvesting and habitat degradation. The diversity and availability of these plant species must be protected in order to guarantee their long-term sustainability. Establishing and enforcing laws and rules that control the gathering and exchange of wild edible plants is a key component of conservation efforts. By establishing sustainable harvest quotas and seasons as well as stiff fines for illegal exploitation, these regulations seek to avoid overharvesting. In order to monitor and manage the harvest, they could also need permits or licenses for collection (Moraes et al., 2021).

Initiatives for conservation also put an emphasis on preserving and restoring habitat. The sustainability of wild edible plants depends on safeguarding their natural environments. In order for these plants to survive undisturbed, protected areas, national parks, and reserves must be established (Winter et al., 2020). Reforestation or wetland restoration are examples of habitat restoration practices that aim to restore ecosystems that have been damaged by human activity so that these plants can grow and reproduce in a healthy setting. Furthermore, education and awareness campaigns are frequently a part of conservation efforts. These programs seek to raise awareness among the general public, decision-makers, and industry stakeholders of the value of preserving these plant species and their ecosystems as well as the significance of sustainable harvesting methods. These efforts aid in the long-term preservation of wild edible plants by encouraging a sense of accountability and stewardship (Gann et al., 2019).

8.3.2   Safety and Toxicity Issues

8.3.2.1   Potential Side Effects

Safety and toxicity issues are pivotal considerations when exploring wild edible plants for cancer drug development. While these plants offer a plethora of bioactive compounds with potential medicinal benefits, they also carry the risk of adverse effects and toxicities, necessitating careful evaluation (Soave et al., 2017). One significant concern is the potential side effects associated with the consumption or use of wild edible plants for therapeutic purposes. These side effects can range from mild gastrointestinal discomfort to severe allergic reactions or organ damage, depending on the plant and its bioactive constituents. For instance, some wild edible plants may contain compounds that irritate the digestive system or cause skin rashes when ingested or applied topically (Kuete, 2014).

Another significant problem involves toxicity concerns. If not carefully detected and reduced, the presence of poisonous chemicals in some wild food plants can have detrimental effects on health. When consumed, certain plants may contain alkaloids, glycosides, or other toxic chemicals that can cause vomiting, nausea, and in severe cases, organ failure. The safety and toxicity concerns highlight how crucial it is to thoroughly assess wild food plants as potential sources of cancer therapy. Despite the potential of these plants, care must be taken while using them to make sure that any medicinal advantages outweigh the dangers of toxicities and side effects (Serrano, 2018).

8.3.2.2   Preclinical and Clinical Testing

The preclinical and clinical testing of safety and toxicity issues are crucial steps in the assessment of wild edible plants for potential use in the development of cancer drugs. These testing phases are essential to confirming that any therapeutic advantages balance the dangers of toxicities and side effects (Seca and Pinto, 2018).

Preclinical Testing: Preclinical testing entails a series of investigations carried out in lab settings and on animals to evaluate the toxicity and safety of bioactive substances produced from wild food plants. Cells are exposed to these substances in in vitro research to examine how cell viability, proliferation, and potential toxic responses are affected. The effects of the chemicals on living beings, including any negative impacts on organ function, behavior, or general health, are assessed in vivo using animal models (Aware et al., 2022). Investigations into the substances' mechanisms of action, pharmacokinetics (absorption, distribution, metabolism, and excretion), and potential interactions with other medications are also a part of preclinical testing. Before moving on to human trials, the aim is to determine safe dosage ranges and collect information on the pharmacological characteristics of the substances (Soave et al., 2017).

Clinical Testing: Clinical trials are an essential step in the process of evaluating the safety and effectiveness of chemicals produced from wild edible plants in human subjects after preclinical research. Phase I trials, which concentrate on safety and dose, are often the first of these trials. A small group of healthy volunteers or patients participate in these trials to ascertain the compound's safety profile, potential side effects, and ideal dosage levels (Artaud et al., 2019). Larger patient populations are included in later Phase II and Phase III trials, which concentrate on efficacy, safety, and comparisons to currently available treatments. Critical information on the therapeutic advantages and long-term consequences of the substances in actual patient populations is provided by these trials (Schadendorf et al., 2015).

8.4   LACK OF SYSTEMATIC RESEARCH

8.4.1   FRAGMENTED KNOWLEDGE

The problem of a fragmented knowledge base or lack of systematic study in the area of developing cancer drugs from wild edible plants is characterized by the scattered and frequently discontinuous nature of the material that is currently available. There is a lack of comprehensive and standardized data due to the historically dispersed and disorganized research activities in this field (Heinrich et al., 2017). The ability to fully comprehend and utilize the potential of wild food plants for cancer therapy is hampered by this fragmentation. To fill in information gaps, compile findings, and build a firm foundation for the systematic investigation of these beneficial botanical resources in cancer treatment, a more coherent and interdisciplinary approach is necessary (Panieri et al., 2020).

8.4.2   NEED FOR INTERDISCIPLINARY COLLABORATION

In order to effectively address the issue of a lack of systematic research on wild edible plants for the development of cancer drugs, interdisciplinary cooperation is essential. Interdisciplinary collaboration promotes the synthesis of information and skills from other disciplines, including ethnobotany, pharmacology, cancer, and botany. This teamwork encourages a more comprehensive understanding of the plants, their bioactive components, and their potential in cancer treatment. It encourages methodical exploration, fills in gaps in fragmented research, and ensures a thorough evaluation of the advantages and difficulties related to these botanical resources. In the end, interdisciplinary cooperation is essential to releasing the full potential of wild edible plants in the war against cancer (Samuels and Ben-Arye, 2020).

8.5   CURRENT RESEARCH AND CASE STUDIES

8.5.1   NOTABLE WILD EDIBLE PLANTS WITH ANTICANCER POTENTIAL

The promising stars of this sector are well-known wild food plants with anticancer potential, according to recent research and case studies. These plants, which come from many parts of the world, have demonstrated strong evidence of containing bioactive components that help fight cancer (Hamedi et al., 2022). They are the focus of continuing research, with laboratory studies, preclinical trials, and clinical trials shedding light on their efficacy against various cancer types. By focusing on these plants, researchers hope to find cutting-edge cancer treatments that take advantage of nature's gifts, perhaps revolutionizing cancer care while honoring the sustainability and cultural significance of these floral riches (Schadendorf et al., 2015).

8.5.2 Laboratory Studies and Preclinical Trials

Current investigations and case studies involving wild food plants with putative anticancer properties heavily rely on laboratory tests and preclinical trials. In controlled environments and using animal models, these studies are conducted to evaluate the efficacy and safety of the bioactive chemicals obtained from these plants. The mechanisms of action are investigated in laboratory experiments, looking at how these substances affect cancer cells at the cellular and molecular levels (Hostettmann et al., 2000). Building on this understanding, preclinical studies examine the drugs' therapeutic potential in living things before moving on to human trials. These phases direct the compounds' ongoing development as potential cancer therapeutics and offer crucial information into the anticancer activities of the compounds (Caruso et al., 2019).

8.5.3 Clinical Trials and Outcomes

The critical phases of the ongoing research and case studies involving wild edible plants with anticancer potential are clinical trials and results. Clinical trials entail the methodical testing of these plant-derived substances in human patients to gauge their efficacy and safety. They want to know if these substances are effective against cancer and, if so, how they stack up against already used treatments (Wasser, 2011). Clinical trial results offer essential information on the compounds' actual performance, assisting in determining their clinical efficacy and potential negative effects. Positive results from these trials carry the prospect of introducing fresh, nature-inspired cancer treatments for patients, ultimately influencing cancer treatment (Panossian et al., 2010).

8.6 EXTRACTION AND FORMULATION TECHNIQUES

8.6.1 Methods for Extracting Bioactive Compounds

Methods for extracting bioactive compounds are fundamental in the process of harnessing the therapeutic potential of wild edible plants for cancer drug development. These methods involve the separation of bioactive compounds from plant materials to create concentrated extracts for further study or formulation (Ahmad et al., 2008).

Various extraction techniques are employed, including:

- Solvent Extraction: This method uses solvents like ethanol, methanol, or water to dissolve and extract bioactive compounds. It's versatile and widely used, but the choice of solvent can impact compound selectivity and yield (Lefebvre et al., 2021).
- Supercritical Fluid Extraction: This technique employs supercritical fluids, often carbon dioxide, to extract compounds under controlled pressure and temperature conditions. It's efficient and avoids the use of potentially harmful solvents (Goyeneche et al., 2018).
- Ultrasound-Assisted Extraction: Ultrasonic waves are used to disrupt plant cell walls and enhance compound release. This method is known for its speed and effectiveness (Pico, 2013).

Effective extraction methods are crucial to obtaining high-quality extracts enriched with bioactive compounds as seen in Figure 8.2. These extracts can then be further formulated into drug delivery systems or pharmaceutical formulations, ensuring their stability, bioavailability, and targeted delivery to cancer cells, thus optimizing their therapeutic potential in clinical applications.

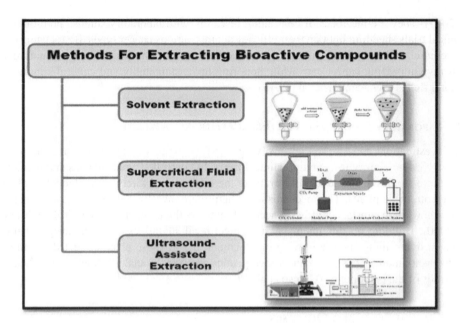

FIGURE 8.2 Methods for extracting bioactive compounds.

8.6.2 Formulation Strategies for Drug Development

Formulation strategies for drug development are integral when considering wild edible plants' bioactive compounds for cancer therapy. After successful extraction, these compounds require formulation into drug delivery systems to optimize their effectiveness and safety (Singh and Singh, 2023).

Common formulation strategies include:

- Nanoencapsulation: This technique involves enclosing bioactive compounds in nanoscale carriers, such as liposomes or nanoparticles. Nanoencapsulation enhances compound stability, solubility, and controlled release, improving their bioavailability and targeting to cancer cells (Pateiro et al., 2021).
- Targeted Drug Delivery: Formulations can be designed to specifically deliver bioactive compounds to cancer cells while sparing healthy tissues. Ligand-functionalized nanoparticles, for instance, can recognize and bind to cancer cell receptors, increasing compound uptake and minimizing side effects (Pathak et al., 2022).
- Sustained-Release Formulations: To maintain therapeutic concentrations over extended periods, sustained-release formulations are developed. These release bioactive compounds gradually, reducing the need for frequent dosing and improving patient compliance (Kumar et al., 2012).

By employing these formulation strategies, researchers aim to enhance the therapeutic potential of wild edible plant-derived bioactive compounds, making them more effective, safe, and practical for cancer treatment. These formulations contribute to the advancement of cancer drug development and its potential integration into mainstream oncology practices.

8.6.3 Ensuring Bioavailability and Stability

The extraction and formulation of bioactive components from wild edible plants for the development of cancer drugs must ensure bioavailability and stability. Stability is the ability of these substances to retain their chemical integrity and efficacy over time, whereas bioavailability is the amount and rate at which they are absorbed and reach their intended destinations in the body (Hadidi et al., 2023). Formulations are made to increase these chemicals' solubility and ease of absorption in the body in order to increase bioavailability. Compounds can be protected against degradation in the hostile gastrointestinal environment using strategies like nanoencapsulation and micellar systems, enabling improved absorption into the bloodstream (Gupta et al., 2013).

In order to prevent deterioration, oxidation, or loss of potency, stability is achieved by choosing the proper excipients, storage settings, and packaging. This is essential to keep the chemicals therapeutically effective for the duration of their shelf life and until the patient receives them. The overall effectiveness of bioactive compounds is improved by incorporating bioavailability and stability techniques into the formulation process. This raises their potential as cancer treatment agents and increases the possibility that their clinical applications will be successful (Dulińska-Litewka et al., 2019).

8.7 REGULATORY AND ETHICAL CONSIDERATIONS

8.7.1 REGULATORY PATHWAYS FOR NATURAL PRODUCTS

There are many legal and moral issues to take into account on the path from foraging for edible wild plants to developing cancer drugs. The Food and Drug Administration (FDA) in the United States is mostly in charge of regulating natural products, including those made from edible wild plants (Poe et al., 2013).

8.7.1.1 FDA Approval Process

The FDA approval process for natural products, often categorized as botanical drugs, is a rigorous and multiphase procedure designed to ensure safety, efficacy, and quality (Frestedt, 2017). The process involves several key steps as seen in Figure 8.3:

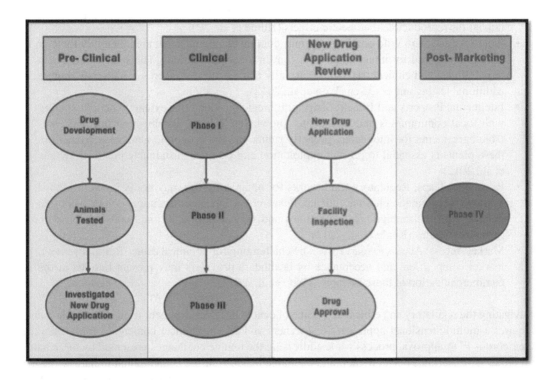

FIGURE 8.3 FDA approval process.

- Preclinical Testing: Before human trials, researchers conduct preclinical studies to evaluate the safety and efficacy of the botanical drug in laboratory settings and animal models. This phase aims to establish a solid foundation of evidence regarding the compound's potential (Wang et al., 2023).
- Investigational New Drug (IND) Application: Researchers must submit an IND application to the FDA, outlining their plans for clinical trials. The FDA reviews this application to assess the safety of the proposed trials. Approval of the IND allows researchers to proceed with human trials (Zhao et al., 2021).
- Clinical Trials: Clinical trials occur in three phases: Phase I focuses on safety and dosage in a small group of participants; Phase II expands to a larger group to assess efficacy and side effects; and Phase III involves a larger, diverse population to confirm efficacy, monitor side effects, and compare the drug to existing treatments (Dulińska-Litewka et al., 2019).
- New Drug Application (NDA): If clinical trials are successful, researchers submit an NDA to the FDA, summarizing the drug's safety and effectiveness data. The FDA reviews this application meticulously, considering the risk-benefit ratio before granting approval (Khin et al., 2011).
- Post-Marketing Surveillance: After approval, the drug is closely monitored in the post-marketing phase for any unexpected side effects or long-term safety concerns (Selvan et al., 2013).

8.7.1.2 Challenges in Registration

While the FDA's approval process for natural products is well established, unique challenges arise when registering therapies derived from wild edible plants as seen in Figure 8.4.

- Standardization: Wild edible plants can exhibit variability in their chemical composition due to factors such as geographical location, climate, and soil conditions. This variability complicates the standardization of botanical drugs, as consistency in active compounds is crucial for reproducible therapeutic effects (Kunle et al., 2012).
- Safety Assessment: Wild edible plants may contain compounds that are inherently toxic or have the potential for interactions with conventional medications. The safety assessment in preclinical and clinical trials must carefully address these concerns, which may require additional testing and research (Woo et al., 2012).
- Intellectual Property and Benefit-Sharing: Ethical considerations extend to benefit-sharing with local communities and indigenous knowledge holders. Establishing fair and equitable agreements for intellectual property rights and the economic benefits derived from these plants is essential to prevent exploitation and promote sustainable practices (Kunle et al., 2012).
- Regulatory Gaps: Regulatory frameworks for botanical drugs may not be as well-defined as those for synthetic pharmaceuticals. Researchers and pharmaceutical companies often grapple with the ambiguity of guidelines and may require additional consultation with regulatory bodies (Kunle et al., 2012).
- Market Access: Access to markets can be challenging for botanical drugs. Reimbursement, market competition, and acceptance by healthcare providers may present hurdles in the commercialization of these therapies (Woo et al., 2012).

Navigating the regulatory and ethical landscape of cancer drug development from wild edible plants requires a multidimensional approach. Researchers and pharmaceutical companies must embrace the rigorous FDA approval process while addressing the unique challenges presented by these natural products. Ensuring safety, efficacy, and equitable benefit sharing is paramount to harnessing the potential of wild edible plants as a valuable resource in the fight against cancer (Saad et al., 2006).

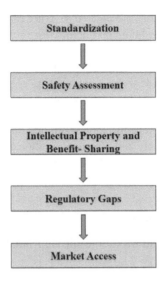

FIGURE 8.4   Challenges in registration.

8.8   FUTURE DIRECTIONS

8.8.1   THE POTENTIAL OF PERSONALIZED MEDICINE

Personalized medicine has more and more potential as the landscape of developing cancer drugs from wild edible plants changes. Precision medicine, commonly referred to as personalized medicine, is a paradigm change in healthcare in which each patient's therapies are catered to their particular genetic, molecular, and clinical characteristics (Krzyszczyk et al., 2018).

8.8.2   TAILORING TREATMENTS TO INDIVIDUAL PATIENTS

According to the theory behind personalized medicine, no two people are alike, and as a result, everyone responds to cancer and its therapies differently. Researchers can examine the complex molecular profiles of both the patient and the tumor by merging genomes, proteomics, metabolomics, and other omics tools (Janssens et al., 2018). Specific genetic mutations or changes that fuel a patient's cancer can be identified through genetic testing. These genetic indicators can help in the selection of chemicals or formulations produced from wild edible plants that are specifically designed to target the disease's molecular processes (Dagogo-Jack and Shaw, 2018).

Identifying biomarkers – molecules that can help with patient classification and treatment choices – can show the existence or behavior of a disease. For instance, in individuals with particular biomarker profiles, certain chemicals from wild edible plants may be very helpful, improving treatment outcomes (Abi-Darghamb and Horga, 2016). Combination therapies that combine conventional treatments with chemicals derived from wild edible plants can be created thanks to personalized medicine. Depending on the unique characteristics of each patient, these combinations can be adjusted, potentially improving therapeutic effectiveness while reducing adverse effects. Genetic variations in patients may impact how well their bodies metabolize medication. To ensure that medicines derived from wild edible plants are given at doses that are most effective for each patient, personalized medicine can direct the optimization of dosing regimens (Krzyszczyk et al., 2018).

The dynamic nature of cancer necessitates ongoing observation. Real-time monitoring of a patient's response to treatment is made possible by personalized medicine, permitting necessary modifications. In the event that the original course of treatment is unsuccessful or has unfavorable effects, this can entail switching to alternate wild edible plant components or formulations

(Ferguson et al., 2013). As medications are created to target cancer cells specifically while protecting healthy organs, personalized medicine can help reduce unneeded toxicity and side effects. A patient-centered approach to cancer therapy is fostered by personalized medicine. Patients actively participate in making decisions about their own care, and treatment plans are created to reflect their preferences, values, and objectives (de Belvis et al., 2021).

Personalized medicine represents a significant advance toward more efficient and patient-focused cancer therapy despite these obstacles. The possibility of customizing therapies to each patient's particular needs, along with the therapeutic toolkit provided by wild edible plants, opens up new avenues in the effort to more precisely and effectively fight cancer. Personalized medicine has the potential to revolutionize cancer treatment and enhance the lives of countless patients as research in this area advances (Patel, 2016).

8.8.3 Advances in Omics Technologies

8.8.3.1 Genomic, Proteomic, and Metabolomic Approaches

The development of cancer drugs from wild edible plants is closely related to the amazing advancements in omics technologies. The potential of these botanical resources can be fully realized by using genomic, proteomic, and metabolomic methods, which have the potential to completely alter the landscape of cancer therapy.

Genomic approaches include genomics involves the comprehensive study of an individual's entire genetic makeup, including their genes and DNA sequences. In the context of wild edible plants and cancer drug development, genomic approaches hold the promise of targeted therapy, and genomic profiling allows the identification of genetic mutations or alterations driving cancer growth. Researchers can pinpoint wild edible plant-derived compounds that specifically target these genetic aberrations, paving the way for more precise and effective treatments (Green et al., 2011).

Genomic data can help identify predictive biomarkers that indicate how a patient is likely to respond to a particular treatment. This knowledge aids in treatment selection and optimization, reducing the trial-and-error approach often associated with cancer therapies (Schmidt et al., 2016).

8.8.4 Integrating Wild Edible Plants into Mainstream Cancer Therapy

A fascinating oncological frontier is the incorporation of edible wild plants into conventional cancer treatment. This integration entails the creation of combination therapies and the investigation of the interactions between conventional medications and chemicals produced from wild edible plants as seen in Figure 8.5.

1. **Combination Therapies**: There is a lot of potential in combining chemicals from wild food plants with well-known cancer therapies like chemotherapy or radiation therapy. These combo therapy have the following benefits:
   - Enhanced Efficacy: Wild edible plant compounds can complement the mechanisms of conventional treatments, increasing their effectiveness against cancer cells.
   - Reduced Side Effects: Combining therapies may allow for lower doses of conventional treatments, reducing their toxic side effects while maintaining therapeutic benefits.
   - Overcoming Resistance: Some cancer cells develop resistance to single treatments. Combination therapies can target multiple pathways, making it more difficult for cancer cells to evade treatment (Esmeeta et al., 2022).
2. **Synergistic Effects**: When two or more agents work together, their combined influence is greater than the sum of their separate effects. Wild edible plant-derived substances may interact synergistically with one another or with established medical procedures:
   - Multi-Target Approach: Different compounds from wild edible plants can target various aspects of cancer development simultaneously, leading to a more comprehensive treatment approach (Pezzani et al., 2019).

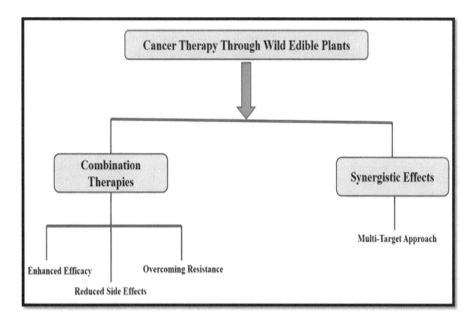

FIGURE 8.5 Cancer therapy through wild edible plants.

The integration of these botanical resources into conventional cancer therapy and advancements in omics technology shed light on the future of cancer medication discovery from wild edible plants. The tools to personalize treatment and enhance therapeutic regimens are provided by genomic, proteomic, and metabolomic methods. Additionally, synergistic effects of conventional cancer medicines and chemicals originating from wild food plants hold the prospect of more potent, specialized, and less toxic cancer therapeutics. These advancements have the potential to change the way cancer is treated, giving patients new hope and better outcomes (Zhu et al., 2022).

8.9 CONCLUSION

In this exploration of cancer drug development from wild edible plants, several key findings and takeaways have emerged, shedding light on both the promise and challenges within this burgeoning field. First off, a substantial pool of bioactive chemicals with tremendous potential in the fight against cancer may be found in the natural world, which is brimming with botanical diversity. A rich reservoir of novel anticancer drugs can be found in the complex chemistry of wild edible plants, offering hope for the creation of groundbreaking treatments. The route from nature to the clinic is difficult, though. These floral riches are in danger of going extinct due to overharvesting and habitat damage, underscoring the urgent need for sustainable practices and conservation initiatives. A strong data infrastructure, equitable benefit sharing, and prudent resource management are essential, and these factors are all heavily weighed by ethical and regulatory concerns.

The promise of harnessing wild edible plants in cancer drug development lies in the potential for highly effective, personalized therapies. These botanical resources, when integrated into mainstream cancer treatment, hold the prospect of reducing side effects, overcoming drug resistance, and enhancing treatment outcomes. Moreover, the synergy between wild edible plant-derived compounds and conventional treatments presents an exciting avenue for research. However, the challenges are substantial. Standardizing these natural products, assuring their safety, and negotiating complex regulatory frameworks are challenges that the scientific community must overcome. To make these medicines available to a large patient population, further obstacles including market accessibility and pricing must be removed. In conclusion, the journey of cancer drug development from wild edible plants is an ongoing odyssey. It calls for continued research, innovation, and collaboration across disciplines. Sustainability must be at the forefront of our efforts, ensuring that the healing potential of these plants endures for future generations. In addition to being moral requirements, ethical benefit sharing and respect for indigenous knowledge are crucial for this field's long-term survival.

REFERENCES

Abd El-Hack, M.E., El-Saadony, M.T., Swelum, A.A., Arif, M., Abo Ghanima, M.M., Shukry, M., Noreldin, A., Taha, A.E. and El-Tarabily, K.A., 2021. Curcumin, the active substance of turmeric: Its effects on health and ways to improve its bioavailability. *Journal of the Science of Food and Agriculture, 101*(14), pp.5747–5762.

Abi-Dargham, A. and Horga, G., 2016. The search for imaging biomarkers in psychiatric disorders. *Nature Medicine, 22*(11), pp.1248–1255.

Ahmad, I., Zahin, M., Aqil, F., Hasan, S., Khan, M.S.A. and Owais, M., 2008. Bioactive compounds from Punica granatum, Curcuma longa and Zingiber officinale and their therapeutic potential. *Drugs of the Future, 33*(4), p.329.

Ahmed, E., Arshad, M., Khan, M.Z., Amjad, M.S., Sadaf, H.M., Riaz, I., Sabir, S. and Ahmad, N., 2017. Secondary metabolites and their multidimensional prospective in plant life. *Journal of Pharmacognosy and Phytochemistry, 6*(2), pp.205–214.

Antonelli, A., Smith, R.J., Fry, C., Simmonds, M.S., Kersey, P.J., Pritchard, H.W., Abbo, M.S., Acedo, C., Adams, J., Ainsworth, A.M. and Allkin, B., 2020. *State of the World's Plants and Fungi* (Doctoral dissertation, Royal Botanic Gardens (Kew); Sfumato Foundation).

Artaud, C., Kara, L. and Launay, O., 2019. Vaccine development: from preclinical studies to phase 1/2 clinical trials. *Malaria Control and Elimination*, pp.165–176.

Aware, C.B., Patil, D.N., Suryawanshi, S.S., Mali, P.R., Rane, M.R., Gurav, R.G. and Jadhav, J.P., 2022. Natural bioactive products as promising therapeutics: A review of natural product-based drug development. *South African Journal of Botany, 151*, pp.512–528.

Caruso, G., Caraci, F. and Jolivet, R.B., 2019. Pivotal role of carnosine in the modulation of brain cells activity: Multimodal mechanism of action and therapeutic potential in neurodegenerative disorders. *Progress in neurobiology, 175*, pp.35–53.

Chen, S.L., Yu, H., Luo, H.M., Wu, Q., Li, C.F. and Steinmetz, A., 2016. Conservation and sustainable use of medicinal plants: Problems, progress, and prospects. *Chinese Medicine, 11*, pp.1–10.

Dagogo-Jack, I. and Shaw, A.T., 2018. Tumour heterogeneity and resistance to cancer therapies. *Nature Reviews Clinical Oncology, 15*(2), pp.81–94.

Das, S. and Sharangi, A.B., 2017. Madagascar periwinkle (Catharanthus roseus L.): Diverse medicinal and therapeutic benefits to humankind. *Journal of Pharmacognosy and Phytochemistry, 6*(5), pp.1695–1701.

Dasari, S., Njiki, S., Mbemi, A., Yedjou, C.G. and Tchounwou, P.B., 2022. Pharmacological effects of cisplatin combination with natural products in cancer chemotherapy. *International Journal of Molecular Sciences, 23*(3), p.1532.

de Belvis, A.G., Pellegrino, R., Castagna, C., Morsella, A., Pastorino, R. and Boccia, S., 2021. Success factors and barriers in combining personalized medicine and patient centered care in breast cancer. Results from a systematic review and proposal of conceptual framework. *Journal of Personalized Medicine, 11*(7), p.654.

Dulińska-Litewka, J., Łazarczyk, A., Hałubiec, P., Szafrański, O., Karnas, K. and Karewicz, A., 2019. Superparamagnetic iron oxide nanoparticles—Current and prospective medical applications. *Materials, 12*(4), p.617.

Esmeeta, A., Adhikary, S., Dharshnaa, V., Swarnamughi, P., Maqsummiya, Z.U., Banerjee, A., Pathak, S. and Duttaroy, A.K., 2022. Plant-derived bioactive compounds in colon cancer treatment: An updated review. *Biomedicine & Pharmacotherapy, 153*, p.113384.

Ferguson, B.S., Hoggarth, D.A., Maliniak, D., Ploense, K., White, R.J., Woodward, N., Hsieh, K., Bonham, A.J., Eisenstein, M., Kippin, T.E. and Plaxco, K.W., 2013. Real-time, aptamer-based tracking of circulating therapeutic agents in living animals. *Science Translational Medicine, 5*(213), pp.213ra165–213ra165.

Frestedt, J.L., 2017. Similarities and difference between clinical trials for foods and drugs. *Austin Journal of Nutrition and Food Sciences, 5*(1), p.1086.

Gann, G.D., McDonald, T., Walder, B., Aronson, J., Nelson, C.R., Jonson, J., Hallett, J.G., Eisenberg, C., Guariguata, M.R., Liu, J. and Hua, F., 2019. International principles and standards for the practice of ecological restoration. *Restoration Ecology, 27*(S1), pp.S1–S46.

Goyeneche, R., Fanovich, A., Rodrigues, C.R., Nicolao, M.C. and Di Scala, K., 2018. Supercritical CO2 extraction of bioactive compounds from radish leaves: Yield, antioxidant capacity and cytotoxicity. *The Journal of Supercritical Fluids*, *135*, pp.78–83.

Green, E.D., Guyer, M.S., National Human Genome Research Institute Overall leadership Green Eric D. Guyer Mark S. and Coordination of writing contributions (see Acknowledgements for list of other contributors) Manolio Teri A. Peterson Jane L., 2011. Charting a course for genomic medicine from base pairs to bedside. *Nature*, *470*(7333), pp.204–213.

Gullett, N.P., Amin, A.R., Bayraktar, S., Pezzuto, J.M., Shin, D.M., Khuri, F.R., Aggarwal, B.B., Surh, Y.J. and Kucuk, O., 2010, June. Cancer prevention with natural compounds. In *Seminars in oncology* (Vol. 37, No. 3, pp. 258–281). WB Saunders.

Gupta, S., Kesarla, R. and Omri, A., 2013. Formulation strategies to improve the bioavailability of poorly absorbed drugs with special emphasis on self-emulsifying systems. *International Scholarly Research Notices*.

Hadidi, M., Tan, C., Assadpour, E., Kharazmi, M.S. and Jafari, S.M., 2023. Emerging plant proteins as nanocarriers of bioactive compounds. *Journal of Controlled Release*, *355*, pp.327–342.

Hamedi, A., Bayat, M., Asemani, Y. and Amirghofran, Z., 2022. A review of potential anti-cancer properties of some selected medicinal plants grown in Iran. *Journal of Herbal Medicine*, *33*, p.100557.

Heinrich, M., Barnes, J., Prieto-Garcia, J., Gibbons, S. and Williamson, E.M., 2017. *Fundamentals of pharmacognosy and phytotherapy E-BOOK*. Elsevier Health Sciences.

Hostettmann, K., Marston, A., Ndjoko, K. and Wolfender, J.L., 2000. The potential of African plants as a source of drugs. *Current Organic Chemistry*, *4*(10), pp.973–1010.

Huntley, B., Collingham, Y.C., Green, R.E., Hilton, G.M., Rahbek, C. and Willis, S.G., 2006. Potential impacts of climatic change upon geographical distributions of birds. *Ibis*, *148*, pp.8–28.

Hussain, Y., Islam, L., Khan, H., Filosa, R., Aschner, M. and Javed, S., 2021. Curcumin–cisplatin chemotherapy: A novel strategy in promoting chemotherapy efficacy and reducing side effects. *Phytotherapy Research*, *35*(12), pp.6514–6529.

Janssens, J.P., Schuster, K. and Voss, A., 2018. Preventive, predictive, and personalized medicine for effective and affordable cancer care. *EPMA Journal*, *9*(2), pp.113–123.

Jasovský, D., Littmann, J., Zorzet, A. and Cars, O., 2016. Antimicrobial resistance—a threat to the world's sustainable development. *Upsala Journal of Medical Sciences*, *121*(3), pp.159–164.

Jiang, L., Lin, J., Taggart, C.C., Bengoechea, J.A. and Scott, C.J., 2018. Nanodelivery strategies for the treatment of multidrug-resistant bacterial infections. *Journal of Interdisciplinary Nanomedicine*, *3*(3), pp.111–121.

Karjalainen, E., Sarjala, T. and Raitio, H., 2010. Promoting human health through forests: Overview and major challenges. *Environmental Health and Preventive Medicine*, *15*, pp.1–8.

Khin, N.A., Chen, Y.F., Yang, Y., Yang, P. and Laughren, T.P., 2011. Exploratory analyses of efficacy data from major depressive disorder trials submitted to the US Food and Drug Administration in support of new drug applications. *The Journal of Clinical Psychiatry*, *72*(4), p.6970.

Krzyszczyk, P., Acevedo, A., Davidoff, E.J., Timmins, L.M., Marrero-Berrios, I., Patel, M., White, C., Lowe, C., Sherba, J.J., Hartmanshenn, C. and O'Neill, K.M., 2018. The growing role of precision and personalized medicine for cancer treatment. *Technology*, *6*(03n04), pp.79–100.

Kuete, V., 2014. Physical, hematological, and histopathological signs of toxicity induced by African medicinal plants. In *Toxicological survey of African medicinal plants* (pp. 635–657). Elsevier.

Kumar, K.S., Bhowmik, D., Srivastava, S., Paswan, S. and Dutta, A.S., 2012. Sustained release drug delivery system potential. *The Pharma Innovation*, *1*(2).

Kunle, Oluyemisi F., Egharevba, Henry O., and Ahmadu, Peter O. 2012. Standardization of herbal medicines-A review. *International Journal of Biodiversity and Conservation*, *4*(3), pp.101–112.

Lefebvre, T., Destandau, E. and Lesellier, E., 2021. Selective extraction of bioactive compounds from plants using recent extraction techniques: A review. *Journal of Chromatography A*, *1635*, p.461770.

Liu, H., Gale, S.W., Cheuk, M.L. and Fischer, G.A., 2019. Conservation impacts of commercial cultivation of endangered and overharvested plants. *Conservation Biology*, *33*(2), pp.288–299.

Moraes, R.M., Cerdeira, A.L. and Lourenço, M.V., 2021. Using micropropagation to develop medicinal plants into crops. *Molecules*, *26*(6), p.1752.

Padam, B.S., Tin, H.S., Chye, F.Y. and Abdullah, M.I., 2014. Banana by-products: An under-utilized renewable food biomass with great potential. *Journal of Food Science and Technology*, *51*, pp.3527–3545.

Panieri, E., Buha, A., Telkoparan-Akillilar, P., Cevik, D., Kouretas, D., Veskoukis, A., Skaperda, Z., Tsatsakis, A., Wallace, D., Suzen, S. and Saso, L., 2020. Potential applications of NRF2 modulators in cancer therapy. *Antioxidants*, *9*(3), p.193.

Panossian, A., Wikman, G. and Sarris, J., 2010. Rosenroot (Rhodiola rosea): Traditional use, chemical composition, pharmacology and clinical efficacy. *Phytomedicine*, *17*(7), pp.481–493.

Pateiro, M., Gómez, B., Munekata, P.E., Barba, F.J., Putnik, P., Kovačević, D.B. and Lorenzo, J.M., 2021. Nanoencapsulation of promising bioactive compounds to improve their absorption, stability, functionality and the appearance of the final food products. *Molecules*, *26*(6), p.1547.

Patel, J.N., 2016. Cancer pharmacogenomics, challenges in implementation, and patient-focused perspectives. *Pharmacogenomics and Personalized Medicine*, pp.65–77.

Pathak, N., Singh, P., Singh, P.K., Sharma, S., Singh, R.P., Gupta, A., Mishra, R., Mishra, V.K. and Tripathi, M., 2022. Biopolymeric nanoparticles based effective delivery of bioactive compounds toward the sustainable development of anticancerous therapeutics. *Frontiers in Nutrition*, *9*, p.963413.

Peduruhewa, P.S., Jayathunge, K.G.L.R. and Liyanage, R., 2021. Potential of underutilized wild edible plants as the food for the future–a review. *Journal of Food Security*, *9*(4), pp.136–147.

Pezzani, R., Salehi, B., Vitalini, S., Iriti, M., Zuñiga, F.A., Sharifi-Rad, J., Martorell, M. and Martins, N., 2019. Synergistic effects of plant derivatives and conventional chemotherapeutic agents: An update on the cancer perspective. *Medicina*, *55*(4), p.110.

Piawah, S. and Venook, A.P., 2019. Targeted therapy for colorectal cancer metastases: A review of current methods of molecularly targeted therapy and the use of tumor biomarkers in the treatment of metastatic colorectal cancer. *Cancer*, *125*(23), pp.4139–4147.

Pico, Y., 2013. Ultrasound-assisted extraction for food and environmental samples. *TrAC Trends in Analytical Chemistry*, *43*, pp.84–99.

Poe, M.R., McLain, R.J., Emery, M. and Hurley, P.T., 2013. Urban forest justice and the rights to wild foods, medicines, and materials in the city. *Human Ecology*, *41*, pp.409–422.

Saad, B., Azaizeh, H., Abu-Hijleh, G. and Said, O., 2006. Safety of traditional Arab herbal medicine. *Evidence-Based Complementary and Alternative Medicine*, *3*, pp.433–439.

Samuels, N. and Ben-Arye, E., 2020. Exploring herbal medicine use during palliative cancer care: The integrative physician as a facilitator of pharmacist–patient–oncologist communication. *Pharmaceuticals*, *13*(12), p.455.

Savoia, D., 2012. Plant-derived antimicrobial compounds: Alternatives to antibiotics. *Future Microbiology*, *7*(8), pp.979–990.

Schadendorf, D., Hodi, F.S., Robert, C., Weber, J.S., Margolin, K., Hamid, O., Patt, D., Chen, T.T., Berman, D.M. and Wolchok, J.D., 2015. Pooled analysis of long-term survival data from phase II and phase III trials of ipilimumab in unresectable or metastatic melanoma. *Journal of Clinical Oncology*, *33*(17), p.1889.

Schmidt, K.T., Chau, C.H., Price, D.K. and Figg, W.D., 2016. Precision oncology medicine: The clinical relevance of patient-specific biomarkers used to optimize cancer treatment. *The Journal of Clinical Pharmacology*, *56*(12), pp.1484–1499.

Seca, A.M. and Pinto, D.C., 2018. Plant secondary metabolites as anticancer agents: Successes in clinical trials and therapeutic application. *International Journal of Molecular Sciences*, *19*(1), p.263.

Selvan, A., Mohan, C.J., Sundari, S. and Suthakaran, R., 2013. Study on role of postmarketing surveillance in new drug development. *Asian Journal of Management*, *4*(1), pp.12–15.

Serrano, R., 2018. Toxic plants: Knowledge, medicinal uses and potential human health risks. *Environment and Ecology Research*, *6*(5), pp.487–492.

Singh, A. and Singh, B., 2023. Bioactive compounds in cancer care and prevention. In *Role of Nutrigenomics in Modern-Day Healthcare and Drug Discovery* (pp. 439–468). Elsevier.

Soave, C.L., Guerin, T., Liu, J. and Dou, Q.P., 2017. Targeting the ubiquitin-proteasome system for cancer treatment: Discovering novel inhibitors from nature and drug repurposing. *Cancer and Metastasis Reviews*, *36*, pp.717–736.

Sugumaran, A., Pandiyan, R., Kandasamy, P., Antoniraj, M.G., Navabshan, I., Sakthivel, B., Dharmaraj, S., Chinnaiyan, S.K., Ashokkumar, V. and Ngamcharussrivichai, C., 2022. Marine biome-derived secondary metabolites, a class of promising antineoplastic agents: A systematic review on their classification, mechanism of action and future perspectives. *Science of The Total Environment*, *836*, p.155445.

Wang, X., Chan, Y.S., Wong, K., Yoshitake, R., Sadava, D., Synold, T.W., Frankel, P., Twardowski, P.W., Lau, C. and Chen, S., 2023. Mechanism-driven and clinically focused development of botanical foods as multitarget anticancer medicine: Collective perspectives and insights from preclinical studies, IND applications and early-phase clinical trials. *Cancers*, *15*(3), p.701.

Wasser, S.P., 2011. Current findings, future trends, and unsolved problems in studies of medicinal mushrooms. *Applied Microbiology and Biotechnology*, *89*, pp.1323–1332.

Winter, K., Ticktin, T. and Quazi, S., 2020. Biocultural restoration in Hawai'i also achieves core conservation goals. *Ecology and Society*, *25*(1).

Woo, C.S.J., Lau, J.S.H. and El-Nezami, H., 2012. Herbal medicine: Toxicity and recent trends in assessing their potential toxic effects. In *Advances in botanical research* (Vol. 62, pp. 365–384). Academic Press.

Yao, R., Heinrich, M. and Weckerle, C.S., 2018. The genus Lycium as food and medicine: A botanical, ethno-botanical and historical review. *Journal of Ethnopharmacology*, *212*, pp.50–66.

Zhang, H., Mittal, N., Leamy, L.J., Barazani, O. and Song, B.H., 2017. Back into the wild—Apply untapped genetic diversity of wild relatives for crop improvement. *Evolutionary Applications*, *10*(1), pp.5–24.

Zhao, Q., Han, Z., Wang, J. and Han, Z., 2021. Development and investigational new drug application of mesenchymal stem/stromal cells products in China. *Stem Cells Translational Medicine*, *10*, pp.S18–S30.

Zhu, X., Yao, Q., Yang, P., Zhao, D., Yang, R., Bai, H. and Ning, K., 2022. Multi-omics approaches for in-depth understanding of therapeutic mechanism for Traditional Chinese Medicine. *Frontiers in Pharmacology*, *13*, p.1031051.

9 Dandelions

Ethnopharmacology, Phytochemistry, and Pharmacological Activities for Human Wellness

Ari Satia Nugraha, Watchara Sangsopha,
Ria Cahyaningsih, Reza Yuridian Purwoko, Lilla Nur Firli
and Phurpa Wangchuk

9.1 INTRODUCTION

Mankind and plants have a complex relationship throughout history in which the plants have shaped human society by providing food, clothes, shelter, cosmetics, and medicine. Dandelion is one fine example to this. It is a wild invasive plant, referred to *Taraxacum* species. Many species mostly grow in the temperate zones. There are about 2336 species of *Taraxacum*. Of these numbers, 20 *Taraxacum* species were reported to be traditionally used as medicine by several ethnic peoples across the globe Some examples are: *T. androssovii* Schischk, *T. crepidiforme* DC. *T. cyprium* H.Lind., *T. dissectum* (Ledeb.) Ledeb., *T. erythrospermum* Andrz. ex Besser, *T. fedtschenkoi* Hand-Mazz, *T. hybernum* Stev., *T. macrolepium* Schischk., *T. megalorrhizon* (Forsskal) Hand.-Mazz, *T. mongolicum* Hand.-Mazz., *T. obovatum* (Willd.) DC., *T. officinale* Weber ex F.H. Wigg., *T. palustre* (Lyons) Symons., *T. panalpinum* van Soest., *T. platycarpum* Dahlst., *T. pseudobrachyglossum* van Soest, *T. pyropappum* Boiss. & Reut., *T. sikkimense* Hand.-Mazz., *T. stenolepium* Hand.-Mazz., *T. tibetanum* Handel-Mazzett. These species are used for treating various diseases including gout, diarrhea, blisters, and spleen and liver complaints. The earliest record of using these plants in medicine dates back to the tenth century when the herbs were described in Arabian medicine to alleviate liver and spleen diseases. Since then, this plant has generated interest in many herbalists and scientific scholars of the past and present. Many *Taraxacum* species have been screened for several pharmacological activities including diuretic, antioxidant, anti-rheumatic, anti-allergic, anti-inflammatory, analgesic, anticoagulant, antibiotic, choleretic, angiogenic and anticarcinogenic properties. This chapter describes the biology, ethnobotany, phytochemicals, and pharmacological properties of *Taraxacum* species.

9.2 BIOLOGY, TAXONOMY, ECOLOGY, DIVERSITY, AND DISTRIBUTION

The dandelion is a perennial herbaceous plant that can reach heights of 20 to 60 cm. The plant typically has a single distinctive flower structure with bright yellow blooms throughout the year and unique cotton-like fruits with lots of seeds that are dispersed by the wind. The leaves have light green to green color and toothed edges resembling a lion's tooth, or dent-de-lion (from French) where the term "dandelion" may have originated. The root is succulent and fragile, but the taproot is strong, 2 to 3 cm wide, and 15 to 100 cm long. The plant can thrive in any type of soil and requires full sun to light shade and low to moderate water (Jalili et al. 2020; Janke 2004), thus commonly prevalent as a weed in pastures, crops, gardens, and wasteland (Di Napoli and Zucchetti 2021).

DOI: 10.1201/9781003395935-9

Dandelion is a common English name and its botanical/latin name is known as *Taraxacum officinale* F.H.Wigg. in Prim. Fl. Holsat.: 56 (1780) (MPNS 2021). *Taraxacum* genus is a member of Asteraceae, a family that has 23,000 species distributed in 1600 genera. The flowering plant family appears in diverse morphological traits from 1 cm to 30 m heights (Gutiérrez-Grijalva et al. 2020). There are at least 521 species that are recorded in the *Taraxacum* genus alone. However, the local names are diverse and there is ambiguity on some of the local dandelion's names, which lead to false dandelion identification. Local nomenclature can identify the plant only at the genus level and not at the species level. Consequently, due to similar physical features, some plants that belong to different families (Fabaceae, Apocynaceae) have been misplaced or misidentified as dandelions. For example, *Araujia gardensericifera* is known locally as cape-dandelion (Wiersema et al. 1999), however according to Plants of the World Online (POWO) information and the herbarium, it belongs to the Apocynaceae family, not the Asteraceae, and so has a different floral structure. *Senna occidentalis* (L.), which is known as the dandelion by Jamaicans (Mitchell 2011), similarly has a different flower form because it is a member of the Fabaceae family. As a result, we don't consider these two species as dandelions.

The eight species of dandelions from the Asteraceae family are the subjects discussed in this section.

Their taxonomy classification (National Museum of Natural History… 2023) is as follows:

Kingdom: Plantae
Subkingdom: Viridiplantae
Infrakingdom: Streptophyta – land plants
Superdivision: Embryophyta
Division: Tracheophyta – vascular plants, tracheophytes
Subdivision: Spermatophytina – spermatophytes, seed plants
Class: Magnoliopsida
Superorder: Asteranae
Order: Asterales
Family: Asteraceae

9.2.1 Taraxacum alaskanum

Author: Rydb
Genus: *Taraxacum*
Heterotypic Synonyms: *Taraxacum pseudokamtschaticum* Jurtzev in Bot. Zhurn. (Moscow & Leningrad) 82(1): 114 (1997). Distribution: Native to Alaska, Washington, Yukon (POWO 2023).

9.2.2 Taraxacum besarabicum

Author: (Hornem.) Hand.-Mazz.
Genus: *Taraxacum*
Homotypic Synonyms: *Leontodon besarabicus* Hornem. in Suppl. Hort. Bot. Hafn.: 88 (1819); *Taraxacum serotinum* subsp. *besarabicum* (Hornem.) Hegi in Ill. Fl. Mitt.-Eur. 6: 1081 (1928); Heterotypic Synonyms: *Taraxacum dens-leonis* subsp. *salsugineum* (Lamotte) R.C.V.Douin in G.E.M.Bonnier, Fl. Ill. France 6: 78 (1923); *Taraxacum fulvipile* Harv. in Fl. Cap. 3: 527 (1865); *Taraxacum leptocephalum* Rchb. in Fl. Germ. Excurs. 1: 270 (1831); *Taraxacum officinale* var. *leptocephalum* (Rchb.) W.D.J.Koch in Syn. Fl. Germ. Helv., ed. 2: 492 (1844); *Taraxacum officinale* proles *leptocephalum* (Rchb.) Rouy in

G.Rouy & J.Foucaud, Fl. France 9: 191 (1905); *Taraxacum procumbens* Less. in Linnaea 9: 181 (1834); *Taraxacum salinum* Besser in Enum. Pl. Volh.: 31 (1821); *Taraxacum salsugineum* Lamotte in Bull. Soc. Bot. France 21: 123 (1874) (POWO 2023).

Distribution: native to Afghanistan, Altay, Austria, Bulgaria, Central European Russia, Czechoslovakia, East European Russia, France, Hungary, Inner Mongolia, Iran, Irkutsk, Kazakhstan, Kirgizstan, Krasnoyarsk, Krym, Lebanon-Syria, Libya, Mongolia, North Caucasus, Poland, Romania, South European Russi, Tadzhikistan, Transcaucasus, Turkey, Turkmenistan, Tuva, Ukraine, Uzbekistan, West Himalaya, West Siberia, Xinjiang, Yugoslavia. Introduced into Cape Provinces, Northern Provinces (POWO 2023).

9.2.3 Taraxacum erythrospermum

Author: Andrz. ex Besser

Genus: *Taraxacum*

Local names: re-seed dandelion or rock dandelion as a local name (FDA 2016)

Synonyms: *Leontodon erythrospermum* (Andrz. ex Besser) Eichw. in Skizze: 150 (1830), *Taraxacum laevigatum* var. *erythrospermum* (Andrz. ex Besser) J.Weiss in W.D.J.Koch, Syn. Deut. Schweiz. Fl., ed. 3: 1656 (1900), *Taraxacum officinale* subsp. *erythrospermum* (Andrz. ex Besser) Berher in L.Louis, Fl. Vosges, éd. 2: 146 (1887), *Taraxacum officinale* var. *erythrospermum* (Andrz. ex Besser) Bab. in Man. Brit. Bot.: 179 (1843), *Taraxacum officinale* f. *erythrospermum* (Andrz. ex Besser) Merino in Fl. Galicia 2: 466 (1906), *Taraxacum vulgare* subsp. *erythrospermum* (Andrz. ex Besser) Arcang. in Comp. Fl. Ital.: 428 (1882) (POWO 2023).

Heterotypic Synonyms: *Leontodon erythrospermus* Britton in N.L.Britton & A.Brown, Ill. Fl. N. U.S., ed. 2, 3: 316 (1913), *Leontodon laevigatus* Willd. in Sp. Pl. ed. 4. 3: 1546 (1803), *Leontodon taraxacum* subsp. *laevigatus* (Willd.) Gaudin in Fl. Helv. 5: 61 (1829), *Leontodon taraxacum* var. *laevigatus* (Willd.) Benth. in Cat. Pl. Pyrénées: 94 (1826), *Taraxacum austriacum* Soest in Proc. Kon. Ned. Akad. Wetensch. C 69: 434 (1966), *Taraxacum austriacum* var. *denubium* (Richards) R.Doll in Feddes Repert. 84: 21 (1973), *Taraxacum austriacum* var. *punctatum* (Richards) R.Doll in Feddes Repert. 84: 22 (1973), *Taraxacum laevigatum* (Willd.) DC. in Cat. Pl. Horti Monsp.: 149 (1813), *Taraxacum laevigatum* f. *scapifolium* F.C.Gates & S.F.Prince in Trans. Kansas Acad. Sci. 41: 119 (1938), *Taraxacum officinale* subsp. *laevigatum* (Willd.) Cout. in Fl. Portugal: 673 (1913), *Taraxacum officinale* var. *laevigatum* (Willd.) Bouvier in Fl. Alpes, ed. 2: 394 (1882), *Taraxacum punctatum* A.J.Richards in Acta Fac. Rerum Nat. Univ. Comen., Bot. 18: 111 (1970), *Taraxacum scanicum* Dahlst. in Ark. Bot. 10(11): 21 (1911), *Taraxacum taraxacoides* var. *laevigatum* (Willd.) Willk. in M.Willkomm & J.M.C.Lange, Prodr. Fl. Hispan. 2: 231 (1865), *Taraxacum taraxacum* var. *laevigatum* (Willd.) Kuntze in Revis. Gen. Pl. 3(3): 181 (1898), not validly publ., *Taraxacum tauricum* Kotov in V.L.Komarov (ed.), Fl. URSS 29: 736 (1964) (POWO 2023).

Distribution: Native to Europe continent, introduced and distributed to North America (Figure 9.1) (POWO 2023).

9.2.4 Taraxacum gracilens

Author: Dahlst.

Genus: *Taraxacum*

Distribution: Native to Greece, Turkey-in-Europe, Yugoslavia. Introduced into New South Wales, South Australia, Victoria (POWO 2023).

Taraxacum erythrospermum

FIGURE 9.1   *Taraxacum erythrospermum* Andrz. ex Besser.

9.2.5   Taraxacum kok-saghyz

Author: Rodin

Genus: *Taraxacum*

Heterotypic Synonyms: *Taraxacum brevicorniculatum* Korol. in Bot. Mater. Gerb. Bot. Inst. Komarova Akad. Nauk S.S.S.R. 8: 93 (1940) (POWO 2023).

Distribution: native to Kazakhstan, Kirgizstan, Mongolia, Xinjiang. Introduced into Austria, Czechoslovakia, Germany, Northwest European R, Romania, Tasmania, Ukraine (POWO 2023).

9.2.6   Taraxacum mongolicum

Author: Hand.-Mazz.

Genus: *Taraxacum*

Local names: Dandelion or Mongolian dandelion (McGuffin et al. 2000)

Synonyms, heterotypic Synonyms: *Taraxacum argute-denticulatum* Nakai & H.Koidz. in Bot. Mag. (Tokyo) 50: 142 (1936), *Taraxacum formosanum* Kitam. in Acta Phytotax. Geobot. 2: 48 (1933), *Taraxacum hangchouense* H.Koidz. in Bot. Mag. (Tokyo) 50: 144 (1936), *Taraxacum hondae* Nakai & H.Koidz. in Bot. Mag. (Tokyo) 50: 143 (1936), *Taraxacum huhhoticum* Z.Xu & H.C.Fu in Fl. Intramongolica 6: 329 (1982), *Taraxacum kansuense* Nakai ex H.Koidz. in Bot. Mag. (Tokyo) 50: 91 (1936), *Taraxacum liaotungense* Kitag. in Bot. Mag. (Tokyo) 47: 825 (1933), *Taraxacum liaotungense* f. *lobulatum* Kitag. in Bot. Mag. (Tokyo) 47: 1933 (1933), *Taraxacum mongolicum* var. *formosanum* (Kitam.) Kitam. in Compos. Nov. Jap. 6: 42 (1957), *Taraxacum pseudodissectum* Nakai & H.Koidz. in Bot. Mag. (Tokyo) 50: 92 (1936), *Taraxacum quelpaertense* Kitam. in Acta Phytotax. Geobot. 2: 184 (1933) (POWO 2023).

Distribution: Native to East Asia and Central Eastern Europe (Figure 9.2) (POWO 2023).

Taraxacum mongolicum

FIGURE 9.2   *Taraxacum mongolicum* Hand.-Mazz.

9.2.7 TARAXACUM SECT. TARAXACUM

Author: F.H.Wigg
Genus: *Taraxacum*
Local names: common dandelion or dandelion (Mitchell 2011)
Synonyms: *Chondrilla taraxacum* (L.) Stokes in Bot. Mat. Med. 4: 122 (1812), *Crepis taraxacum* (L.) Stokes in W.Withering, Bot. Arr. Brit. Pl. ed. 2. 2: 853 (1787), *Leontodon officinalis* (F.H.Wigg.) J.F.Gmel. in Syst. Nat. ed. 13bis. 2(2): 1174 (1792), nom. superfl., *Leontodon taraxacum* L. in Sp. Pl.: 798 (1753), *Leontodon taraxacum* var. *villosum* Lej. in Rev. Fl. Spa: 167 (1825), *Leontodon taraxacum* var. *vulgare* Benth. in Cat. Pl. Pyrénées: 94 (1826), not validly publ., *Leontodon vulgaris* Lam. in Fl. Franç. 2: 113 (1779), nom. superfl., *Taraxacum dens-leonis* Desf. in Fl. Atlant. 2: 228 (1799), *Taraxacum dens-leonis* subsp. *officinale* (Lyons) R.C.V.Douin in G.E.M.Bonnier, Fl. Ill. France 6: 78 (1923), nom. illeg., *Taraxacum officinale* F.H.Wigg. in Prim. Fl. Holsat.: 56 (1780), *Taraxacum officinale* var. *angustifolium* Gray in Nat. Arr. Brit. Pl. 2: 426 (1821 publ. 1822), *Taraxacum officinale* subsp. *dens-leonis* (Desf.) Cout. in Fl. Portugal 2: 793 (1939), *Taraxacum officinale* var. *genuinum* Willk. in M.Willkomm & J.M.C.Lange, Prodr. Fl. Hispan. 2: 230 (1865), not validly publ., *Taraxacum officinale* subsp. *vulgare* Schinz & R.Keller in H.Schinz & R.Keller, Fl. Schweiz ed. 2, 2: 543 (1905), not validly publ., *Taraxacum palustre* var. *vulgare* (Benth.) Fernald in Rhodora 35: 380 (1933), *Taraxacum taraxacum* (L.) H.Karst. in Deut. Fl.: 1138 (1883), not validly publ., *Taraxacum vulgare* Schrank in Baier. Reise: 11 (1786), nom. superfl. *Taraxacum vulgare* var. *dens-leonis* (Desf.) Samp. in Fl. Port., ed. 2: 618 (1947), nom. superfl. (POWO 2023).
Distribution: Native to Europe and northern Africa, and introduced and distributed around the globe (Figure 9.3) (POWO 2023).

9.2.8 TARAXACUM PALUSTRE

Author: (Lyons) Symons
Genus: *Taraxacum*
Local names: marsh dandelion (FDA 2016)
Synonyms:
Homotypic Synonyms: *Leontodon palustris* Lyons in Fasc. Pl. Cantabr.: 48 (1763), *Leontodon taraxacum* var. *paludosus* Lej. in Rev. Fl. Spa: 167 (1825), *Leontodon taraxacum* subsp. *palustris* (Lyons) Schübl. & G.Martens in Fl. Würtemberg: 511 (1834), *Taraxacum officinale* proles *palustre* (Lyons) Rouy in G.Rouy & J.Foucaud, Fl. France 9: 190 (1905), *Taraxacum officinale* subsp. *palustre* (Lyons) Hartm. in Sv. Norsk Exc.-Fl.: 110 (1846) (POWO 2023).
Heterotypic Synonyms: *Leontodon salinus* Pollich in Hist. Pl. Palat. 2: 380 (1777), *Leontodon taraxacum* var. *salinus* (Pollich) E.Mey. in Pl. Labrador.: 58 (1830), *Taraxacum commutatum* Jord. in Mém. Acad. Roy. Sci. Lyon, Sect. Lett., sér. 2, 1: 327 (1851), *Taraxacum dens-leonis* subsp. *palustre* (Willd.) R.C.V.Douin in G.E.M.Bonnier, Fl. Ill. France 6: 78 (1923), *Taraxacum gremlii* Appel ex Naegeli & Wehrli in Mitth. Thurgauischen Naturf. Ges. 9: 150 (1890), *Taraxacum lanceolatum* Poir. in J.B.A.M.de Lamarck, Encycl. 5: 349 (1804), *Taraxacum limnanthes* subsp. *limnanthoides* Soest in Acta Bot. Neerl. 14: 38 (1965), *Taraxacum maritimum* Hagend., Soest & Zevenb. in Gorteria 5: 86 (1970), *Taraxacum officinale* f. *commutatum* (Jord.) Merino in Fl. Galicia 2: 465 (1906), *Taraxacum officinale* proles *lanceolatum* (Poir.) Rouy in G.Rouy & J.Foucaud, Fl. France 9: 190 (1905),

Taraxacum sect. Taraxacum

FIGURE 9.3   *Taraxacum* sect. *Taraxacum* F.H.Wigg.

Taraxacum palustre var. *latifolium* A.Gray in Mem. Amer. Acad. Arts, n.s., 4(1): 115 (1849), *Taraxacum palustre* f. *spurium* Beck in Fl. Nieder-Österreich 2: 1314 (1893), *Taraxacum scorzonera* Rchb. in Fl. Germ. Excurs. 1: 270 (1831), *Taraxacum westhoffii* Hagend., Soest & Zevenb. in Gorteria 5: 84 (1970) (POWO 2023).

Distribution: Native to Eastern Part of Europe and Introduced to western part of North America (Figure 9.4) (POWO 2023).

9.2.9   TARAXACUM PLATYCARPUM

Author: Dahlst.
Genus: *Taraxacum*
Local names: dandelion (Safety 2002)
Synonyms: -
Distribution: Native to Japan and Korea (POWO 2023).

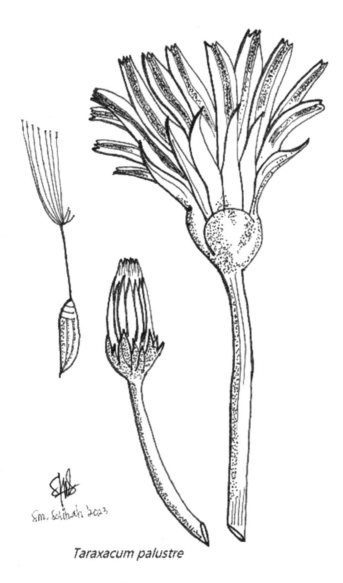

Taraxacum palustre

FIGURE 9.4 *Taraxacum palustre* (Lyons) Symons.

9.2.10 Taraxacum portentosum

Author: Kirschner & Štěpánek
Genus: Taraxacum
Distribution: native to Czechoslovakia, Poland (POWO 2023).

9.2.11 Taraxacum siphonanthum

Author: X.D.Sun, X.J.Ge, Kirschner & Štěpánek
Genus: *Taraxacum*
Heterotypic Synonyms: *Neo-taraxacum siphonanthum* Y.R.Ling & X.D.Sun in Bull. Bot. Res., Harbin 21: 176 (2001), no holotype indicated. Distribution: native to Inner Mongolia (POWO 2023).

9.2.12 TARAXACUM SINICUM

Author: Kitag.

Genus: *Taraxacum*

Local names: dandelion or Chinese dandelion (McGuffin et al. 2000)

Synonyms: *Taraxacum armeriifolium* Soest in Feddes Repert. 70: 61 (1965), *Taraxacum czuense* Schischk. in Sist. Zametki Mater. Gerb. Krylova Tomsk. Gosud. Univ. Kuybysheva 1949(1–2): 6 (1949), *Taraxacum sinicum* var. *armeriifolium* (Soest) Tzvelev in Rast. Tsentral. Azii 14b: 170 (2008) (POWO 2023).

Distribution: Native to China to Center-eastern Europe. Introduced to northwest Europe and eastern Europe (POWO 2023).

9.2.13 TARAXACUM UDUM

Author: Jord.

Genus: *Taraxacum*

Homotypic Synonyms: *Taraxacum palustre* var. *udum* (Jord.) Corb. in Nouv. Fl. Normandie: 360 (1894). Heterotypic Synonyms: *Taraxacum crassiceps* Soest in Acta Bot. Neerl. 14: 26 (1965); *Taraxacum hagendijkii* Soest in Acta Bot. Neerl. 20: 141 (1971); *Taraxacum laeticolorifrons* Soest in Acta Bot. Neerl. 14: 36 (1965); Distribution: native to France, Germany, Netherlands, Poland, Switzerland (POWO 2023).

9.3 TRADITIONAL USES ACROSS THE GLOBE

Dandelions are not only well known as medicinal and food plants but also as weed plants which are distributed commonly in temperate regions across the globe (Figure 9.5). Each plant part (leaf, stem, flower, and root) or whole plant contains some phytochemical components that might be beneficial for health (Honek et al. 2011). Dandelions' leaves (*Taraxacum* spp.) were used in soups during times of scarcity in some parts of the Polish Carpathians, but this kind of use disappeared at the beginning of the 20th century (Łuczaj 2010).

Taraxacum sect. *Taraxacum* F.H.Wigg has the most records in medicinal uses as it is a more widespread and more commonly used dandelion among other dandelions (MPNS 2021). It is frequently picked for the root, leaf, and flower, all of which can be consumed fresh or dried. The leaves can be eaten raw in salads, the root is consumed as a substitute for cereal coffee, and the flowers are used to make syrups (Kania-Dobrowolska et al. 2022). In Europe, it is approved for use to treat appetite loss, hepatic and gallbladder problems, urinary tract infections, and indigestion (Janke 2004). Algerian and North African people harvest its root, process it into infusion or decoction, and use it for antidiabetic and tonic (Boudjelal et al. 2013), while southern Africans consume the extract of its flowers, leaves, roots, and whole plant orally to treat tuberculosis (TB) (Cock et al. 2020). South Kosovo locals drink dandelion tea as a diuretic and use the syrup as a general tonic (Pieroni et al. 2017). Romania and other Eastern European people use the roots, aerial parts, flowers, and sap for curing internal acne, varicose veins, varicose ulcers, dermatitis, scurvy, external acne, verrucae, tinea, freckles, and sensitive skin, as well as other conditions (Gilca et al. 2018). South-Eastern Albania's villagers that are Albanian, Aromanian, and Macedonian consume it as a wild vegetable (Pieroni and Sõukand 2017).

There were very few authoritative sources of information on the medicinal uses of other common dandelions. *Taraxacum mongolicum* have been used since ancient time for curing mastitis, breast abscess, and hyperplasia of mammary glands by Chinese people (MPNS 2021; Deng et al. 2021). *Taraxacum sinicum*'s whole plant is used for medicinal purposes (MPNS 2021). *Taraxacum*

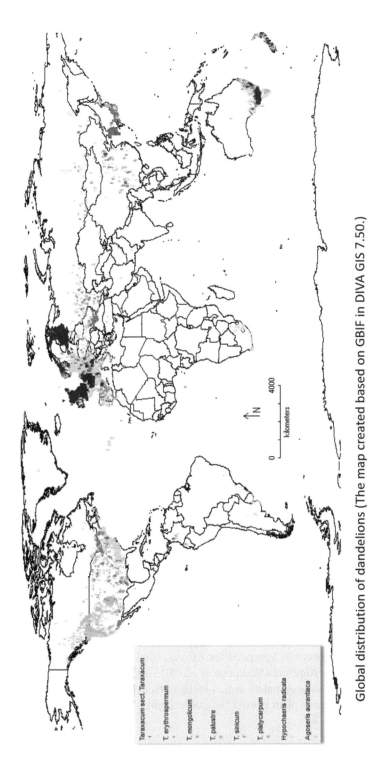

Global distribution of dandelions (The map created based on GBIF in DIVA GIS 7.50.)

FIGURE 9.5 Global distribution of eight most common dandelion species (The map created in DIVA GIS 7.50, based on GBIF occurrence data) (GBIF 2023).

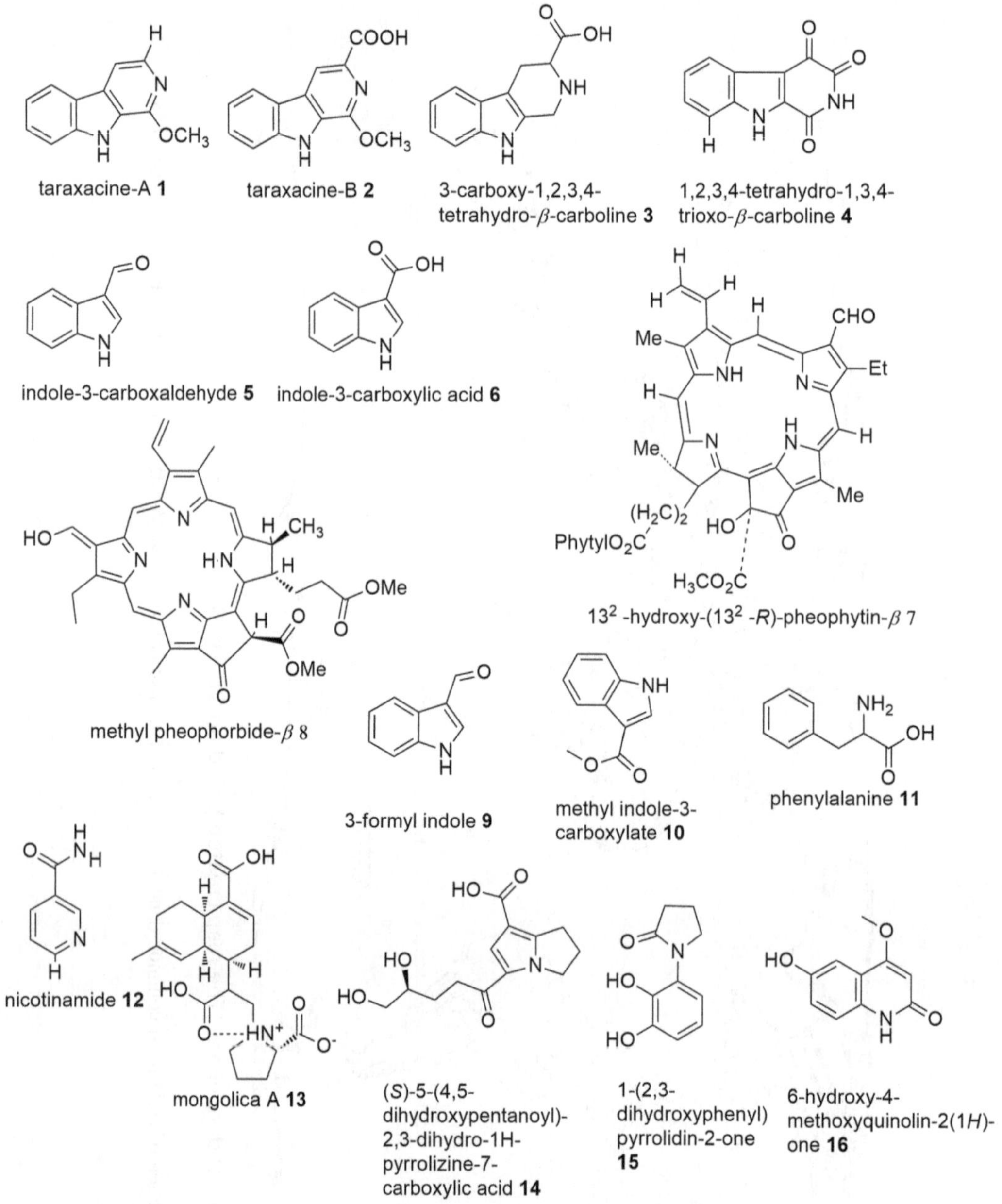

FIGURE 9.6 Alkaloids from *T. formossum* and *T. mongolicum*.

erythrospermum Andrz. ex Besser is harvested for its root, used freshly or processed into infusion for sudorific and diuretic by Algerian (Boudjelal et al. 2013). For *Taraxacum palustre*, there was no documentation of medicinal use, unless it was recognized as medicinal plant in MPNS (2021). *Taraxacum platycarpum*'s whole plant is used for medical purposes in Korea (MPNS 2021).

9.4 PHYTOCHEMISTRY

The oldest attempt on natural product purification is distillation or preparing decoction. In order to study the secondary metabolites, isolation and characterization is required. The isolation requires tedious efforts as the molecules are small in size and have diverse physical-chemical properties (Cannell 1998). In the secondary metabolites isolation, there is a general target to obtain all possible compounds or a specific target to gain specific bioactive compounds, which are responsible for certain bioactivity (Tsuda 2004). In general, isolation starts from sample preparation (pulverization), extractions-fractionation, and separation. About 16 *Taraxacum* species are phytochemically studied and reported that the dandelions are rich in alkaloids, terpenes, and phenolic group of compounds.

9.4.1 Alkaloids

The alkaloids represent a group of naturally occurring compounds which contain mostly basic nitrogen atoms. Alkaloids from plants have been used from the ancient to modern era across cultures and continents with its diverse capability for medicinal purposes. Although alkaloids containing plants have been used since ancient time, the term alkaloid was introduced in the early 19th century. This nitrogen containing secondary metabolites is rich in molecular structure diversities derived from amino acids and distinct biosynthetic pathways (Lichman 2021). Chemical scaffold of alkaloids is commonly used for classification despite the diversities across the plants. The compound names are often derived from their plant's taxonomical classification. For example, anonaine and atropine have derived their names from *Annona squamosa* and *Atropa belladonna*, respectively (Gutiérrez-Grijalva et al. 2020).

There are more than 40,000 reported alkaloids with several well-known plant families in producing alkaloids, including Amaryllidaceae, Apocynaceae, Papaveraceae, Solanaceae, Rutaceae, and Asteraceae families. Phytochemical studies on *Taraxacum* genera revealed several alkaloids including indole-containing molecules (Figure 9.7). The indole backbone is a fused ring formed by a six-membered benzene ring and a five-membered pyrrole ring which contributed to its basicity (Lichman 2021). Seven indoles were isolated from Taiwanese *T. formosum*, taraxacine-A **1**, taraxacine-B **2**, 3-carboxy-1,2,3,4-tetrahydro-β-carboline **3**, 1,2,3,4-tetrahydro-1,3,4-trioxo-β-carboline **4**, Indole-3-carboxaldehyde **5**, Indole-3-carboxylic acid **6**, 13²-hydroxy-(13²-*R*)-pheophytin-b **7**, methyl pheophorbide-β **8**, 3-formyl indole **9**, methyl indole-3-carboxylate **10** (Leu et al. 2003; Leu et al. 2005). The study also produce protoalkaloid compounds, phenylalanine **11** and nicotinamide **12**. The secondary amine present in the molecules produce signal at 3300 cm⁻¹ under IR spectroscopy. Another studies on Chinese *T. mongolicum* Hand.-Mazz successfully isolated four true alkaloids, mongolica A **13**, (*S*)-5-(4,5-dihydroxypentanoyl)-2,3-dihydro-1*H*-pyrrolizine-7-carboxylicacid **14**,1-(2,3-dihydroxyphenyl) pyrrolidine-2-one **15** and 6-hydroxy-4-methoxyquinolin-2(1*H*)-one **16** (Xie et al. 2022).

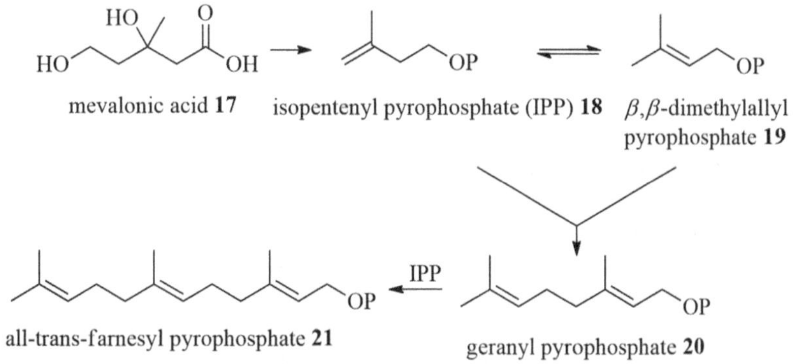

mevalonic acid 17   isopentenyl pyrophosphate (IPP) 18 β,β-dimethylallyl pyrophosphate 19

all-trans-farnesyl pyrophosphate 21   geranyl pyrophosphate 20

FIGURE 9.7 Biosynthesis of important terpene building blocks (geranyl pyrophosphate and farnesyl pyrophosphate) from mevalonic acid.

9.4.2 TERPENES

There are 30,000 recorded terpenes produced by plants. The compounds are generated from two important building blocks, geranyl pyrophosphate and *trans*-farnesyl pyrophosphate through mevalonic acid pathways (Hanson 2003). The mevalonic acid **17** is firstly converted into isopentenyl pyrophosphate **18** (reactive intermediate) (Hanson 2003). This reactive intermediate could undergo isomerization into β,β-dimethylallyl pyrophosphate **19** in which a combination of its isomers (isopentenyl pyrophosphate) further produce two important terpenes building blocks, geranyl pyrophosphate **20** and *trans*-farnesyl pyrophosphate **21** (Figure 9.7). In the biosynthesis of monoterpenes and diterpene, geranyl pyrophosphate and geranylgeranyl pyrophosphate, respectively, act as parent compounds. Farnesyl pyrophosphate is related to sesquiterpenes biosynthesis. Nevertheless, folding differentiation, mixed condensation and other enzymatic reactions contribute to the large derivatives in the terpenoids classes of compounds. For example, the simple geranol could undergo a slightly different folding to form the cyclopentene skeleton of iridoids (Torssell 1982). A metabolic engineering study of *T. brevicorniculatum* Korol (accepted name *T. kok-saghyz* L.E.Rodin) through the main enzyme evolved in mevalonic pathway including 3-hydroxy-methyl-glutaryl-CoA reductase (HMGR), acetoacetyl-CoA thiolase (AACT), and adenosine triphosphate citrate lyase (AAL) has caused significant terpenes production in the latex biomass (Putter et al. 2017).

9.4.2.1 Sesquiterpenes

Investigation on aerial part of *T. mongolicum* of Korean origins produced 1β,3β-dihydroxyeudesman-11(13)-en-6α,12-olide **22**, ainslioside **23**, 1β,3β-dihydroxyeudesman-6α,12-olide **24** and 11β,13-dihydrotaraxinic acid **25** (W. Li, Lee, et al. 2017). Phytochemical study on root of *T. platycarpum* disclosed seven sesquiterpenes with half as glycosylated form, namely sonchuside A **26**, cichorioside C **27**, deacetylmatricarin 8-*O*-β-D-glucopyranoside **28**, 11β-hydroxydeacetylmatricarin 8-*O*-β-D-glucopyranoside **29**, 11β-hydroxyleucodin-11-O-β-D-glucopyranoside **30**, 11β,13-dihydroxydeacetylmatricarin 8-O-β-D-glucopyranoside **31**, ixerin D **32** (Figure 9.8) (Warashina et al. 2012). Another sesquiterpene was also isolated from latex of *T. officinale* grown in Europe, taraxinic acid β-D-glucopyranosyl ester **29** (Figure 9.8) (Huber et al. 2015).

9.4.2.2 Triterpenes

Taraxacum species were able to produce arrays type of triterpenes, including taraxastane, lupane, oleanane, and ursane types (Figure 9.9). Extensive chromatographic separation of ether fraction of root extract of Japanese *T. platycarpum* yielded 18 tritrepenes, taraxasteryl acetate **30**, 21α-hydroperoxy-taraxasteryl acetate **31**, ptiloepoxyl acetate **32**, φ-taraxasteryl acetate **33**, 30-hydroperoxy-φ-taraxasteryl acetate **34**, α-amyrin acetate **35**, 3β-acetoxy-11α-hydroperoxy-12-ursene **36**, neoilexonol acetate **37**, β-amyrin acetate **38**, 3β-acetoxy-11α-hydroperoxy-12-oleanene **39**, 3β-acetoxy-12-oleanen-11-one **40**, and lupenyl acetate **41**, 3β-acetoxy-lup-18-en-21-one **42**, 3-acetoxy-11α-hydroperoxy-neolup-12-ene **43**, 3β-acetoxy neolup-12-en-11-one **44**, 3β-acetoxy-19α, 21α-epoxyl-neolup-12-ene **45**, 3β-acetoxy19β, 21β-epoxy-19-*epi*-lactuc-14-ene **46** and one 17,18-seco-lupanyl acetate-type triterpene **47** (Warashina, Umehara, and Miyase 2012). Constituent isolation of non-saponifiable lipids of flower of *T. platycarpum* obtained from Chiba-Japan successfully obtained ursane type triterpene, uvaol (urs-12-ene-3β, 28-diol) **48** (Yasukawa et al. 1996). Huber et al. conducted phytochemical study on latex of *T. officinale* and revealed several triterpenes including β-amyrin acetate **38**, α-amyrin acetate **35**, lupeol/lupenyl acetate **41**. Intensive phytochemical study on Chilean-sourced *T. officinalle* revealed several triterpenes including lupeol **49**, lupeol acetate **41**, betulin **50** and α-amyrin **51** (Díaz et al. 2018). Intensive phytochemical study on *Taraxacum japonicum* Koidz. isolated 11 triterpenoids, taraxasterol **52**, taraxsteryl acetate **30**, φ-taraxasteryl acetate **33**, α-amyrin acetate **35**, β-amyrin **53**, β-amyrin acetate **38**, taraxerol **54**, taraxeryl acetate **55**, lupeol **49**, lupenyl acetate **41**, lupenone **56** (Takasaki et al. 1999). Korean *T. mongolicum* also produced triterpenoids from its aerial part, gigantursenol A **57** and taraxasterol **52** (W. Li, Lee, et al. 2017).

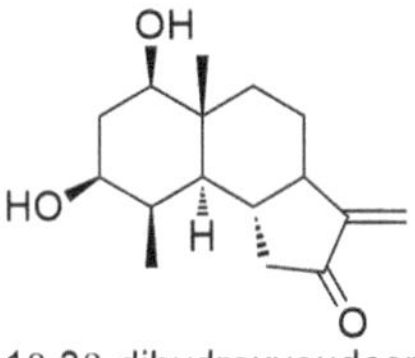

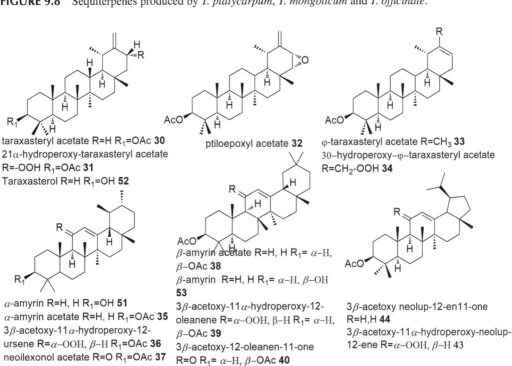

FIGURE 9.8 Sequiterpenes produced by *T. platycarpum*, *T. mongolicum* and *T. officinale*.

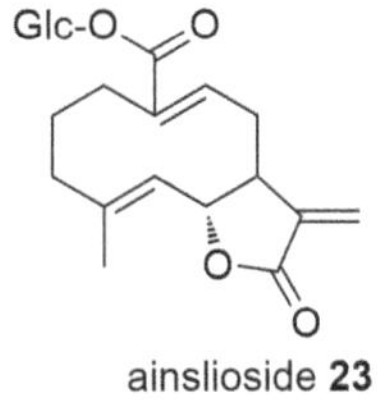

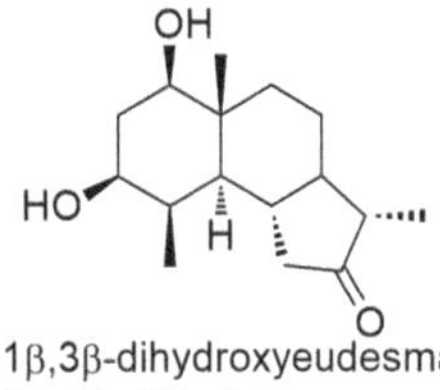

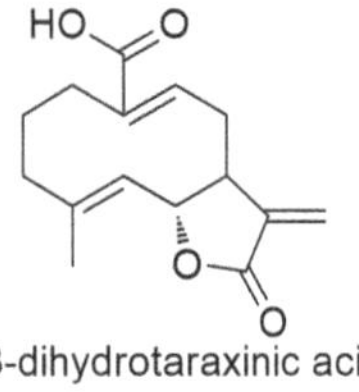

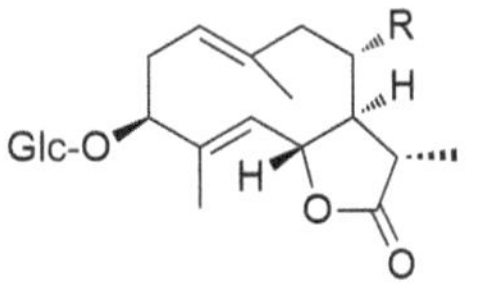

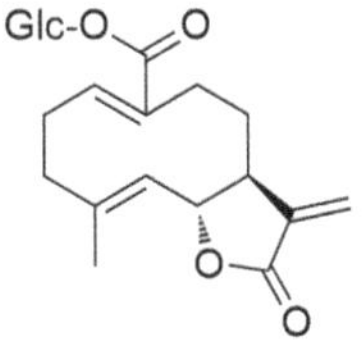

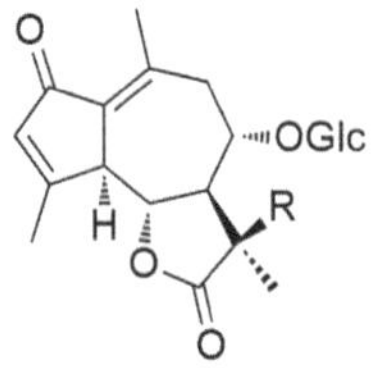

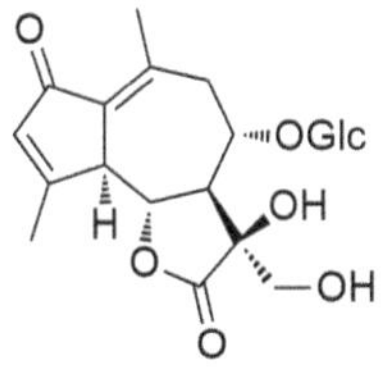

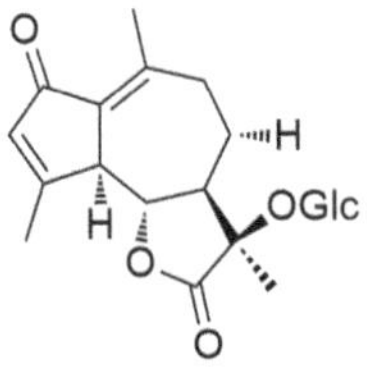

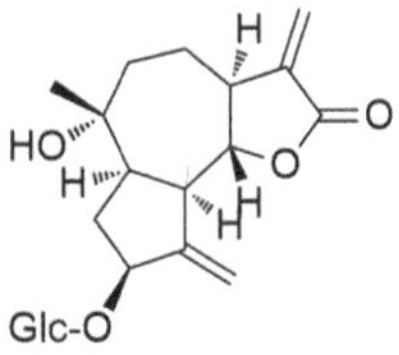

FIGURE 9.9 Triterprenoids secondary metabolite of *Taraxacum* species.

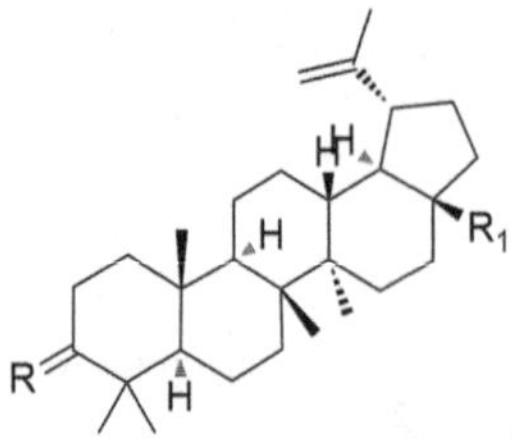

lupenyl acetate R=–H, β–OAc , R$_1$=CH$_3$ **41**
lupeol R=α–H, β–OH , R$_1$=CH$_3$ **49**
betulin R=α–H, β–OH , R$_1$=CH$_2$OH **50**
lupenone R=O, R$_1$=CH$_3$ **56**

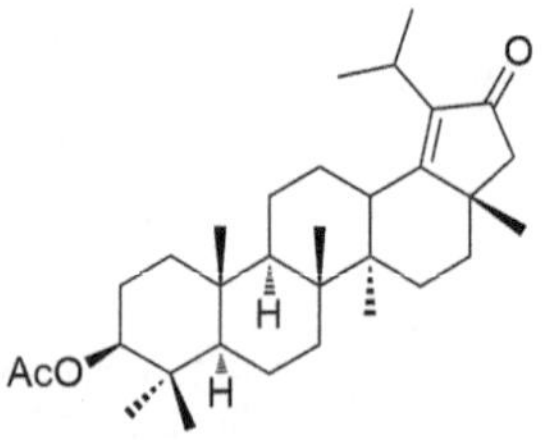

3β-acetoxy-lup-18- en-21-
one **42**

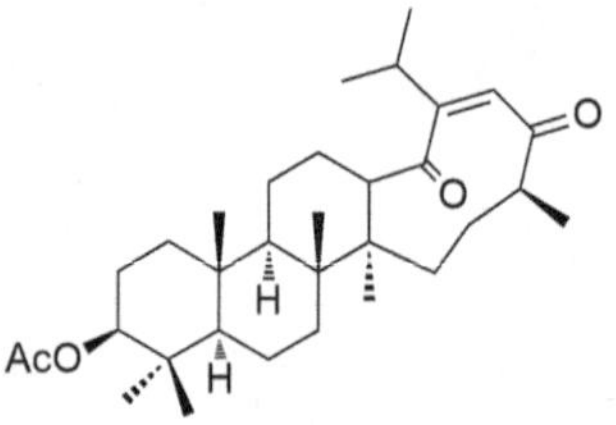

17,18-seco-lupanyl acetate-type
triterpene **47**

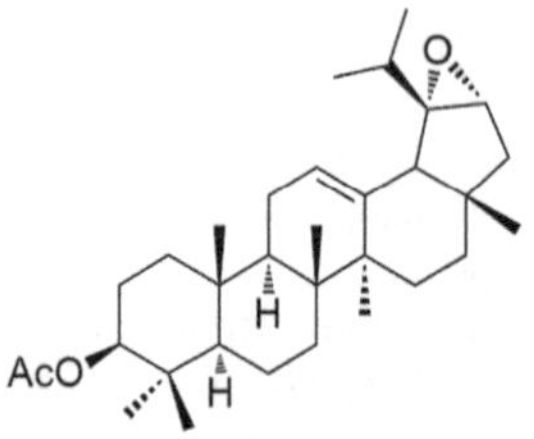

3β-acetoxy-19α,21αepoxyl-
neolup-12-ene **45**

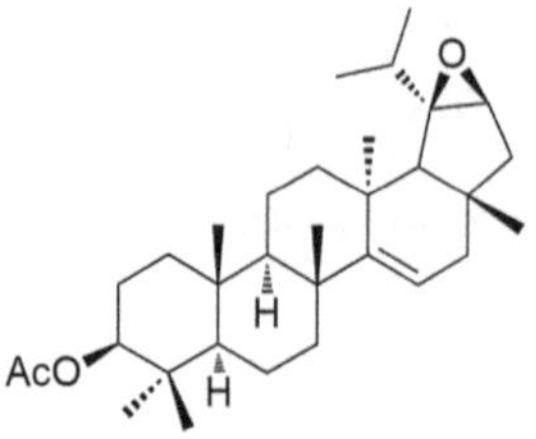

3β-acetoxy 19β,21β-epoxyl-
19-epi-lactuc-14-ene **46**

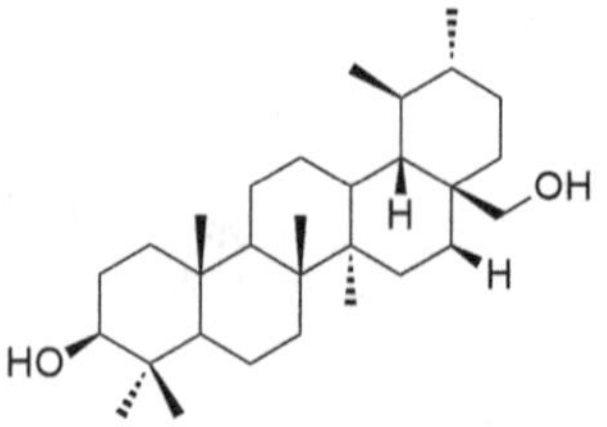

urs-12-ene-3β,28-diol **48**

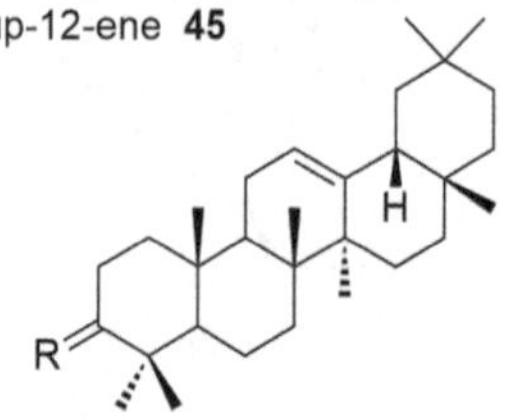

taraxerol R=α–H, β–OH 54
taraxeryl acetate R=–H, β–OAc **55**

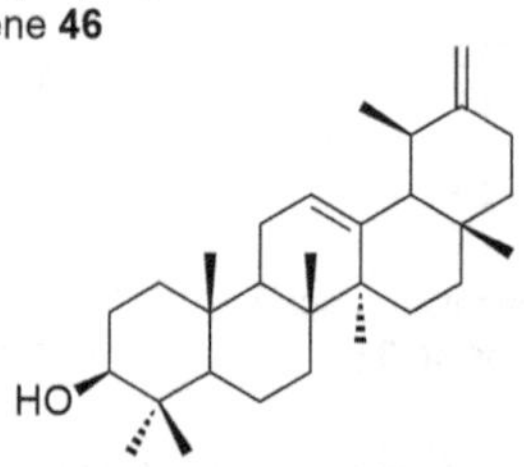

gigantursenol A **57**

FIGURE 9.9   (Continued)

9.4.2.3   Sterols

Common sterol of higher plant Artocarpus, cycloartenol acetate **58**, was also presented in *T. offi-cinale* with a low yield compared to other terpenes especially ursane type triterpenes (Huber et al. 2015). Other common sterols were also presented in Chilean-originated *T. officinale*, β-sitosterol **59**, (22Z)-stigmasta-5,22-diene-3β-ol acetate **60**, 3-ethyl-3-hydroxy-5β-androstan-17-one **61** (Figure 9.10) (Díaz et al. 2018). Li et al. reported sterol constituents of *T. mongolicum* grown in Korea, β-sitosterol 59, β-sitosterol-3-*O*-β-D-glucoside **62**, stigmasterol **63** and β-sigmasterol-3-*O*-β-D-glucoside **64** (W. Li, Lee, et al. 2017).

9.4.3   Phenolic Compounds

Phenolic compounds are the most widely distributed secondary metabolites across species in the plant kingdom; they are biosynthetically produced through shikimate or malonate pathways to provide phenylpropanoids or simple phenols (Cheynier et al. 2013). The aromatic systems present in the molecular structures enable the molecules to act as antioxidants, which are the basic survival components against abiotic stress (M. Kumar et al. 2020). Phenolics are categorized based on their skeleton from simple C6 into more complex skeleton evolving condensation of phenolic monomers such as lignins and tannins.

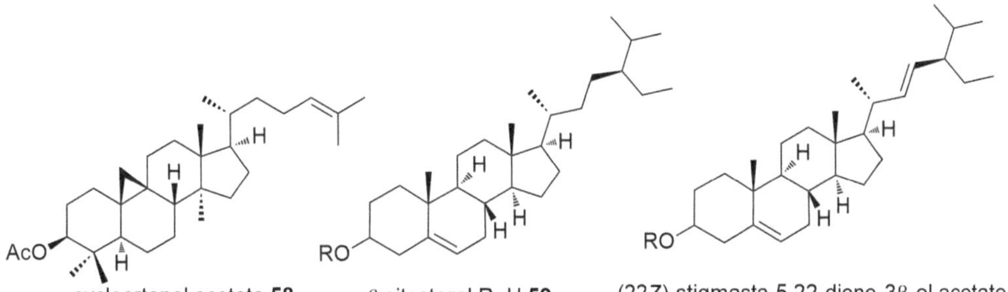

cycloartenol acetate **58**

β-sitosterol R=H **59**
β-sitosterol R=Glc **62**

(22Z)-stigmasta-5,22-diene-3β-ol acetate R=Ac **60**
stigmasterol R=H **63**
β-sigmasterol-3-O-β-D-glucoside R=glc **64**

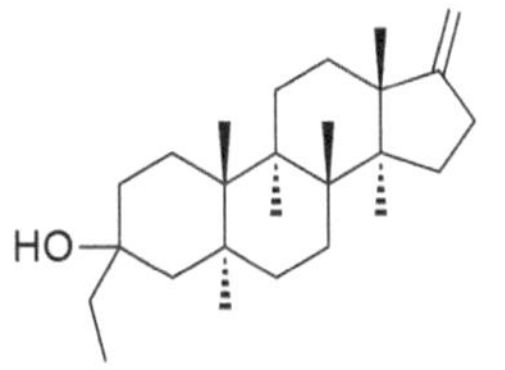

3-ethyl-3-hydroxy-5 -androstan-17-one **61**

FIGURE 9.10   Sterols of *Taraxacum* genera.

9.4.3.1 Flavonoids

The flavonoids comprised of a ubiquitous group of phenolics compounds with C6-C3-C6 molecular skeleton in which a total of 10,000 reported flavonoids are present as a aglycone or glycosylated form (Ekalu et al. 2020). Flavonoids study on Asteraceae tribes revealed flavonoid diversity as possible tools for taxonomical study, in which their production balanced against other secondary metabolite biosynthesis (Emerenciano et al. 2001). Phytochemical experiment revealed several specific flavonoids present across the *Taraxacum* genus including luteolin and quercetin (Figure 9.11). Methanol extract of Neo-*Taraxacum siphonanthum* X.D.Sun et al. (accepted name *T. siphonanthum* X.D.Sun, X.J.Ge, Kirschner & Štěpánek) from Anhui Province China are rich of luteolin (luteolin **65**, luteolin-3′-O-β-D-glucopyranoside **66**, luteolin-4′-O-β-D-glucopyranoside **67**, luteolin-7-O-β-D-glucopyranoside **68**) and quecertin (quercetin **69**, quercetin-3-O-β-D-glucopyranoside **70**, quercetin-3-O-α-D-arabinofuranoside **71**, quercetin-3-*O*-α-L-rhamnoside **72**) (S.Y. Shi et al. 2010). The plant sample also produced hydrogenated methyl quercetin, (2R,3R)-(+)-4′-*O*-methyldihydroquercetin **73** and (2R,3R)-(+)-4′,7-di-O-methyldihydroquercetin **74**as well as methylated flavone, genkwanin **75**, genkwanin-4′-O-β-D-rutinoside **76** (S.Y. Shi et al. 2010). Interestingly, despite luteolin and quercetin are major derivatives the hydrogenated forms present in higher abundant compare to the rest of the flavonoids. The luteolin **65** and luteolin-7-O-β-D-glucopyranoside **68** were also isolated from methanol extract of *T. bessarabicum* (Hornem.) Hand.-Mazz. grown in Anatolia-Turkey along with gossypetin **77**, a unique quercetin derivative with high energy hydroxylation at C-8 (Sari et al. 2019). Phytochemical study on Chinese *T. alaskanum* Rydb also isolated luteolin-7-O-β-D-glucopyranoside **68** along with quercetin **69**, apigenin **78**, and C4' methoxylated luteolin, diosmetin **79** (Fang et al. 2019). Luteolin **65** was reported as major flavonoids reported in Korean dandelions, *T. coreanum* Nakai and *T. gracilens* Dahlst. from Turkey (S. Lee et al. 2012) (Karahuseyin et al. 2022). Luteolin-7-O-β-D-glucopyranoside **68** were also reported from Korean dandelions, *T. coreanum* and Taiwanese *T. formosanum* (accepted name *T. mongolicum* Hand.-Mazz.) Kitam (S. Lee et al. 2012) (H.-J. Chen et al. 2012). Luteolin **65** was also reported as major constituents of polar exudates of *T. gracilens* Dahlst. From Turkey (Karahuseyin, Sari, and Ozsoy 2022).

Study on notorious Asteraceae medicinal herbs in China, *T. mongolicum* Hand-Mazz, revealed arrays of flavonoids including luteolin **65**, luteolin-7-O-β-D-glucopyranoside **68**, luteolin-7-O-β-D-galactopyranoside **80**, quercetin **69**, quercetin-7-O-[β-D-glucopyranosyl(1→6)-

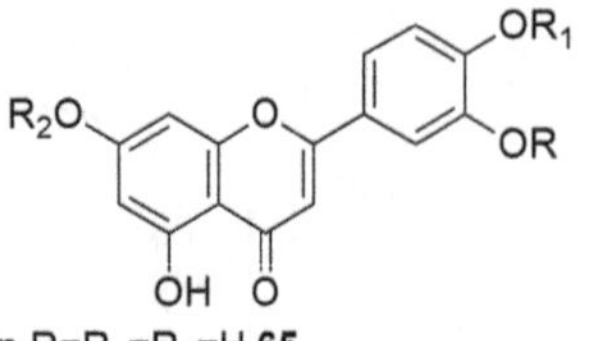

luteolin R=R$_1$=R$_2$=H **65**
luteolin-3'-O-β-D-glucopyranoside R=glu, R$_1$=H, R$_2$=H **66**
luteolin-4'-O-β-D-glucopyranoside R=H, R$_1$=glu, R$_2$=H **67**
luteolin-7-O-β-D-glucopyranoside R=H, R$_1$=H, R$_2$=glu **68**
luteolin-7-O-β-D-galactopyranoside R=H, R$_1$=H, R$_2$=gal **80**

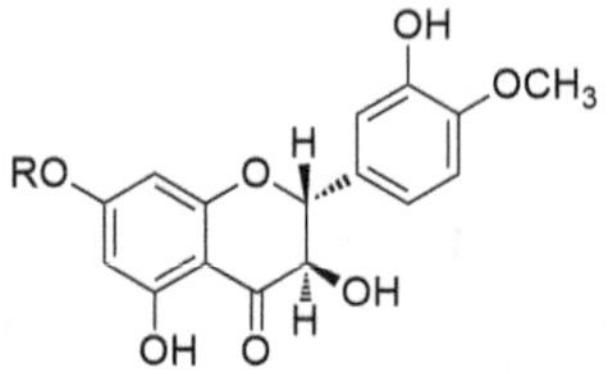

(2R,3R)-(+)-4'-O-methyldihydroquercetin
R=H **73**
(2R,3R)-(+)-4',7-di-O-
methyldihydroquercetin R=OCH$_3$ **74**

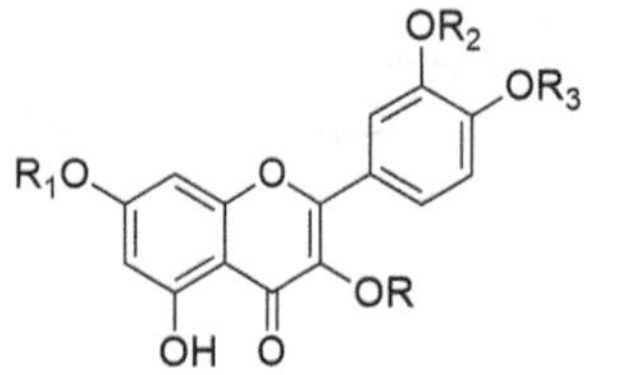

quercetin R=R$_1$=R$_2$=R$_3$=H **69**
quercetin-3-O-β-D-glucopyranoside R=glc R$_1$=R$_2$=R$_3$=H **70**
quercetin-3-O-β-D-arabinofuranoside R=arab R$_1$=R$_2$=R$_3$=H **71**
quercetin-3-O-α-L-rhamnoside R=rham R$_1$=R$_2$=R$_3$=H **72**
quercetin-7-O-β-D-glucopyranoside R$_1$=glc R=R$_2$=R$_3$=H **92**
quercetin-7-O-[β-D-glucopyranosyl(1-->6)-β-D-glucopyranoside]
R$_1$= β-D-glu(1-->6)-β-D-glu R=R$_2$=R$_3$=H **81**
quercetin-3,7-O-β-D-diglucopyranoside R=R$_1$=β-D-glu R$_2$=R$_3$=H **82**
quercetin-3',4',7-trimethyl ether R=H R$_1$=R$_2$=R$_3$=CH$_3$ **90**

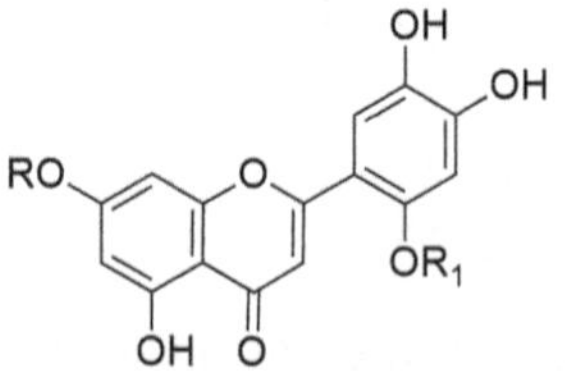

isoetin R=R$_1$=H **83**
isoetin-7-O-β-D-glucopyranosyl-2'-O-α-L-
arabinopyranoside R=β-D-glu R$_1$=α-L-ara **84**
isoetin-7-O-β-D-glucopyranosyl-2'-O-β-D-
xyloypyranoside R=β-D-glu R$_1$=β-D-xyl **86**
isoetin-7-O-β-D-glucopyranosyl-2'-O-β-D-
glucopyranoside R=R$_1$=β-D-glu **85**

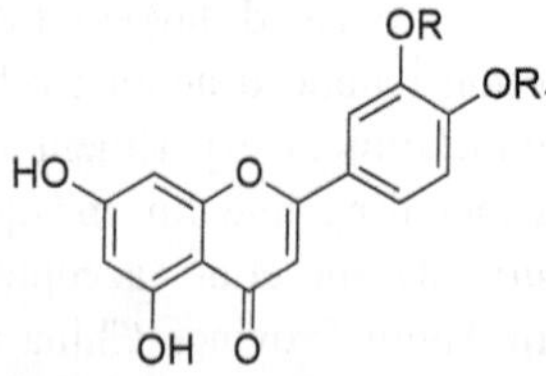

diosmetin R=H R$_1$=CH$_3$ **79**
chrysoeriol R=CH$_3$ R$_1$=H **93**

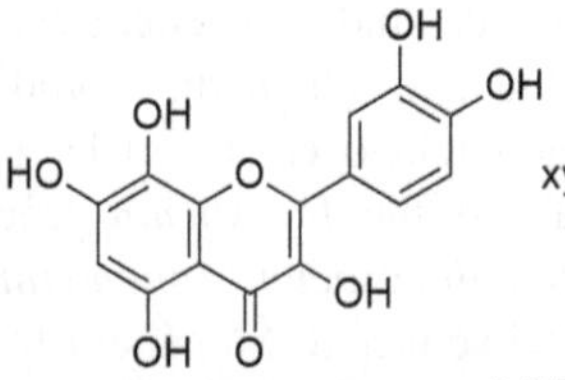

gossypetin **77**

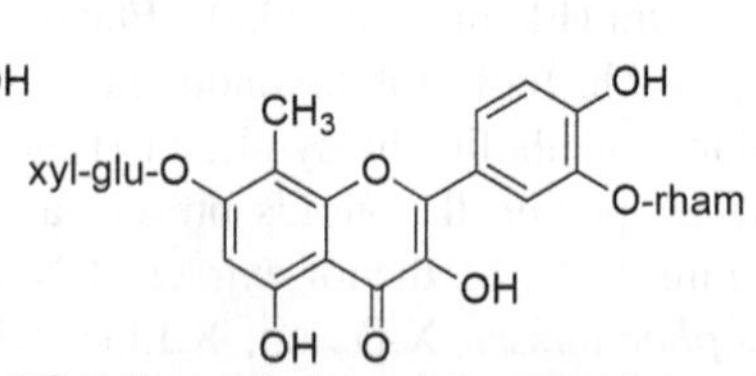

3,5,7,3',4'-pentahydroxy 8-C methyl flavone 7-
O-β-D-xylopyranosyl (1-->4)O-β-D
glucopyranosyl 3'-O-α-L-rhamnopyranoside **95**

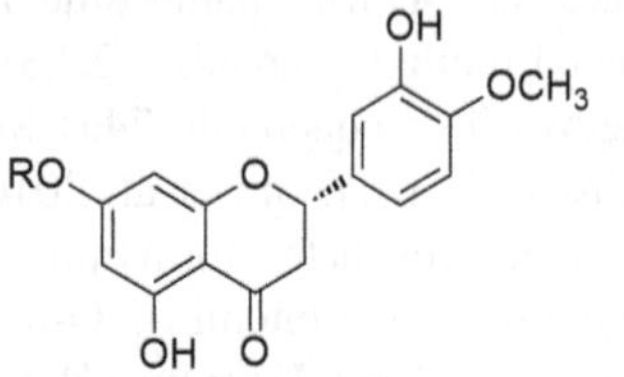

3',5,7-trihydroxy-4'-
methoxyflavanone R=H **87**
hesperidin R=rha-(1-->6)-glu **88**

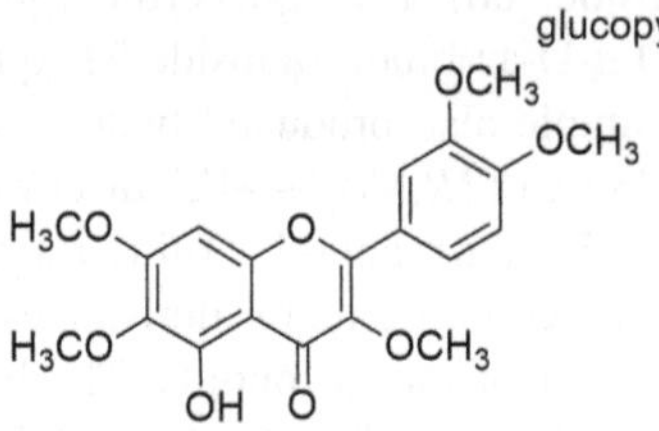

artemitin **89**

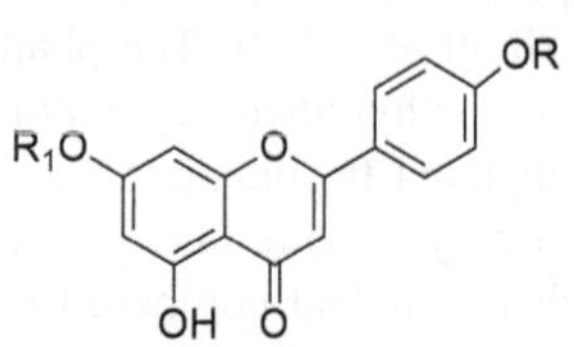

genkwanin R=H R$_1$=CH$_3$ **75**
genkwanin-4'-O-β-D-rutinoside R=Rut
R$_1$=CH$_3$ **76**
apigenin R=R$_1$=H **78**

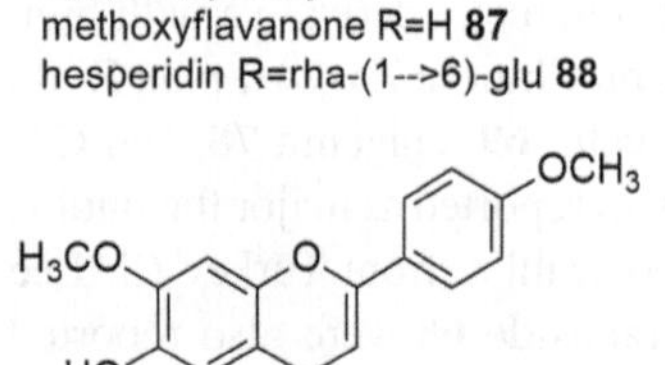

ladanein B **96**

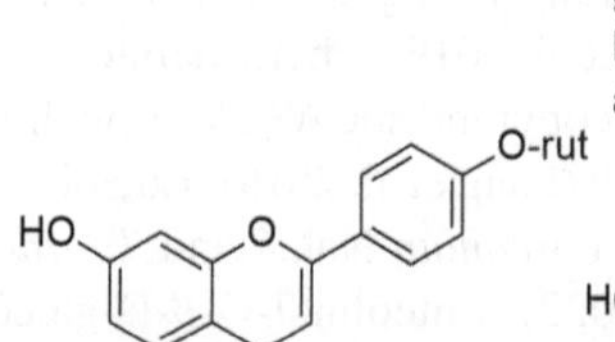

alquds **91**

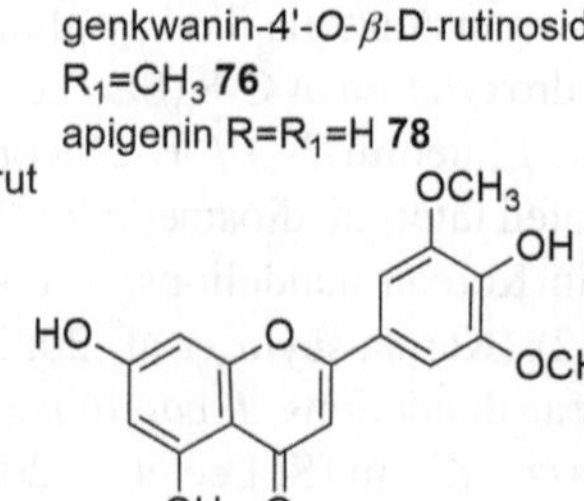

tricin **94**

FIGURE 9.11 Flavonoids isolated from *Taraxacum* medicinal plants across the globe.

β-D-glucopyranoside] **81**, quercetin-3,7-O-β-D-diglucopyranoside **82**, genkwanin **75**, genkwanin-4'-O-β-D-rutinoside **76**, isoetin **83**, isoetin-7-O-β-D-glucopyranosyl-2'-O-α-L-arabinopyranoside **84**, isoetin-7-O-β-D-glucopyranosyl-2'-O-α-D-glucopyranoside **85**, isoetin-7-O-β-D-glucopyranosyl-2'-O-β-D-xyloypyranoside **86**, 3',5,7-trihydroxy-4'–methoxyflavanone **87**. The herbs also produced hydrogenated luteolin, hesperidin **88** as well as methoxylated quercetin derivates, artemitin **89** and quercetin-3',4',7-trimethyl ether **90** (S.-Y. Shi, Zhang, Huang, et al. 2008). Further study on more polar water fraction produced main flavonoids presented as isoetin-7-O-β-D-glucopyranosyl-2'-O-α-L-arabinopyranoside **84**, isoetin-7-*O*-β-D-glucopyranosyl-2'-O-α-D-glucopyranoside **85** and Isoetin-7-O-β-D-glucopyranosyl-2'-O-α-D-xyloypyranoside **86**, quercetin **69** and diosmetin **79** (S. Shi, Zhang, Zhao, et al. 2008) (Fang et al. 2019). Preparative isolation technique development on traditional Chinese medicine using high-speed counter-current chromatography was able to produce 85 mg alquds **91** and 45 mg hesperidin **88** from single operation of 600 mg extract *T. mongolicum* (S. Shi, Huang, et al. 2008). Study on Korean *T. mongolicum* revealed several flavonoids, apigenin **78**, luteolin **65**, quercetin **69**, luteolin-7-*O*-β-D-glucopyranoside **65**, quercetin-7-*O*-β-D-glucopyranoside **92**, quercetin-3,7-O-β-D-diglucopyranoside **82** (W. Li, Lee, et al. 2017).

In the European continent, dandelion is commonly referred to as *T. officinale*. The phytochemical study of a French sample revealed several flavonoids such as luteolin-7-*O*-β-D-glucopyranoside **65**, chrysoeriol **93**, tricin **94** and luteolin **65** (Williams et al. 1996) (Kurkin and Aznagulova 2016). A study on Indian-originated *T. officinale* produced 3,5,7,3',4'-pentahydroxy 8-C methyl flavone 7-O-β-D-xylopyranosyl (1→4)O-β-D glucopyranosyl 3'- O-α-L-rhamnopyranoside **95** and ladanein B **96** (Yadava and Khan 2013). More flavonoids were reported from water extract of Chinese *T. officinale*, apigenin **78**, luteolin-7-*O*-β-D-glucopyranoside **68**, quercetin **69**, diosmetin **79** (Fang et al. 2019).

9.4.3.2 Cinnamic Acids

The cinnamic acid derivatives are natural product phenolics with C6–C3 skeleton, which include *p*-coumaric acid, caffeic acid, and ferulic acid (Padmanabhan et al. 2016). Several cinnamic acids were isolated from arrays of *Taraxacum* species (Figure 9.12). Phytochemical study on *Neo-Taraxacum siphonanthum* (accepted name *T. siphonanthum* X.D.Sun, X.J.Ge, Kirschner & Štěpánek) from China revealed caffeic acid **97**, ferulic acid **98** and ester condensate with tartaric acid (*trans*-caftaric acid **99**), quinic acid (4-*O*-caffeoylquinic acid **100**, 3-*O*-caffeoylquinic acid or chlorogenic acid **101**) (S.Y. Shi et al. 2010). The caffeic acid **97**, ferulic acid **98**, and chlorogenic acid **101** were isolated from water extract of *T. alaskanum* that originated from China (Fang et al. 2019). The caffeic acid **97** and ferulic acid **98** were also isolated from *T. bessarabicum* (Hornem.) Hand.- Mazz. subsp. bessarabicum (Hornem.) Hand.-Mazz. along with *p*-coumaric acid **102**, chlorogenic acid methyl ester **103**, 3,5-di-*O*-caffeoylquinic acid **104**, 3,5-di-O-caffeoylquinic acid methyl ester **105** (Sari and Kececi 2019). Interestingly, *cis*-caftaric acid **106** presented in Taiwanese *T. formosanum* (accepted name *T. mongolicum Hand.-Mazz.*) indicated asymmetric enzymatic synthesis differentiation to *Neo-Taraxacum siphonanthum* (accepted name *T. siphonanthum* X.D.Sun, X.J.Ge, Kirschner & Štěpánek) (H.-J. Chen, Inbaraj, and Chen 2012). The study also revealed other cinnamic acid derivates, chlorogenic acid **101**, caffeic acid **97**, 3,5-di-*O*-caffeoylquinic acid **104**, chicoric acid **107** (H.-J. Chen, Inbaraj, and Chen 2012). Caffeic acid **97**was also reported in ethanolic extract of *T. gracilens* Dahlts from Turkey (Karahuseyin, Sari, and Ozsoy 2022). Chicoric acid **107**, chlorogenic acid **101** and *trans*-caftaric acid **99** were isolated from French orginiated dandelions, *T. officinale* (Williams, Goldstone, and Greenham 1996). Additional study of *T. officinalle* from Samara revealed the isolation of *trans*-caftaric acid **99** and caffeic acid **97** (Kurkin and Aznagulova 2016). The caffeic acid **97**, ferulic acid **98**, chlorogenic acid **101**were also reported from Chinese origin *T. officinale* (Fang et al. 2019). The isoferulic acid **108** was reported from Korean dandelions, *T. coreanum* (Mo et al. 2017).

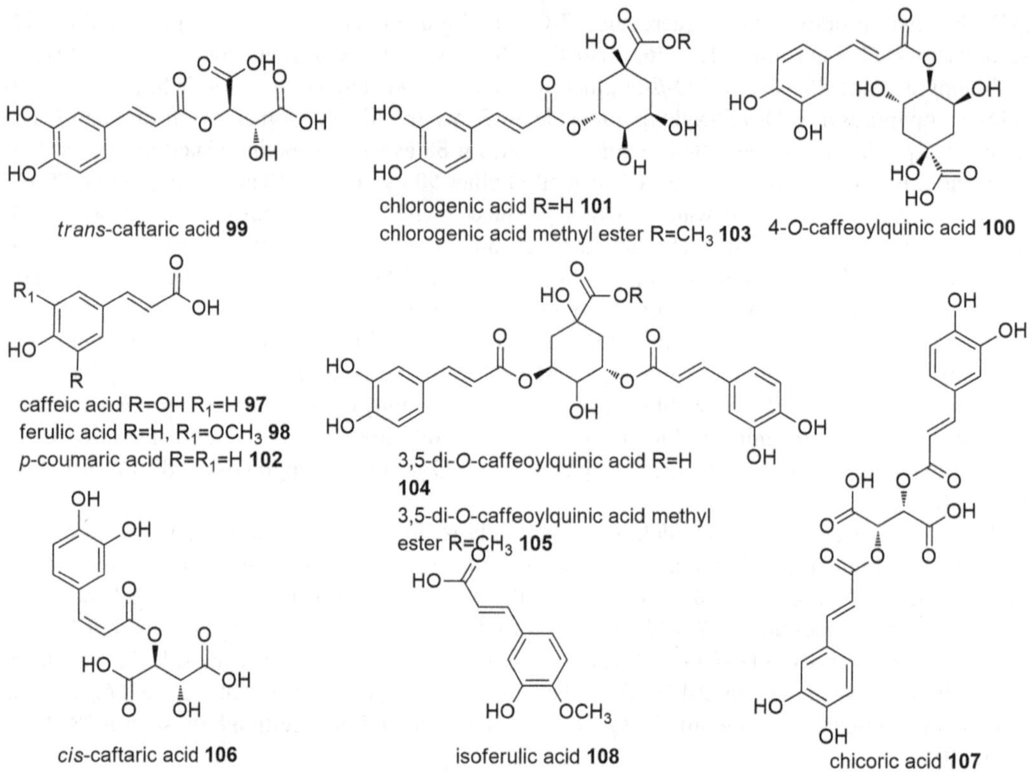

FIGURE 9.12 Cinnamic acids were isolated from arrays of *Taraxacum* species.

9.4.3.3 Coumarins

Coumarin derivatives were isolated form Korean *T. coreanum*, nodakenetin **109**, decursinol **110**, prangol **111**, isobyakangelicin **112** (Mo et al. 2017). Glycosilated coumarins were reported from Frenhc methanolic extract of, *T. officinale*, cichoriin **113**, aesculin **114** (Williams, Goldstone, and Greenham 1996). Cichoriin **113** was also reported from *T. bessarabicum* (Hornem.) Hand.- Mazz. subsp. bessarabicum (Hornem.) Hand.-Mazz. Along with esculetin **115** (Sari and Kececi 2019). Methoxylated esculetin, scopoletin **116**, was reported from *T. portentosum* (Figure 9.13) (Michalska et al. 2021).

9.4.3.4 Lignan

Common pinoresinol **117**, a furanofuran lignan was isolated from Korean *T. coreanum* along with syringaresinol-4′ -O-β-D-glucoside **118** and syringaresinol **119** (Figure 9.14) (Mo et al. 2017).

9.4.3.5 Phenyl Acetics

Phenyl acetic derivates were isolated from polar extract of *T. coreana*, 4-methoxyphenylacetic acid **120**, 4-hydroxyphenylacetic acid methyl ester **121**, taraxinositols A **122**, taraxinositols B **123**, tarax-inol **124**, neo-inositol-1,4-bis (4-hydroxybenzeneacetate) **125**, chiro-inositol-1,5-bis (4- ydroxy-benzeneacetate) **126**, chiro-inositol-2,3-bis (4-hydroxybenzeneacetate) **127**, chiro-inositol1,2,3-tris (4-hydroxybenzeneacetate) **125** (Figure 9.15) (Mo et al. 2017).

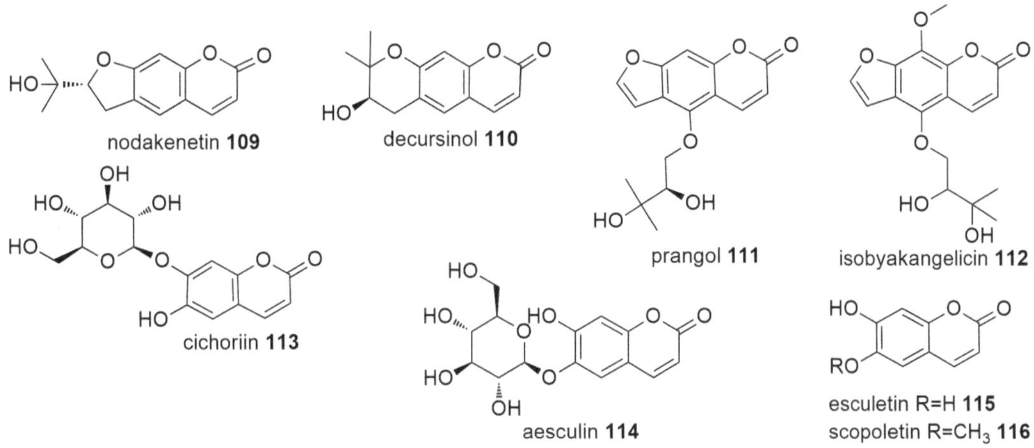

FIGURE 9.13 Coumarine derivates of *Taraxacum* species.

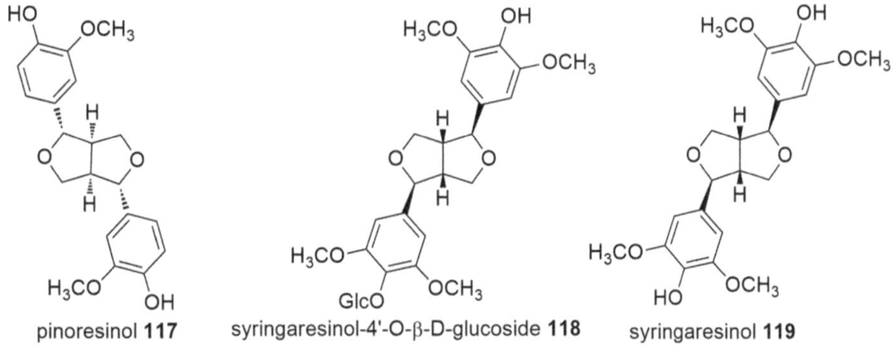

FIGURE 9.14 Lignan isolated from *T. coreanum*.

9.4.3.6 Other Phenolic Compounds

Phenolic acids and aldehydes are commonly present in plants with potential antioxidant activities including *Taraxacum* (Figure 9.16). These simple phenolics were isolated from *T. coreanum*, *p*-hydroxybenzaldehyde **129**, vanillin **130**, syringaldehyde **131**, vanillic acid **132** (Mo et al. 2017). The vanillic acid **132** was also reported from *T. gracilens* Dhalst of Turkey origins (Karahuseyin, Sari, and Ozsoy 2022). De-methoxylated vanillic acid, protocatechuic acid **133** was isolated from *T. officinale* water extract obtained from China (Fang et al. 2019). Other unique phenolics such as syringine derivatives were reported from *Taraxacum* genera, syringin **134**, dihydrosyringin **135** which are reported from *T. portentosum* and *T. udum* of Poland origins (Michalska, Marciniuk, and Stojakowska 2021; Michalska et al. 2010). This simple aromatic phenols, dihydrosyringin **135**, was also isolated from polar fraction of root extract of Japanese *T. platycarpum* (Warashina, Umehara, and Miyase 2012). Other unique phenolics were reported from *Taraxacum portentosum* Kirschner & Štěpánek and *Taraxacum udum* Jord. were 3-hydroxy-1-(4-hydroxy-3-methoxyphenyl)-1-propano ne **136**, methyl *p*-hydroxyphenylacetate **137**, 5-methoxy-eugenyl-4-*O*-β-glucopyranoside **138**, dihy-drodehydrodiconiferyl alcohol 9-*O*-β-glucopyranoside **139**, syringaresinol-4′-*O*-β-glucopyranoside **140**, dihydroconiferin **141** (Michalska, Marciniuk, and Stojakowska 2021) (Michalska, Marciniuk, and Kisiel 2010).

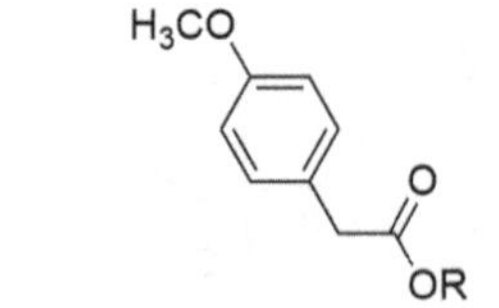

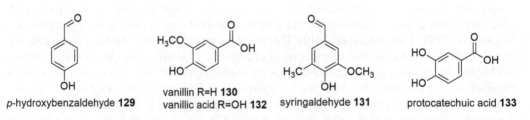

4-methoxyphenylacetic acid R=H **120**
4-hydroxyphenylacetic acid methyl
ester R=CH₃ **121**

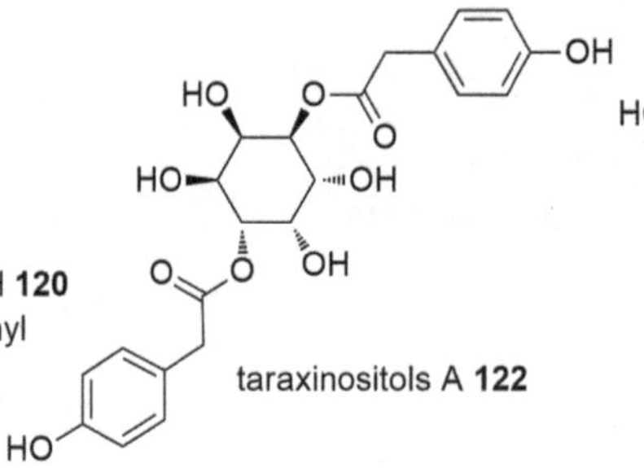

taraxinositols A **122**

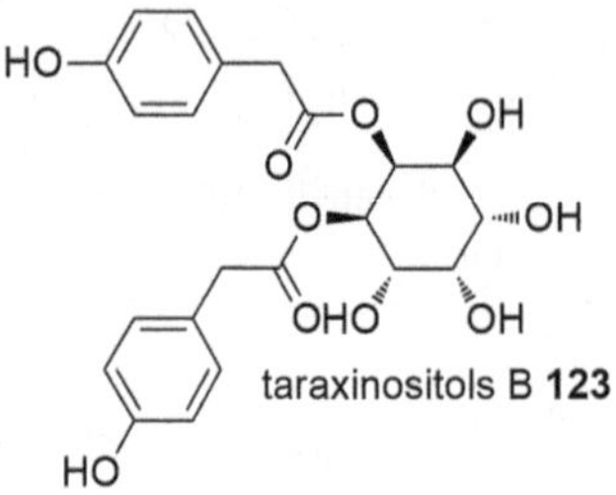

taraxinositols B **123**

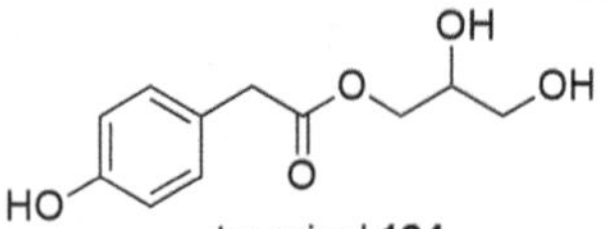

taraxinol **124**

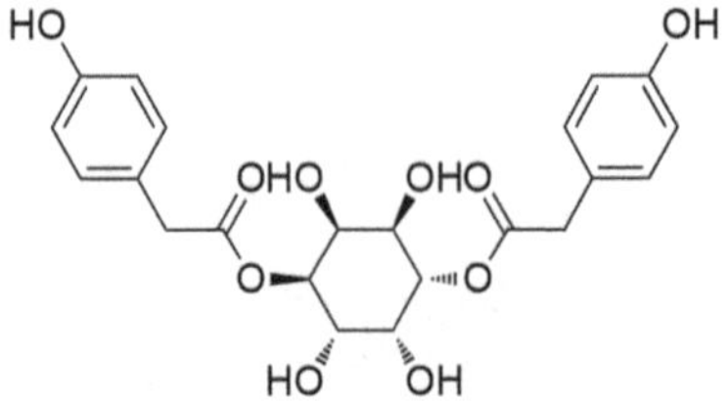

neo-inositol-1,4-bis (4-
hydroxybenzeneacetate) **125**

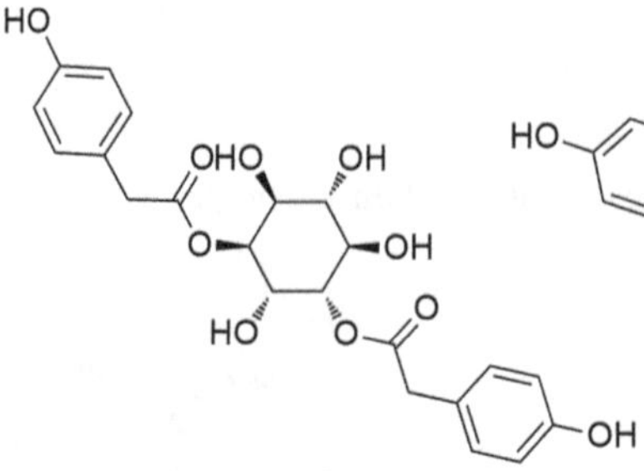

chiro-inositol-1,5-bis(4-
hydroxybenzeneacetate) **126**

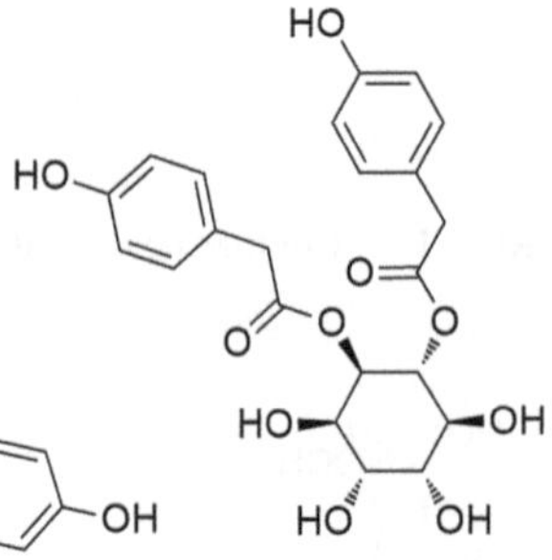

chiro-inositol-2,3-bis (4-
hydroxybenzeneacetate) **127**

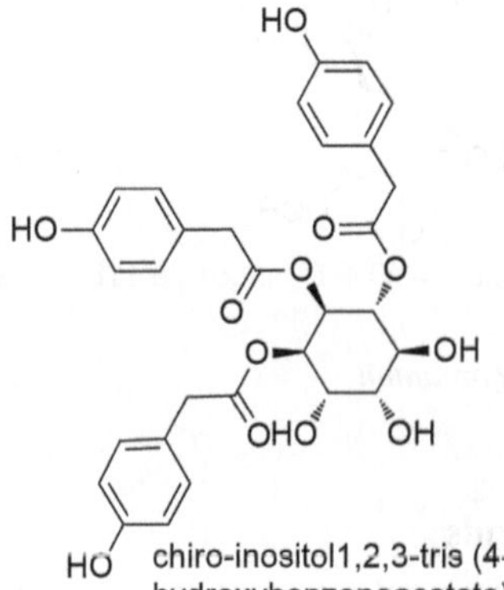

chiro-inositol1,2,3-tris (4-
hydroxybenzeneacetate) **128**

FIGURE 9.15   Phenyl acetic derivates produced by *Taraxacum* species.

p-hydroxybenzaldehyde **129**

vanillin R=H **130**
vanillic acid R=OH **132**

syringaldehyde **131**

protocatechuic acid **133**

FIGURE 9.16   Unique phenols produced by *Taraxacum* species.

FIGURE 9.16 (Continued)

9.4.4 Medicinal Uses and Biological Activities against Non-infectious Diseases

The dandelion species *T. officinale* is a common herbaceous plant that has been traditionally used for medicinal purposes for centuries. Recent studies have highlighted the potential anticancer, antidiabetes, and antioxidant properties of dandelion, making it a promising source of natural compounds for the prevention and treatment of various diseases. Studies revealed distribution of a medicinally significant phytochemical of *T. officinalle* from terpenoids to phenolics compounds (Wirngo et al. 2016). These constituents contributed to their pharmacological activities against noncommunicable diseases.

The anticancer properties of dandelion extracts and compounds have been studied in different types of cancer, including hepatocellular carcinoma, esophageal squamous cell carcinoma, triple-negative breast cancer, colon cancer, ovarian cancer, and leukemia (Hou et al. 2019). Dandelion extracts were found to inhibit angiogenesis, induce cell apoptosis, trigger cell stress-related signals, and selectively induce programmed cell death in cancer cells (Ovadje et al. 2016; Ovadje et al. 2011; Ovadje, Chochkeh, et al. 2012; Ovadje, Hamm, et al. 2012). Additionally, dandelion compounds were found to inhibit the migration, invasion, and epithelial-mesenchymal transition of cancer cells (Y. Li et al. 2022). These findings suggest that dandelion extracts and compounds may be used as alternative or complementary therapies for cancer treatment.

Dandelion has also shown potential in treating and preventing metabolic disorders, such as diabetes, non-alcoholic fatty liver disease (NAFLD), and obesity (Kania-Dobrowolska and Baraniak 2022). Flavonoids from dandelion were found to inhibit pancreatic alpha-amylase and form complexes that quench the protein's intrinsic fluorescence (Huang et al. 2021). Dandelion extracts were also found to improve antioxidant enzyme activity, reduce lipid peroxidation, and enhance glucose uptake, contributing to the treatment of type 2 diabetes and atherosclerosis. Additionally, dandelion extracts were found to inhibit adipogenesis and demonstrate alpha-amylase, alpha-glucosidase inhibition, and advanced glycation end-product (AGE) formation, suggesting their potential as anti-diabetic and anti-obesity agents. These findings indicate that dandelion has potential as a treatment option for metabolic disorders (Kania-Dobrowolska and Baraniak 2022).

Studies on dandelion have also shown that it contains numerous bioactive compounds that exhibit antioxidant properties, making it a promising source of natural compounds for the prevention and treatment of diseases associated with oxidative stress. Dandelion extracts were found to inhibit lipid peroxidation and oxidation of proteins in plasma, suggesting their potential as a source of natural compounds with antioxidant properties that could benefit diseases associated with oxidative stress and changes in hemostasis (Olas 2022).

9.4.4.1 Dandelion as Anticancer

Feng Ren et al. found that dandelion polysaccharide (DP) inhibited angiogenesis in hepatocellular carcinoma (HCC) by suppressing key factors involved in angiogenesis (Ren et al. 2020). Yuxi Li et al. examined the effects of aqueous dandelion seed extract (DSE) on esophageal squamous cell carcinoma (ESCC) and found that DSE selectively inhibited cell growth, proliferation, migration, invasion, angiogenesis, and induced cell apoptosis in ESCC cells (Y. Li et al. 2022). Kumar et al. synthesized self-assembled ZnO "dandelion" capsules, which have potential for use in sustained drug delivery and gene delivery due to their simplicity, feasibility, and cost effectiveness (V.B. Kumar et al. 2014). Xiao-Hong Li et al. found that dandelion extract decreased TNBC cell viability, triggered cell apoptosis and G2/M phase arrest, and activated several cell stress-related signals, including the PERK/p-eIF2α/ATF4/CHOP axis (X.-H. Li, He, et al. 2017). Pei Chen et al. extracted a novel polysaccharide, DLP120, from dandelion leaves, which has potential as an anticancer agent (P. Chen et al. 2022). Pamela Ovadje et al. found that an aqueous dandelion root extract (DRE) selectively induced programmed cell death in more than 95% of colon cancer cells within 48 hours of treatment and was effective in reducing the occurrence of cancer cells' drug resistance. Additionally, they found that DRE effectively induced apoptosis in human leukemia cell lines in a dose- and time-dependent manner but did not affect noncancerous cells (Ovadje et al. 2011, 2016). Lastly, J. Zhu et al. found that taraxasterol (TAR) **52**, a compound extracted from dandelion, inhibited the migration, invasion, and EMT of ovarian cancer cells (J. Zhu, Li, et al. 2021; S. Zhu and Viejo-Borbolla 2021). In addition, taraxasterol **52** and taraxerol **54** generated from *T. japonium* showed valuable chemo-preventive agents based on animal model experiments (Takasaki et al. 1999).

Overall, these studies suggest that dandelion extracts and compounds may have promising anti-cancer properties and could be used as alternative or complementary therapies for cancer treatment.

9.4.4.2 Dandelion as Antidiabetes

Dandelion has been found to have potential in treating and preventing metabolic disorders such as diabetes, NAFLD, and obesity. Studies have revealed that flavonoids from dandelion inhibit the pancreatic α-amylase in a noncompetitive manner, and the hydrophobic interactions of these flavonoids and the alpha-amylase lead to the formation of complexes that quench the protein's intrinsic fluorescence (Kania-Dobrowolska and Baraniak 2022). A combination of metformin and lemon balm and dandelion was found to be a promising solution for preventing and treating metabolic diseases involving insulin resistance. Prebiotics found in foods such as almonds, artichoke, and dandelion greens are being explored for their benefits in mineral absorption, metabolite production, gut microbiota modulation, and the prevention of diseases such as diabetes, allergies, metabolic disorders, and necrotizing enterocolitis (J.Y. Choi et al. 2022). Dandelion water extract (DWE) supplementation decreased serum glucose levels, improved antioxidant enzyme activity, reduced lipid peroxidation, and improved lipid metabolism in diabetic rats compared to nondiabetic rats, suggesting that DWE supplementation may be beneficial in preventing diabetic complications (Cho et al. 2002). Dandelion chloroform extract could be a potential hypoglycemic agent for the treatment of type 2 diabetes mellitus (T2DM), as it was found to activate the AMPK signaling pathway, increase GLUT4 expression, promote its translocation to the cell membrane, and enhance glucose uptake in L6 cells (Zhao et al. 2018). Ethanol extract of sweet gale, roseroot, sheep sorrel, stinging nettles, and dandelion showed potential antidiabetic and anti-obesity properties, inhibiting adipogenesis and demonstrating α-amylase, α-glucosidase inhibition, and advanced glycation end product formation (Sekhon-Loodu and Vasantha Rupasinghe 2019). Dandelion was also found to regulate lipid and carbohydrate metabolism, thus contributing to the treatment of type 2 diabetes and atherosclerosis (Kania-Dobrowolska and Baraniak 2022). Dandelion sterol treatment has also been found to significantly improve diabetes-induced renal injury, through regulation of the miR-181a/TNF-α/NF-κB pathway (Tian et al. 2021). These findings suggest that dandelion has potential as a treatment option for metabolic disorders, and that further research into the benefits and mechanisms of action of dandelion could lead to its widespread use in functional foods and pharmaceuticals.

9.4.4.3   Dandelion as Antioxidant

Studies on dandelion (*T. officinale*) have shown that it contains numerous bioactive compounds that exhibit antioxidant properties, making it a promising source of natural compounds that could benefit various diseases associated with oxidative stress. Molinu et al. conducted a study on the potential use of the leaves of *T. kok-saghyz* (TKS), a dandelion species, as a source of phenolic compounds with antioxidant properties (Molinu et al. 2019). The leaves of TKS018 showed the highest antioxidant capacity, suggesting the potential for exploiting TKS leaves as a source of antioxidant compounds for use in forage, nutraceutical, and pharmacological fields.

Dariusz Je et al. focused on the antioxidant properties of four different phenolic fractions from dandelion leaves and petals (Jedrejek et al. 2019). The study investigated the effect of these fractions on the production of oxidative stress markers in human plasma. The results showed that the tested dandelion fractions could inhibit lipid peroxidation and oxidation of proteins in plasma, with the phenolic fractions from petals having better antioxidant properties compared to those from leaves. The findings suggest that these dandelion fractions could be a potential source of natural compounds with antioxidant properties that could benefit diseases associated with oxidative stress and changes in homeostasis.

Majewski et al. evaluated the effects of alcoholic leaf and petal fractions of *T. officinale* on Wistar rats over a four-week period (Majewski et al. 2020). The results showed that the dandelion fractions exerted antioxidant activities, as demonstrated by decreased levels of TBARS and increased thiol levels in the blood plasma. Additionally, the leaf fraction showed a beneficial effect on the lipid profile and modified the participation of cyclooxygenase products in noradrenaline-induced vascular contractions of thoracic arteries. The findings suggest that dandelion leaf and petal phenolic fractions, enriched with chicoric acid **98**, may have promising in vivo antioxidant effects.

Tajner-Czopek et al. investigated the antioxidant activity of six medicinal plants, including dandelion, by analyzing the content of caffeic acid derivatives in water and water-ethanolic extracts (Tajner-Czopek et al. 2020). The study found a positive correlation between caffeic acid derivatives and antioxidant activity as measured by radical cation scavenging activity (ABTS) and radical scavenging activity (DPPH).

Cai et al. isolated two novel polysaccharides (DRP-2b, DRP-3a) from dandelion root and found that DRP-3a showed higher radical scavenging activity against DPPH, hydroxyl, and superoxide anions, as well as strong protective effects against H_2O_2-induced damage in liver cells (Cai et al. 2019). DRP-3a was found to be a potential antioxidant and functional food ingredient.

Colle et al. evaluated the potential of *Taraxacum officinale* leaf extract as a treatment for acetaminophen-induced hepatotoxicity. The extract was shown to decrease oxidative stress and prevent liver damage, as well as having antioxidant activity and scavenging abilities against reactive oxygen and nitrogen species.

Gomez et al. identified and quantified phytochemicals in *T. officinale* var. Garnet Stem harvested from Texas and New Jersey, USA (Gomez et al. 2018). The study revealed the presence of four anthocyanins (cyanidin-3-glucoside, cyanidin-3-(6-malonyl)-glucoside, cyanidin-3-(6-malonyl)-glucoside, and peonidin-3-(malonyl)-glucoside) for the first time based on liquid chromatography–mass spectrometry (LCMS) data analysis. The combination of methanol with water and formic acid showed the highest level of DPPH free radical scavenging activity, and sodium chenodeoxycholate was found to be the most bound bile acid in both Texas and New Jersey samples.

Choi et al. investigated the potential hypolipidemic and antioxidant effects of dandelion root and leaf in rabbits fed with a high-cholesterol diet. Results showed that dandelion root and leaf improved (U.-K. Choi et al. 2010).

9.4.4.4   Dandelion as Anti-inflammatory

Dandelion (*T. officinale*) has been studied for its potential therapeutic effects in various diseases. Majewski et al. studied the effect of dandelion flower water syrup (TOFS) on obese male albino Wistar rats and found that TOFS had beneficial effects in regulating blood lipids, improving antioxidant status, and decreasing markers for liver damage (Majewski et al. 2021). Choi et al. investigated the effects of a mixture of lemon balm and dandelion extracts on ethanol-induced liver injury and found that the mixture attenuated hepatic accumulation of triacyl glycerides, reduced oxidative stress and inflammation, and enhanced antioxidant activity (B.-R. Choi et al. 2020). Badr et al. investigated the protective effects of dandelion leaf extract (DLE) against cisplatin-induced nephrotoxicity in rats and found that DLE pretreatment helped alleviate kidney damage by reducing oxidative stress, inflammation, and apoptosis (Badr et al. 2019). Zhou et al. studied the mechanism by which dandelion root extract (DRE) protects mice against ulcerative colitis induced by dextran sodium sulphate (DSS) and found that DRE-H (high dose) attenuated colonic mucosal damage and inhibited DSS-induced inflammatory responses and oxidative stress in the bloodstream and colon tissues (Zhou et al. 2022). Ge Hu et al. investigated the potential therapeutic effects of dandelion leaf aqueous extracts (DAE) on mastitis in a mouse model and found that DAE can inhibit the expression of inflammatory mediators TNF-α and ICAM-1 in a time-dependent manner, suggesting that dandelion may have therapeutic potential for mastitis treatment by targeting the endothelium (Hu et al. 2017). Finally, Liu et al. characterized 22 phenolic compounds in dandelion and identified cichoric acid **107**, caffeic acid **97**, and luteolin **65** as quality markers based on their correlations with dandelion's bioactivities (Liu et al. 2018). These markers were quantified in two frequently used forms of dandelion, providing a strategy for quality control of dandelion by identifying bioactive markers using chemometrics and in silico pharmacology. Overall, these studies provide evidence for the potential therapeutic effects of dandelion in various diseases and suggest that it may be a promising natural remedy.

Rolnik and Olas' research highlights the beneficial effects of Asteraceae's chemical composition, including inulin, a natural polysaccharide, and phytochemical compounds such as polyphenols, phenolic acids, flavonoids, acetylenes, and triterpenes (Rolnik and Olas 2021). Dandelion extract has been found to be useful in identifying Compositae allergy and is a safe supplement to sesquiterpene lactone mix testing. Lee et al.'s study found *T. coreanum* chloroform fraction to be a potent anti-inflammatory agent (M.-H. Lee et al. 2013). Bernadetta Lis et al. found that dandelion preparations have promising antioxidant and anticoagulant activities, particularly those based on aerial parts rich in hydroxycinnamic acid derivatives (Majewski et al. 2021). Kikuchi et al. discovered that ten triterpenoids from *Taraxacum officinale* can inhibit the production of nitric oxide, suggesting potential as anti-inflammatory agents (Kikuchi et al. 2016). Lastly, Heo et al. found *T. officinale's* inhibitory effects on osteoclast genesis and OVX-induced bone loss, indicating its potential as an anti-osteoporotic agent (Heo et al. 2022). Park et al. found the synergistic anti-inflammatory effects of luteolin and chicoric acid **107**, two constituents of *Taraxacum officinale*, suggesting their central role in ameliorating LPS–induced inflammatory cascades (Park et al. 2011). These studies demonstrate the promising medicinal and nutritional properties of the Asteraceae family, particularly dandelion and its extracts, and their potential use in the prevention and treatment of various diseases.

9.4.5 Medicinal Uses and Biological Activities against Infectious Diseases

Infectious diseases can be caused by harmful agents (pathogens) such as bacteria, fungi, viruses, or parasites. Pathogens can get into the human body through mouth, nose, eyes, urogenital organs, or wounds. Infectious diseases can be spread from one person to another by airborne transmission, contaminated water/food, bites of insects or animals, and so on. A variety of symptoms depend on the microorganism that causes the infection; it can result in mild to severe conditions such as sore throat, fever, chills, or chest pain, resulting in a decrease the quality of life or even death. Nowadays, treating these infections have become more difficult due to the emergence of drug resistance of pathogens toward the existing drugs. Natural products used either as whole plants or crude extracts show a wide range of biological activity against pathogens and are widely used in different ethnomedicinal systems worldwide. Medicinal plants, as one of the most important sources of biologically active compounds, have demonstrated therapeutic potential throughout human history. Plant-based natural products are still being used nowadays with expanding popularity for mitigation and preventing various ailments (Atanasov et al. 2021; van Seventer and Hochberg 2017). The following paragraphs provide a compilation of scientific research on the bioactivity and medicinal uses of natural products that act against infectious diseases.

9.4.5.1 Antibacterial Activity

In recent years, the antimicrobial activity of various plants including from the Asteraceae family were evaluated; despite major studies were limited to crude extract antibacterial efficacies without further purification of the bioactive antibiotics (Bessada et al. 2015). Nevertheless, several studies on *Taraxacum* species revealed significant antibacterial activities from small to macromolecules exudated by the plants. Crude ethanol extracts of aerial part of Chinese origins *T. mongolicum* Hand-Mazz exerted antimicrobial activities against *Staphylococcus aureus*, *Eschericia coli*, and *Pseudomonas aeruginosa* with MIC value of 50, 50, and 100 µg/mL respectively, in which the study suggested its phenylpropanoid and sesquiterpene constituents were responsible for the activity (Gao 2010). Further study on transcriptomic response of *Escherichia coli* to *T. mongolicum* revealed that 3962 differentially global gene expression between control and treated group suggested the antibacterial activity evolved stress response, DNA damage and repair, protein translation and secretion (Jianming Zhu et al. 2017). These studies have supported clinical uses of *T. mongolicum* in Chinese medication to treat abscesses, eye inflammation, and viral infectious diseases.

In the other hand, bioprospecting evaluation on leaves of Chilean-originated dandelion (*T. officinale*) showed less polar component (*n*-hexane fraction) to possess better antibacterial activities than the moderate polar fraction, ethyl acetate, in which the hexane fraction showed significant activity against *S. aureus*, *E. coli*, *Kelbsiella pneumoniae*, and *Proetus mirabilis* with MIC value of 200, 400, 400, and 800 µg/mL, respectively (positive control chloramphenicol with MIC range from 25–200 µg/mL) (Díaz et al. 2018). Further phytochemical experiments on the fraction successfully isolated and identified several terpenoids, lupeol **49**, lupeol acetate **41**, *α*-Amyrin **51**, and stigmasta-5-en-3-ol (*β*-Sitosterol) **59** (Díaz et al. 2018). In a similar study conducted on a sample from Bulgaria, hexane fraction was inactive against *Bacillus subtilis ATCC 6633, Proteus vulgaris, Salmonella* sp*, Candida albicans, Aspergillus niger*, and *Fusarium moliniforme*. In addition, the fraction possessed moderate activity against *S. aureus, Listeria monocytogenes, Enterococcus faecalis*, and *E. coli* ATCC 8739 with inhibition zone value range from 8–10 mm at 2.5 mg/mL. The GCMS protocol detected some similar components compared to Chilean origins (Ivanov et al. 2018). Antimicrobial study on the root of Irish *T. officinale* revealed the fraction produced from chromatographic separation of hydrophobic crude extract (ethyl acetate) indicated strong antimicrobial activity against *S. aureus* NCTC 8178, methicillin resistant *S. aureus* (clinical), and *B. cereus* NCTC 7464 with MIC value of 93.75, 62.5, and 93.75 µg/mL respectively, in which further work detected hydroxylated fatty acids: 9-hydroxyoctadecatrienoic acid (9-HOTE) and 9-hydroxyoctadecadienoic acid as the bioactive components based on LCMS data analysis (Kenny et al. 2015). A study conducted by Yang et al. suggested that root extract and taraxasterol **52** was able to reverse bacterial resistance to antibiotics using heavy metal–induced resistance study in *E. coli* (Yang & Zhang 2020). The study showed that low concentration of nickel (0.5 μg mL^{-1}), cadmium (0.05 μg mL^{-1}), and arsenic (0.05 μg mL^{-1}) possessed no effect on bacterial growth, but induced the bacteria to become resistant to kanamycin and ampicillin in which presence of water extract of *T. officinale* suppressed the β-lactamase and acetyltransferase protein expression (Yang and Zhang 2020). Another study revealed virulence genes expression of *S. aureus, Salmonella sp, E. coli*, and cell membrane permeability were reduced on petroleum extract of *T. officinale* treated group. Increased concentration of the extract was able to cause intracellular ATP concentration to decrease which led to the *T. officinale* constituents ability to inhibit food-borne bacteria growth both Gram negative and Gram positive (Wu 2022).

Apart from small molecule components, study on flowers of *T. officinale* discovered unusual peptides, namely ToAMP1, ToAMP2, and ToAMP3 (cysteine rich peptide with molecular mass size of 4000 to 6000 Da), which displayed strong antibacterial activity against *Pseudomonas syringae, B. subtillis*, and *Xanthomonas campestris* with inhibition zone range 0.8–1.3 cm (kanamycin inhibition zone range 1.4–2.3) at concentration of 6 μg mL^{-1}. In addition, the fungi *Botrytis cinerea, A. niger, Bipolaris sorokiniana*, and *Pythium debaryanum* were sensitive to ToAMP2 with IC$_{50}$ value of 5.2, 2.6, 5.2, 2.6, μM, respectively. (A. A. Astafieva et al. 2012). Further peptide exploration led to the discovery of other peptides, ToHyp1 and ToHyp2, which are composed of cysteine free peptide and proline hydroxyproline-rich and pentoses. These peptides indicated a significant antibacterial activity against bacterial pathogens (*P. syringae, X. campestris, Clavibacter michiganense, Erwinia coratovora*, and *B. subtilis*) (Alexandra A. Astafieva et al. 2015).

9.4.5.2   Anti-fungal Activity

The essential oils, containing (-)-α-thujone, fenchone, (+)-β-thujone, and (+)-hibaene from the plants *T. mongolicum* exhibited valuable antifungal activities against pathogenic fungi, including *Fusarium graminearum, Curvularia lunata*, and *Bipolaris maydis*, as well as *C. albicans which is pathogenic to humans*. Moreover, quantitative analysis of GCMS data revealed high concentrations of (-)-α-thujone in the leaves of *T. occidentalis* (Bai et al. 2020).

The hexane and ethyl acetate extracts of *T. officinale* exhibited antimicrobial activities against *S. aureus, E. coli*, and *Pseudomonas aeruginosa* and two fungal strains, namely *A. niger* and *Aspergilus flavus* (Gani & Alam 2022).

Albumin 2S proteins, ToA1, ToA2, and ToA3 from *Taraxacum* seeds are active against *Helminthosperium sativum, Pleospora betae*, and *Verticillium albo-atrum*. However, *Furarium oxysporum* and *Verticillium albo-atrum* were only inhibited by ToA2 and ToA3, respectively (Odintsova et al. 2010).

Three novel peptides, ToAMP1, ToAMP2, and ToAMP3 from flowers of *T. Officinale* exhibited antifungal activity against *Botrytis cinerea, Bipolaris sorokiniana, A. niger, Pythium debaryanum, F. Oxysporum* and *Phytophthora Infestans* with IC$_{50}$ values ranging of 1.2 to 5.8 μM. The ToAMPs were also active against *P. syringae, B. subtilis*, and *Xanthomas campestris*. In addition, ToAMP2 was active against *Clavibacter michiganensis* and *Phytophthora infestans* (A. A. Astafieva et al. 2012).

In 2013, Astrafieva and coworkers published a study on flowers of *T. officinale* Wigg., which led to the isolation of ToAMP4, a novel peptide which showed antifungal activity against the fungi *Aternaria alternata, A. niger, Furarium avenaceum*, and *Pseudomonas betae* with IC$_{50}$ values of 2.9, 4.2, 13.1, and 10.7 mM, respectively (MIC values in the range of 1.0 to 8.0 mM).

Additionally, in 2015, the Astrafieva team research also reported a novel peptide, ToHyp2 from *T. officinale* Wigg., consisting of 35 amino acids. The peptide showed antifungal activity and inhibits growth of Gram-positive and Gram-negative bacteria (Astafieva et al. 2015).

9.4.5.3   Antiviral Activity

Viruses are very tiny particles that can cause a lot of infectious diseases. For example, Coronavirus (COVID-19) is caused by coronavirus 2 (SARS-CoV-2), which became a public health epidemic, which led to the discovery of innovative antiviral vaccines. By early 2020, more than 107 million have confirmed infections, resulting in over 2 million deaths worldwide due to COVID-19.

In 2021, De Pellegrin and team reported five extracts of herbs showing the potential to restrict SARS-CoV-2 replication in vitro: Bronchipret thyme-ivy, Bronchipret thyme-primrose, Imupret, Sinupret extract, and Tonsipret (De Pellegrin et al. 2021). Additionally, in the same year, Jan and co-workers reported 15 chemical compounds from a library of 2,855 molecules showed the anti-SARS-CoV-2 activity in this Vero E6 cell-based study. Additionally, several extracts of Chinese plants showed anti-SARS-CoV-2 effects in Vero E6 cell-based assays, and the plants from Asteraceae, Theaceae, Mentheae, and Lamiaceae families are of interest to scientific communities. These medicinal plants could also be promising sources for a new drug discovery as inhibitors of the virus (Jan et al. 2021).

One of the most severe diseases in the history of mankind is acquired immunodeficiency syndrome, caused by the human immunodeficiency virus (HIV). Dandelion extract showed strong activity against HIV-1 RT and inhibited both the replication of HIV-1 vector and the hybrid-MoMuLV/MoMuSV retrovirus. These results showed additional support for the potential therapeutic efficacy of the extracts of *Taraxacum officinale*, which may be a good start for the development of an antiretroviral therapy with fewer side effects (Han et al. 2011).

In addition, the investigation on *T. mongolicum* extract confirms the strong antiviral activity against hepatitis B virus in cell culture. The protective effect on hepatocytes and antiviral effects were stronger than those of the reference drugs, silybin and lamivudine, providing scientific evidence for the use of *T. mongolicum* in treating hepatitis (Jia et al. 2014).

Methanol extracts of *T. officinale* (containing luteolin and caffeoylquinic acids derivatives and quercertin diclycosides) showed inhibitory activities on the replication dengue virus serotype (DENV2) replication. The higher activity fraction was found for *T. officinale* with IC_{50} ranging of 165.7 ± 3.85 and 126.1 ± 2.80 µg/mL, respectively; each fraction showed SI values were 5.59 and 6.01 (Flores-Ocelotl et al. 2018). In addition, the methanol extracts of *T. officinale* has been tested for antiviral activity against *Herpes simplex* virus 1. *E. lancifolia* exhibited the most antiviral effect with EC_{50} values of 6.31 µg/mL (SI = 51.82) (Franco-Espínola et al. 2022).

9.4.5.4 Antiparasitic Activity

Parasites, which includes protozoa, worm, and ectoparasites, live in or on other living organisms through consuming nutrients from their host. These parasites have caused serious health burden across the world, for example, plasmodium protozoa caused 229 million malarial cases globally, in 2019, leading to approximately 409,000 deaths (Nugraha et al. 2022). Ineffective treatment due to drug resistance led to drug efficacy loss, from which emerged the need for new and effective drugs against parasite disorders.

The biological activity potential of 53 legally approved and marketed herbal medicinal products (HMPs) in Germany were investigated for their bioactivities against antiprotozoal agents Neglected tropical diseases, including sleeping disorders, caused by the species *Trypanosoma brucei*, leishmaniasis (caused by the parasite *Leishmania donovani*), Chagas illness (coming from spices of the genus *Trypanosoma*), and malaria (caused by the *Plasmodium* species), which are infectious diseases caused by parasites. The study showed dichloromethane extract of Naturreiner Heilpflanzensaft Löwenzahn (containing *T. officinale* F.H.Wigg) indicated significant activity against *T. brucei*, *L. donovani*, and *P. falciparum* with percentage inhibition value of 35.3, 0.0, 32.2, 53.0 at concentration of 10 µg/mL, respectively. Despite the extract being not active against *Trypanosoma* parasite, the overall results confirm the remarkable potential of the HMPs in the search for new antiprotozoal agents (Montesino et al. 2015). Computational-based study using molecular docking (MD) indicated potential phytochemicals from *T. officinale* that provided excellent docking results and binding affinity. Luteolin **65** and taraxasterol **52** have superior binding affinities. These phytochemicals may promote as potential medicines against cerebral malaria, a severe display of parasite infection (Shaikh et al. 2022). Additionally, dichloromethane and *n*-butanol extracts from *T. officinale*, containing flavonoids (polyphenols), provided the reduction of methanogenesis (Bhatta et al. 2013). Moreover, plant extracts and *T. officinale*–provided flavonoids (mostly quercetin derivates) as well as volatile compounds, eucalyptol, and chavicol-methyl-ether help honeybee populations from protozoan disease nosemosis caused by *Nosema apis* Zander and *Nosema caranae* Fries. (Cristina et al. 2020).

9.5 CONCLUSION

Dandelions have contributed to human society as food delicacies and medicine despite the plants being categorized as wild plants and the lawn culture label the plants as weed. The health benefit of dandelion is numerous and varied across cultures. Dandelion crude extracts and its chemical constituents have shown promise as anticancer, antidiabetes, antioxidant agents, antibacterial, antifungal, antivirus and antiparasite, making them a promising source of natural compounds for the prevention and treatment of various diseases. Nevertheless, further research is needed to fully understand the mechanisms of action of dandelion and its potential therapeutic uses. The study is necessary to provide evidence-based medicinal uses for treating modern ailments.

ACKNOWLEDGMENT

The botanical illustration of the dandelions was provided by Saniyatun Mar'atus Solihah, a botanist from the Bogor Botanic Garden, Mitra Natura Raya.

REFERENCES

Astafieva, A. A., E. A. Rogozhin, T. I. Odintsova, N. V. Khadeeva, E. V. Grishin, and Ts A. Egorov. 2012. "Discovery of novel antimicrobial peptides with unusual cysteine motifs in dandelion *Taraxacum officinale* Wigg. flowers." *Peptides* 36 (2):266–271.

Astafieva, Alexandra A., Atim A. Enyenihi, Eugene A. Rogozhin, Sergey A. Kozlov, Eugene V. Grishin, Tatyana I. Odintsova, Roman A. Zubarev, and Tsezi A. Egorov. 2015. "Novel proline-hydroxyproline glycopeptides from the dandelion (*Taraxacum officinale* Wigg.) flowers: De novo sequencing and biological activity." *Plant Science* 238:323–329.

Atanasov, Atanas G., Sergey B. Zotchev, Verena M. Dirsch, and Claudiu T. Supuran. 2021. "Natural products in drug discovery: Advances and opportunities." *Nature Reviews Drug Discovery* 20 (3):200–216.

Badr, Amira, Dalia Fouad, and Hala Attia. 2019. "Insights into protective mechanisms of dandelion leaf extract against cisplatin-induced nephrotoxicity in rats: role of inhibitory effect on inflammatory and apoptotic pathways." *Dose-response* 17 (3):1–11.

Bai, L., W. Wang, J. Hua, Z. Guo, and S. Auid-Orcid Luo. 2020. "Defensive functions of volatile organic compounds and essential oils from northern white-cedar in China." *BMC Plant Biology* 20 (1):500.

Bessada, Sílvia M. F., João C. M. Barreira, and M. Beatriz P. P. Oliveira. 2015. "Asteraceae species with most prominent bioactivity and their potential applications: A review." *Industrial Crops and Products* 76:604–615.

Bhatta, Ragavendra, L. Baruah, Mani Agila Saravanan, K. P. Suresh, and K. T. Sampath. 2013. "Effect of medicinal and aromatic plants on rumen fermentation, protozoa population and methanogenesis in vitro." *Journal of Animal Physiology and Animal Nutrition* 97 3:446–56.

Boudjelal, Amel, Cherifa Henchiri, Madani Sari, Djamel Sarri, Noui Hendel, Abderrahim Benkhaled, and Giuseppe Ruberto. 2013. "Herbalists and wild medicinal plants in M'Sila (North Algeria): An ethnopharmacology survey." *Journal of Ethnopharmacology* 148 (2):395–402.

Cai, Liangliang, Bohua Chen, Fanglian Yi, and Shanshan Zou. 2019. "Optimization of extraction of polysaccharide from dandelion root by response surface methodology: Structural characterization and antioxidant activity." *International Journal of Biological Macromolecules* 140:907–919.

Cannell, R. J. P. 1998. *Natural Products Isolation*: Humana Press.

Chen, Hung-Ju, Baskaran Stephen Inbaraj, and Bing-Huei Chen. 2012. "Determination of phenolic acids and flavonoids in *Taraxacum formosanum* Kitam by liquid chromatography-tandem mass spectrometry coupled with a post-column derivatization technique." *International Journal of Molecular Sciences* 13:260–285.

Chen, Pei, Suyun Ding, Zhiqian Yan, Huiping Liu, Jianqiu Tu, Yi Chen, and Xiaowei Zhang. 2022. "Structural characteristic and in-vitro Anticancer activities of dandelion leaf polysaccharides from pressurized hot water extraction." *Nutrients* 15 (1):1–19.

Cheynier, Véronique, Gilles Comte, Kevin M. Davies, Vincenzo Lattanzio, and Stefan Martens. 2013. "Plant phenolics: Recent advances on their biosynthesis, genetics, and ecophysiology." *Plant Physiology and Biochemistry* 72:1–20.

Cho, S. Y., Ji Yeun Park, Eun Mi Park, Myung Sook Choi, Mi Kyung Lee, Seon Min Jeon, Moon Kyoo Jang, Myung Joo Kim, Yong Bok Park. 2002. "Alternation of hepatic antioxidant enzyme activities and lipid profile in streptozotocin-induced diabetic rats by supplementation of dandelion water extract." *Clinica Chimica Acta* 317 (1–2):109–117.

Choi, Beom-Rak, Il Je Cho, Su-Jin Jung, Jae Kwang Kim, Sang Mi Park, Dae Geon Lee, Sae Kwang Ku, and Ki-Moon Park. 2020. "Lemon balm and dandelion leaf extract synergistically alleviate ethanol-induced hepatotoxicity by enhancing antioxidant and anti-inflammatory activity." *Journal of Food Biochemistry* 44 (8):1–16.

Choi, Jae Young, Tae-Woo Jang, Phil Hyun Song, Seong Hoon Choi, Sae Kwang Ku, and Chang-Hyun Song. 2022. "Combination effects of metformin and a mixture of lemon balm and dandelion on high-fat diet-induced metabolic alterations in mice." *Antioxidants* 11 (3):1–17.

Choi, Ung-Kyu, Ok-Hwan Lee, Joo Hyuk Yim, Chang-Won Cho, Young Kyung Rhee, Seong-Il Lim, and Young-Chan Kim. 2010. "Hypolipidemic and antioxidant effects of dandelion (Taraxacum officinale) root and leaf on cholesterol-fed rabbits." *International Journal of Molecular Sciences* 11 (1):67–78.

Cock, Ian E., and Sandy F. Van Vuuren. 2020. "The traditional use of southern African medicinal plants for the treatment of bacterial respiratory diseases: A review of the ethnobotany and scientific evaluations." *Journal of Ethnopharmacology* 263:113204.

Cristina, Romeo T., Zorana Kovačević, Marko Cincović, Eugenia Dumitrescu, Florin Muselin, Kalman Imre, Dumitru Militaru, Narcisa Mederle, Isidora Radulov, Nicoleta Hădărugă, and Nikola Puvača. 2020. "Composition and efficacy of a natural phytotherapeutic blend against nosemosis in honey bees." *Sustainability* 12:1–16.

De Pellegrin, Michela Luisa, Anette Rohrhofer, Philipp Schuster, Barbara Schmidt, Philipp Peterburs, and André Gessner. 2021. "The potential of herbal extracts to inhibit SARS-CoV-2: A pilot study." *Clinical Phytoscience* 7 (1):1–7.

Deng, Xin-Xin, Yan-Na Jiao, Hui-Feng Hao, Dong Xue, Chang-Cai Bai, and Shu-Yan Han. 2021. "Taraxacum mongolicum extract inhibited malignant phenotype of triple-negative breast cancer cells in tumor-associated macrophages microenvironment through suppressing IL-10 / STAT3 / PD-L1 signaling pathways." *Journal of Ethnopharmacology* 274:113978.

Di Napoli, Agnese, and Pietro Zucchetti. 2021. "A comprehensive review of the benefits of Taraxacum officinale on human health." *Bulletin of the National Research Centre* 45 (1):110.

Díaz, Katy, Luis Espinoza, Alejandro Madrid, Leonardo Pizarro, and Rolando Chamy. 2018. "Isolation and identification of compounds from bioactive extracts of *Taraxacum officinale* Weber ex F. H. Wigg. (Dandelion) as a potential source of antibacterial agents." *Evidence-Based Complementary and Alternative Medicine* 2018:1–8.

Ekalu, Abiche, and James Dama Habila. 2020. "Flavonoids: isolation, characterization, and health benefits." *Beni-Suef University Journal of Basic and Applied Sciences* 9 (1):1–14.

Emerenciano, V. P., J. S. L. T. Militão, C. C. Campos, P. Romoff, M. A. C. Kaplan, M. Zambon, and A. J. C. Brant. 2001. "Flavonoids as chemotaxonomic markers for Asteraceae." *Biochemical Systematics and Ecology* 29 (9):947–957.

Fang, Mingyue, Shuangyue Liu, Qingqing Wang, Xuan Gu, Pengmin Ding, Weihua Wang, Yi Ding, Junxiu Liu, and Rufeng Wang. 2019. "Qualitative and quantitative analysis of 24 components in Jinlianhua decoction by UPLC-MS/MS." *Chromatographia* 82 (12):1801–1825.

FDA, U.S. 2016. *United States Food and Drug Administration Substance Registration System*, Silver Spring, USA: U.S. Food and Drug Administration.

Flores-Ocelotl, María del Rosario, Nora Hilda Rosas-Murrieta, Diego A. Moreno, Verónica Vallejo-Ruiz, Julio Reyes-Leyva, Fabiola Domínguez, and Gerardo Santos-López. 2018. "*Taraxacum officinale* and *Urtica dioica* extracts inhibit dengue virus serotype 2 replication in vitro." *BMC Complementary and Alternative Medicine* 18:1–10.

Franco-Espínola, José, Marvin J. Núñez, Y. Mabel Sanabria-Ramírez, Carlos Fabio Villar-Duarte, Ulises G. Castillo, Guadalupe Cantero-González, Marcos Florentín-Pavía, Patricia Langjahr, and Pablo Hernán Sotelo Torres. 2022. "Screening of medicinal plants from El Salvador for anti-viral activity against Herpes simplex 1." *Natural Product Research* 14:1–5.

Gani, Murtaza , and Tanveer Alam. 2022. "Biological activity of the indigenous fruits and vegetables of the Indian Himalayan region." *International Journal of Pharmaceutical Sciences and Research* 13 (7):2697–2702.

Gao, Demin. 2010. "Analysis of nutritional components of *Taraxacum mongolicum* and its antibacterial activity." *Pharmacognosy Journal* 2 (12):502–505.

GBIF. 2023. GBIF occurence.

Gilca, Marilena, George Sorin Tiplica, and Carmen Maria Salavastru. 2018. "Traditional and ethnobotanical dermatology practices in Romania and other Eastern European countries." *Clinics in Dermatology* 36 (3):338–352.

Gomez, Maricella K., Jashbir Singh, Pratibha Acharya, G. K. Jayaprakasha, and Bhimanagouda S. Patil. 2018. "Identification and Quantification of Phytochemicals, Antioxidant Activity, and Bile Acid-Binding Capacity of Garnet Stem Dandelion (Taraxacum officinale)." *Journal of Food Science* 83 (6): 1569–1578.

Gutiérrez-Grijalva, Erick Paul, Leticia Xochitl López-Martínez, Laura Aracely Contreras-Angulo, Cristina Alicia Elizalde-Romero, and José Basilio Heredia. 2020. "Plant alkaloids: Structures and bioactive properties." In *Plant-derived bioactives: Chemistry and mode of action*, edited by Mallappa Kumara Swamy, 85–117. Singapore: Springer Singapore.

Han, Huamin, Wen He, Wei Wang, and Bin Gao. 2011. "Inhibitory effect of aqueous dandelion extract on HIV-1 replication and reverse transcriptase activity." *BMC Complementary and Alternative Medicine* 11 (1):112.

Hanson, James Ralph. 2003. *Natural products : the secondary metabolites*. Cambridge: Royal Society of Chemistry.

Heo, Jun, Minsun Kim, Jae-Hyun Kim, Hwajeong Shin, Seo-Eun Lim, Hyuk-Sang Jung, Youngjoo Sohn, and Jaseung Ku. 2022. "Effect of Taraxaci herba on bone loss in an OVX-induced model through the regulation of osteoclast differentiation." *Nutrients* 14 (20):1–20.

Honek, Alois, Zdenka Martinkova, and Pavel Saska. 2011. "Effect of size, taxonomic affiliation and geographic origin of dandelion (Taraxacum agg.) seeds on predation by ground beetles (Carabidae, Coleoptera)." *Basic and Applied Ecology* 12 (1):89–96.

Hou, Y. N., Jun J. Deng, G. Fau-Mao, and J. J. Mao. 2019. "Practical application of "about herbs" website: Herbs and dietary supplement use in oncology settings." *Cancer Journal* 25 (5):357–366.

Hu, Ge, Junjie Wang, Dong Hong, Tao Zhang, Huiqin Duan, Xiang Mu, and Zuojun Yang. 2017. "Effects of aqueous extracts of Taraxacum Officinale on expression of tumor necrosis factor-alpha and intracellular adhesion molecule 1 in LPS-stimulated RMMVECs." *BMC Complementary and Alternative Medicine* 17 (1):38–38.

Huang, Yanmei, Peng Wu, Jian Ying, Zhizhong Dong, and Xiao Dong Chen. 2021. "Mechanistic study on inhibition of porcine pancreatic α-amylase using the flavonoids from dandelion." *Food Chemistry* 344:128610.

Huber, Meret, Daniella Triebwasser-Freese, Michael Reichelt, Sven Heiling, Christian Paetz, Jima N. Chandran, Stefan Bartram, Bernd Schneider, Jonathan Gershenzon, and Matthias Erb. 2015. "Identification, quantification, spatiotemporal distribution and genetic variation of major latex secondary metabolites in the common dandelion (*Taraxacum officinale* agg.)." *Phytochemistry* 115:89–98.

Ivanov, Ivan, Nadezhda Petkova, Julian Tumbarski, Ivayla Dincheva, Ilian Badjakov, Panteley Denev, and Atanas Pavlov. 2018. "GC-MS characterization of *n*-hexane soluble fraction from dandelion (*Taraxacum officinale* Weber ex F.H. Wigg.) aerial parts and its antioxidant and antimicrobial properties." *A Journal of Biosciences: Zeitschrift für Naturforschung C.* 73 (1–2):41–47.

Jalili, C., M. Taghadosi, M. Pazhouhi, F. Bahrehmand, S. S. Miraghaee, D. Pourmand, and I. Rashid. 2020. "An overview of therapeutic potentials of *Taraxacum officinale* (dandelion): A traditionally valuable herb with a reach historical background." *World Cancer Research Journal* 7:1–19.

Jan, J. T., Tr Auid-Orcid X. Cheng, Yp Auid-Orcid Juang, H. H. Ma, Yt Auid-Orcid X. Wu, W. B. Yang, Cw Auid-Orcid Cheng, X. Auid-Orcid Chen, Th Auid-Orcid Chou, Jj Auid-Orcid X. Shie, W. C. Cheng, Rj Auid-Orcid Chein, S. S. Mao, Ph Auid-Orcid Liang, C. Auid-Orcid Ma, Sc Auid-Orcid X. Hung, and C. H. Wong. 2021. "Identification of existing pharmaceuticals and herbal medicines as inhibitors of SARS-CoV-2 infection." *Proceedings of the National Academy of Sciences of the United States of America* 118 (5):1–8.

Janke, Rhonda. 2004. *Farming a few acres of herbs: Dandelion*: Kansas State University.

Jedrejek, Dariusz, Bernadetta Lis, Agata Rolnik, Anna Stochmal, and Beata Olas. 2019. "Comparative phytochemical, cytotoxicity, antioxidant and haemostatic studies of *Taraxacum officinale* root preparations." *Food and Chemical Toxicology* 126:233–247.

Jia, Y. Y., R. F. Guan, Y. H. Wu, X. P. Yu, W. Y. Lin, Y. Y. Zhang, T. Liu, J. Zhao, S. Y. Shi, and Y. Zhao. 2014. "Taraxacum mongolicum extract exhibits a protective effect on hepatocytes and an antiviral effect against hepatitis B virus in animal and human cells." *Molecular Medicine Reports* 9 (4):1381–1387.

Kania-Dobrowolska, Malgorzata, and Justyna Baraniak. 2022. "Dandelion (*Taraxacum officinale* L.) as a source of biologically active compounds supporting the therapy of co-existing diseases in metabolic syndrome." *Foods* 11 (18):1–17.

Karahuseyin, Secil, Aynur Sari, and Nurten Ozsoy. 2022. "Antioxidant activity and three phenolic compounds from the roots of *Taraxacum gracilens* Dahlst." *Istanbul Journal of Pharmacy* 52 (1):69–72.

Kenny, O., N. P. Brunton, D. Walsh, C. M. Hewage, P. McLoughlin, and T. J. Smyth. 2015. "Characterisation of antimicrobial extracts from Dandelion root (*Taraxacum officinale*) using LC-SPE-NMR." *Phytotherapy Research* 29 (4):526–532.

Kikuchi, Takashi, Ayaka Tanaka, Mayu Uriuda, Takeshi Yamada, and Reiko Tanaka. 2016. "Three novel triterpenoids from *Taraxacum officinale* roots." *Molecules* 21 (9):1–11.

Kumar, Manoj, Yamini Tak, Jayashree Potkule, Prince Choyal, Maharishi Tomar, Nand Lal Meena, and Charanjit Kaur. 2020. "Phenolics as plant protective companion against abiotic stress." In *Plant phenolics in sustainable agriculture* edited by R. Lone, R. Shuab and A. Kamili, 277–308. Singapore: Springer.

Kumar, Vijay Bhooshan, Koushi Kumar, Aharon Gedanken, and Pradip Paik. 2014. "Facile synthesis of self-assembled spherical and mesoporous dandelion capsules of ZnO: efficient carrier for DNA and anticancer drugs." *Journal of Materials Chemistry* 2 (25):3956–3964.

Kurkin, V. A., and A. V. Aznagulova. 2016. "Constituents of the aerial part of *Taraxacum officinale*." *Chemistry of Natural Compounds* 52 (4):711–712.

Lee, Mi-Hwa, Hee Kang, Kyungjin Lee, Gabsik Yang, Inhye Ham, Youngmin Bu, Hocheol Kim, and Ho-Young Choi. 2013. "The aerial part of Taraxacum coreanum extract has an anti-inflammatory effect on peritoneal macrophages in vitro and increases survival in a mouse model of septic shock." *Journal of Ethnopharmacology* 146 (1):1–8.

Lee, Sanghyun, Mi Jin Choi, Ji Myung Choi, Sullim Lee, Hyun Young Kim, and Eun Ju Cho. 2012. "Flavonoids from *Taraxacum coreanum* protect from radical-induced oxidative damage." *Journal of Medicinal Plants Research* 6 (40):5377–5384.

Leu, Yann-Lii, Li-Shian Shi, and Amooru Gangaiah Damu. 2003. "Chemical constituents of *Taraxacum formosanum*." *Chemical and Pharmaceutical Bulletin* 51 (5):599–601.

Leu, Yann-Lii, Yu-Li Wang, Shih-Chin Huang, and Li-Shian Shi. 2005. "Chemical constituents from roots of *Taraxacum formosanum*." *Chemical and Pharmaceutical Bulletin* 53 (7):853–855.

Li, Wei, Changyeol Lee, Young Ho Kim, Jin Yeul Ma, and Sang Hee Shim. 2017. "Chemical constituents of the aerial part of *Taraxacum mongolicum* and their chemotaxonomic significance." *Natural Product Research* 31 (19):2303–2307.

Li, Xiao-Hong, Xi-Ran He, Yan-Yan Zhou, Hai-Yu Zhao, Wen-Xian Zheng, Shan-Tong Jiang, Qun Zhou, Ping-Ping Li, and Shu-Yan Han. 2017. "Taraxacum mongolicum extract induced endoplasmic reticulum stress associated-apoptosis in triple-negative breast cancer cells." *Journal of Ethnopharmacology* 206:55–64.

Li, Yuxi, Yuying Deng, Xiuli Zhang, Han Fu, Xue Han, Wenqing Guo, Wei Zhao, Xuening Zhao, Chunxue Yu, Hui Li, Kaijian Lei, and Tianxiao Wang. 2022. "Dandelion Seed extract affects tumor progression and enhances the sensitivity of Cisplatin in esophageal squamous cell Carcinoma." *Frontiers in Pharmacology* 13:897465–897465.

Lichman, Benjamin R. 2021. "The scaffold-forming steps of plant alkaloid biosynthesis." *Natural Product Reports* 38 (1):103–129.

Liu, Qiang, Heng Zhao, Yue Gao, Yan Meng, Xiang-Xuan Zhao, and Shi-Nong Pan. 2018. "Effects of Dandelion extract on the proliferation of rat skeletal muscle cells and the inhibition of a Lipopolysaccharide-induced inflammatory reaction." *Chinese Medical Journal* 131 (14):1724–1731.

Łuczaj, Łukasz. 2010. "Changes in the utilization of wild green vegetables in Poland since the 19th century: A comparison of four ethnobotanical surveys." *Journal of Ethnopharmacology* 128 (2):395–404.

Majewski, Michał, Bernadetta Lis, Jerzy Juśkiewicz, Katarzyna Ognik, Małgorzata Borkowska-Sztachańska, Dariusz Jedrejek, Anna Stochmal, and Beata Olas. 2020. "Phenolic fractions from Dandelion leaves and petals as modulators of the antioxidant status and lipid profile in an *in vivo* study." *Antioxidants* 9 (2):1–13.

Majewski, Michał, Bernadetta Lis, Jerzy Juśkiewicz, Katarzyna Ognik, Dariusz Jedrejek, Anna Stochmal, and Beata Olas. 2021. "The composition and vascular/antioxidant properties of *Taraxacum officinale* flower water syrup in a normal-fat diet using an obese rat model." *Journal of Ethnopharmacology* 265:1–10.

McGuffin, M., J. T. Kartesz, A. Y. Leung, and A. O. Tucker. 2000 *Herbs of Commerce*. 2nd ed. Silver Spring, USA: AHPA.

Michalska, Klaudia, Jolanta Marciniuk, and Wanda Kisiel. 2010. "Sesquiterpenoids and phenolics from roots of *Taraxacum udum*." *Fitoterapia* 81 (5):434–436.

Michalska, Klaudia, Jolanta Marciniuk, and Anna Stojakowska. 2021. "A new sesquiterpenoid and further natural products from *Taraxacum portentosum* Kirschner & Stepanek, an endangered species." *Natural Product Research* 35 (21):4058–4062.

Mitchell, Sylvia A. 2011. "The Jamaican root tonics: A botanical reference." *Focus on Alternative and Complementary Therapies* 16 (4):271–280.

Mo, Eun Jin, Jong Hoon Ahn, Yang Hee Jo, Seon Beom Kim, Bang Yeon Hwang, and Mi Kyeong Lee. 2017. "Inositol derivatives and phenolic compounds from the roots of *Taraxacum coreanum*." *Molecules* 22 (8):1–8.

Molinu, Maria Giovanna, Giovanna Piluzza, Giuseppe Campesi, Leonardo Sulas, and Giovanni Antonio Re. 2019. "Antioxidant sources from leaves of Russian Dandelion." *Chemistry & Biodiversity* 16 (8):1–10.

Montesino, N. L., M. Kaiser, R. Brun, and T. J. Schmidt. 2015. "Search for antiprotozoal activity in herbal medicinal preparations; New natural leads against neglected tropical diseases." *Molecules* 20 (8):14118–14138.

MPNS. 2021. The Medicinal Plant Names Services (MPNS) Resource is V11. Kew Gardens.

National Museum of Natural History, Smithsonian Institution. 2023. Integrated Taxonomic Information System (ITIS).

Nugraha, Ari S., Yoshinta D. Purnomo, Antonius N. Widhi Pratama, Bawon Triatmoko, Rudi Hendra, Hendris Wongso, Vicky M. Avery, and Paul A. Keller. 2022. "Isolation of antimalarial agents from Indonesian medicinal plants: *Swietenia mahagoni* and *Pluchea indica*." *Natural Product Communications* 17 (1):1–5.

Odintsova, I. T., A. E. Rogozhin, V. I. Sklyar, K. A. Musolyamov, M. A. Kudryavtsev, A. V. Pukhalsky, N. A. Smirnov, V. E. Grishin, and A. T. Egorov. 2010. "Antifungal activity of storage 2S albumins from seeds of the Iinvasive weed Dandelion *Taraxacum officinale* Wigg." *Protein & Peptide Letters* 17 (4):522–529.

Olas, Beata. 2022. "New perspectives on the effect of Dandelion, Its food products and other preparations on the cardiovascular system and its diseases." *Nutrients* 14 (7):1–11.

Ovadje, P., S. Chatterjee, C. Griffin, C. Tran, C. Hamm, and S. Pandey. 2011. "Selective induction of apoptosis through activation of caspase-8 in human leukemia cells (Jurkat) by dandelion root extract." *Journal of Ethnopharmacology* 133 (1):86–91.

Ovadje, Pamela, Saleem Ammar, Jose-Antonio Guerrero, John Thor Arnason, and Siyaram Pandey. 2016. "Dandelion root extract affects colorectal cancer proliferation and survival through the activation of multiple death signalling pathways." *Oncotarget* 7 (45):73080–73100.

Ovadje, Pamela, Madona Chochkeh, Pardis Akbari-Asl, Caroline Hamm, and Siyaram Pandey. 2012. "Selective induction of apoptosis and autophagy through treatment with dandelion root extract in human pancreatic cancer cells." *Pancreas* 41 (7):1039–1047.

Ovadje, Pamela, Caroline Hamm, and Siyaram Pandey. 2012. "Efficient induction of extrinsic cell death by Dandelion root extract in human chronic myelomonocytic leukemia (CMML) cells." *PLoS One* 7 (2):1–10.

Padmanabhan, P., J. Correa-Betanzo, and G. Paliyath. 2016. "Berries and related fruits." In *Encyclopedia of food and health*, edited by Benjamin Caballero, Paul M. Finglas and Fidel Toldrá, 364–371. Oxford: Academic Press.

Park, Chung Mu, Ji Young Park, Kyung Hee Noh, Jin Hyuk Shin, and Young Sun Song. 2011. "Taraxacum officinale Weber extracts inhibit LPS-induced oxidative stress and nitric oxide production via the NF-κB modulation in RAW 264.7 cells." *Journal of Ethnopharmacology* 133 (2):834–42.

Pieroni, Andrea, and Renata Sõukand. 2017. "The disappearing wild food and medicinal plant knowledge in a few mountain villages of North-Eastern Albania." *Journal of Applied Botany and Food Quality* 90:58–67.

POWO. 2023. *Plants of the World Online*. The Royal Botanic Gardens, Kew.

Putter, Katharina M., Nicole van Deenen, Kristina Unland, Dirk Prufer, and Christian Schulze Gronover. 2017. "Isoprenoid biosynthesis in dandelion latex is enhanced by the overexpression of three key enzymes involved in the mevalonate pathway." *BMC Plant Biology* 17:1–13.

Ren, Feng, Kaixuan Wu, Yun Yang, Yingying Yang, Yuxia Wang, and Jian Li. 2020. "Dandelion polysaccharide exerts anti-angiogenesis effect on hepatocellular carcinoma by regulating VEGF/HIF-1α expression." *Frontiers in Pharmacology* 11:460–460.

Rolnik, Agata, and Beata Olas. 2021. "The plants of the asteraceae family as agents in the protection of human health." *International Journal of Molecular Sciences* 22 (6):1–10.

Safety, Ministry of Food and Drug. 2002. *The Korean Herbal Pharmacopoeia* 8th ed. Cheonju-si: Korea Food and Drug Administration.

Sari, Aynur, and Zeynep Kececi. 2019. "Phytochemical investigations on chemical constituents of *Taraxacum bessarabicum* (Hornem.) Hand.-Mazz. subsp. bessarabicum (Hornem.) Hand.-Mazz." *Iranian Journal of Pharmaceutical Research* 18 (1):400–405.

Sekhon-Loodu, Satvir, and H. P. Vasantha Rupasinghe. 2019. "Evaluation of antioxidant, antidiabetic and anti-obesity potential of selected traditional medicinal plants." *Frontiers in Nutrition* 6:1–11.

Shaikh, Mohd S., Fahadul Islam, Parag P. Gargote, Rutuja R. Gaikwad, Kalpana C. Dhupe, Sharuk L. Khan, Falak A. Siddiqui, Ganesh G. Tapadiya, Syed S. Ali, Abhijit Dey, and Talha B. Emran. 2022. Potential epha2 receptor blockers involved in cerebral malaria from *Taraxacum officinale*, *Tinospora cordifolia*, *Rosmarinus officinalis* and *Ocimum basilicum*: A computational approach. *Pathogens* 11 (11):1–22. doi:10.3390/pathogens11111296.

Shi, Shu-Yun, Yu-Ping Zhang, Ke-Long Huang, Yu Zhao, and Su-Qin Liu. 2008. "Flavonoids from *Taraxacum mongolicum*." *Biochemical Systematics and Ecology* 36 (5–6):437–440.

Shi, Shu Yun, Yu Ping Zhang, Hong Hao Zhou, Ke Long Huang, and Xin Yu Jiang. 2010. "Screening and identification of radical scavengers from *Neo-taraxacum siphonanthum* by online rapid screening method and nuclear magnetic resonance experiments." *Journal of Immunoassay and Immunochemistry* 31 (3):233–249.

Shi, Shuyun, Kelong Huang, Yuping Zhang, and Suqin Liu. 2008. "Preparative isolation and purification of two flavonoid glycosides from *Taraxacum mongolicum* by high-speed counter-current chromatography." *Separation and Purification Technology* 60 (1):81–85.

Shi, Shuyun, Yuping Zhang, Yu Zhao, and Kelong Huang. 2008. "Preparative isolation and purification of three flavonoid glycosides from Taraxacum mongolicum by high-speed counter-current chromatography." *Journal of Separation Science* 31 (4):683–688.

Tajner-Czopek, Agnieszka, Mateusz Gertchen, Elżbieta Rytel, Agnieszka Kita, Alicja Z. Kucharska, and Anna Sokół-Łętowska. 2020. "Study of antioxidant activity of some medicinal plants having high content of caffeic acid derivatives." *Antioxidants* 9 (5):1–21.

Takasaki, Midori, Takao Konoshima, Harukuni Tokuda, Kazuo Masuda, Yōko Arai, K. Shiojima, and Hiroyuki Ageta. 1999. "Anti-carcinogenic activity of Taraxacum plant. II." *Biological & Pharmaceutical Bulletin* 22 (6):606–610.

Tian, Lin, Peng Fu, Min Zhou, and Jiping Qi. 2021. "Dandelion sterol improves diabetes mellitus-induced renal injury in *in vitro* and *in vivo* study." *Food Science and Nutrition* 9 (9):5183–5197.

Torssell, K. 1982. *Natural products chemistry*. Wiley.

Tsuda, Y. 2004. *Isolation of natural products Japan Analytical* Industry Co., Ltd.

van Seventer, Jean Maguire, and Natasha S. Hochberg. 2017. "Principles of infectious diseases: Transmission, diagnosis, prevention, and control." In *International encyclopedia of public health* (2nd edition), edited by Stella R. Quah, 22–39. Oxford: Academic Press.

Warashina, Tsutomu, Kaoru Umehara, and Toshio Miyase. 2012. "Constituents from the roots of *Taraxacum platycarpum* and their effect on proliferation of human skin fibroblasts." *Chemical and Pharmaceutical Bulletin* 60 (2):205–212.

Wiersema, J. H., and B. León. 1999. *World economic plants: A standard reference*. 1st ed: CRC Press.

Williams, Christine A., Fiona Goldstone, and Jenny Greenham. 1996. "Flavonoids, cinnamic acids and coumarins from the different tissues and medicinal preparations of *Taraxacum officinale*." *Phytochemistry* 42 (1):121.

Wirngo, F. E., M. N. Lambert, and P. B. Jeppesen. 2016. "The physiological effects of Dandelion (*Taraxacum Officinale*) in type 2 diabetes." *The Review of Diabetic Studies* 13 (2–3):113–131.

Wu, Jiang. 2022. "Antibacterial activity of *Taraxacum officinale* against foodborne pathogens." *Pakistan Journal of Zoology* 54 (5):2021–2028.

Xie, Jun, Hongqing Wang, Chaoxuan Dong, Shengtian Lai, Jianbo Liu, Ruoyun Chen, and Jie Kang. 2022. "Benzobicyclic ketones, cycloheptenone oxide derivatives, guaiane-type sesquiterpenes, and alkaloids isolated from *Taraxacum mongolicum* Hand.-Mazz." *Phytochemistry* 201:113277.

Yadava, Raj Nath, and Shirin Khan. 2013. "A new flavonoidal constituent from *Taraxacum officinale* (L.) Weber." *Asian Journal of Chemistry* 25 (7):4117–4118.

Yang, Kerry, and Yanjie Zhang. 2020. "Reversal of heavy metal-induced antibiotic resistance by dandelion root extracts and taraxasterol." *Journal of Medical Microbiology* 69 (8):1049–1061.

Yasukawa, Ken, Toshihiro Akihisa, Hirotoshi Oinuma, Yoshimasa Kasahara, Yumiko Kimura, Sakae Yamanouchi, Kunio Kumaki, Toshitake Tamura, and Michio Takido. 1996. "Inhibitory effect of di- and trihydroxy triterpenes from the flowers of Compositae on 12-*O*-tetradecanoylphorbol-13-acetate-induced inflammation in mice." *Biological and Pharmaceutical Bulletin* 19 (10):1329–1331.

Zhao, Ping, Qian Ming, Mingrui Xiong, Guanjun Song, Li Tan, Di Tian, Jia Liu, Zhao Huang, Jingyi Ma, Jinhua Shen, Qing-Hua Liu, and Xinzhou Yang. 2018. "Dandelion chloroform extract promotes glucose uptake via the AMPK/GLUT4 pathway in L6 cells." *Evidence-Based Complementary and Alternative Medicine* 2018:1–10.

Zhou, Anfu, Shuqing Zhang, Chengliang Yang, Nansheng Liao, and Yan Zhang. 2022. "Dandelion root extracts abolish MAPK pathways to ameliorate experimental mouse ulcerative colitis." *Advances in Cinical and Experimental Medicine* 31 (5):529–538.

Zhu, J., X. Li, S. Zhang, J. Liu, X. Yao, Q. Zhao, B. Kou, P. Han, X. Wang, Z. Bai, Z. Zheng, and C. Xu. 2021. "Taraxasterol inhibits TGF-β1-induced epithelial-to-mesenchymal transition in papillary thyroid cancer cells through regulating the Wnt/β-catenin signaling." *Human & Experimental Toxicology* 40 (12):S87–S95.

Zhu, Jianming, Xingbei Weng, Rujin Jiang, Jinlan Wu, and Zuhuang Mi. 2017. "Transcriptomic response of uropathogenic *Escherichia coli* to *Taraxacum mongolicum*." *International Journal of Clinical and Experimental Medicine* 10 (8):12030–12043.

Zhu, Shuyong, and Abel Viejo-Borbolla. 2021. "Pathogenesis and virulence of herpes simplex virus." *Virulence* 12 (1):2670–2702.

10 Consumption of Mushroom as a Food Source
Overview, Health Benefits and Safety

Özge Cemali and Duygu Ağagündüz

INTRODUCTION

For centuries, mushrooms have gained widespread popularity as a result of their recognized nutritional value and pharmacological properties, leading to their prevalent consumption. Protein, fiber, bioactive peptides, polysaccharides, vitamins, minerals, and a variety of other bioactive substances are all present in mushrooms. It also has low sodium, fat, cholesterol, and calorie levels and is low in energy density. Besides being easily edible, it also has pleasant organoleptic characteristics. Mushrooms are commonly accessible because of their cost and ability to be grown throughout the world. Because of these qualities, mushrooms can be a possible member of a balanced and healthy diet (Boin and Nunes 2018). In addition to the various research on the chemical characteristics of mushrooms, they have gained recognition as healthcare due to their effects on antimicrobials, antioxidants, inflammation, possible positive effects on cancer and metabolic diseases and modulation of gut microbiota (Chou, Sheih, and Fang 2013, Gong et al. 2020, Mishra et al. 2021, Rathore, Prasad, and Sharma 2017, Li et al. 2021). On each continent, edible mushrooms are in terms of commerce grown across more than 100 nations using various techniques and scales. The number of mushrooms produced worldwide is rising at a pace of roughly 7.0% annually. There are already 13 million tons of mushrooms consumed worldwide, and more mushrooms are projected to be produced – 20.8 million tons are predicted by 2026 (Shirur, Barh, and Annepu 2021). The majority of the species are found in subtropical moist woodlands and temperate areas such as Australia, North America, Asia, and Europe (Andrew et al. 2013, Sande et al. 2019). The Asian Continent, with a 69% share of mushroom production, holds the top position globally in terms of quantity, followed by the Chinese mainland specifically. *Agaricus bisporus* (white button mushroom) takes the lead as the most widely cultivated species, followed by *Pleurotus ostreatus* (oyster mushroom), *Flammulina Velutipes* (golden needle mushroom), and *Lentinula edodes* (shiitake) in terms of production volume. Unwanted color and texture changes, mass loss, and so on, can occur quickly after harvest. Mushrooms have a brief shelf life, lasting between 1 and 3 days at room temperature and approximately 8 to 10 days when refrigerated (Castellanos-Reyes, Villalobos-Carvajal, and Beldarrain-Iznaga 2021, Dawadi et al. 2022, Lin and Sun 2019).

Mushrooms require optimal heat and humidity levels for their growth to occur. The optimal temperature is 20°C or higher, while the ideal soil and air humidity levels are 60% and higher. Extreme dryness and frost events prevent fungus from procreating. It can grow in meadows, woods, tree trunks, field soils, sandy plains, stumps, and branches when the correct circumstances are present. Through commercial cultivation, mushrooms are predominantly grown on agricultural waste, effectively repurposing these discarded materials into a valuable and sustainable source of nutrition for humans (Das et al. 2021, Zhang, Li, and Fadel 2002).

DOI: 10.1201/9781003395935-10

Based on their characteristics, mushrooms can be categorized into three distinct groups: ecological habitat, taxonomic position, and extent of use. Edible mushrooms are those that have fruit parts that can be consumed fresh or dried (for example, *Lentinus edodes*), whereas medicinal mushrooms are those that are used in pharmaceutical applications but cannot be consumed directly (for example, *Ganoderma lucidum*) because of the substantial number of bioactive components in the plant's cell walls (Azieana et al. 2015).

Mushrooms that are produced in locations with anthropogenic activities such as radiogeological anomalies, pesticide use, and mining can collect heavy metals and radioactive toxic elements that are damaging to the organism (Barea-Sepúlveda et al. 2022). Because of the reality that edible wild mushroom species are produced in more unfamiliar areas than cultured mushrooms (Dowlati et al. 2021, Ernst et al. 2022), the accumulation of toxic elements may be higher in these species. It is therefore crucial to get mushrooms that will be consumed from reliable sources. Mushrooms are viewed as a remarkable food in terms of sustainability given the natural resources consumed in their production and consumption in terms of other nutrients, particularly protein content (Poore and Nemecek 2018).

In this chapter, the mushroom family, which has taken on a number of novel interests, was examined in detail in order to assess its nutritional value, potential health benefits, potential mechanisms of action, and safety concerns.

MUSHROOM TYPES

Although mushrooms share several similarities with plants and animals, mycology is the field in which they have since been studied. The fleshy, umbrella-shaped fruit parts of macrofungal species are known as fungus. Fungi are typically thought of as macrofungal species with identifiable fruit portions that can be selected by hand and seen with the naked eye. Despite having a cell wall similar to that of a plant cell, fungi that reproduce sexually and asexually with spores lack chlorophyll organelles and are therefore unable to photosynthesize. They are unable to produce organic materials on their own, including starch, sugar, and oil, due to this characteristic. They obtain their nourishment from waste material and other living things. These are living organisms with mycelium making up their primary bodies and fruiting organs. There are several shapes for mushrooms. Mushrooms typically have the following structures: a cap, gills or lamella, a stalk, an annulus or ring, a volva, and scales. When it comes to cultivated mushrooms, both external and internal factors play a significant role in influencing stem height, stem diameter, and cap size. Among the crucial variables, temperature, humidity, fresh air, and substrate composition hold utmost importance (Das et al. 2021). According to their taxonomic position, ecological environment, and human usage, mushrooms can be categorized in three different ways.

TAXONOMIC CLASSIFICATION

Taxonomically, some mushrooms belong under the *Basidiomycetes* class, some the *Ascomycetes* class, and yet others the *Deuteromycetes* class (Mwangi et al. 2022), the class that is subject to the majority of clinical research (Figure 10.1).

NATURAL HABITAT CLASSIFICATION

According to their natural habitats, humus (saprotic – *Lepista nuda*, *Volvarielle* spp, *Marasmium* spp, *Polyporuz*, *Tuberaster* – and symbiotic – *Boletus*, *Lacterius*, *Tricholoma*, *Tuber*, *Morchella*), wood (saprophytic – *Agrocybei Pleurotus*, *Auricularia* spp., *Auricularia* spp., and parasitic – *Armillaria mellea*, *Cyttaria*) and dung (*Agaricus* spp., *Coprinus* spp.), which are their native habitats. According

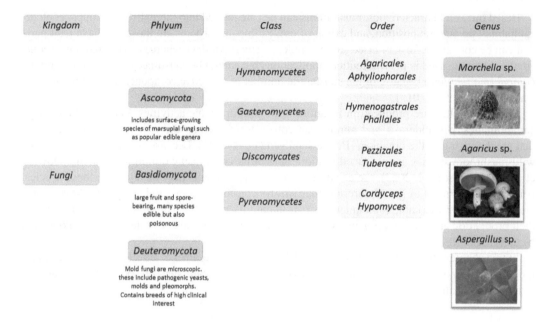

FIGURE 10.1   Taxonomic classification of mushrooms.

to their eating habits, there are three categories into which mushroom species can be divided. The host plant and mycorrhizal or symbiotic organisms interact closely and mutualistically. Saprotrophic animals rely on decomposing organic matter for their nutrition. The last group consists of parasitic organisms, which can parasitize other species in some ways (Kalač 2013).

HUMAN CONSUMPTION CLASSIFICATION

While the majority of edible mushrooms have healing value, it can be hard to differentiate between therapeutic and edible mushrooms because certain species with healing properties are also edible (Azieana et al. 2015). Mushrooms can be divided into four groups when it comes to human consumption: cultured mushrooms, wild mushrooms, pharmacological mushrooms, and poisonous mushrooms (Korman 2019). Although it can be difficult to differentiate between the various kinds of mushrooms, we examine edible (cultivated, wild, and pharmacological) and nonedible mushrooms (poisonous, wild, pharmacological) in this chapter.

Edible Mushrooms

Mushrooms that can be eaten are particularly nutrient dense. An edible mushroom is delicious and doesn't leave a nasty taste on the tongue. It is also a sign that something is edible if it has worms on it and the cap has no scales (Shirur, Barh, and Annepu 2021). Among the most significant of mushrooms grown for commerce include *Agaricus bisporus* (button/agaric mushroom), *Pleurotus ostreatus* (oyster mushroom), *Lentinula edodes* (shiitake), *Flammulina velutipes* (enoki or winter mushroom), *Pleurotus eryngii* (king trumpet mushroom), *Volvariella volvacea* (paddy straw mushroom), *Hericium erinaceus* (lion's mane mushroom), *Boletus edulis* (king bolete mushroom), *Calocybe indica* (milk mushroom), *Agrocybe aegerita* (pioppini), and *Grifola frondosa* (maitake or hen-of-the-woods mushroom) (Li et al. 2022). About 85% of the world's cultivated edible mushrooms are represented by only five genera: *Lentinula, Agaricus, Pleurotus, Auricularia*, and *Flammulina*. Interestingly, these five genera dominate the global production of cultivated edible mushrooms (Royse, Baars, and Tan 2017).

Nonedible Mushrooms

In some cultures and nations, a wide variety of wild or agricultural mushrooms are regarded as appetizing. A significant proportion of mushrooms, however, have unpleasant tastes and odors or can be dangerous if consumed. However, most have more intense colors. Those with poisons include amanitin, muscarine, ibotenic acid, phallotoxin, bolesatine, gyromitrin, arabitol, and ergotamine in them. They smell bad, have a sour taste that burns the tongue, do not contain worms, and have scale on the cap (Shirur, Barh, and Annepu 2021).

The following are some instances of poisonous mushrooms: *Amanita phalloides*, *Galerina* spp., *Gyromitra esculenta*, *Coprinus* spp., *Inocybe* spp., *Amanita muscaria*, *Psilocybe* spp. (Rumack and Spoerke 1994). They include the potential for gastrointestinal toxicity, hepatotoxicity, nephrotoxicity, neurotoxicity, rhabdomyolysis, bone marrow toxicity, hypoglycemia, cutaneous response, and psychotropic or complicated combination effects (White et al. 2019).

NUTRITIONAL COMPONENTS OF MUSHROOM

The nutritional content of mushrooms varies substantially based on a variety of characteristics, including species, intraspecific genetic diversity, age, growth conditions, location, environment, and post-harvest conditions (Marçal et al. 2021). Typically, mushrooms have 90% moisture and 10% dry substance. Nucleic acids (3–8%), carbs (35–70%), protein (15.0–34.7%), fat (10%), minerals (6.0–10.9%), and moisture (85–95%) KAL are all present in mushrooms. Mushrooms include few calories, minimal fat, no cholesterol, no gluten, and very low sodium (Rahi and Malik 2016). Mushrooms are characterized by their low-calorie content, absence of fat and cholesterol, gluten-free nature, and remarkably low sodium levels. Bioactive substances generated from mushrooms called triterpenes and triterpenoids are useful to human health (Das and Prakash 2022). This section discusses the contents of mushrooms', especially the edible ones, important macro- and micronutrients and bioactive components.

NUTRITIONAL VALUES

Proteins and Amino Acids

Protein, an essential element in human nutrition, is a complex macromolecule composed of extended chains of amino acids (Gong et al., 2020). According to Wani, Bodha, and Wani (2010), mushrooms have a protein content of 19–35% by dry weight, which is significantly greater than that of typical vegetables and on pace with that of meat and dairy products (Wang and Zhao 2023; Wani, Bodha, and Wani 2010). Mushrooms provide adequate amounts of essential amino acids (Rathore, Prasad, and Sharma 2017). Many elements, including the type of mushrooms, the carbon and nitrogen source, the pH, the temperature, and the administration, have an impact on the protein content (Chan et al. 2015).

In terms of protein content, certain species like *Flammulina velutipes, Agaricus bisporus, Pleurotus eryngii*, and *Tricholoma matsutake* (Wang and Zhao 2023). *Agaricus bisporus* had the highest protein content, at 4.70 g/L, than *Pleurotus ostreatus* at 3.99 g/L, *Pleurotus eryngii* at 1.91 g/L, and *Agaricus subrufescens* at 0.612 g/L in a study on the protein content of several fungal strains (Argyropoulos et al. 2022).The most frequently cultivated mushroom types, including as *Pleurotus*, *Lentiluna edodes*, and *Agaricus*, have high protein/biomass ratios between 10.5% to 42.0% (Smiderle et al. 2012, Wu, Siu, and Geng 2021). The typical range of protein digestibility in mushrooms is between 72 and 83% (compared to 74 and 82% for rice, 82 and 94% for meat, 98 and 100% for eggs, and 97 and 100% for dairy products) (Wang and Zhao 2023, Wani, Bodha, and Wani 2010).

Given to these features, edible mushrooms are a possible sustainable resource of protein that could be added to a variety of cuisines (Hamza, Ghanekar, and Kumar 2022). Because of its considerable sulfur-containing amino acid and glutamic acid content, mushroom protein has a meaty flavor and can potentially reduce meat consumption in the context of sustainable nutrition (Poore and

Nemecek 2018). Because of their efficient digestion, mushrooms are viewed as possible replacements for muscle protein (González et al. 2020). Therefore, they can be administered as an alternative to protein derived from animals or as a component of dietary fortifications (Yu et al. 2021).

Fats and Fatty Acids

Mushrooms have a modest energy (calorie) content and a 4–6% fat and cholesterol level. Mushrooms provide between 2–6 g fat / 100 g dry weight and contain any cholesterol (Das et al. 2021, Wang and Zhao 2023). Lipids, which contain triglycerides, phospholipids, sterols, and fatty acids, belong to another class of bioactive metabolites produced by mushrooms. The ratio of unsaturated/saturated fatty acids in mushroom lipids is substantial because mono- and polyunsaturated fatty acid components dominated. Linoleic acid ranges from 0.0% to 81.1%, oleic acid 1.0%–60.3%, and linolenic acid 0.0%–28.8% in 100 g serving of mushrooms (Sande et al. 2019). The precursor to 1-octen-3-ol, a crucial aromatic molecule which gives mushrooms their characteristic smells, is linoleic acid (Rathore, Prasad, and Sharma 2017). Ergosterol and tocopherol are additional components of lipid fractions that function as powerful antioxidants (Valverde, Hernández-Pérez, and Paredes-López 2015). Considering fat content, mushrooms are a food that's capable of being consumed in sufficient amounts in a balanced diet plan because they don't contain cholesterol, do include important fatty acids, though rather in little amounts, and also do contain antioxidant chemicals.

CHOs and Dietary Fibers

According to Rathore et al. (2017), mushrooms have a dry weight ratio of approximately 50–65 g of carbs per 100 g of their constituent parts (Rathore, Prasad, and Sharma 2017). In addition to insoluble carbohydrates like chitin, mannans, and beta-glucan, mushrooms also contain soluble carbohydrates such trehalose, glycogen, mannitol, and glucose (Yadav and Negi 2021). The dominant fibers in mushrooms, chitin and beta-glucans, make up over 50% of the dietary fiber content, while water-soluble fiber generally accounts for less than 10% of the dry weight. Mushroom micelles contain other polysaccharides that constitute 90% of them, including chitin (Nitschke et al. 2011). A remarkable source of dietary fiber that is insoluble is mushroom stems. Of the fibre in mushrooms, 4% to 9% is soluble, and 22% to 30% is insoluble. 100 g of *Pleurotus ostreatus* mushrooms contain 4.1 g of dietary fiber (Ma et al. 2020).

Glucose, fructose, trehalose, arabinose, and mannitol are the monosaccharides most frequently found in mushrooms (Samsudin and Abdullah 2019). Mannitol, commonly referred to as mushroom sugar, accounts for approximately 80% of all free sugars present in mushrooms. However, variables including mushroom type, substrate (also contains cellulose, hemicellulose, and lignin-containing agricultural waste), and ambient circumstances during growth, which all can affect the content of carbohydrates in mushrooms (Rathore, Prasad, and Sharma 2017). Mushrooms, depending on the variety, can serve as an excellent dietary fiber source. However, they often go unnoticed compared to other fiber-rich foods like grains, legumes, fruits, and vegetables. Consuming edible mushrooms can help you achieve the necessary daily intake of dietary fiber of about 25% (Cheung 2013). For illustration, it has been noted that the *Pleurotus* is a significant amount of carbohydrates and dietary fiber (Alam et al. 2008). Mushroom stems are mostly made of glucans and insoluble dietary fiber. The mushroom's stem in particular can be used to make biologically active polysaccharide complexes for dietary supplements. *Flammulina velutipes* (winter or enoki mushroom) stems were shown to provide 32% dietary fiber in one study (Banerjee et al. 2020, Das et al. 2021). Variations in mushrooms can lead to differing outcomes, and mushrooms can be a significant source of carbs, particularly insoluble fiber, which is frequently mentioned for its health benefits.

Vitamins

Mushrooms exhibit a range of vitamin content per kg of dry matter, including 2.6 to 9.0 mg riboflavin, 1.7 to 6.3 mg thiamine, 0.5 to 3 mg tocopherol, 63.8 to 83.7 mg niacin, 1.4 to 5.6 mg pyridoxine, 124 to 442 mcg folate -, 150–300 mg ascorbic acid (Kalač 2013). Mushrooms are commonly consumed sources of B vitamins (thiamine, riboflavin, pyridoxine etc.). The vitamin D content in mushrooms can be influenced by their exposure to sunlight or ultraviolet (UV)-B light during the growing process (Cardwell et al. 2018, Das et al. 2021, Mattila et al. 2002, Taofiq et al. 2017, Wang and Zhao 2023), however, it should be noted that all dietary sources are inadequate, including mushrooms, where the main source of vitamin D is sunlight. *Schizophyllum commune* is recognized for its elevated levels of 2711.30 mg/g of vitamin A and 85.08 mg/g vitamin E (ratio of fresh weight) (Chye, Wong, and Lee 2008). Mushrooms can provide a sufficient and balanced contribution in terms of vitamins, which are micronutrients necessary for the organism. It can be considered that adding mushrooms to the diet in the appropriate amount and frequency, especially in vegetarians, may contribute to the intake of some B group vitamins that they cannot get from animal-derived products.

Minerals

According to Kalač (2013), mushrooms frequently have high quantities of minerals such as potassium, phosphorus, and magnesium and contain moderate levels of calcium and low levels sodium (Kalač 2013). When compared to other vegetables, mushrooms are thought to be a safe option for those with hypertension because of their favorable nutrient profile. Low sodium is present in *Lentinula edodes*. *Marasmius oreades* is known to contain notable amounts of copper, zinc, iron, and folic acid. *Pleurotus ostreatus* is higher in essential minerals, with potassium (466.24 mg/100 g dry matter per stem and cap), phosphorus (497.35 mg/100 g per cap), and magnesium (340.59 mg/100 g per stem) being particularly abundant (in dry matter) (Oluwafemi, Seidu, and Fagbemi 2016). The major ingredient in edible mushrooms, potassium, is unevenly distributed throughout the fruiting bodies; the cap has the highest quantity, followed by the stem. The second significant mineral that can be naturally present in wild edible mushrooms is magnesium (Kalač 2013). *Boletus edulis*, which has a content of roughly 20 g per 100 g 1 dm, has the greatest concentration of dietary selenium across all mushrooms (Falandysz 2013).

When the macro and micro nutrient composition is evaluated, it differs according to the type of mushroom (Table 10.1) (Ahlawat, Manikandan, and Singh 2016; Tsai, Tsai, and Mau 2008). Although the nutritional content of mushrooms varies based on their species, they may appear to be a healthy food when measured for protein, necessary amino acids, essential fatty acids, carbs, insoluble fiber, and vitamin and mineral contents.

BIOACTIVE COMPONENTS

Along with the previously listed necessary nutrients, mushrooms also include a variety of substances that have bioactive properties in the human body.

Bioactive peptides: Lectins

Bioactive peptides, with their diverse array of primary and secondary amino acid composition, exhibit a broad spectrum of physiological functions that contribute to human health (Moussa et al. 2020). Peptides made from proteins can be found in mushrooms in the required balance (Zhou et al. 2020).

Mushrooms consist of wide vary of proteins and peptides, including, antimicrobial proteins, ribonucleases, laccases, lectins, fungal immunomodulatory proteins (FIPs), and ribosome-inactivating proteins (RIPs) that have particular biological capabilities (Kalač 2013). Mushrooms include aminobutyric acid, which enhances liver function and promotes the detoxification of harmful substances (Rathore, Prasad, and Sharma 2017). Several mushroom species, including *Flammulina velutipes* and other *Ganoderma* species, produce FIPs, a unique class of bioactive proteins with prospective medical applications (Bao et al. 2018). Furthermore, by deleting one or more adenosine residues from rRNA, RIPs can deactivate ribosomes (Jayachandran, Xiao, and Xu 2017). Confluenines, oxidized isoleucine derivatives, were made using the fruiting stems of the *Albatrellaceae* species *Albatrellus confluens* (Zhang et al. 2018).

TABLE 10.1

Nutrional Value of Some Edible Mushroom Types

Scientific Name	Energy and Nutrients						
(dry weight)	Energy (kcal/100 g)	Moisture (%)	Carb (%)	Protein (%)	Fat (%)	Ash (%)	Ref.
Agaricus bisporus	-	-	51.05	29.14	1.56	-	(Ahlawat, Manikandan, & Singh, 2016)
Agaricus bisporus	184.37	-	62.20	26.49	2.53	8.78	(Tsai, Tsai, & Mau, 2008)
Flammulina velutipes (wild)	272.9–449.5	88.05–90.68	42.6–70.85	23.2–27.5	1.73–7.0	7.25–7.4	(Tang et al., 2016)
Flammulina velutipes (Cultivated)	315.65–519.34	88.0–89.7	58.0–87.14	17.89–27.95	1.84–7.33	7.39–10.4	(Mau, Lin, & Chen, 2001)
Grifola frondosa	155.06	-	66.83	21.8	3.10	6.99	(Ahlawat et al., 2016)
Lentinula edodes	-	-	63.6	18.85	1.22	-	(Sharif et al., 2016)
Lentinula edodes	374.3	-	70.62	22.61	0.78	5.99	(Sharif et al., 2016)
Pleurotus ostreatus	378.51	-	69.86	21.14	2.02	7.02	(Ahlawat et al., 2016)
Volvariella vovacea	-		42.30	38.10	0.97	-	(Srikram & Supapvanich, 2016)
Volvariella vovacea	-	90.21	39.14	32.57	1.43	11,54	(Sharif et al., 2016)

In the fruiting body and vegetative mycelium of fungus, lectins are abundantly located. A wide range of lectins, including intracellular, extracellular, and surface mycelial forms, have been identified in *Basidiomycetes*. These lectins have been found in a number of fungal genera, including *Agrocybe, Coprinus, Lactarius, Lentinus, Ganoderma, Panaeolus, Gymnopilus, Pycnoporus, Punctularia, Termitomyces, Schizophyllum*, and *Volvariella*. This suggests that these fungi contain both intracellular and surface mycelial lectins (Nikitina, Loshchinina, and Vetchinkina 2017). In the *Basidiomycetes* class, lectins from genera like *Lactarius, Boletus, Russula*, and *Phallus* as well as families like *Hygrophoraceae* and *Tricholomataceae* have been identified in quantity (Konska 2006). Recent research has revealed that the Russulaceae family, notably the species *Lactarius* and *Russula*, contain a wide range of fungal lectins (Panchak 2019). *Pleurotus ostreatus, Agaricus bisporus*, and *Ganoderma applanatum*, known for their lectin content, exhibit the ability to bind to cell surfaces and stimulate antitumoral, antibacterial, and antiproliferative activities (Singh et al. 2014). Although it varies according to the type of mushroom, mushrooms come to the fore with their lectin content, which has various health benefits.

Bioactive polysaccharides: Glucans

Using different extraction methods, over 47 bioactive polysaccharide fractions from parts of mushrooms have been described over the past 30 years (Maity et al. 2021).

Mushrooms contain heteropolysaccharides, high in galactose, glucose, mannose, fucose, rhamnose, and xylose. The 80 species of the genus *Ganoderma* comprise an important class of polysaccharide-rich curative mushrooms. Biologically active polysaccharides of more than 100 different varieties have been discovered in the sporophores and micelles of *Ganoderma lucidum* (Sonawane et al. 2014). Glucans are polysaccharides with the general formula of glucose as a single monomer unit (C6H12O5). Glucan, a major constituent of the fungal cell, is present in mushrooms at notable concentrations, typically ranging from 0.21 to 0.53 grams/100 grams (dry weight).

According to research on various medicinal mushrooms, the primary active polysaccharide that can offer a variety of health advantages is glucans, which constitute a significant portion of the fungal cell wall. There are three types of mushroom glucans: α-glucans, β-glucans, and α, β-glucans (De Silva et al. 2012). Both heteropolysaccharides and β-glucans, which are recognized as the primary bioactive polysaccharides in mushrooms, demonstrate significant biological activities (Ren, Perera, and Hemar 2012). β-glucans, specifically polysaccharides such as β(1 → 3) and β(1 → 6), are abundant in several mushroom varieties, including *Agaricus bisporus, Lentinula edodes, Pleurotus ostreatus*, and *Ganoderma lucidum*.

Lentinula edodes and the genus *Pleurotus* both contain between 0.6% and 1.5% of soluble β-glucans (Sari et al. 2017). *Sparassis crispa* (the cauliflower mushroom) is another potent source of β-glucans; up to 40% soluble β-glucans were found (Kimura 2013). Mushrooms have a wide variety of glucans, which constitute about 10 to 50% of their dry content (Wasser 2005). Certain species, such as *Agrocybe aegerita, Collybia dryophila, Inonotus obliquus, Calocybe indica, Lyophyllum decastes, Hirneola auricula-judae, Trametes versicolor, Schizophyllum commune*, and *Stropharia aeruginosa*, are recognized for their significant β-glucan content, making them valuable for potential pharmaceutical applications (Klaus et al. 2016, Ukawa, Ito, and Hisamatsu 2000). It is therefore reasonable to suggest that mushrooms are a source of β-glucan, even though it varies depending on the type of mushroom.

Phenolic Compounds

The main properties of the varied class of compounds known as phenols are a number of hydroxyl groups that are linked to a benzene ring. They include salicylic acid, quinones, lignans, lignins, stilbenes, tocopherols, oxidized polyphenols, curcuminoids, phenolic acids including hydroxybenzoic acids, and flavonoids. The phenolic content of *Morchella esculenta, Hericium erinaceus, Lentinula edodes*, and *Ganoderma lucidum* were analyzed (Mitra, Mandal, and Acharya 2016). The genus *Agaricus* is rich in phenolic compounds (Gąsecka et al. 2018).

Agaricus bitorquis, Agaricus silvaticus, Agaricus bisporus, Agaricus campestris Agaricus blazei, and *Agaricus arvensis* were among the *Agaricus* species whose phenolic derivative concentration was observed. The significance of *Agaricus species'* antioxidant potential as a provider of phenolic compounds cannot be overstated (Froufe, Abreu, and Ferreira 2009). Extracts from the Bosnian market plants *Boletus edulis, Lentinus edodes*, and *Pleurotus ostreatus* were tested for phenolic constituents and antioxidant activity (Butkhup, Samappito, and Jorjong 2018). The species *Boletus edulis* had the largest concentration of phenolic chemicals. In an independent study, eight edible mushroom species, namely *Agaricus bisporus, Craterellus cornucopioides, Boletus edulis, Calocybe gambosa, Hygrophorus marzuolus, Cantharellus cibarius Pleurotus ostreatus*, and *Lactarius deliciosus* were analyzed, showing that the dry substance's total phenolic and flavonoid concentrations range from 3.0 to 10.9 mg. The major flavonoids identified were myricetin and catechin. *Boletus edulis* and *Agaricus bisporus* showed the greatest quantities of phenolic compounds and flavonoids, whereas *Pleurotus ostreatus* and *Calocybe gambosa* had the smallest concentrations, respectively (Palacios et al. 2011). The ability of oxalic acid to eliminate bacteria extracted from *Lentula edodes* was observed against various other bacteria and *S. aureus* (Bender et al. 2003). *Laetiporus sulphureus* shows antioxidant, cancer modulation, anti-inflammatory, and antimicrobial effects with its flavonoids and phenolic compounds, which are lycopene and ascorbic acid content (Zięba et al. 2020).

According to Nowacka et al. (2015), hyphodontia paradoxa is abundant in a number of phenolic acids, such as salicylic acid, 4-OH-benzoic acid, proto-catechuic acid, p-coumaric acid, caffeic acid, sinapic acid, syringic acid, and ferulic acid (Nowacka et al. 2015). In a similar vein, research by Sabino Ferrari et al. (2021) and Bach et al. (2019) found phenolic acid compounds in

Agaricus brasiliensis and *Agaricus subrufescens*, including ferulic acid, p-hydroxybenzoic acid, trans-cinnamic acid, gallic acid, fumaric benzoic acid, p-coumaric acid, catechol, and gentisic (Bach et al. 2019, Sabino Ferrari et al. 2021).

According to Liu, Xiao, Wang, Chen, and Hu (2017), *Agrocybe aegerita* contains chlorogenic acid, gallic acid, sinapic acid, and ferulic acid (Liu et al. 2017). *Calocybe* (Alam et al. 2019) and *Amanita crocea* (Alkan et al. 2020) catechol with fumaric acids and flavonoids including naringin, naringenin, homogentisic acid, hesperetin, and formononetin, as well as p-coumaric production have been described.

Terpenoids

Some of the most significant and bioactive substances is terpene. This is a result of their adaptability, which makes them powerful against microbial illnesses. While it was once believed that terpenoids were primarily produced by plants, a potential source has gained attention in recent decades; these are mushrooms or *Basidiomycetes* (Moussa et al. 2020).

Terpenes are categorized as monoterpenes, sesquiterpenes, diterpenes, and triterpenes based on the isoprene unit they contain. Terpenoids are a subclass of modified terpenes that have undergone numerous functional group changes as well as the repositioning or removal of an oxidized methyl group (Fukushima-Sakuno 2020). The triterpenoids are classified as the lanostan type, whereas the diterpenoids derived from the fungus are classified as the cyatan type (Kour et al. 2022). *Ganoderma lucidum, Inonotus obliquus*, and *Antrodia camphorata* are among the mushrooms known to possess significant levels of it. In the case of *Cyathus africans*, three diterpenes, namely cyathatriol, neosarcodonin, and 11O-acetylcyathetriol, have been isolated. Additionally, a study by Han et al. (2013) has identified five novel cyatan diterpenes in *Cyathus africans*, referred to as cyathi (Han et al. 2013). The edible umbrella mushroom *Macrolepiota procera* was found to have the lanostan triterpene lepiotaprocerin I (Chen et al. 2018). According to their type, mushrooms have various terpene molecules and exhibit various effects; it is crucial to mention that their terpene content is remarkable.

POTENTIAL HEALTH EFFECTS OF MUSHROOMS

Edible mushrooms contain a diverse range of bioactive compounds that offer a plethora of advantageous health effects alongside their nutritional value. This segment will delve into the potential positive impacts on well-being and elucidate the mechanisms by which mushroom constituents act (Figure 10.2).

ANTIVIRAL AND ANTIMICROBIAL EFFECT

The antiviral effects of lectin have been effective against a wide range of viruses, including the Japanese encephalitis virus, severe acute respiratory syndrome (SARS) virus, hepatitis C, influenza A/B, hepatitis C, herpes simplex types 1 and 2, and HIV. The range of lectin's effective concentrations is 1.6 nanomolar (nM) to 1.3 millimolar (M) (Mazalovska and Kouokam 2018). In order to prevent viral infection, mushrooms' bioactive metabolites focus on reaching viral entrance and genome replication while also influencing the immune system.

Sarcodon imbricatus, Tricholoma portentosum, and *Lactarius deliciosus* are among the edible mushrooms with antibacterial properties. Gram-positive (like *S. aureus, B. subtilis, S. epidermidis and M. luteus*) and Gram-negative (like *P. aeruginosa, E. coli, P. mirabailis, and K. pneumonia*) bacteria can both be killed by phenolic extracts from *Hyphodontia paradoxa* (Nowacka et al. 2015).

The erinacine cyathane, the fermentation broth culture of some chemicals, was separated from *Dentipellis fragilis* and demonstrated activity against *Propionibacterium acnes, Bacillus subtilis, Staphylococcus epidermidis, Bacillus atrophaeus*, and *Bacillus cereus* (Ha et al. 2021). The identified substances 3-acetyl-4-methoxybenzoic acid and (4-chloro-3, 5-dimethoxyphenyl) methanol,

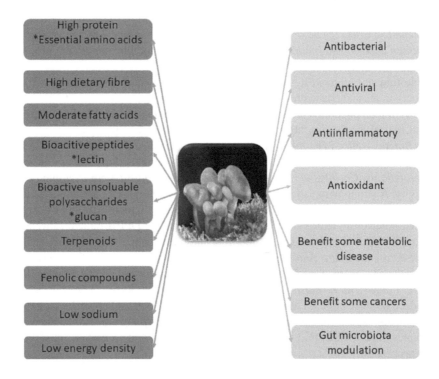

FIGURE 10.2 Health benefits of edible mushroom ingredients.

along with the unidentified diterpene discovered as the erinacine sugar diastereomer in *Hericium erinaceous*, exhibited moderate efficacy against *H. pylori*. Furthermore, 1-(5-chloro-2-hydroxyphen yl)-3-methylbutan-1-one displayed significant activity in the same investigation (Zhang et al. 2015). Two benzoate derivatives were extracted from mycelial cultures of *Stereum hirsutum*, both of which exhibited antibacterial properties against *S. aureus* (Ma et al. 2014). In the case of *Wolfiporia cocos*, the diterpenes 7p-hydroxyabieta-8,11,13-triene-18-oic acid, 7-oxy-8,11,13-abietatriene-18-oic acid, and the steroid demethylincisterol were isolated and found to effectively combat *S. aureus* (Baosong et al. 2020). From *Lentinus fasciatus*, connatusin A and B (hirsutan sesquiterpene) was extracted and exhibited antibacterial activity (Helaly et al. 2016).

Besides that, *Agaricus brasiliensis*'s β-glucan has been discovered to be an antiviral agent that possesses the ability to inhibit the proliferation of bovine herpes virus (Zhao et al. 2008). Numerous studies have shown that lectin, phenolic compounds, terpenoids, and B-glucans derived from different mushroom species have antiviral and antibacterial activities.

ANTIOXIDANT EFFECT

Mushrooms act as a natural wellspring of antioxidants, containing phenolic acids like P-hydroxybenzoic acid, tyrosine, catechol, and p-coumaric acid (Stojanova et al. 2021). Reactive oxygen species (ROS) are becoming more prevalent, which damages cells and tissues and causes oxidative stress. This stress leads to metabolic imbalances, metabolic disorders, malignancies, and neurodegeneration (Askarova et al. 2020). In addition to the organisms' natural defense systems against free radicals, eating is a significant additional source of antioxidants that may support oxidative equilibrium (Mwangi et al. 2022). According to a study conducted on the *Pleurotus ostreatus* species, saponin was the strongest phytochemical, followed by alkaloids, flavonoids, tannin, and phenolic compounds, all of which were said to have antioxidant characteristics (Rahimah et al. 2019).

Research has indicated that the polysaccharides sourced from mushrooms like *Agaricus brasiliensis Agraicus bisporus, Tricholoma mongolicum Imai, Trichoderma harzianum, Volvariella volvacea, Lentinula edodes*, and *Pleurotus ostreatus* possess remarkable antioxidant properties (Maity et al. 2014, Zhao et al. 2016).

The edible fungus *Entoloma lividoalbum's* branching β-glucan has been shown to exhibit antioxidant effects in an alkaline environment (Kosanić, Ranković, and Dašić 2012). Kosanić et al. used measures of total phenolic content, whole flavonoids content, and 2,2diphenyl-1picrylhydrazyl (DPPH) free radical scavenging ability to examine the antioxidant potential of extracts from the mushrooms *Boletus edulis, Leccinum carpini*, and *Boletus aestivalis*. According to the study's findings, Boletus edulis' acetonic extract has more potent antioxidant activity.

Suillus granulatus, Coriolus versicolor, and *Fuscoporia torulosa* were examined for their antioxidant activity. A strong association between their antioxidant activity and the total phenolic and flavonoid contents was discovered (Stojanova et al. 2021). The aqueous extract of *Fuscoporia torulosa* exhibited noteworthy antioxidant activity, as evaluated by DPPH radical scavenging activity (Shafiqah 2019). Similar outcomes were reported in studies where ethanolic extracts of various edible mushrooms, including *Amanita caesarea, Tricholoma giganteum Coprinus comatus*, and *Volvariella volvacea*, were used for antioxidant investigations (Alispahić et al. 2015, Rai and Acharya 2012). *Agaricus bisporus, Flammulina velutipes*, and *Auricularia auricula-judae* have been recognized because of their capacity to scavenge free radicals and powerful antioxidants, making them promising natural antioxidants for human nutrition (Zhao et al. 2008). Research suggests that compounds such as phenolic compounds and glucan present in certain mushrooms possess antioxidant properties.

ANTIINFLAMMATORY EFFECT

By inhibiting the NF-kB pathway, which is mediated by ROS, in murine macrophage cells, hispidin, a phenolic molecule isolated from *Phellinus species*, has shown particular anti-inflammatory capabilities (Shao et al. 2015). According to a study, *P. ostreatus* has the ability to reduce inflammation through a variety of mechanisms, including the antihistamine action, the suppression of nitric oxide production, inhibition of cell migration to the site of inflammation, and in vitro membrane stabilizing activity (Jayasuriya et al. 2020). In several ways, including modulating cytokines, preventing oxidative stress, controlling the gut microbiota, and affecting the human immune system, mushroom polysaccharides can also have immunomodulatory effects (Yin et al. 2021). Numerous varieties of mushrooms have been discovered to successfully limit the generation of inflammatory mediators such nitric oxide, prostaglandins, and cytokines due to the presence of phenolic compounds and polysaccharides, decreasing macrophage activity and lowering cellular inflammation. Additionally, possible implications for the treatment and prevention of cancer have been linked to their anti-inflammatory characteristics.

POSSIBLE EFFECTS ON CANCER

The bioactive components found in edible mushrooms have been shown in the literature to have some possible anti-carcinogenic effects, particularly in cases of liver, uterus, breast, and pancreatic malignancies. Mushrooms activate lymphocytes, which are considered to be immune cells that fight cancer, to display their therapeutic benefit potential. Polysaccharides, in particular *Lentinan schizophyllan*, are the bioactive metabolites from *Pleurotus ostreatus* that have drawn the most attention for the potential for treating cancer (Mishra et al. 2021). In contrast to having no effect on healthy witness cells, ganoderic acid, a triterpene produced by *Ganoderma lucidum*, can cause cancer cells to go through cell death (Kolniak-Ostek et al. 2022).

Mushroom phenolic compounds have some anti-inflammatory and cancer fighting activities through promoting apoptosis, blocking lipopolysaccharide-induced nitric oxide, and deactivating

ROS in the NF-Kb pathway (Venturella et al. 2021). They also help with improving macrophage performance and enhancing immunity (Yin et al. 2021). According to reports, polysaccharides' ability to limit tumor growth is mediated by a thymus-dependent immunological process, not by constantly targeting cancer cells (Smiderle et al. 2012). Thus, *Agaricus bisporus* lectins and lectin-like proteins show promising results for both care and protection of cancer according to the literature (Tirta Ismaya, Tjandrawinata, and Rachmawati 2020). When monoterpenes and sesquiterpenoids from *Pleurotus cornicopiae* were extracted for cytotoxicity against the cancer line, encouraging results were obtained (Wang et al. 2013). An effective inhibitor of lanostane triterpenoid (ganoderol B (3,24E)-lanosta-7,9(11),24-triene-3,26-diol) from *Ganoderma lucidum* has been reported (Xu et al. 2010).

Flammulina velutipes produced several types of bioactive sesquiterpenoids, some of which have been shown to have anticancer potential (Wang et al. 2012). Inotodiol and tramethenolic acid, two triterpenoids isolated from *Inonotus obliquus*, were examined in a different study in order to assess their antiproliferative effects (Kim et al. 2020).

Mushroom polysaccharides are thought to mediate their cytotoxic effects on cancer cells through the control of the host defense system. In addition to increasing the relative abundance of *Bacteroides* and *Prevotellaceae* it has been observed that *Ganoderma* spp. polysaccharides also prevent tumor growth. This process could be connected to the activation of the TLR-4-associated MAPK/NF-B signaling pathway (Li et al. 2018). Additionally, it is suggested that polysaccharides can alter the gut microbiota, as will be covered in more detail under the topics that follow, to explain their impacts on cancer cells.

Possible Effects on Metabolic Diseases

Several species of mushrooms, including *Agaricus subrufescens, Agaricus bisporus, Inonotus obliquus, Coprinus comatus, Ganoderma lucidum, Cordyceps sinensis, Phellinus linteus, and Pleurotu*s spp., have been shown to be advantageous for controlling blood sugar and reducing diabetic problems (De Silva et al. 2012). Mushroom homopolysaccharides modulate insulin release in the hormone signaling system, which effects insulin metabolism (Kumar et al. 2021). Through regulating appetite, digestion, and absorption of nutrients, mushroom components have anti-obesogenic and antidiabetic benefits. Patients with type 2 diabetes (T2DM) received 50 g/kg of *Pleurotus ostreatus* orally every day for two weeks. The results demonstrated their efficacy as functional meals for diabetes, showing a significant decrease in plasma glucose levels during fasting and postprandial plasma glucose (Jayasuriya et al. 2015).

It is common practice to swap out high-energy-density foods with low-energy-density foods in overweight and obese individuals in order to improve effective weight loss and their body composition. In clinical research, replacing red meat with mushrooms resulted in a decrease in caloric and fat intake, which resulted in increased weight loss and lower levels of a pro-inflammatory marker linked to obesity (Poddar et al. 2013). Scientific research has shown that increasing mushroom consumption can enhance markers of body mass index (BMI), triglycerides, serum alanine aminotransferase (ALT), high-density lipoprotein (HDL), and fasting blood glucose levels (Zhang et al. 2020).

Through affecting the gut microbiota, fungal polysaccharides have the potential to affect metabolic diseases. The metabolic syndrome, particularly prevalent in cases of obesity and diabetes, is often characterized by moderate inflammation and elevated levels of lipopolysaccharides (LPS) in the body (Chang et al. 2015). LPS, an ingredient of Gram-negative bacteria's cell walls, can enter the circulation when glucagon-like peptide 2 affects the tight junctions of the intestinal epithelium. This can lead to an increased release of chemokines and proinflammatory cytokines through association with the LPS innate immune receptor complex (Di Lorenzo et al. 2019). According to a study by Chang et al. (2015), intestinal dysbiosis brought on by a high-fat diet in mice might be reversed by high molecular weight polysaccharides generated from *Ganoderma lucidum mycelia* (Chang et al. 2015). This was accomplished by lowering the numbers of *Proteobacteria* and *Firmicutes* to

Bacteroidetes balance. Additionally, these polysaccharides helped maintain the intestinal barrier's integrity and mitigated metabolic decline. In a different investigation, S. Xu et al. (2017) discovered that mice eating a diet with 200 mg/kg of *Ganoderma lucidum* polysaccharides had a significantly higher abundance of fecal *Enterococcus* (Xu et al. 2017). Improved insulin sensitivity and anti-inflammatory benefits were linked to this rise. Further supporting evidence from studies conducted by M. Chen et al. (2020) and Lv, Guo, Li, Yu, and Liu (2019) demonstrated that *Ganoderma lucidum* polysaccharides could restore imbalances in the gut microbiota of mice with type 2 diabetes (Chen et al. 2020, Lv et al. 2019). Short-chain fatty acids (SCFAs), which are thought to be essential for reestablishing energy balance, are produced by these polysaccharides, also lowered the quantity of dangerous bacteria. In vitro studies have also shown that extracts of *Ganoderma lucidum* can modify the composition of the gut microbiota in a manner that may help prevent obesity (Delzenne and Bindels 2015). In summary, mushroom polysaccharides, specifically those derived from *Ganoderma lucidum*, have the capacity to affect the gut microbial population, maintain intestinal barrier integrity, reduce inflammation, and alleviate metabolic disturbances associated with conditions such as obesity and diabetes. However, it is important to note that further research is required to fully understand the underlying mechanisms involved and confirm these effects in human subjects.

It is suggested that consuming mushrooms has a relation to metabolic illnesses indirectly through reduced dietary energy intake, specifically with the presence of polysaccharide, modified blood pressure and body weight, or affected gut microbiota.

GUT MICROBIOTA MODULATION

Combined with glucose, energy, and lipid metabolism, the gut microbiota consists of different physiological processes such as immunology, cancer, and inflammation. The large intestine produces SCFAs, such as butyric, acetic, and propionic acid, as a result of the fermentation of food fibers. SCFAs are crucial for preserving human health (Danneskiold-Samsøe et al. 2019). The chemical and physical properties of dietary fiber are essential for the formation of SCFAs, as well as for the development of a range of gut bacteria (Wang et al. 2019). Prebiotics, which are nondigestible and fermentable food components that improve health by altering the composition and/or activity of the gut microbiota, include dietary fiber, nondigestible starches, and oligosaccharides (Cheung 2013, Yin et al. 2020). Important prebiotic sources include mushroom compounds such chitin, α- and β-glucans, mannans, xylans, and galactans (Vamanu and Gatea 2020). Polysaccharides have prebiotic potential because they raise the body's level of SCFAs, which enhances the microbiota in the gut (Moumita and Das 2022). The prebiotics in question significantly increase the amount of good bacteria in *Firmicutes*, such *Lactobacillus* (Inyod et al. 2022). Because of the *Ganoderma lucidum* components, it serves as a crucial prebiotic for the development of bacterial flora. According to research, *Ganoderma lucidum* polysaccharides change the gut microbiota mostly by flipping the *Bacteroides/Firmicutes* ratio, which increases the production of SCFA–producing and anti-inflammatory bacteria while reducing the amount of disease-causing bacteria ((Khan et al. 2018). Important mushrooms with high prebiotic effect include *Lentinula edodes*, *Trametes versicolor*, and *Ganoderma lucidum*. Prebiotics derived from mushrooms, according to a study (Chou, Sheih, and Fang 2013), pass undigested through the human small intestine and stomach before entering the colon, where they encourage the formation of healthy bacteria (*Lactobaccilus acidophilus* and *Bifidobacterium longum*). According to Synytsya et al. (2009) and Khan et al. (2018), *Bifidobacterium* also contains glucans from *Pleurotus ostreatus and Pleurotus eryngii* as well as *Ganoderma lucidum* (Khan et al. 2018, Synytsya et al. 2009). By modifying the host's energy homeostasis and plasma glucose levels, prebiotic fungal polysaccharides have also been demonstrated to have effects against diabetes and obesity (Friedman 2016). The mushrooms *Hericium erinaceus, Grifola frondose, Lentinula edodes*, and *Ganoderma lucidum* have been found to contain bioactive compounds that alter the gut flora and improve health. A rise in *Allobaculum, Bifidobacterium*, and *Bacteroides* in the gut is associated with improved liver function and decreased obesity, according to studies on *Grifola frondosa* polysaccharides (Li et al. 2021). *Flammulina velutipes* polysaccharides alter the distribution of *Bifidobacteraceae* and *Bacteroidaceae* in an in vitro fermentation test, while reducing the amounts of *Enterococcaceae* and *Lachnospiraceae genera* (Su et al. 2019). According to another study, *Lentinula edodes* polysaccharides can reduce the *Firmicutes/Bacteroidetes* ratio as well as the diversity and homogeneity of the gut microbiota throughout the entire gut, particularly in the cecum and colon (Xu and Zhang 2015). Polysaccharides derived from *Flammulina velutypes* can modify the makeup of the gut microbiota by decreasing the *Firmicutes/Bacteroidetes* balance and the corresponding abundances of *Akkermansia* and *Lactobacillus* in vivo (Zhao et al. 2019). *Auricularia auricular* polysaccharides supplementation appears to increase the variety of the fecal microbiota and decrease the *Firmicutes/Bacteroidetes* balance (Ma et al. 2021, Zhao et al. 2019).

Intestinal microbiota composition can be changed by polysaccharides in two different ways. The first step in the digestion of polysaccharides is the formation of SCFAs, which lower intestinal pH and change the balance of microbial metabolites as well as the make-up of the intestinal microbiota. The generation of SCFAs has the ability to inhibit pathogenic bacteria and improve the absorption of particular substances and mineral ingredients, such as magnesium, calcium, and phosphorus (Wallace et al. 2011). Second, certain types of gut microbiota exploit polysaccharide degradation to get carbon sources and energy, making them the dominant microbiota in the gut. The fact that some gut bacteria have a cross-feeding effect that encourages the growth of other bacteria is interesting to notice. It has been shown that SCFAs, particularly butyrate, provide intestinal epithelial cells with energy and increase the production of mucin, which enhances intestinal barrier function (Jung et al. 2015).

While most polyphenols are converted by bacteria in the colon after consumption, some are taken in in the small intestine during digestion. They actually operate as a prebiotic, promoting the development of particular bacterial strains, and change the microbiota in a variety of ways. In the microbiome of hypertension patients, curcumin from the *Curcuma longa* plant has a detrimental effect on the *Bacteroides-Prevotella Porphyromona*s group and *Enterobacteriaceae* group (Kour et al. 2022, Vamanu et al. 2019). The human intestinal mucosa, bacterial metabolite synthesis, and altered metabolite profiles are all significantly impacted by mushroom proteins due to altered amino acid breakdown in bacterial metabolism (Beaumont et al. 2017).

The bioactive components that mushrooms contain are summarized in Table 10.2, and new health advantages are constantly being discovered as a result. It would be hard to reap the health benefits of mushrooms without a sufficient and well-balanced diet. Because these effects may differ depending on the species of mushroom, the consumption dose, and many other factors based on the disease-health condition of the individual, it is vital to pay attention to the appropriate amount and frequency of intake.

TABLE 10.2
Edible Mushrooms, Bioactive Components and Health Effects

Mushrooms	Bioactive Compounds	Health Effects	Reference
Lentinula edodes	Polysaccharides	Gut microbiota modulation	(Chou, Sheih, & Fang,
Trametes versicolor	prebiotic	Lactobacillus	2013)
Ganoderma lucidum		Bifidobacterium	
Pleurotus ostreatus	Glucan	Gut microbiota modulation	(Synytsya et al., 2009)
Pleurotus eryngii		Lactobacillus	(Khan et al., 2018)
Ganoderma lucidum		Bifidobacterium	
Grifola frondosa	Polysaccharides	Gut microbiota modulation	(Li et al., 2021)
		Allobaculum	
		Bacteroides	
		Bifidobacterium	
		Antiobesity	
		Liver protective	

(*Continued*)

TABLE 10.2   (Continued)
Edible Mushrooms, Bioactive Components and Health Effects

Mushrooms	Bioactive Compounds	Health Effects	Reference
Ganoderma lucidum		Gut microbiota modulation *Bacteroides/Firmicutes* *SCFA* Anti-inflammatory	(Khan et al., 2018)
Flammulina velutiplerinden	Polysaccharides	Gut microbiota modulation *Firmicutes/Bacteroidetes azaleaAkkermansia Lactobacillus azalea*	(Zhao et al., 2019)
Auricularia auricular	Polysaccharides	Gut microbiota modulation *Firmicutes/Bacteroidetes Fecalmicrobiota modulation*	(Ma et al., 2021)
Curcuma longa	Curcumin	Gut microbiota modulation	(Vamanu, Gatea, Sârbu, & Pelinescu, 2019)
Pleurotus ostreatus	Polysaccharides	Benefit some cancers	(Mishra, Tomar, Yadav, & Singh, 2021)
Ganoderma lucidum	Triterpenoid ganoderic acid	Antitumor Benefit some cancers	(Kolniak-Ostek, Oszmiański, Szyjka, Moreira, & Barg, 2022)
Cordyceps sinensis	Polysaccharides	Immunomodulator Antibacterial Antioxidant Antitumor Anti-inflammatory Antilipidemic	(Gong et al., 2020; Rathore, Prasad, & Sharma, 2017)
Agaricus bisporus	Lectin	Antibacterial Antidiabetic Anti-inflammatory	(Tirta Ismaya, Tjandrawinata, & Rachmawati, 2020)
Pleurotus cornicopiae	Monoterpenler Seskiterpenoidler	Benefit some cancers	(Wang et al., 2013)
Ganoderma lucidum, Inonotus obliquus Antrodia camphorata	Triterpenoidler	Anti-inflammatory	(Ríos 2010)
Inonotus obliquus	Triterpenoidler Inotodiol trametenolik acid	Antiproliferative	(Kim, Yang, Hwang, Cho, & Hwang, 2020)
Phellinus spp.	Hispidinin	Anti-inflammatory	(Shao, Jeong, Kim, & Lee, 2015)
Cyathus africans	Terpenoid cyathatriol, neosarcodonin 11-O-acetylcyathetriol	Anti-inflammatory	(Han et al., 2013)
Pleurotus Ostreatus	Saponin Alkaloids, flavonoids tannin phenolic compounds	Antioxidant	(Rahimah, Djunaedi, Soeroto, & Bisri, 2019)
Lentula edodes	Oxalic acid	Antimicrobial	(Bender et al., 2003)
Laetiporus Sulphuerus	Flavaniides, phenolic compounds, lycopene, Ascorbic acid	Antioxidant Benefit some cancers, Anti-inflammatory Antimicrobial	(Zięba et al., 2018)

SAFETY ISSUES OF MUSHROOM CONSUMPTION

The main feature controlling a mushroom's edibility is its safety. Since this food has been investigated and some of its possible health effects have been shown, consumer demand for both cultivated and wild mushrooms has increased. The ingredients in a mushroom's composition are significantly affected by its surroundings. It is obvious that cultivated mushrooms grow in a more known and reliable environment. When it comes to wild edible mushrooms, though, the situation is distinct. Some such edible wild mushroom species have been shown to collect toxic substances more frequently, according to the literature. Human-related actions, including the application of pesticides, mining operations, and the burning of fossil fuels, generate substantial quantities of contaminants. These pollutants, upon entering the atmosphere, water bodies, or soil, disrupt the ecological balance in areas characterized by radiogeological irregularities (Barea-Sepúlveda et al. 2022). Previous literature have shown that mushrooms growing on polluted soils – soils containing heavy metals, organic hazardous compounds, radioactive contaminants, or radionuclides – threaten the health of humans (Turkey (Keskin et al. 2021), China (Ernst et al. 2022), Poland (Ronda et al. 2022), Iran (Dowlati et al. 2021), Spain (Melgar & García, 2021)). Metallic elements and metalloids found in the environment, such as cadmium (Cd), mercury (Hg) (Rahgo, Mojerlou, and Jahanbin 2019), chromium, lead (Jiang et al. 2007), and arsenic, are thought to be hazardous to human health, even at low concentrations, as they lack any beneficial biological functions within the human system (Barea-Sepúlveda et al. 2022). Cooking and digestion tend to diminish the affects of heavy metals, according to in vitro research. In order to evaluate the potential toxicological risk associated with edible mushrooms, it is essential to validate the impact of these factors in vivo, particularly considering that certain fungi may exceed the permissible limits for Hg and Cd levels. Determining whether the amount of these elements entering the circulatory system after processing and ingestion would reach a level that creates a toxicological risk is the most important factor to consider (Chiocchetti et al. 2020).

Comparatively, mushrooms possess a higher protein content compared to other plant sources, considering the investment of time and resources required for their growth. Nevertheless, due to their increased respiration rate and lack of a protective cuticles layer that prevents water loss, mushrooms are vulnerable to rapid disintegration. The possibility of microbial infection is also increased by the high water content of mushrooms. Nonetheless, the categorization of edible mushrooms as such a sustainable and nutritious food heavily relies on specific factors.

It should also be noted that consuming edible wild mushrooms rather than cultivated ones can have adverse affects on anyone's health. As a result, mushrooms should be consumed in moderation and at appropriate frequency. At this point, it might prove more reliable to prefer cultivated species rather than wild species.

CONCLUSION AND FUTURE DIRECTIONS

The nutritional value of edible mushrooms is well-documented, and they provide a variety of bioactive elements that may help to improve health. These consist of amino acids, dietary fiber, polyunsaturated fatty acids, proteins, polysaccharides, vitamins, and minerals. These nutrients and bioactive ingredients have been related to a number of health advantages, including antioxidant properties, antimicrobial and anti-inflammatory effects, and potential advantages for cancer prevention and management, as well as the ability to influence gut microbiota.

Mushrooms also stand out in the context of sustainable nutrition due to their ability to be produced more sustainably than meat products. In summary, their nutritional and sustainability characteristics have improved in recent years. Nonetheless, a careful differentiation between edible species should be formed, taking into account the classification in terms of human consumption as a result of thorough screening. Due to toxic and radioactive compounds they may be exposed to depending on where they are grown, some edible wild mushroom species rather than cultivated mushrooms present a health risk. Making a clear identification between types is crucial to explore further into both the positive and negative properties and components of mushroom species as well as to accurately display health effects. Furthermore, because of their high protein content and the growing consumer demand for healthier and more sustainable food options, the usage of mushrooms as a substitute for animal products is anticipated to grow in the future.

The main objective of the study seems to revolve around enhancing techniques for extracting secondary metabolites from edible mushrooms, exploring new edible mushroom types, and extracting their vital bioactive components, with the aim of utilizing them in the prevention and protection of specific diseases due to their diverse health-promoting effects.

REFERENCES

Ahlawat, OP, K Manikandan, and Manjit Singh. 2016. "Proximate composition of different mushroom varieties and effect of UV light exposure on vitamin D content in Agaricus bisporus and Volvariella volvacea." *Mushroom Res* 25 (1):1–8.

Alam, Nuhu, Ruhul Amin, Asaduzzaman Khan, Ismot Ara, Mi Ja Shim, Min Woong Lee, and Tae Soo Lee. 2008. "Nutritional analysis of cultivated mushrooms in Bangladesh–Pleurotus ostreatus, Pleurotus sajor-caju, Pleurotus florida and Calocybe indica." *Mycobiology* 36 (4):228–232.

Alam, Nuhu, Md Maniruzzaman Sikder, Md Abdul Karim, and Sheikh Md Ruhul Amin. 2019. "Antioxidant and antityrosinase activities of milky white mushroom." *Bangladesh Journal of Botany* 48 (4):1065–1073.

Alispahić, A, A Šapčanin, M Salihović, E Ramić, A Dedić, and M Pazalja. 2015. "Phenolic content and antioxidant activity of mushroom extracts from Bosnian market." *Bulletin of the Chemists and Technologist of Bosnia and Herzegovina* 44:5–8.

Alkan, Sinan, Ahmet Uysal, Giyasettin Kasik, Sanja Vlaisavljevic, Sanja Berežni, and Gokhan Zengin. 2020. "Chemical characterization, antioxidant, enzyme inhibition and antimutagenic properties of eight mushroom species: A comparative study." *Journal of Fungi* 6 (3):166.

Andrew, Egbe Enow, Tonjock Rosemary Kinge, Ebai Maureen Tabi, Nji Thiobal, and Afui Mathias Mih. 2013. "Diversity and distribution of macrofungi (mushrooms) in the Mount Cameroon Region." *Journal of Ecology and The Natural Environment* 5 (10):318–334.

Argyropoulos, Dimitrios, Charoula Psallida, Paraskevi Sitareniou, Emmanouil Flemetakis, and Panagiota Diamantopoulou. 2022. "Biochemical evaluation of Agaricus and Pleurotus strains in batch cultures for production optimization of valuable metabolites." *Microorganisms* 10 (5):964.

Askarova, Sholpan, Bauyrzhan Umbayev, Abdul-Razak Masoud, Aiym Kaiyrlykyzy, Yuliya Safarova, Andrey Tsoy, Farkhad Olzhayev, and Almagul Kushugulova. 2020. "The links between the gut microbiome, aging, modern lifestyle and Alzheimer's disease." *Frontiers in Cellular and Infection Microbiology* 10:104.

Azieana, J, MN Zainon, A Noriham, and MN Rohana. 2015. "Morphological identification and toxicity evaluation of two selected wild mushrooms." *Advances in Environmental Biology* 9 (22 S3):6–11.

Bach, Fabiane, Acácio Antonio Ferreira Zielinski, Cristiane Vieira Helm, Giselle Maria Maciel, Alessandra Cristina Pedro, Ana Paula Stafussa, Suelen Ávila, and Charles Windson Isidoro Haminiuk. 2019. "Bio compounds of edible mushrooms: In vitro antioxidant and antimicrobial activities." *Lwt* 107:214–220.

Banerjee, Dipak Kumar, Arun K Das, Rituparna Banerjee, Mirian Pateiro, Pramod Kumar Nanda, Yogesh P Gadekar, Subhasish Biswas, David Julian McClements, and Jose M Lorenzo. 2020. "Application of enoki mushroom (Flammulina Velutipes) stem wastes as functional ingredients in goat meat nuggets." *Foods* 9 (4):432.

Bao, Da-Peng, Rui Bai, Ying-Nv Gao, Yingying Wu, and Ying Wang. 2018. "Computational insights into the molecular mechanism of the high immunomodulatory activity of LZ-8 protein isolated from the Lingzhi or Reishi medicinal mushroom Ganoderma lucidum (Agaricomycetes)." *International Journal of Medicinal Mushrooms* 20 (6):537–548.

Baosong, Chen, Wang Sixian, Liu Gaoqiang, Bao Li, Huang Ying, Zhao Ruilin, and Liu Hongwei. 2020. "Anti-inflammatory diterpenes and steroids from peels of the cultivated edible mushroom Wolfiporia cocos." *Phytochemistry Letters* 36:11–16.

Barea-Sepúlveda, Marta, Estrella Espada-Bellido, Marta Ferreiro-González, Hassan Bouziane, José Gerardo López-Castillo, Miguel Palma, and Gerardo F. Barbero. 2022. "Exposure to essential and toxic elements via consumption of Agaricaceae, Amanitaceae, Boletaceae, and Russulaceae mushrooms from Southern Spain and Northern Morocco." *Journal of Fungi* 8 (5):545.

Beaumont, Martin, Kevin Joseph Portune, Nils Steuer, Annaïg Lan, Victor Cerrudo, Marc Audebert, Florent Dumont, Giulia Mancano, Nadezda Khodorova, and Mireille Andriamihaja. 2017. "Quantity and source of dietary protein influence metabolite production by gut microbiota and rectal mucosa gene expression: A randomized, parallel, double-blind trial in overweight humans." *The American Journal of Clinical Nutrition* 106 (4):1005–1019.

Bender, Stefan, Cristina N Dumitrache-Anghel, Juergen Backhaus, Gregor Christie, Reg F Cross, Greg T Lonergan, and Warren L Baker. 2003. "A case for caution in assessing the antibiotic activity of extracts of culinary-medicinal Shiitake mushroom [Lentinus edodes (Berk.) Singer](Agaricomycetideae)." *International Journal of Medicinal Mushrooms* 5 (1):6.

Boin, Elisa, and João Nunes. 2018. "Mushroom consumption behavior and influencing factors in a sample of the Portuguese population." *Journal of International Food & Agribusiness Marketing* 30 (1):35–48.

Butkhup, Luchai, Wannee Samappito, and Sujitar Jorjong. 2018. "Evaluation of bioactivities and phenolic contents of wild edible mushrooms from northeastern Thailand." *Food Science and Biotechnology* 27:193–202.

Cardwell, Glenn, Janet F Bornman, Anthony P James, and Lucinda J Black. 2018. "A review of mushrooms as a potential source of dietary vitamin D." *Nutrients* 10 (10):1498.

Castellanos-Reyes, Katy, Ricardo Villalobos-Carvajal, and Tatiana Beldarrain-Iznaga. 2021. "Fresh Mushroom Preservation Techniques." *Foods* 10 (9):2126.

Chan, Jannie Siew Lee, Mikheil D Asatiani, Lital E Sharvit, Beny Trabelcy, Gayane S Barseghyan, and Solomon P Wasser. 2015. "Chemical composition and medicinal value of the new Ganoderma tsugae var. jannieae CBS-120304 medicinal higher basidiomycete mushroom." *International Journal of Medicinal Mushrooms* 17 (8):735–47.

Chang, Chih-Jung, Chuan-Sheng Lin, Chia-Chen Lu, Jan Martel, Yun-Fei Ko, David M Ojcius, Shun-Fu Tseng, Tsung-Ru Wu, Yi-Yuan Margaret Chen, and John D Young. 2015. "Ganoderma lucidum reduces obesity in mice by modulating the composition of the gut microbiota." *Nature Communications* 6 (1):7489.

Chen, He-Ping, Zhen-Zhu Zhao, Zheng-Hui Li, Ying Huang, Shuai-Bing Zhang, Yang Tang, Jian-Neng Yao, Lin Chen, Masahiko Isaka, and Tao Feng. 2018. "Anti-proliferative and anti-inflammatory lanostane triterpenoids from the Polish edible mushroom Macrolepiota procera." *Journal of Agricultural and Food Chemistry* 66 (12):3146–3154.

Chen, Mingyi, Dan Xiao, Wen Liu, Yunfei Song, Baorong Zou, Lin Li, Pei Li, Ying Cai, Deliang Liu, and Qiongfeng Liao. 2020. "Intake of Ganoderma lucidum polysaccharides reverses the disturbed gut microbiota and metabolism in type 2 diabetic rats." *International Journal of Biological Macromolecules* 155:890–902.

Cheung, Peter CK. 2013. "Mini-review on edible mushrooms as source of dietary fiber: Preparation and health benefits." *Food Science and Human Wellness* 2 (3-4):162–166.

Chiocchetti, Gabriela M, Teresa Latorre, María Jesús Clemente, Carlos Jadan-Piedra, Vicenta Devesa, and Dinoraz Vélez. 2020. "Toxic trace elements in dried mushrooms: Effects of cooking and gastrointestinal digestion on food safety." *Food Chemistry* 306:125478.

Chou, Wei-Ting, I-Chuan Sheih, and Tony J. Fang. 2013. "The applications of polysaccharides from various mushroom wastes as prebiotics in different systems." *Journal of Food Science* 78 (7):M1041–M1048.

Chye, Fook Yee, Jin Yi Wong, and Jau-Shya Lee. 2008. "Nutritional quality and antioxidant activity of selected edible wild mushrooms." *Food Science and Technology International* 14 (4):375–384.

Danneskiold-Samsøe, Niels Banhos, Helena Dias de Freitas Queiroz Barros, Rosangela Santos, Juliano Lemos Bicas, Cinthia Baú Betim Cazarin, Lise Madsen, Karsten Kristiansen, Glaucia Maria Pastore, Susanne Brix, and Mário Roberto Maróstica Júnior. 2019. "Interplay between food and gut microbiota in health and disease." *Food Research International* 115:23–31.

Das, Arun K, Pramod K Nanda, Premanshu Dandapat, Samiran Bandyopadhyay, Patricia Gullón, Gopalan Krishnan Sivaraman, David Julian McClements, Beatriz Gullón, and José M Lorenzo. 2021. "Edible mushrooms as functional ingredients for development of healthier and more sustainable muscle foods: A flexitarian approach." *Molecules* 26 (9):2463.

Das, Somenath, and Bhanu Prakash. 2022. "Edible mushrooms: Nutritional composition and medicinal benefits for improvement in quality life." In *Research and Technological Advances in Food Science*, 269–300. Elsevier.

Dawadi, Ebha, Prem Bahadur Magar, Sagar Bhandari, Subash Subedi, Suraj Shrestha, and Jiban Shrestha. 2022. "Nutritional and post-harvest quality preservation of mushrooms: A review." *Heliyon*:e12093.

De Silva, Dilani D, Sylvie Rapior, Kevin D Hyde, and Ali H Bahkali. 2012. "Medicinal mushrooms in prevention and control of diabetes mellitus." *Fungal Diversity* 56:1–29.

Delzenne, Nathalie M, and Laure B Bindels. 2015. "Ganoderma lucidum, a new prebiotic agent to treat obesity?" *Nature Reviews Gastroenterology & Hepatology* 12 (10):553–554.

Di Lorenzo, Flaviana, Cristina De Castro, Alba Silipo, and Antonio Molinaro. 2019. "Lipopolysaccharide structures of Gram-negative populations in the gut microbiota and effects on host interactions." *FEMS Microbiology Reviews* 43 (3):257–272.

Dowlati, Mohsen, Hamid Reza Sobhi, Ali Esrafili, Mahdi FarzadKia, and Mojtaba Yeganeh. 2021. "Heavy metals content in edible mushrooms: A systematic review, meta-analysis and health risk assessment." *Trends in Food Science & Technology* 109:527–535.

Ernst, Anna-Lena, Gerald Reiter, Meike Piepenbring, and Claus Bässler. 2022. "Spatial risk assessment of radiocesium contamination of edible mushrooms–Lessons from a highly frequented recreational area." *Science of the Total Environment* 807:150861.

Falandysz, Jerzy. 2013. "On published data and methods for selenium in mushrooms." *Food Chemistry* 138 (1):242–250.

Friedman, Mendel. 2016. "Mushroom polysaccharides: chemistry and antiobesity, antidiabetes, anticancer, and antibiotic properties in cells, rodents, and humans." *Foods* 5 (4):80.

Froufe, Hugo JC, RMV Abreu, and Isabel CFR Ferreira. 2009. "A QCAR model for predicting antioxidant activity of wild mushrooms." *SAR and QSAR in Environmental Research* 20 (5-6):579-590.

Fukushima-Sakuno, Emi. 2020. "Bioactive small secondary metabolites from the mushrooms Lentinula edodes and Flammulina velutipes." *The Journal of Antibiotics* 73 (10):687–696.

Gąsecka, Monika, Zuzanna Magdziak, Marek Siwulski, and Mirosław Mleczek. 2018. "Profile of phenolic and organic acids, antioxidant properties and ergosterol content in cultivated and wild growing species of Agaricus." *European Food Research and Technology* 244:259–268.

Gong, Pin, Siyuan Wang, Meng Liu, Fuxin Chen, Wenjuan Yang, Xiangna Chang, Ning Liu, Yuanyuan Zhao, Jing Wang, and Xuefeng Chen. 2020. "Extraction methods, chemical characterizations and biological activities of mushroom polysaccharides: A mini-review." *Carbohydrate Research* 494:108037.

González, Abigail, Mario Cruz, Carolina Losoya, Clarisse Nobre, Araceli Loredo, Rosa Rodríguez, Juan Contreras, and Ruth Belmares. 2020. "Edible mushrooms as a novel protein source for functional foods." *Food & Function* 11 (9):7400–7414.

Ha, Lee Su, Dae-Won Ki, Ji-Yul Kim, Dae-Cheol Choi, In-Kyoung Lee, and Bong-Sik Yun. 2021. "Dentipellin, a new antibiotic from culture broth of Dentipellis fragilis." *The Journal of Antibiotics* 74 (8):538–541.

Hamza, Arman, Shreya Ghanekar, and Devarai Santhosh Kumar. 2022. "Current trends in health-promoting potential and biomaterial applications of edible mushrooms for human wellness." *Food Bioscience*:102290.

Han, JunJie, YuHui Chen, Li Bao, XiaoLi Yang, Dailin Liu, ShaoJie Li, Feng Zhao, and Hongwei Liu. 2013. "Anti-inflammatory and cytotoxic cyathane diterpenoids from the medicinal fungus Cyathus africanus." *Fitoterapia* 84:22–31.

Helaly, Soleiman E, Christian Richter, Benjarong Thongbai, Kevin D Hyde, and Marc Stadler. 2016. "Lentinulactam, a hirsutane sesquiterpene with an unprecedented lactam modification." *Tetrahedron Letters* 57 (52):5911–5913.

Inyod, Tanapak, Francis Ayimbila, Achara Payapanon, and Suttipun Keawsompong. 2022. "Antioxidant activities and prebiotic properties of the tropical mushroom Macrocybe crassa." *Bioactive Carbohydrates and Dietary Fibre* 27:100298.

Jayachandran, Muthukumaran, Jianbo Xiao, and Baojun Xu. 2017. "A critical review on health promoting benefits of edible mushrooms through gut microbiota." *International Journal of Molecular Sciences* 18 (9):1934.

Jayasuriya, WJA Banukie N, Chandanie A Wanigatunge, Gita H Fernando, D Thusitha, U Abeytunga, and T Sugandhika Suresh. 2015. "Hypoglycaemic activity of culinary Pleurotus ostreatus and P. cystidiosus mushrooms in healthy volunteers and type 2 diabetic patients on diet control and the possible mechanisms of action." *Phytotherapy Research* 29 (2):303–309.

Jayasuriya, WJA, Shiroma M Handunnetti, Chandanie A Wanigatunge, Gita H Fernando, D Thusitha U Abeytunga, and T Sugandhika Suresh. 2020. "Anti-inflammatory activity of Pleurotus ostreatus, a culinary medicinal mushroom, in wistar rats." *Evidence-Based Complementary and Alternative Medicine* 2020.

Jiang, P, F Burczynski, C Campbell, G Pierce, JA Austria, and CJ Briggs. 2007. "Rutin and flavonoid contents in three buckwheat species Fagopyrum esculentum, F. tataricum, and F. homotropicum and their protective effects against lipid peroxidation." *Food Research International* 40 (3):356–364.

Jung, Tae-Hwan, Jeong Hyeon Park, Woo-Min Jeon, and Kyoung-Sik Han. 2015. "Butyrate modulates bacterial adherence on LS174T human colorectal cells by stimulating mucin secretion and MAPK signaling pathway." *Nutrition Research and Practice* 9 (4):343–349.

Kalač, Pavel. 2013. "A review of chemical composition and nutritional value of wild-growing and cultivated mushrooms." *Journal of the Science of Food and Agriculture* 93 (2):209–218.

Keskin, Feyyaz, Cengiz Sarikurkcu, Ilgaz Akata, and Bektas Tepe. 2021. "Element concentration, daily intake of elements, and health risk indices of wild mushrooms collected from Belgrad Forest and Ilgaz Mountain National Park (Turkey)." *Environmental Science and Pollution Research* 28 (37):51544–51555.

Khan, Imran, Guoxin Huang, Xiaoang Li, Waikit Leong, Wenrui Xia, and WL Wendy Hsiao. 2018. "Mushroom polysaccharides from Ganoderma lucidum and Poria cocos reveal prebiotic functions." *Journal of Functional Foods* 41:191–201.

Kim, Jaecheol, Si Chang Yang, Ah Young Hwang, Hyunnho Cho, and Keum Taek Hwang. 2020. "Composition of triterpenoids in Inonotus obliquus and their anti-proliferative activity on cancer cell lines." *Molecules* 25 (18):4066.

Kimura, Takashi. 2013. "Natural products and biological activity of the pharmacologically active cauliflower mushroom Sparassis crispa." *BioMed Research International* 2013.

Klaus, Anita, Maja Kozarski, Jovana Vunduk, Predrag Petrović, and Miomir Nikšić. 2016. "Antibacterial and antifungal potential of wild basidiomycete mushroom Ganoderma applanatum." *Lekovite Sirovine* (36):37–46.

Kolniak-Ostek, Joanna, Jan Oszmiański, Anna Szyjka, Helena Moreira, and Ewa Barg. 2022. "Anticancer and antioxidant activities in Ganoderma lucidum wild mushrooms in Poland, as well as their phenolic and triterpenoid compounds." *International Journal of Molecular Sciences* 23 (16):9359.

Konska, Grazyna. 2006. "Lectins of higher fungi (Macromycetes)—their occurrence, physiological role, and biological activity." *International Journal of Medicinal Mushrooms* 8 (1).

Korman, Richard. 2019. Growing Mushrooms: The Complete Grower's Guide to Becoming a Mushroom Expert and Starting Cultivation at Home: Richard Korman.

Kosanić, Marijana, Branislav Ranković, and Marko Dašić. 2012. "Mushrooms as possible antioxidant and antimicrobial agents." *Iranian Journal of Pharmaceutical Research: IJPR* 11 (4):1095.

Kour, Harpreet, Divjot Kour, Satvinder Kour, Shaveta Singh, Syed Azhar Jawad Hashmi, Ajar Nath Yadav, Krishan Kumar, Yash Pal Sharma, and Amrik Singh Ahluwalia. 2022. "Bioactive compounds from mushrooms: An emerging bioresources of food and nutraceuticals." *Food Bioscience*: 102124.

Kumar, Krishan, Rahul Mehra, Raquel PF Guiné, Maria João Lima, Naveen Kumar, Ravinder Kaushik, Naseer Ahmed, Ajar Nath Yadav, and Harish Kumar. 2021. "Edible Mushrooms: A comprehensive review on bioactive compounds with health benefits and processing aspects." *Foods* 10 (12):2996.

Li, Li-Feng, Hong-Bing Liu, Quan-Wei Zhang, Zhi-Peng Li, Tin-Long Wong, Hau-Yee Fung, Ji-Xia Zhang, Su-Ping Bai, Ai-Ping Lu, and Quan-Bin Han. 2018. "Comprehensive comparison of polysaccharides from Ganoderma lucidum and G. sinense: chemical, antitumor, immunomodulating and gut-microbiota modulatory properties." *Scientific Reports* 8 (1):6172.

Li, Miaoyu, Leilei Yu, Jianxin Zhao, Hao Zhang, Wei Chen, Qixiao Zhai, and Fengwei Tian. 2021. "Role of dietary edible mushrooms in the modulation of gut microbiota." *Journal of Functional Foods* 83:104538.

Li, Yongbo, Guofu Qin, Fengrui He, Keting Zou, Bei Zuo, Ruixiao Liu, Wei Zhang, Bixia Yang, Guipeng Zhao, and Guangfeng Jia. 2022. "Investigation and analysis of pesticide residues in edible fungi produced in the mid-western region of China." *Food Control* 136:108857.

Lin, Xiaohui, and Da-Wen Sun. 2019. "Research advances in browning of button mushroom (Agaricus bisporus): Affecting factors and controlling methods." *Trends in Food Science & Technology* 90:63–75.

Liu, Kun, Xuan Xiao, Junli Wang, C-Y Oliver Chen, and Huagang Hu. 2017. "Polyphenolic composition and antioxidant, antiproliferative, and antimicrobial activities of mushroom Inonotus sanghuang." *LWT-Food Science and Technology* 82:154–161.

Lv, Xu-Cong, Wei-Ling Guo, Lu Li, Xiao-Dan Yu, and Bin Liu. 2019. "Polysaccharide peptides from Ganoderma lucidum ameliorate lipid metabolic disorders and gut microbiota dysbiosis in high-fat diet-fed rats." *Journal of Functional Foods* 57:48–58.

Ma, Gaoxing, Hengjun Du, Qiuhui Hu, Wenjian Yang, Fei Pei, and Hang Xiao. 2021. "Health benefits of edible mushroom polysaccharides and associated gut microbiota regulation." *Critical Reviews in Food Science and Nutrition* 62 (24):6646–6663.

Ma, Ke, Li Bao, Junjie Han, Tao Jin, Xiaoli Yang, Feng Zhao, Shaifei Li, Fuhang Song, Miaomiao Liu, and Hongwei Liu. 2014. "New benzoate derivatives and hirsutane type sesquiterpenoids with antimicrobial activity and cytotoxicity from the solid-state fermented rice by the medicinal mushroom Stereum hirsutum." *Food Chemistry* 143:239–245.

Ma, Qiu-Lan, Cansheng Zhu, Marco Morselli, Trent Su, Matteo Pelligrini, Zhengqi Lu, Mychica Jones, Paul Denver, Daniel Castro, and Xuelin Gu. 2020. "The Novel Omega-6 Fatty Acid Docosapentaenoic Acid Positively Modulates Brain Innate Immune Response for Resolving Neuroinflammation at Early and Late Stages of Humanized APOE-Based Alzheimer's Disease Models." *Frontiers in Immunology*:2364.

Maity, Prasenjit, Surajit Samanta, Ashis K Nandi, Ipsita K Sen, Soumitra Paloi, Krishnendu Acharya, and Syed S Islam. 2014. "Structure elucidation and antioxidant properties of a soluble β-d-glucan from mushroom Entoloma lividoalbum." *International Journal of Biological Macromolecules* 63:140–149.

Maity, Prasenjit, Ipsita K Sen, Indranil Chakraborty, Soumitra Mondal, Harekrishna Bar, Sunil K Bhanja, Soumitra Mandal, and Gajendra Nath Maity. 2021. "Biologically active polysaccharide from edible mushrooms: A review." *International Journal of Biological Macromolecules* 172:408–417.

Marçal, Sara, Ana Sofia Sousa, Oludemi Taofiq, Filipa Antunes, Alcina MMB Morais, Ana Cristina Freitas, Lillian Barros, Isabel CFR Ferreira, and Manuela Pintado. 2021. "Impact of postharvest preservation methods on nutritional value and bioactive properties of mushrooms." *Trends in Food Science & Technology* 110:418–431.

Mattila, Pirjo, Anna-Maija Lampi, Riitta Ronkainen, Jari Toivo, and Vieno Piironen. 2002. "Sterol and vitamin D2 contents in some wild and cultivated mushrooms." *Food Chemistry* 76 (3):293–298.

Mau, J. L., H. C. Lin, and C. C. Chen. 2001. "Non-volatile components of several medicinal mushrooms." *Food Research International* 34(6):521–526.

Mazalovska, Milena, and J Calvin Kouokam. 2018. "Lectins as promising therapeutics for the prevention and treatment of HIV and other potential coinfections." *BioMed Research International* 2018.

Melgar, María Julia, and María Ángeles García. 2021. "Natural radioactivity and total K content in wild-growing or cultivated edible mushrooms and soils from Galicia (NW, Spain)." *Environmental Science and Pollution Research* 28:52925–52935.

Mishra, Vartika, Sarika Tomar, Priyanka Yadav, and MP Singh. 2021. "Promising anticancer activity of polysaccharides and other macromolecules derived from oyster mushroom (Pleurotus sp.): An updated review." *International Journal of Biological Macromolecules* 182:1628–1637.

Mitra, Payel, Narayan Chandra Mandal, and Krishnendu Acharya. 2016. "Polyphenolic extract of Termitomyces heimii: Antioxidant activity and phytochemical constituents." *Journal für Verbraucherschutz und Lebensmittelsicherheit* 11:25–31.

Moumita, Sahoo, and Bhaskar Das. 2022. "Assessment of the prebiotic potential and bioactive components of common edible mushrooms in India and formulation of synbiotic microcapsules." *LWT* 156:113050.

Moussa, Ashaimaa Y, Christopher Lambert, Theresia EB Stradal, Samad Ashrafi, Wolfgang Maier, Marc Stadler, and Soleiman E Helaly. 2020. "New peptaibiotics and a cyclodepsipeptide from Ijuhya vitellina: Isolation, identification, cytotoxic and nematicidal activities." *Antibiotics* 9 (3):132.

Mwangi, Ruth W, John M Macharia, Isabel N Wagara, and Raposa L Bence. 2022. "The antioxidant potential of different edible and medicinal mushrooms." *Biomedicine & Pharmacotherapy* 147:112621.

Nikitina, Valentina E, Ekaterina A Loshchinina, and Elena P Vetchinkina. 2017. "Lectins from mycelia of basidiomycetes." *International Journal of Molecular Sciences* 18 (7):1334.

Nitschke, Jörg, Hans-Josef Altenbach, Tim Malolepszy, and Helga Mölleken. 2011. "A new method for the quantification of chitin and chitosan in edible mushrooms." *Carbohydrate Research* 346 (11):1307–1310.

Nowacka, Natalia, Renata Nowak, Marta Drozd, Marta Olech, Renata Los, and Anna Malm. 2015. "Antibacterial, antiradical potential and phenolic compounds of thirty-one polish mushrooms." *PloS one* 10 (10):e0140355.

Oluwafemi, GI, KT Seidu, and TN Fagbemi. 2016. "Chemical composition, functional properties and protein fractionation of edible oyster mushroom (Pleurotus ostreatus)." *Annals Food Science and Technology* 17 (1):218–223.

Palacios, I, M Lozano, C Moro, M D'Arrigo, MA Rostagno, JA Martínez, A García-Lafuente, E Guillamón, and A Villares. 2011. "Antioxidant properties of phenolic compounds occurring in edible mushrooms." *Food Chemistry* 128 (3):674–678.

Panchak, LV. 2019. "Russulaceae family mushrooms lectins: Function, purification, structural features and possibilities of practical applications." *Biotechnologia Acta* 12 (1):29–38.

Poddar, Kavita H, Meghan Ames, Chen Hsin-Jen, Mary Jo Feeney, Youfa Wang, and Lawrence J Cheskin. 2013. "Positive effect of mushrooms substituted for meat on body weight, body composition, and health parameters. A 1-year randomized clinical trial." *Appetite* 71:379–387.

Poore, Joseph, and Thomas Nemecek. 2018. "Reducing food's environmental impacts through producers and consumers." *Science* 360 (6392):987–992.

Rahgo, Zahra, Shideh Mojerlou, and Kambiz Jahanbin. 2019. "Statistical optimization of culture conditions for protein production by a newly isolated Morchella fluvialis." *BioMed Research International* 2019.

Rahi, Deepak K, and Deepika Malik. 2016. "Diversity of mushrooms and their metabolites of nutraceutical and therapeutic significance." *Journal of Mycology* 2016.

Rahimah, Santun Bhekti, Dhiah Dianawaty Djunaedi, Arto Yuwono Soeroto, and Tatang Bisri. 2019. "The phytochemical screening, total phenolic contents and antioxidant activities in vitro of white oyster mushroom (Pleurotus ostreatus) preparations." *Open Access Macedonian Journal of Medical Sciences* 7 (15):2404.

Rai, Manjula, and Krishnendu Acharya. 2012. "Proximate composition, free radical scavenging and NOS activation properties of Ramaria aurea." *Research Journal of Pharmacy and Technology* 5 (11):1421–1427.

Rathore, Himanshi, Shalinee Prasad, and Satyawati Sharma. 2017. "Mushroom nutraceuticals for improved nutrition and better human health: A review." *PharmaNutrition* 5 (2):35–46.

Ren, Lu, Conrad Perera, and Yacine Hemar. 2012. "Antitumor activity of mushroom polysaccharides: A review." *Food & Function* 3 (11):1118–1130.

Ríos, José-Luis. 2010. "Effects of triterpenes on the immune system." *Journal of Ethnopharmacology* 128 (1):1–14.

Ronda, Oskar, Elżbieta Grządka, Iwona Ostolska, Jolanta Orzeł, and Bartłomiej Michał Cieślik. 2022. "Accumulation of radioisotopes and heavy metals in selected species of mushrooms." *Food Chemistry* 367:130670.

Royse, Daniel J, Johan Baars, and Qi Tan. 2017. "Current overview of mushroom production in the world." *Edible and Medicinal Mushrooms: Technology and Applications*:5–13.

Rumack, Barry H, and David G Spoerke. 1994. *Handbook of mushroom poisoning: diagnosis and treatment*: CRC Press.

Sabino Ferrari, Anna Beatriz, George Azevedo de Oliveira, Helena Mannochio Russo, Luiza de Carvalho Bertozo, Vanderlan da Silva Bolzani, Diego Cunha Zied, Valdecir Farias Ximenes, and Maria Luiza Zeraik. 2021. "Pleurotusostreatus and Agaricus subrufescens: Investigation of chemical composition and antioxidant properties of these mushrooms cultivated with different handmade and commercial supplements." *International Journal of Food Science & Technology* 56 (1):452–460.

Samsudin, NIP, and N Abdullah. 2019. "Edible mushrooms from Malaysia; a literature review on their nutritional and medicinal properties." *International Food Research Journal* 26 (1):11–31.

Sande, Denise, Geane Pereira de Oliveira, Marília Aparecida Fidelis e Moura, Bruna de Almeida Martins, Matheus Thomaz Nogueira Silva Lima, and Jacqueline Aparecida Takahashi. 2019. "Edible mushrooms as a ubiquitous source of essential fatty acids." *Food Research International* 125:108524.

Sari, Miriam, Alexander Prange, Jan I Lelley, and Reinhard Hambitzer. 2017. "Screening of beta-glucan contents in commercially cultivated and wild growing mushrooms." *Food Chemistry* 216:45–51.

Shafiqah, H. 2019. "Primary and secondary antioxidant activities of nine edible mushrooms species." *Food Research* 3 (1):14–20.

Shao, Hong Jun, Jin Boo Jeong, Kui-Jin Kim, and Seong-Ho Lee. 2015. "Anti-inflammatory activity of mushroom-derived hispidin through blocking of NF-κB activation." *Journal of the Science of Food and Agriculture* 95 (12):2482–2486.

Sharif, S., G. Mustafa, H. Munir, C. M. Weaver, Y. Jamil, and M. Shahid. 2016. "Proximate composition and micronutrient mineral profile of wild Ganoderma lucidum and four commercial exotic mushrooms by ICP-OES and LIBS." *Journal of Food and Nutrition Research* 4(11):703–708.

Shirur, Mahantesh, Anupam Barh, and Sudheer Kumar Annepu. 2021. "Sustainable Production of Edible and Medicinal Mushrooms: Implications on Mushroom Consumption." *Climate Change and Resilient Food Systems: Issues, Challenges, and Way Forward* 315–346.

Singh, Senjam Sunil, Hexiang Wang, Yau Sang Chan, Wenliang Pan, Xiuli Dan, Cui Ming Yin, Ouafae Akkouh, and Tzi Bun Ng. 2014. "Lectins from edible mushrooms." *Molecules* 20 (1):446–469.

Smiderle, FR, LM Olsen, AC Ruthes, PA Czelusniak, AP Santana-Filho, GL Sassaki, PAJ Gorin, and M Iacomini. 2012. "Exopolysaccharides, proteins and lipids in Pleurotus pulmonarius submerged culture using different carbon sources." *Carbohydrate Polymers* 87 (1):368–376.

Sonawane, Hiralal, H Bhosle, Gauri Bapat, and Ghole Vikram. 2014. "Pharmaceutical metabolites with potent bioactivity from mushrooms." *Journal of Pharmacy Research* 8 (7):969–972.

Srikram, A., and S. Supapvanich. 2016. "Proximate compositions and bioactive compounds of edible wild and cultivated mushrooms from Northeast Thailand." *Agriculture and Natural Resources* 50(6):432–436.

Stojanova, Monika, Milena Pantić, Mitko Karadelev, Biljana Čuleva, and Miomir Nikšić. 2021. "Antioxidant potential of extracts of three mushroom species collected from the Republic of North Macedonia." *Journal of Food Processing and Preservation* 45 (2):e15155.

Su, Anxiang, Gaoxing Ma, Minhao Xie, Yang Ji, Xiangfei Li, Liyan Zhao, and Qiuhui Hu. 2019. "Characteristic of polysaccharides from Flammulina velutipes in vitro digestion under salivary, simulated gastric and small intestinal conditions and fermentation by human gut microbiota." *International Journal of Food Science & Technology* 54 (6):2277–2287.

Synytsya, Andriy, Kateřina Míčková, Alla Synytsya, Ivan Jablonský, Jiří Spěváček, Vladimír Erban, Eliška Kováříková, and Jana Čopíková. 2009. "Glucans from fruit bodies of cultivated mushrooms Pleurotus ostreatus and Pleurotus eryngii: Structure and potential prebiotic activity." *Carbohydrate Polymers* 76 (4):548–556.

Tang, C., P. C. X. Hoo, L. T. H. Tan, P. Pusparajah, T. M. Khan, L. H. Lee, … K. G. Chan. 2016. "Golden needle mushroom: A culinary medicine with evidenced-based biological activities and health promoting properties." *Frontiers in Pharmacology*, 7:474.

Taofiq, Oludemi, Angela Fernandes, Lillian Barros, Maria Filomena Barreiro, and Isabel CFR Ferreira. 2017. "UV-irradiated mushrooms as a source of vitamin D2: A review." *Trends in Food Science & Technology* 70:82–94.

Tirta Ismaya, Wangsa, Raymond Rubianto Tjandrawinata, and Heni Rachmawati. 2020. "Lectins from the edible mushroom Agaricus bisporus and their therapeutic potentials." *Molecules* 25 (10):2368.

Tsai, Shu-Yao, Hui-Li Tsai, and Jeng-Leun Mau. 2008. "Non-volatile taste components of Agaricus blazei, Agrocybe cylindracea and Boletus edulis." *Food Chemistry* 107 (3):977–983.

Ukawa, Yuuichi, Hitoshi Ito, and Makoto Hisamatsu. 2000. "Antitumor effects of $(1 \rightarrow 3)$-β-D-glucan and $(1 \rightarrow 6)$-β-D-glucan purified from newly cultivated mushroom, Hatakeshimeji (Lyophyllum decastes Sing.)." *Journal of Bioscience and Bioengineering* 90 (1):98–104.

Valverde, María Elena, Talía Hernández-Pérez, and Octavio Paredes-López. 2015. "Edible mushrooms: improving human health and promoting quality life." *International Journal of Microbiology* 2015.

Vamanu, Emanuel, and Florentina Gatea. 2020. "Correlations between microbiota bioactivity and bioavailability of functional compounds: A mini-review." *Biomedicines* 8 (2):39.

Vamanu, Emanuel, Florentina Gatea, Ionela Sârbu, and Diana Pelinescu. 2019. "An in vitro study of the influence of Curcuma longa extracts on the microbiota modulation process, in patients with hypertension." *Pharmaceutics* 11 (4):191.

Venturella, Giuseppe, Valeria Ferraro, Fortunato Cirlincione, and Maria Letizia Gargano. 2021. "Medicinal mushrooms: bioactive compounds, use, and clinical trials." *International Journal of Molecular Sciences* 22 (2):634.

Wallace, Taylor C, Francisco Guarner, Karen Madsen, Michael D Cabana, Glenn Gibson, Eric Hentges, and Mary Ellen Sanders. 2011. "Human gut microbiota and its relationship to health and disease." *Nutrition Reviews* 69 (7):392–403.

Wang, Meiqi, and Ruilin Zhao. 2023. "A review on nutritional advantages of edible mushrooms and its industrialization development situation in protein meat analogues." *Journal of Future Foods* 3 (1):1–7.

Wang, Miaomiao, Santad Wichienchot, Xiaowei He, Xiong Fu, Qiang Huang, and Bin Zhang. 2019. "In vitro colonic fermentation of dietary fibers: Fermentation rate, short-chain fatty acid production and changes in microbiota." *Trends in Food Science & Technology* 88:1–9.

Wang, Shaojuan, Li Bao, Feng Zhao, Quanxin Wang, Shaojie Li, Jinwei Ren, Li Li, Huaan Wen, Liangdong Guo, and Hongwei Liu. 2013. "Isolation, identification, and bioactivity of monoterpenoids and sesquiterpenoids from the mycelia of edible mushroom Pleurotus cornucopiae." *Journal of Agricultural and Food Chemistry* 61 (21):5122–5129.

Wang, Yaqi, Li Bao, Dailin Liu, Xiaoli Yang, Shaifei Li, Hao Gao, Xinsheng Yao, Huaan Wen, and Hongwei Liu. 2012. "Two new sesquiterpenes and six norsesquiterpenes from the solid culture of the edible mushroom Flammulina velutipes." *Tetrahedron* 68 (14):3012–3018.

Wani, Bilal Ahmad, RH Bodha, and AH Wani. 2010. "Nutritional and medicinal importance of mushrooms." *Journal of Medicinal Plants Research* 4 (24):2598–2604.

Wasser, Solomon P. 2005. "Reishi or ling zhi (Ganoderma lucidum)." *Encyclopedia of Dietary Supplements* 1:603–622.

White, Julian, Scott A Weinstein, Luc De Haro, Regis Bédry, Andreas Schaper, Barry H Rumack, and Thomas Zilker. 2019. "Mushroom poisoning: A proposed new clinical classification." *Toxicon* 157:53–65.

Wu, Jian-Yong, Ka-Chai Siu, and Ping Geng. 2021. "Bioactive ingredients and medicinal values of Grifola frondosa (Maitake)." *Foods* 10 (1):95.

Xu, Ke, Xin Liang, Feng Gao, Jianjiang Zhong, and Jianwen Liu. 2010. "Antimetastatic effect of ganoderic acid T in vitro through inhibition of cancer cell invasion." *Process Biochemistry* 45 (8):1261–1267.

Xu, S, Y Dou, B Ye, Q Wu, Y Wang, M Hu, F Ma, X Rong, and J Guo. 2017. "Ganoderma lucidum polysaccharides improve insulin sensitivity by regulating inflammatory cytokines and gut microbiota composition in mice." *Journal of Functional Foods* 38:545–552.

Xu, Xiaofei, and Xuewu Zhang. 2015. "Lentinula edodes-derived polysaccharide alters the spatial structure of gut microbiota in mice." *PloS one* 10 (1):e0115037.

Yadav, Divya, and Pradeep Singh Negi. 2021. "Bioactive components of mushrooms: Processing effects and health benefits." *Food Research International* 148:110599.

Yin, Chaomin, Giuliana D Noratto, Xiuzhi Fan, Zheya Chen, Fen Yao, Defang Shi, and Hong Gao. 2020. "The impact of mushroom polysaccharides on gut microbiota and its beneficial effects to host: A review." *Carbohydrate Polymers* 250:116942.

Yin, Zhenhua, Zhenhua Liang, Changqin Li, Jinmei Wang, Changyang Ma, and Wenyi Kang. 2021. "Immunomodulatory effects of polysaccharides from edible fungus: a review." *Food Science and Human Wellness* 10 (4):393–400.

Yu, Xiao-Ying, Yuan Zou, Qian-Wang Zheng, Feng-Xian Lu, De-Huai Li, Li-Qiong Guo, and Jun-Fang Lin. 2021. "Physicochemical, functional and structural properties of the major protein fractions extracted from Cordyceps militaris fruit body." *Food Research International* 142:110211.

Zhang, Ruihong, Xiujin Li, and J.G. Fadel. 2002. "Oyster mushroom cultivation with rice and wheat straw." *Bioresource Technology* 82 (3):277–284.

Zhang, Shuai-Bing, Ying Huang, He-Ping Chen, Zheng-Hui Li, Bin Wu, Tao Feng, and Ji-Kai Liu. 2018. "Confluenines A–F, N-oxidized l-isoleucine derivatives from the edible mushroom Albatrellus confluens." *Tetrahedron Letters* 59 (34):3262–3266.

Zhang, Shunming, Yeqing Gu, Min Lu, Jingzhu Fu, Qing Zhang, Li Liu, Ge Meng, Zhanxin Yao, Hongmei Wu, and Xue Bao. 2020. "Association between edible mushroom intake and the prevalence of newly diagnosed non-alcoholic fatty liver disease: results from the Tianjin Chronic Low-Grade Systemic Inflammation and Health Cohort Study in China." *British Journal of Nutrition* 123 (1):104–112.

Zhang, Zhong, Rui-Na Liu, Qing-Jiu Tang, Jing-Song Zhang, Yan Yang, and Xiao-Dong Shang. 2015. "A new diterpene from the fungal mycelia of Hericium erinaceus." *Phytochemistry Letters* 11:151–156.

Zhao, Lei, Guanghua Zhao, Ming Du, Zhengdong Zhao, Lixia Xiao, and Xiaosong Hu. 2008. "Effect of selenium on increasing free radical scavenging activities of polysaccharide extracts from a Se-enriched mushroom species of the genus Ganoderma." *European Food Research and Technology* 226:499–505.

Zhao, Ruiqiu, Qiuhui Hu, Gaoxing Ma, Anxiang Su, Minhao Xie, Xiangfei Li, Guitang Chen, and Liyan Zhao. 2019. "Effects of Flammulina velutipes polysaccharide on immune response and intestinal microbiota in mice." *Journal of Functional Foods* 56:255–264.

Zhao, Yong-Ming, Jin-Hui Song, Jin Wang, Jian-Ming Yang, Zhi-Bao Wang, and Ying-Hui Liu. 2016. "Optimization of cellulase-assisted extraction process and antioxidant activities of polysaccharides from Tricholoma mongolicum Imai." *Journal of the Science of Food and Agriculture* 96 (13):4484–4491.

Zhou, Juanjuan, Mengfei Chen, Shujian Wu, Xiyu Liao, Juan Wang, Qingping Wu, Mingzhu Zhuang, and Yu Ding. 2020. "A review on mushroom-derived bioactive peptides: Preparation and biological activities." *Food Research International* 134:109230.

Zięba, P., Sękara, A., Sułkowska-Ziaja, K., & Muszyńska, B. 2020. "Culinary and medicinal mushrooms: Insight into growing technologies." *Acta Mycologica*, 55(2). doi:10.5586/am.5526

11 *Mimosa caesalpiniifolia* Benth., a Brazilian Caatinga Species with Anti-Hypertensive Properties
Developing a Phytomedicine from Its Inflorescences

Sara Léa Fortes Barbosa, Angélica Gomes Coêlho,
Francisco Valmor Macedo Cunha, Brenda Nayranne Gomes
dos Santos, Iolanda Souza do Carmo, Bernardo Melo Neto,
José de Sousa Lima-Neto, Lívio César Cunha Nunes,
Daniel Dias Rufino Arcanjo and Antônia Maria das Graças
Lopes Citó

11.1 INTRODUCTION

Mimosa caesalpiniifolia Benth. (Fabaceae) is a species from northeastern Brazil characterized by its fast growth, high regeneration capacity, and resistance to drought. Considered as a multiuse type of plant, it is source of food for cattle; also, it is used as living fence and its wood raw material for the manufacturing of stakes and charcoal. Besides those uses, some parts of the plant are employed in the traditional medical practices of the region (Monção et al. 2014; Albuquerque et al. 2007; Márcio E P Santos et al. 2015).

The tea from the inflorescences of this species is used by some communities for the treatment of arterial hypertension; its efficacy has been confirmed with in vivo studies, in which it was observed to induce tachycardia and hypotension, whereas the ethanolic extract induces bradycardia and hypotension via muscarinic and ganglionar pathways. For its vasorelaxant effect, the blockade of the calcium channels is reported as a possible pharmacological mechanism involved (Santos et al. 2015; Albuquerque et al. 2007). In addition to this activity, the present research group confirmed the antioxidant activity of the extract toward *Saccharomyces cerevisiae*, wherein the extract attenuated the oxidant effect of the hydrogen peroxide.

As for the search for new medications, these findings foster the development of an herbal remedy from a standardized extract of *M. caesalpiniifolia* inflorescences that can be utilized for treating arterial hypertension. They also meet the goals of Política Nacional de Plantas Medicinais e Fitoterápicos (National Policy for Medicinal Plants and Herbal Remedies), which aims at granting the Brazilian population safe access to and rational use of medicinal plants and herbal remedies, promoting the sustainable use of biodiversity, as well as the development of both the productive chain and the national industry (BRASIL et al. 2016).

DOI: 10.1201/9781003395935-11

The release of new herbal products with the same efficacy and safety as allopathic medicines ranks them as promising substitutes for the synthetic products of the pharmaceutical industry. The advantages of the herbal remedies include action through the synergism of different molecules, association of mechanisms by compounds acting on different molecular targets, lower risk of side effects, and lower research costs (Yunes, Pedrosa, and Cechinel Filho 2001).

The production of herbal remedies requires the transformation of raw materials into a product suitable for industrial use and other therapeutic applications. However, being a complex operation, it requires the use of specific processing technologies (Couto et al. 2011). The spray-dried plant extracts, for instance, have been used as intermediate products for obtaining different pharmaceutical forms (Soares and Petrovick 1999).

This study targeted the outlining of at least one formulation of conventional-release tablets from the standardized extract from *M. caesalpiniifolia* inflorescences that meets the protocols of the "General methods applied to medicines" (Brazilian pharmacopeia) for solid preparations.

11.2 MATERIAL AND METHODS

11.2.1 OBTAINING AND STANDARDIZATION OF THE EXTRACT

The plant material for the study was inflorescences of *Mimosa caesalpiniifolia* Benth., collected in June 2015 in the municipality of Teresina-Pi, in the native forest of the Federal University of Piauí (UFPI). An exsiccate was deposited at the "Graziela Barroso" herbarium of UFPI with the number: TEPB 26824.

The inflorescences were dried at room temperature, submitted to spraying on a Marconi knife mill (Model: MA 680 and serial 98 2422; LPN-UFPI). The extraction conditions were optimized, seeking the best yield in the extraction via factorial planning, which pointed to the following significant factors: maceration with hydroethanolic solvent (ethanol/water ratio 7:3) and filtration and renewal of the solvent daily for 3 days without the need for the use of a sonicator. The excess ethanol was evaporated, under reduced pressure, with heating up to 50°C, through a Heidolph Laborota 4000 rotary evaporator coupled to a Kohlbach vacuum pump.

The level of gallic acid, chosen as marking substance, was quantified in the dried extract using a calibration curve obtained through High-Performance Liquid Chromatography.

11.2.2 SPRAY DRYING

After removing the excess ethanol, the residual volume was spray dried on a Büchi B-290 Mini Spray Dryer (Büchi Labortechnik AG; LITE-UFPI). The high hygroscopicity of the extract harms the powder flow. In order to reduce this problem, colloidal silicon dioxide was added to the extract solution (Aerosil 200®) as a drying adjuvant, in a proportion of 23.1% in relation to the mass of the extract's solid residue which was determined following the guidelines of the Brazilian Pharmacopeia 5th ed., item 5.4.3.2.2 (Determination of the dry residue in fluid and soft extracts) (BRASIL and Agência Nacional de Vigilância Sanitária – ANVISA 2019). The configurations for the drying on the device were: input temperature at 150°C, output temperature at 97°C, peristaltic pump flow of 3 ml/minute, pressure of 55 bar, air aspiration at 80% of the maximal capacity. The resulting powder showed theoretical constitution of 76.9% of extract and 23.1% of Aerosil 200®.

11.2.3 OUTLINE OF THE TABLET FORMULATIONS

Considering the extract content in the powder yielded through spray drying, for a formulation with dose of 50 mg of standardized extract, 65 mg of the powder was necessary. By means of the 2^3-factorial planning (assessment of three factors varying in 2 levels), we analyzed the factor 1 – diluent influence: pregelatinized starch (+1) × microcrystalline cellulose (−1); factor 2 – the use of the agglutinating agent PVP at 1%: with PVP (+1) × without PVP (−1); and factor 3 – production method: direct compression production (+1) or through humid pathway (−1). Eight formulations, named F1, F2, F3, F4, F5, F6, F7, and F8, were outlined for tablets of 125 mg of weight. The lots produced in the humid method went through previous granulation before compression. In order to humidify the post mixture, the 1:1 ethanol and water mixture was used. When there was the use of agglutinant in these lots, PVP was added to the ethanol and water mixture. The humid mass was granulated through a No. 16 tamis and dried in an oven at 40 °C for 3 hours. Next, the size of the granules was standardized in No. 20 tamis and received the lubricant. The lots produced by direct compression had the mixing of the powders performed through grade and pistillate.

The following pharmaceutical excipients were used in the formulations: PH 101 microcrystalline cellulose (supplier: Genix®, Lot: 90652), pregelatinized starch (Starch 1500®, supplier: Colorcon®), and magnesium stearate (supplier: Natural Pharma®).

11.2.4 QUALITY CONTROL OF THE TABLET FORMULATIONS

Assessment tests were applied for average weight, hardness, friability, disintegration, and dissolution following the instructions preconized in the Brazilian Pharmacopeia, 5th Ed. (BRASIL and Agência Nacional de Vigilância Sanitária – ANVISA 2019).

On account of the fact that the study is dealing with uncoated tablets, for the determination of the average weight, each lot was assessed separately through the weighing of 20 tablets on a Sartorius model BL 210S analytical scale and the means of each lot was calculated.

In the hardness test, a portable digital hardness meter (Ethik Technology) was used to assess the resistance of 10 tablets of each formulation toward rupture under radial pressure, measured in Newtons, and expressed by the mean hardness of each formulation.

For the disintegration test, a tablet was placed followed by the addition of the appropriate disk in each six tubes of the disintegrator (Model 301 AC, Nova Ética, Brazil). The immersion liquid used was water kept at 37°C. The total disintegration time of each formulation was recorded, and the time of the experiment was 30 minutes, which is the maximal time allowed for uncoated tablets.

Following the recommendation from the Brazilian Pharmacopeia (BRASIL and Agência Nacional de Vigilância Sanitária – ANVISA 2019) for tablets with average weight of 0.65 g or lower, we used in the test of friability 20 tablets of each formulation previously weighed. They underwent 25 rotations per minute during 4 minutes on a friabilometer. After the test, the tablets were weighed again and the percentage of loss of each one was calculated.

The results of the hardness and disintegration tests were analyzed with a spreadsheet developed by Lima et al. 2016, on the Microsoft Excel® Software version 2007 for the calculus of the effects of the factors over the response in study and the assessment of their statistical significance in the factorial planning. After plotting and analysis of the results, we performed the profile of dissolution of the two formulations with characteristics that were mostly favorable to the industry.

For so, the dissolution test was performed on a Nova Ética® dissolutor tester (Model 299). The rotation parameters at 75 RPM were adopted, as well as temperature at 37°C in 900 mL of HCl 0.1 M medium per bath. The collection times for the reading of the samples were 5, 15, 30, 45, and 60 minutes, following the guidelines from the Brazilian Pharmacopeia (BRASIL and Agência Nacional de Vigilância Sanitária – ANVISA 2019). Each aliquot went through filtration on a quantitative sieve that retained the insoluble excipients. The maximal wavelength of the extract was obtained through previous scanning on the Visible UV (UV-1800 SHIMADZU) of filtered HCl 0.1 M. An analytical curve of the extract was developed in the concentrations of 10.4, 15.2, 24.0, 32.0, 40.0, 56.0, 72.0, and 144.00 µg ml⁻¹, at wavelength of 269 nm, according to the previous scanning. Through this curve, the concentrations of the dissolutor tester aliquots removed at each time were quantified.

The means and standard error of the average weight and dissolution tests were calculated and analyzed on the Software GraphPad Prism 6 software. The Pareto and surface charts were designed using the softwares Microsof Excel® version 2013 and Statistica® version 12, respectively.

11.3 RESULTS AND DISCUSSION

The dried extract of the inflorescences of *M. caesalpiniifolia* obtained with the extractive technique was standardized, showing content of 7.9 mg of gallic acid per gram of extract. Beginning from the drying process, we used the spray-drying technique, which is one of the most frequently used in the industry of herbal remedies thanks to its stability and the possibility to control the characteristics of the final product (Fahr 2021). Several substances are used as co-adjuvants, such as: starch, cyclodextrins, colloidal silicon dioxide (Aerosil 200®), tricalcium phosphate, gelatin, Arabic gum, lactose, and maltodextrin, among others (Oliveira and Petrovick 2010). Aerosil 200® stands out due to its greater suitability regarding the stability of the dried extracts, reducing their residual humidity (Vasconcelos et al. 2005). Besides, its small particle size and large specific surface area provide interesting flow characteristics that are explored to improve its flow properties of dried powders (Rowe, Sheskey, and Quinn 2009).

The drying of the standardized extract of *Mimosa caesalpiniifolia* Benth. with Aerosil® in a spray dryer showed yield of 74.4 % over the partition mass. The process improved the aspect of the extract, which was no longer very hygroscopic and low flow and became fine powder, fluid and homogeneous when observed macroscopically. Such improvements enabled the use of the extract in the pharmaceutical operations employed in the production of tablets.

A stable and efficient pharmaceutical formulation relies on the careful selection of the excipients that will be added to facilitate the administration, promote consistent release, and protect from degradation (Aulton and Taylor 2016). In addition to that, turning plant-based raw material into a suitable product for industrial use and other therapeutic applications is a complex operation that requires the use of specific processing technologies in agreement with strict norms defined in official compendiums (Couto et al. 2011). In order to obtain the homogeneity of the production lots, one must employ analytical methods aiming at the quantification of either active or reference substances, as well as methods that assess the aspects regarding the pharmaceutical form (Cristina et al. 2003).

Generally, the formulation of tablets contains, besides the active substances, adjuvants that aim at both allowing the compression operation to occur satisfactorily and ensuring that the tablets are obtained with the specified quality. The adjuvants employed as diluents in these formulations are inert substances used to achieve the desired mass and improve the flow properties and compression characteristics (Aulton and Taylor 2016; Allen, Popovich, and Ansel 2005). In addition, the choice of the adjuvants and characteristics such as free flow and cohesion properties enable the direct compaction of the medicines without the use of humid or dry granulation. However, only some active substances can be turned into tablets without suffering any processing prior to compression (Toller and Schimdt 2005).

The factorial planning is an important statistical tool, through which it is possible to outline different formulations to produce the tablets with the powder from the spray-dried extract and assess the level of significance of each factor and the possible interactions. In the factorial planning 23, any effect can be understood as the difference between the means, and by using the real repetitions in a particular combination of levels one can estimate the experimental error in this combination. The group variance estimate is then calculated using the means of all the estimates, each of which is defined by its own degree of freedom (Lima et al. 2016).

The formulations outlined with the powder from the extract of *M. caesalpiniifolia* (Figure 11.1) were tested in the factorial analysis using as diluents (factor 1) the pre-gelatinized starch (level +1) or the microcrystalline cellulose (level −1) (Table 11.1; Table 11.2). Starch – also belonging to the category of agglutinants and disintegrants – enhances the fluidity, disintegration, and hardness of the tablets. It has the disintegration property and independent dissolution of the medium pH, which

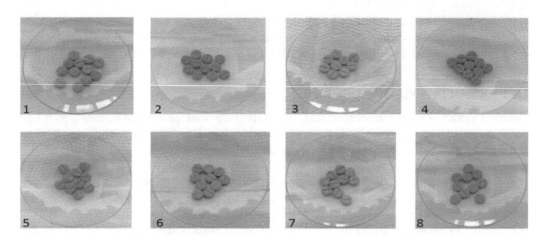

FIGURE 11.1   Images of tablets of each one of the eight formulations outlined by factorial planning 2^3.

TABLE 11.1
Matrix of Factorial Planning 2^3 for the Formulations of Tablets of the Hydroethanolic Extracts of the Inflorescences of *Mimosa caesalpiniifolia*

	Variables		
Formulations	Diluent	Agglutinant	Production mode
F1	+1	+1	+1
F2	+1	+1	−1
F3	+1	−1	+1
F4	+1	−1	−1
F5	−1	+1	+1
F6	−1	+1	−1
F7	−1	−1	+1
F8	−1	−1	−1

Notes: Diluent: starch 1500 (+1), cellulose (−1); PVP agglutinant: Present (+1); Absent (−1); Production mode: Dry way (+1), Humid way (−1).

TABLE 11.2
Masses of the Excipients Employed in the Formulations Outlined for Tablets of Standardized Extract of the Inflorescences of *Mimosa caesalpiniifolia*

	Formulations							
Ingredients	F1	F2	F3	F4	F5	F6	F7	F8
Stand. Extract (52%)	3.625 g	3.625 g	3.625 g	3.625 g	3.625 g	3.625 g	3.625 g	3.625 g
Stand. Mg (0.5%)	0.035 g	0.035 g	0.035 g	0.035 g	0.035 g	0.035 g	0.035 g	0.035 g
PVP (1%)	0.069 g	0.069 g	—	—	0.069 g	0.069 g	—	—
Starch (46.93%)	3.298 g	3.298 g	3.298 g	3.298 g	—	—	—	—
Cellulose (46.93%)	—	—	—	—	3.298 g	3.298 g	3.298 g	3.298 g
Production mode	Dry	Humid	Dry	Humid	Dry	Humid	Dry	Humid

helps promote the disaggregation of the powder mass in primary particles of the drug and speeds up the dissolution rate of the drug (Rowe, Sheskey, and Quinn 2009). Starch 1500® is a form of pre-gelatinized starch that has been modified to make it more fluid and compressible, favoring the direct compression process (Jivraj, Martini, and Thomson 2000). Microcrystalline cellulose is widely used in drugs, mainly as a diluent-agglutinant in formulations of capsules and oral tablets, where it is used in processes of humid granulation and direct compression (Rowe, Sheskey, and Quinn 2009). It is, in addition, one of the most easily compressible adjuvants used in direct compression. That may be explained by the nature of the molecules, which favors the formation of hydrogen ligands – which maintain its characteristic of plasticity (Lima Neto and Petrovick 1997).

The second factor tested (factor 2) was the influence of the use of the PVP agglutinant in the quality tests, where level +1 corresponds to the use of PVP and the level −1 to its absence. The use of agglutinants promotes the adhesion between the powder drug and the pharmaceutical excipients, allows the preparation of granulates, and keeps the integrity of the final tablet (Allen, Popovich, and Ansel 2005).

The mode of production of the tablets was the third factor assessed (factor 3), ranging between the use of the dry method (level +1) and the humid method (−1). In order to produce them, there is generally the need to go for the granulation of the constituents so that the compression in the pharmaceutical form can be obtained, except when in cases of directly compressible substances, in which the granulation operation is dispensable. The use of co-processed excipients (microcrystalline cellulose, pre-gelatinized starch, and spray-dried lactose, as for instance) provides the manufacturing of tablets by direct compression with its material and economic advantages (Sausen and Mayorga 2013; Pessanha et al. 2012). The formulations were prepared following the matrix of factorial planning and outlining presented in Tables 1 and 2. The results of the pharmacopedic assays are presented in Table 11.3.

11.3.1 Weight Assessment

In the weight assessment assay for uncoated tablets weighing more than 80 mg and less than 250 mg, no more than a variation of ±7.5% of the average weight can be accepted. Nevertheless, no

---

TABLE 11.3
Results of the Pharmacopeic Assays of the Formulations of Tablets Outlined from the Standardized Extract of the Inflorescence of *M. caesalpiniifolia*

	ASSAYS				
Formulations	Average Weight (mg)	Friability (%)	Hardness (N)	Disintegration (min)	Dissolution (45 min; %)
F1	122.2 ± 1.5[bcfh]	0.25	35.6 ± 6.4	5.1 ± 0.2	—
F2	125.0 ± 1.3[agce]	1.27	38.2 ± 12.9	4.3 ± 0.2	—
F3	120.6 ± 1.7[abdh]	0.50	60.2 ± 7.5	6.2 ± 0.3	104.3 ± 0.5
F4	125.1 ± 1.9[acge]	0.30	35.5 ± 3.3	9.1 ± 0.2	—
F5	121.9 ± 1.7[bdfh]	0.16	122 ± 17.8	18.0 ± 0.3	—
F6	126.2 ± 2.3[aceg]	0.80	86.3 ± 9.4	35.00*	—
F7	120.1 ± 2.3[bdfh]	0.08	129.2 ± 22	23.2 ± 0.3	82.5 ± 1.6[c]
F8	126.1 ± 3.7[aceg]	0.08	76.2 ± 13.9	23.4 ± 0.5	—

Notes: The numbers correspond to the means ± S.E.M. In the average weight, one-way ANOVA and Tukey´s test of multiple comparisons was employed. Values are considered significant when $P < 0.05$, being "a" the significance regarding F1, "b" regarding F2, "c" regarding F3, "d" regarding F4, "e" regarding F5, "f" regarding F6, "g" regarding F7, "h" regarding F8. (*) The F6 formulation was reproved in the disintegration test because it exceeded the 30-minute limit. (−) Dissolution assays were not performed.

formulation may be above or below the double of this percentage (BRASIL and Agência Nacional de Vigilância Sanitária – ANVISA 2019). All the formulations tested showed satisfactory results for this test, with all units weighing within the accepted variation limit. However, it was observed that those produced with the humid method – F2, F4, F6, and F8 – presented average weight closer to 125 mg, whereas the tablets produced with the dry way – F1, F3, F5 and F7 – presented lower weight (Table 11.3). These formulations formed a mixture of a light and fine powder, which favored losses during the impact of the puncture in the compression process. Such loss was minimal in the compression of the granulated losses once that these were denser; they were better organized within the compressing matrix and did not spread with the impact of the puncture.

11.3.2 FRIABILTY

In the friability test, all the lots of tablets of the extract manufactured presented satisfactory results (Table 11.3), considering that no unit tested could be presented, at the end of the test, broken, chipped, cracked, split, or with loss higher than 1.5% of its weight (BRASIL and Agência Nacional de Vigilância Sanitária – ANVISA 2019). The friability of the tablets is a resistance degree regarding the shock, friction, rolling, agitation, and flexion, being determined by the percentage of debris that are separated in the tablet (Prista, Alves, and Morgado 2008).

11.3.3 HARDNESS

The hardness of an uncoated tablet is its resistance to crushing or rupture under radial pressure caused by friction shocks during the processes of packaging, transportation, and storage (Aulton and Taylor 2016; Allen, Popovich, and Ansel 2005; Prista, Alves, and Morgado 2008). Based on Pareto's chart (Figure 11.2), the influence of each factor on the hardness of the tables was investigated. The threshold for significance of 95% of reliability was calculated and considered only with values higher than 21.2 N, as significant for the answer obtained. The charts were made using a statistical

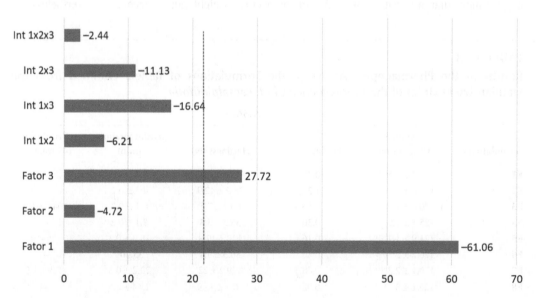

FIGURE 11.2 Pareto's chart for the effects of the factors tested in the factorial planning and their interactions for the response Hardness (Newtons) of the tablets of *M. caesalpiniifolia*.

Legend: Factor 1 (type of diluent); Factor 2 (presence of agglutinant); Factor 3 (production way). The dotted line defines the 21.1 N significance level, with confidence intervals of 95%. Interaction of the variables 1 and 2 (1/2), 1 and 3 (1/3), 2 and 3 (2/3), and 1 and 2 and 3 (1/2/3).

tool developed by (Lima et al. 2016). It was observed that only the factors 1 (diluent) and 3 (compression method) presented results statistically significant for the formulations.

Based on the findings shown in Table 11.3 and Figure 11.2, in a sense, the diluent factor (factor 1) had the greatest influence on the hardness of final tablets, in a way that those produced with the cellulose diluent achieved higher hardness in approximately 61 N, on average, in relation to the lots where starch was used. This finding is according to the literature, which explains that the greater resistance and cohesion of the tablets obtained from cellulose, even under low forces of compression, are due to the proximity of hydrogen groups in its adjacent molecules, allowing the formation of hydrogen bonds. During the compression, the microcrystalline cellulose is plastically deformed and, therefore, maximizes the binding area among particles (Thoorens et al. 2014). As for the compression method (factor 3), through the result of the factorial analysis (Figure 11.2), it was observed that the dry way produced lots of tablets approximately 27 N harder than those produced with the humid way.

The dry way, also called direct compression, is the preparation of tablets from a mixture of ingredients without a process of granulation or preliminary agglomeration. Although it involves few phases, the outlining of products for direct compression can be challenging once that it is directly affected by the properties of the material (Thoorens et al. 2014). The production of tablets through direct compression involves three sequential operations: weighing, mixture of the powders, and compression. Even though it demands filling and agglutinant materials that are usually more expensive, it allows the timesaving, large-scale and low-cost production of tablets. In addition, since water and heat are not employed in the process, the stability of the product can be increased (Aulton and Taylor 2016; Toller and Schimdt 2005).

The humid granulation is also a widely used obtaining method, once the granulates formulated present good flow and cohesiveness characteristics, resulting in tablets with suitable and constant physical characteristics, such as weight, mechanical resistance, and disintegration (Cury et al. 2008). However, the granulation is always a complex operation and calls for a number of steps (humidification, granulation, and drying), some of which are very delicate (Soares and Petrovick 1999).

Still on the factorial analysis of the hardness assay (Figure 11.2), in the interval analyzed no interactions were found among the factors, showing that each factor studied is not dependent on the level of the other factors.

The use of cellulose to produce tablets using dry compression is shown to lead to the production of harder tablets, though the current legislation requires a minimum limit of 30 N for tablets. High levels of hardness are not a basic requirement for the technically satisfactory production of tablets. All the lots studied agreed with the Brazilian pharmacopeia guidelines on this issue (BRASIL and Agência Nacional de Vigilância Sanitária – ANVISA 2019), yet it is worth mentioning that some potential problems related to tablets are bound to the compression method. Among those is the high level of hardness of the tablet, which may hinder the disintegration and dissolution processes (Soares and Petrovick 1999).

Through the surface charts, it was possible to observe the influence of the diluent factor over the hardness of the tablet (Figure 11.3), which increases as it ranges from starch 1500 (+) to microcrystalline cellulose (−) and when the production method ranges from humid granulation (−) to direct compression (+).

11.3.4 Disintegration

According to the statistical analysis performed for hardness, the threshold for significance with 95% reliability was calculated for the disintegration; values above 0.28 min were considered as significant (Table 11.3). It was observed that factor 1 (diluent) was the one that mostly influenced the time of disintegration of the tablets obtained. Those produced with the use of cellulose presented disintegration approximately 18 minutes longer. The fast disintegration of the formulations with starch is since besides being diluent, it has a disintegrating character. It is the most widely employed

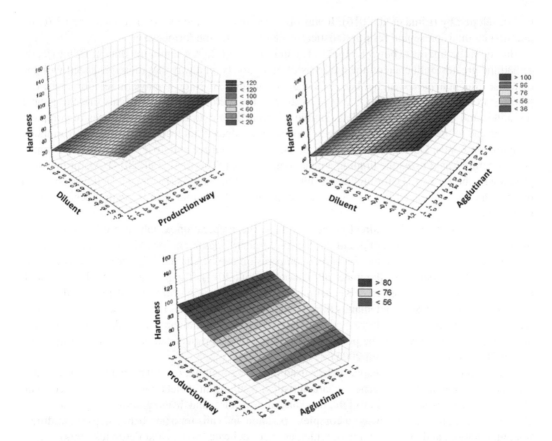

FIGURE 11.3 Response surface of the factorial planning 2^3 produced in the hardness of the tablets of *M. caesalpiniifolia* by the diluent factors (starch+1; cellulose−1), agglutinant (with PVP+1; without PVP−1), and production method (direct compression+1; humid granulation−1).

disintegrant in conventional tablets (Aulton and Taylor 2016). Using or not an agglutinant (factor 2) in the tablet composition did not bring statistically significant influences for the study of the disintegration time.

As for the production method (factor 3), on Pareto's chart (Figure 11.4), one can observe a mild yet significant increase (approximately 5 minutes) in the disintegration time attributed to this factor when produced with the humid granulation. However, it is known that production through direct compression is more related to longer disintegration times (Soares and Petrovick 1999).

Statistically significant double and triple interactions were found among the factors, showing that the effect of each factor over the disintegration time can be modified when the level of the other factors is also modified. The formulation F6, for instance, presented longer disintegration time due to the influence of the responses of the lowest level of factor 1 (cellulose), of the lowest level of factor 3 (humid method), of the 1×3 double interaction, and still of the $1 \times 2 \times 3$ triple interaction (Figure 11.4).

In this test, therefore, only F6 was reproved, exceeding the maximal time allowed of 30 minutes, while the other formulations presented good results in the tests. The tablets produced with pre-gelatinized starch disintegrated more rapidly than did those with cellulose. However, the legal requirements of the Brazilian Pharmacopeia (BRASIL and Agência Nacional de Vigilância Sanitária – ANVISA 2019) demand only that the tablets disintegrate within a maximal limit of up to 30 minutes. The shorter times of disintegration are not a basic requirement for the obtaining of technically satisfactory tablets (Oliveira and Petrovick 2010).

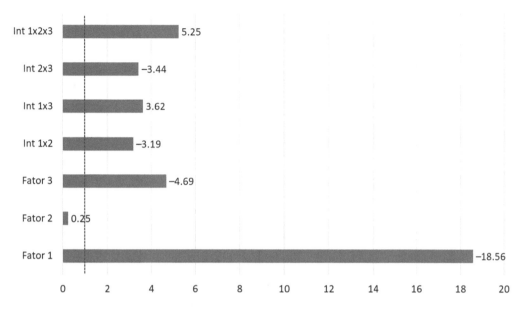

FIGURE 11.4 Pareto chart for the effects of the factors tested and their interactions for the response disintegration time (minutes) of the tablets of *Mimosa caesalpiniifolia*. Captions: Factor 1 (type of diluent); Factor 2 (presence of agglutinant); Factor 3 (production method). The dotted line defines the 0.28 min. significance threshold, with confidence interval of 95%. Interaction of the variables 1 and 2 (1/2), 1 and 3 (1/3), 2 and 3 (2/3), and 1 and 2 and 3 (1/2/3).

In Figure 11.5, one can see in the response surfaces the significant increase in the disintegration time as the diluent ranges from starch 1500 (+) to cellulose (–), while the factors of production method and use of agglutinant were slightly expressive.

Taking into consideration the previous goal of obtaining at least one approved formulation and considering the interests from the pharmaceutical industry, which seeks more efficiency in the production line, low cost, practicality, and easy control (Sausen and Mayorga 2013), it is inferred that the F3 (starch 1500, without PVP, dry way) and F7 (microcrystalline cellulose, without PVP, dry way) formulations are presented as the most attractive for the assessment of the dissolution test.

11.3.5 Dissolution

For the therapeutic effect to appear, tablet dissolution is necessary. From then on, its active principles may be absorbed, thus reaching the blood flow and binding to the action sites. The assessment of the dissolution profile shows the portion of the drug that will be available to be absorbed in the in the gastrointestinal tract (Ngo 2007).

There are three categories of dissolution tests for immediate-release medicines, which allow one to assess and compare the kinetics and efficiency of the dissolution of a certain product: single-point dissolution test, two-point dissolution test, and dissolution profiles. The dissolution profiles, which are obtained from a dissolved percentage of the drug at different sampling times, allow a more conclusive analysis (Serra and Storpirtis 2007).

The dissolution test consists of the use of a device that simulates the conditions of the gastrointestinal tract regarding temperature and agitation to measure the amount of the drug dissolved in the medium within a given time span (Lira, Dornelas, and Cabral 2008). For medicines containing drugs that are slightly soluble in water, which dissolve slowly, at least the two-point dissolution test is recommended, that is, one at 15 minutes and another at 30, 45, or 60 minutes, in order to ensure 85% of dissolution. Moreover, the pharmaceutical forms of immediate release must present average

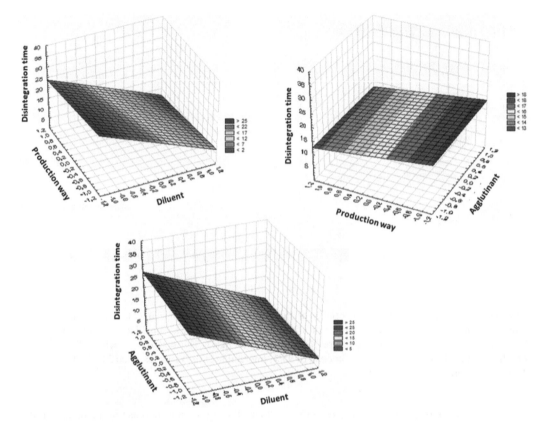

FIGURE 11.5 Response surfaces of the factorial planning 2^3 obtained in the disintegration time of the tablets of *M. caesalpiniifolia* through the factors of diluent (starch+; cellulose–), agglutinant (with PVP+; without PVP–), and production method (direct compression+; humid granulation–).

dissolution of at least 75% of the active substance within 45 minutes (BRASIL and Agência Nacional de Vigilância Sanitária – ANVISA 2019). In order to enable the quantification of the extract of *M. caesalpiniifolia* contained in the tablets, of the F3 and F7 formulations, a scanning of the extract was initially made in UV–visible, which yielded the maximal λ of 269 nm (Figure 11.6), through which it was possible to develop the $A = 0.00789C – 0.01336$ calibration curve (Figure 11.7), where "A" corresponds to the sample absorbance and "C" corresponds to its concentration.

The validation of analytical methodology for herbal remedies must be made according to what is defined on the RE 899/03. However, given the complexity of the plant raw material, the results might conform to the acceptance thresholds set for the bioanalytical procedures. The value of the linear correlation coefficient (r) for the acceptance of the calibration curve for bioanalytical procedures must be 0.98 or higher (BRASIL and Agência Nacional de Vigilância Sanitária – ANVISA 2019). The calibration curve of the hydroethanolic extract of *M. caesalpiniifolia* agrees with the norms of the current legislation for the quality control of plant extracts and herbal remedies once the R value of 0.9986 (Figure 11.7) was achieved.

Using the equation of the line ($A = 0.00789C – 0.01336$), the absorbances of the aliquots of the dissolution test were calculated in a dissolved concentration and released percentage of the extract (Figure 11.8). Forty-five minutes after the beginning of the test, the F3 formulation was shown more than 100% of it dissolved; at the same time, F7 presented 82% of dissolution. Therefore, both formulations agree with the basic requirements of the Brazilian legislation, which states the minimal dissolution of 75% of the drug 45 minutes after the beginning of the dissolution test. The formulation developed with starch 1500, F3, showed a better dissolution profile than did F7, which

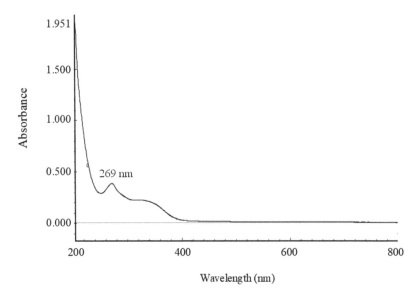

FIGURE 11.6 Absorption spectra in the UV-VIS of the hydroethanolic of *M. caesalpiniifolia*.

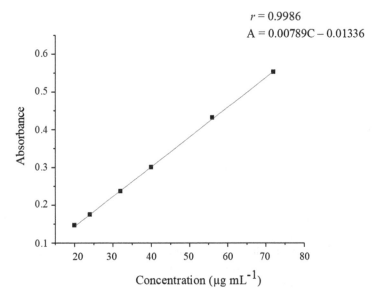

FIGURE 11.7 Calibration curve of the hydroethanolic extract of *M. caesalpiniifolia* in HCl solution 0.1 M (λ maximum 269 nm).

has microcrystalline cellulose as diluent. This difference in performance is due to the disintegrating action of starch, whereas microcrystalline cellulose, if used in high concentrations, may cause a decrease in the dissolution speed of active substances that have low solubility (Rowe, Sheskey, and Quinn 2009; Lima Neto and Petrovick 1997).

Once the outlined tablets F3 and F7 were approved in the pharmacopedic assays – the object of this study – through a factorial planning it was possible to develop two formulations for tablets of the standardized extract of *M. caesalpiniifolia* produced by direct compression, a method that allows fast manufacturing and low cost. These two formulations differ from each other only in the type of diluent. The co-processed diluents showed their potential of direct compression, particularly

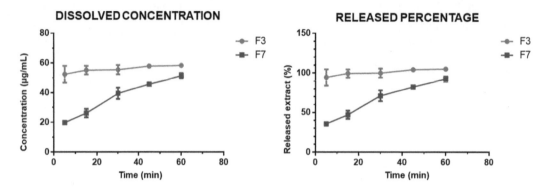

FIGURE 11.8   Release profiles of two formulations of tablets dissolved with hydroethanolic extract of *M. caesalpiniifolia* (F3 and F7).

conveying a plant extract. The diluent factor was responsible for significant differences among the results of the tests of each lot and these differences were not a rejection criterion for the formulations.

11.4   CONCLUSION

Different formulations of tablets based on the standardized extract of *M. caesalpiniifolia* were developed and tested. The spray drying of the extract with the use of a drying adjuvant improved the flow aspect, which favored the pharmaceutical operations of tablet preparation. Through the factorial planning analysis, it was possible to select the two best extract formulations. Different from each other only regarding the type of diluent, both formulations presented hardness, friability, disintegration, and dissolution favorable to the requirements established by the Brazilian Pharmacopeia. However, the use of the diluent pre-gelatinized starch yielded tablets with faster disintegration and dissolution. Both formulations were produced by direct compression, the method that allows lower costs and manufacturing time.

ACKNOWLEDGEMENTS

The author would like to thank UFPI (Federal University of Piauí, Brasil), CNPq (National Board for the Scientific and Technological Development, Brazil), and FAPEPI (Research Supporting Foundation of the State of Piauí, Brazil) for the financial support through the process FAPEPI/ SESAPI/MS/CNPq N° 003/2013 – Research Program for SUS ("PPSUS"). We also thank Prof. Abilio Borghi for English proofreading.

REFERENCES

Albuquerque, Ulysses Paulino De, Medeiros, Patrícia Muniz De, De Almeida, Alyson Luiz S., Monteiro, Marcelino, De Freitas, Ernani Machado, Neto, Lins, and DeMelo, Joabe Gomes. 2007. "Medicinal Plants of the Caatinga (Semi-Arid) Vegetation of NE Brazil: A Quantitative Approach." *Journal of Ethnopharmacology* 114: 325–54. https://doi.org/10.1016/j.jep.2007.08.017.
Allen, Jr, Loyd V., Nicholas G. Popovich, and Howard C. Ansel. 2005. *Ansel's Pharmaceutical Dosage Forms and Drug Delivery Systems*. 8th ed. Baltimore, Md: Lippincott Williams & Wilkins.
Aulton, Michael E, and Kevin M G Taylor. 2016. *Aulton Delineamento de Formas Farmacêuticas*. 4th ed. GEN Guanabara Koogan.
BRASIL, and Agência Nacional de Vigilância Sanitária - ANVISA. 2019. "Brazilian Pharmacopoeia, 6th Edition." https://www.gov.br/anvisa/pt-br/assuntos/farmacopeia/farmacopeia-brasileira/volume-1-ingles.pdf/@@download/file/Volume%201%20-%20Ingl%C3%AAs.pdf.

Brasil, Ministério da Saúde, Secretaria de Ciência Tecnologia e Insumos Estratégicos em Saúde, and Departamento de Assistência Farmacêutica. 2016. *Política e programa nacional de plantas medicinais e fitoterápicos*. 1st ed. Brasília: Ministério da Saúde. https://pesquisa.bvsalud.org/bvsms/resource/pt/mis-41804.

Cristina, Ana, Toledo, Oltramari, Hirata, Lilian Lúcio, Buffon, Marilene Da Cruz M., Miguel, Marilis Dallarmi, and Miguel, Obdulio Gomes. 2003. "Fitoterápicos: Uma Abordagem Farmacotécnica / Herbal Medicines: Pharmacotechnique Approach." *Lecta-USF* 21 (1/2): 7–13.

Couto, R. O., R. R. Araujo, L. A. Tacon, E. C. Conceicão, M. T. F. Bara, J. R. Paula, and L. A. P. Freitas. 2011. "Development of a Phytopharmaceutical Intermediate Product via Spray Drying." *Drying Technology* 29 (6): 709–18. doi:10.1080/07373937.2010.524062.

Cury, B. S. F., Silva Júnior, N. P., and Castro, A. D. 2008. "Influência Das Propriedades de Granulados de Celulose Nas Características Físicas Dos Comprimidos." *Revista de Ciências Farmacêuticas Básica e Aplicada* 29 (1). https://rcfba.fcfar.unesp.br/index.php/ojs/article/view/491.

Fahr, Alfred. 2021. *Voigt Pharmazeutische Technologie*. 13th ed. Deutscher Apotheker Verlag.

Jivraj, Mira, Luigi G. Martini, and Carol M. Thomson. 2000. "An Overview of the Different Excipients Useful for the Direct Compression of Tablets." *Pharmaceutical Science & Technology Today* 3 (2): 58–63. doi:10.1016/S1461-5347(99)00237-0.

Lima, Handerson Rodrigues Silva, Sá, Laisa Lis Fontinele De, Sousa, Francisco Daniel Leal, and Nunes, Lívio César Cunha. 2016. "Development of a Practical Tool for Analysis of Data Obtained by Factorial Design 22 e 23 Using the Software Microsoft Office Excel®2007 (Microsoft Corporation, EUA)." *Boletim Informativo Geum* 7 (2): 39–47. https://revistas.ufpi.br/index.php/geum/article/view/5394.

Lima Neto, Severino Antonio de, and Petrovick, Pedro Ros. 1997. "A Celulose Na Farmácia." *Caderno de Farmácia* 13 (1): 19–23. https://lume.ufrgs.br/handle/10183/19247.

Lira, L. M., C. B. Dornelas, and L. M. Cabral. 2008. "Avaliação de Bentonita Sódica Purificada e Bentonita Sódica Intercalada Como Promotores de Dissolução de Clorpropamida Em Comprimidos Preparados Por Granulação Úmida e Compressão Direta." *Revista de Ciências Farmacêuticas Básica e Aplicada* 29 (2). https://rcfba.fcfar.unesp.br/index.php/ojs/article/view/482.

Monção, Nayana Bruna Nery, Luciana Muratori Costa, Daniel Dias Rufino Arcanjo, Bruno Quirino Araújo, Maria Do Carmo Gomes Lustosa, Klinger Antônio Da França Rodrigues, Fernando Aécio De Amorim Carvalho, Amilton Paulo Raposo Costa, and Antônia Maria Das Graças Lopes Citó. 2014. "Chemical Constituents and Toxicological Studies of Leaves from Mimosa Caesalpiniifolia Benth., a Brazilian Honey Plant." *Pharmacognosy Magazine* 10 (Suppl 3): S456. https://doi.org/10.4103/0973-1296.139773.

Ngo, S. N. T. 2007. "When Do Differences in Dissolution Profiles Predict Clinical Problems?" *Journal of Clinical Pharmacy and Therapeutics* 32 (2): 111–112. https://doi.org/10.1111/J.1365-2710.2007.00807.X.

Oliveira, Olivia Werner, and Pedro Ros Petrovick. 2010. "Secagem Por Aspersão (Spray Drying) de Extratos Vegetais: Bases e Aplicações." *Revista Brasileira de Farmacognosia* 20 (4): 641–650. https://doi.org/10.1590/S0102-695X2010000400026.

Pessanha, Ana Flávia de Vasconcelos, Rolim, Larissa Araújo, Peixoto, Monize Santos, Da Silva, Rosali Maria Ferreira, and Rolim-Neto, Pedro José. 2012. "Influência Dos Excipientes Multifuncionais No Desempenho Dos Fármacos Em Formas Farmacêuticas." *Revista Brasileira de Farmácia* 93 (2): 136–45. https://silo.tips/downloadFile/influencia-dos-excipientes-multifuncionais-no-desempenho-dos-farmacos-em-formas.

Prista, L. Nogueira, Correia Alves, A., and Morgado, Rui. 2008. *Tecnologia Farmacêutica*. 5th ed. Lisboa: Fundação Calouste Gulbenkian.

Rowe, Raymond C., Sheskey, Paul J., and Quinn, Marian E. 2009. *Handbook of Pharmaceutical Excipients*. 6th ed. Grayslake, USA; Washington, USA: Pharmaceutical Press; American Pharmacists Association.

Santos, Márcio E. P., Moura, Lucas H. P., Mendes, Marcelo B., Arcanjo, Daniel D. R., Monção, Nayanna B. N., Araújo, Bruno Q, Lopes, José A. D., et al. 2015. "Hypotensive and Vasorelaxant Effects Induced by the Ethanolic Extract of the Mimosa Caesalpiniifolia Benth. (Mimosaceae) Inflorescences in Normotensive Rats." *Journal of Ethnopharmacology* 164 (April): 120–128. https://doi.org/10.1016/J.JEP.2015.02.008.

Sausen, Tiago Rafael, and Paulo Mayorga. 2013. "Excipientes Para a Produção de Comprimidos Por Compressão Direta." *Infarma - Ciências Farmacêuticas* 25 (4): 199–205. https://doi.org/10.14450/2318-9312.V25.E4.A2013.PP199-205.

Serra, Cristina Helena dos Reis, and Storpirtis, Sílvia. 2007. "Comparação de Perfis de Dissolução Da Cefalexina Através de Estudos de Cinética e Eficiência de Dissolução (ED%)." *Revista Brasileira de Ciências Farmacêuticas* 43 (1): 79–88. https://doi.org/10.1590/S1516-93322007000100010.

Soares, L A L, and P R Petrovick. 1999. "Física Da Compressão." *Cadernos de Farmácia* 15 (2): 65–79. https://edisciplinas.usp.br/pluginfile.php/3851197/mod_folder/content/0/Soares%2C%201999-Fisica-da-compressao.pdf.

Thoorens, Gregory, Krier, Fabrice, Leclercq, Bruno, Carlin, Brian, and Evrard, Brigitte. 2014. "Microcrystalline Cellulose, a Direct Compression Binder in a Quality by Design Environment—A Review." *International Journal of Pharmaceutics* 473 (1–2): 64–72. https://doi.org/10.1016/J.IJPHARM.2014.06.055.

Toller, Aline Brondani, and Cleber Alberto Schimdt. 2005. "Excipientes à Base de Celulose e Lactose Para Compressão Direta." *Disciplinarum Scientia | Saúde* 6 (1): 61–80. https://doi.org/10.37777/877.

Vasconcelos, E. A. F., Medeiros, M. G. F., Raffin, F. N., and Moura, T. F. A. L.. 2005. "Influência Da Temperatura de Secagem e Da Concentração de Aerosil®200 Nas Características Dos Extratos Secos Por Aspersão Da Schinus Terebinthifolius Raddi (Anacardiaceae)." *Revista Brasileira de Farmacognosia* 15 (3): 243–249. https://doi.org/10.1590/S0102-695X2005000300015.

Yunes, Rosendo A., Pedrosa, Rozangela Curi, and Filho, Valdir Cechinel. 2001. "Fármacos e Fitoterápicos: A Necessidade Do Desenvolvimento Da Indústria de Fitoterápicos e Fitofármacos No Brasil." *Química Nova* 24 (1): 147–152. https://doi.org/10.1590/S0100-40422001000100025.

12 Ethnobotanical Review of Wild Edible Plants in Serbia

Milica Aćimović and Biljana Lončar

12.1 INTRODUCTION

Since ancient times, plants that grow and reproduce naturally in their natural habitat, without of human influence, are recognized as wild (Motti 2022). Among them, there are edible, medicinal, and poison plants, plants for animal nutrition, firewood, fiber, and so on. All these plants ensured the existence of humans, and therefore humans have gathered them since ancient times, and they have become part of the human diet and traditional food systems (Cheng et al., 2022). However, different geographical and ecological conditions cause the biodiversity of wild edible plants characteristic of each region (Balemie and Kebebew 2006, Uprety et al., 2012).

The north part of Serbia is a fertile plain known as the Pannonian Basin, which is characterized by well-developed agriculture, while the central and south parts of Serbia consists mainly of hills and low to medium-high mountains (Dinaric Alps, Carpathian, Rila-Rhodope and Balkan Mountains) (Savić et al., 2008, Bogdanov et al., 2017, Grasgruber et al., 2022). In that part, which is characterized by extensive agricultural production (Zlatibor and Pčinja district, Kopaonik, Suva planina and Rtanj Mountains), the traditional application of wild edible plants is much better preserved, as well as the traditional way of preparing old dishes and healing methods (Jarić et al., 2007, Šavikin et al., 2013, Zlatković et al., 2014, Jarić et al., 2015, Živković et al., 2020, Matejić et al., 2020).

In the past, it was food for poor people; today they are delicatess and specific gastronomic offers of rural touristic destinations (Blanco-Salas et al., 2019). Formerly, a remedy for local people, nowadays they're potential for the pharmaceutical industry (Pawera et al., 2020). Autochthonous varieties of cereals, wild fruits and berries, mushrooms, wild medicinal, and other weed plants today are very popular among people who strive to get closer to nature (Turner et al., 2011). Additionally, scientific confirmation of numerous biological activities of old plant varieties and traditional recipes are confirmed daily (Aceituno-Mata et al., 2021, Radovanović et al., 2023).

For these reasons, in this book chapter, we will review more than 70 wild edible plants of Serbia, which belong to different families: Asteraceae (*Achillea* sp., *Arctium lappa*, *Artemisia absinthium*, *Carlina acaulis*, *Cichorium intybus*, *Inula helenium*, *Helianthus tuberosus*, *Matricaria chamomilla*, *Silybum marianum*, *Taraxacum officinale*, *Tussilago farfara*), Acoraceae (*Acorus calamus*), Sapindaceae (*Aesculus hippocastanum*), Rosaceae (*Agrimonia eupatoria*, *Alchemilla vulgaris*, *Filipendula vulgaris*, *Fragaria vesca*, *Rosa canina*, *Rubus* sp.), Amaryllidaceae (*Allium ursinum*), Malvaceae (*Althaea officinalis*), Ericaceae (*Arctostaphylos uva-ursi*, *Vaccinium myrtilis*), Brassicaceae (*Armoracia rusticana*), Betulaceae (*Betula pendula*), Fagaceae (*Castanea sativa*, *Corylus avellana*, *Quercus* sp.), Amaranthaceae (*Chenopodium album*), Cornaceae (*Cornus mas*), Equisetaceae (*Equisetum arvense*), Rubiaceae (*Galium verum*), Gentianaceae (*Gentiana lutea*), Lamiaceae (*Glechoma hederacea*, *Lamium album*, *Leonurus cardiaca*, *Marrubium vulgare*, *Melissa officinalis*, *Mentha* sp., *Origanum vulgare*, *Salvia* sp., *Satureja* sp., *Teucrium chamaedrys*, *Thymus* sp.), Fabaceae (*Glycyrrhiza glabra*, *Medicago sativa*, *Robinia pseudoacacia*), Araliaceae (*Hedera helix*), Cannabaceae (*Humulus lupulus*), Hypericaceae (*Hypericum perforatum*), Juglandaceae (*Juglans regia*), Cupressaceae (*Juniperus communis*), Linaceae (*Linum usitatissimum*), Lythraceae (*Lythrum salicaria*), Malaceae (*Malus sylvestris*, *Crataegus* sp.), Moraceae (*Morus nigra*), Orchidaceae (*Orchis morio*), Papaveraceae (*Papaver rhoeas*), Plantaginaceae (*Plantago major*),

DOI: 10.1201/9781003395935-12

Portulacaceae (*Portulaca oleracea*), Primulaceae (*Primula veris*), Amigdalaceae (*Prunus* sp.,), Boraginaceae (*Pulmonaria officinalis*), Polygonaceae (*Rumex crispus*), Capryfoliaceae (*Sambucus nigra*), Tiliaceae (*Tilia* sp.), Urticaceae (*Urtica diorica*), Scrophulariaceae (*Verbascum phlomoides*) and Violaceae (*Viola odorata*). According to this list, as well as from other studies, plants from Asteraceae (11), Lamiaceae (11), and Rosaceae (5) families are most frequently used (Miskoska-Milevska et al., 2020, Aćimović et al., 2020b, Radovanović et al., 2023).

This chapter is based on folk beliefs (Popović-Radović 2010, Čajkanović 2022), ethnobotany, and ethnopharmacology (Tucakov 2006), but modern scientific confirmations and recommendations for use also support it.

12.1.1 *ACHILLEA* SP. (YARROW)

Achillea L. genus, also known under the generic name of yarrow, in Serbian *hajdučka trava* (in Balkan folkloric tradition, the word *hajduk* derived from the Turkish word *haidut* or *haydut* referred to 'bandit', which was originally used by the Ottomans to refer to Serbian infantry soldiers, who treated wounds with this plant, similar to Greek mythological character Achilles) (Čajkanović 2022). It is one of the most frequent genera that includes approximately 140 herbaceous perennial species (Pleşca and Blaga 2019). These are widespread across southeast Europe and southwest Asia but are also distributed in North America (Radulović et al., 2007). The leaves are lanceolate, highly dissected, and highly variable in size, while the flowers (white, yellow, or pink), formed in a flat-topped corymb, appear from May to October (Chandler et al., 1982). All *Achillea* sp. exhibit a characteristic aromatic odor, and they are highly appreciated for their medicinal and ornamental properties. *A. millefolium* aggregate is a complex group of scarcely separable species that inhabit very diverse habitats. On the other side, a phenomenon of endemism within the genus is also present, 36.7% of the total number of *Achillea* occurs on the territory of the Balkans or in a narrower area (Radulović et al., 2007). Nineteen species are defined in *Flora of Serbia*, and the hybrids within the genus are very common (Josifovic, 1975).

Yarrow, *A. millefolium* (Figure 12.1a), a widespread species, is well known and most widely used in Serbian traditional medicine, known as *hajdučka trava*. It is consumed internally for improving appetite and for stomach disorders as herbal tea before meals, or in combination with other herbs (coltsfoot, lemon balm, mallow) for treating respiratory ailments such as colds, influenza, cough, and bronchial asthma, or externally, in form of dried ground plant directly on wound for wound healing, treating ulcers or hemorrhoids (Jarić et al., 2007, Šavikin et al., 2013). Modern science proved medicinal properties such as wound healing, antioxidative, estrogenic, antispermatogenic, antidiabetic, antiulcer and many other activities of yarrow (Saeidnia et al., 2011, Radovanović et al., 2022).

A. clypeolata, moonshine yarrow, *žuta hajdučka trava* in Serbian, in areas where it grows wild, is also used as a medicinal plant as an antidiabetic infusion (Zlatković et al., 2014), for treating kidney problems, improving appetite, and a soothing tea (Jarić et al., 2015). In addition, the available literature suggests that *A. clypeolata* may have anti-inflammatory, wound healing, digestive health, and potentially anticancer properties (Mueller et al., 2010, Andritoiu et al., 2020, Menković and Tadić, 2021).

12.1.2 *ACORUS CALAMUS* (SWEET FLAG)

Sweet flag is a semi-aquatic, perennial, grass-like monocot aromatic medicinal plant found in moist habitats (bank of ponds, rivers, and swamps) throughout the northern hemisphere (Ibdah et al., 2022). It has a creeping and extensively branched, aromatic rhizome, cylindrical, up to 2.5 cm thick, which is purplish brown to light brown externally and white internally (Balakumbahan et al., 2010). The leaves are long, slender, sword-shaped and simple, arising alternately (Kumar, 2013). It blooms from June to July, the inflorescence is cylindrical, about 12 cm long, and the flowers are

FIGURE 12.1 Plants used as wild edible in Serbia: a) *Achillea millefolium*, b) *Aesculus hyppocastaneum*, c) *Agrimonia eupatoria*, d) *Alchemilla vulgaris*, e) *Allium ursinum*, f) *Althaea officinalis*, g) *Arctium lappa*, h) *Arctostaphylos uvae-ursi*, i) *Artemisia absinthium*. Photographs by the authors.

yellow-green (Igić et al., 2010). The plant has a rich ethnobotanical history dating back possibly to the time of Moses in the Old Testament of the Bible and in early Greek and Roman medicine (Motley, 1994).

The Serbian name for sweet flag is *idirot*. Traditional medicine uses the rhizome (*Calami rhizoma*), which has a characteristic aromatic smell and bitter and pungent taste (Tucakov, 2006). As root decoction sweet flag is used for digestive problems, primarily for improving appetite, and against colic, ulcer, diarrhea, and intestinal gas (Šavikin et al., 2013, Živković et al., 2020). Historically, sweet flag rhizome has been used to address various mental health issues, including but not limited to schizophrenia, psychoneurosis, insomnia, hysteria, epilepsy, and memory loss (Škobić et al., 2019). In southeastern Serbia, traditionally it has been used as an antitumor plant (Kojičić et al., 2022).

Modern science approved that sweet flag rhizome has a wide range of biological properties, including sedative, anticonvulsant, immunosuppressant, antidiabetic, anti-inflammatory effects, and hepatoprotective and nephroprotective activity, as well as anti-cancer properties (Škobić et al., 2019, Das et al., 2019, Nasir, 2021).

12.1.3 *AESCULUS HIPPOCASTANUM* (HORSE CHESTNUT)

The horse chestnut (Figure 12.1b) or conker tree is native to the Balkan Peninsula region, up to 1.200 m asl. It develops an oval crown, bearing large shade-giving leaves composed of 5–7 palmate leaflets. Numerous white hermaphrodite flowers are born in a pyramidal inflorescence that appears in May and June. They have sharp spines and contain one to three seeds. The seed, ripe during September to October, resembles the chestnut fruit in its dark brown colour with a large whitish scar-like mark (Igić et al., 2010, Ravazzi and Caudullo, 2016).

Horse chestnut in Serbian is known as *divlji kesten*. Traditional medicine in Serbia recommends horse chestnut dried ripe seed (*Hippocastani semen*) for tincture preparation, widely used for healing rheumatism, varicose veins, and improving circulation, often in combination with other plants (buttercup herb and lilac flowers) (Šavikin et al., 2013, Jarić et al., 2015, Matejić et al., 2020). Apart from these, seeds are traditionally used to treat symptoms of chronic venous insufficiency, such as pain, heaviness, tension, swelling, and itching in the legs, as well as nocturnal symptoms (Ćalić-Dragosavac, et al., 2010, Vujic et al., 2013). In addition, horse chestnut bark (*Hippocastani cortex*) is collected in December and used as an astringent and for skin diseases (such as lupus) (Tucakov, 2006).

The seeds contain saponins, starch, fatty oil, proteins, and vitamins and are suitable for use for feeding animals, especially horses, justifying the origin of the common name (Ravazzi and Caudullo, 2016). However, the primary active compound found in horse chestnut seeds is escin, a group of chemically related triterpene glycosides, which possess anti-inflammatory, vasoconstrictor, and vasoprotective properties (Vašková et al., 2015). It is successfully used in the management of hemorrhoidal disease in the pharmaceutical industry (Ezberci and Ünal, 2018).

12.1.4 *AGRIMONIA EUPATORIA* (AGRIMONY)

Agrimony (Figure 12.1c) is a perennial herbaceous plant with a short rhizome and an upright, up to 1 m high, hairy stem with few branches. The leaves are leathery and plumose, and the lower ones frequently form a rosette. The flowers are yellow and star-shaped, arranged in thick, spiky bunches; the fruit grows downwards, from June to September (Igić et al., 2010, Muruzović et al., 2017). Agrimony inhabits pasture lands across Europe, and it is common in Serbia.

In the Serbian language, agrimony is known as *petrovac* or *ranjenik*. The plant has gained recognition for its traditional use and has been extensively researched. In traditional medicine, the whole plant is used (*Agrimoniae herba*) internally as an infusion for dyspepsia, urinary incontinence, cystitis and rheumatism, colds, and laryngitis, while externally, a leaf directly on the wound is used for treating wounds and cuts (Jarić et al., 2007, Jarić et al., 2015).

According to previous studies, agrimony is very rich in secondary metabolites and exhibits a range of properties, including anti-inflammatory, antiviral, neuroprotective, antidiabetic, anti-obesity, hepatoprotective, antimutagenic and anticancer effects (Horikawa et al., 1994, Kwon et al., 2005, Lee et al., 2010, Yoon et al., 2012, Ad'hiah et al., 2013, Al-Snafi 2015, Muruzović et al., 2016).

12.1.5 *ALCHEMILLA VULGARIS* (LADY'S MANTLE)

Lady's mantle (Figure 12.1d) is an aggregate species found throughout Europe, especially on upland grassland and verges (Tobyn et al., 2010). It is perennial species, with a woody rhizome and a basal rosette with corrugated and lobed kidney-shaped to semicircular leaves. Characteristic water droplets are exuded by the leaves when air humidity is high. The tiny yellow-green flowers occur in dense, terminal compound cymes from June to September (Vlaisavljević et al., 2019).

In Serbian traditional medicine lady's mantle is known as *virak*, *gospin plašt*, or *vrkuta*. The leaf (*Alchemillae folium*) is collected during the flowering period, and it is used as an infusion against menstrual problems and other women's illnesses (fibroids, cysts, endometriosis, infertility, and reproductive and thyroid hormones balancing), while external use (rinse) is recommended for vaginal itching and bleeding between periods (Šavikin et al., 2013, Jarić et al., 2015). Furthermore, the aerial parts of this plant are frequently employed to address gastrointestinal conditions, nose bleeds, and inflammatory processes. Due to a variety of biological activities, including antimicrobial and anti-inflammatory properties, they are also valued for their ability to enhance wound healing (Vlaisavljević et al., 2019, Tadić et al., 2020).

Lady's mantle is widely used in folk medicine throughout the world (Ghedira et al., 2012, Jelača et al., 2022). Scientists also showed that this plant possesses significant anti-inflammatory, antimicrobial, and antioxidant activities (Boroja et al., 2018, Kanak et al., 2022).

12.1.6 *Allium ursinum* (wild garlic)

Wild garlic (Figure 12.1e), ramsons, broad-leaved garlic, or bear's garlic, is a perennial plant species widespread through Europe and Asia, but rare in the Mediterranean region (Pavlović et al., 2017). Its vegetation starts early in spring, in March, with petioled basal leaves; blade lanceolate to ovate-elliptic (Tutin, 1957). Its bulb is narrow, elongated, about 1.5–6 cm long, surrounded by layers of clear skin with only a few fibers at the base. Sometimes daughter bulbs are formed (Sobolewska et al., 2015). At the top of the stem white flowers are formed, gathered in thyroid inflorescences (Gordanić et al., 2022). The flowering period of wild garlic is between April and May (Voća et al., 2021). The leaves, bulbs, and flowers of wild garlic possess characteristic distinctive smell and taste, and they are edible (Pluta-Kubica et al., 2022).

Wild garlic has been utilized in Serbian folk medicine for centuries and is known as *sremuš* or *medvedi luk*. Wild garlic is traditionally used in nutrition and treatment (Gordanić et al., 2021). Fresh leaves are eaten as salad, spice, or an ingredient in traditional dishes for digestive problems, against anemia, various inflammations, diabetes, obesity, and as a stimulant (Šavikin et al., 2013, Zlatković et al., 2014, Tomšik et al., 2016, Nićetin et al., 2022). In the form of tincture, it is known for lowering cholesterol (Jarić et al., 2015).

The results of numerous studies suggest that wild garlic can be considered as an antioxidant, antimicrobial, cardioprotective, anti-inflammatory and cytotoxic raw material valuable in medicine and nutrition (Ivanova et al., 2009, Pejatović et al., 2017, Rankovic et al., 2021, Voća et al., 2021, Gordanić et al., 2022).

12.1.7 *Althaea officinalis* (marshmallow)

Marshmallow (Figure 12.1f) is a perennial herbaceous plant, with a thick branchy root, whitish yellow outside, white and fibrous within (Rančić et al., 2009). The stem is upright, up to 150 cm high, leaves spirally arranged on the steam, with a triangular-ovate, bluntly toothed or shallowly divided petiole (Igić et al., 2010). It blooms from June to September, and the flowers are light pink or white, located in the axils of the top leaves, single or grouped in an inflorescence (Josifovic, 1975).

Marshmallow in Serbian is well known as *beli slez*. It is a typical mucus plant, widely used in Serbian traditional medicine. Roots, leaves, and flowers (*Althaeae radix, folium et flos*) are used for the same purposes, internally, as a cold infusion of dried catted root for bronchitis, asthma, and whooping cough, while leaves are used for urinary infections and against diarrhea (Jarić et al., 2007, Zlatković et al., 2014, Živković et al., 2020). In addition, the marshmallow root has been found to be particularly effective in treating respiratory illnesses due to its high concentration of mucilage, providing relief for symptoms including coughing, inflammation, and phlegm build-up (Rančić et al., 2009).

Interest in marshmallow in medicine and pharmacy is significant. Investigations showed that it might have been used as an adjunct medication to accelerate wound healing in skin burns (Zabihi et al., 2023). Furthermore, the antidiabetic, neuroprotective, and antimicrobial properties of marshmallow were also detected (Jafari-Sales et al., 2015, Changizi Ashtiyani et al., 2015, Arab et al., 2017).

12.1.8 *Arctium lappa* (greater burdock)

Greater burdock (Figure 12.1g) is a biennial herb. The stem reaches up to 2 m, and has multiple branches, each of which is topped by a flower head. Leaves are large, alternate, and cordiform with a long petiole and are pubescent on the underside. The lower leaves are very large, on long, solid foot-stalks, furrowed above, while the upper leaves are much smaller, more egg-shaped in form and not so densely clothed beneath with the grey down. The flowers are purple and grouped in globular capitula, united in clusters. They appear in midsummer, from July to September. The capitula are surrounded by the involucres of many bracts, each curving to form a hook (Salama and Salama, 2016).

Greater burdock, in Serbia, is known as *veliki čičak* (because of its ability to cling to other objects, greater burdock is used as a magical plant that causes bonding or joining). The leaf is applied directly on the wound for treating ulcers and festering wounds, rheumatism, and herb as an infusion against cold and cough, liver disease, and urinary infections (Jarić et al., 2007, 2015, Zlatković et al., 2014, Matejić et al., 2020). In addition, greater burdock is also applied for the treatment of high blood pressure and diarrhea (Marković et al., 2022a).

Greater burdock has been used for centuries to treat a variety of ailments (Guna 2019). This plant has a vast range of traditional benefits, and its various parts have been used to support overall health and well-being for centuries (Shaheen et al., 2017). In addition, apart from medicinal properties, greater burdock root can be grated or stir-fried and added to soups, stews, and other dishes for its crunchy texture and slightly sweet flavor (Hong and Gruda, 2020).

12.1.9 *Arctostaphylos uva-ursi* (common bearberry)

Common bearberry (Figure 12.1h) is a prostrate evergreen shrub with thick, alternate leaves on short petioles. The leaf surface is dark green, lighter below. As fall comes around, the leaves begin to change in color from their summer dark green to a deep red. In early spring, buds begin to loosen and flower and leaves begin turning dark green. Flowers occur in a drooping apical racemose inflorescence (3–5 flowers per inflorescence). The fruit is a red flattened mealy drupe, resembling miniature apples, which reach full maturity in late July–August and remain red (Young et al., 2012, Shamilov et al., 2021).

In Serbian, common bearberry is known as *medvede grožde*. The leaves are collected in the spring during the flowering season due to their powerful astringent activity (Tucakov, 2006, Radulović et al., 2010). Common bearberry has been traditionally used internally as an infusion to promote diuresis and treat urinary tract infections, kidney stones, and prostate issues (Jarić et al., 2007, Šavikin et al., 2013, Jarić et al., 2015, Živković et al., 2020). It is often combined with other herbs such as birch, licorice, and corn silk for treating urinary disorders (Tucakov, 2006).

Common bearberry leaves and preparations made from them are used in traditional and modern medicine because of their antibacterial and anti-inflammatory activities, widely used in urinary tract infections treatment (Vučić et al., 2013, Dell'Annunziata et al., 2022).

12.1.10 *Armoracia rusticana* (horseradish)

Horseradish is a hardy, perennial herb with white and thickened, pungent roots. The plant forms a rosette of large, entire margined leaves having long flowering stalks with four-merous white flowers that are borne in a terminal panicle and angustiseptate fruits (Sampliner and Miller, 2009, Walters, 2021).

Horseradish in Serbian is known as *ren* or *hren*. The root (*Armoracia rusticana radix*), which has a spicy taste and smell, is traditionally used as food and medicine (Tucakov, 2006). Horseradish is utilized to aid breathing and improve the appetite (eaten fresh), treat stomach ulcers (mixture with honey and grape brandy and wood ash lye), or compress for stomach pain, arthritis, sinusitis, bronchitis, and asthma (Jarić et al., 2015, Živković et al., 2020, Matejić et al., 2020).

The root is a rich sulphur compound source and possesses potent antimicrobial activity against soil-borne and human pathogens. In addition, these compounds have been found to exhibit a range of other health benefits, including anticancer, antioxidative, and detoxification properties (Blagojević et al., 2020).

12.1.11 *Artemisia absinthium* (wormwood)

Wormwood (Figure 12.1i) is a perennial herb native to Europe. The stem and branches are white, very downy, with ridges and furrows and finely silky and hairy, and glandular throughout. The leaves are 2–3 pinnatifidly cut into spreading linear-lanceolate, obtuse segments, hairy on both surfaces. The flower heads are in large cymes. The heads are heterogamous; ray flowers are female and disc flowers are hermaphrodite (Ahamad, 2019).

Wormwood in Serbia, known as *pelin*, is a typical bitter taste, an aromatic remedy used for treating stomach problems, improving appetite, and strengthening the immune system, as a tonic (Jarić et al., 2015, Živković et al., 2020). It is believed to have antimicrobial, antifungal, and antiparasitic properties, which may explain why it is used as a balm against parasitic worms in children with liquorice and aniseed mixed with jam (Tucakov, 2006, Boudreau et al., 2022). It is used to prepare infusions and alcoholic beverages; the most popular in Serbia is wormwood brandy – *pelinkovac*. Worldwide, an alcoholic beverage is made by macerating wormwood in a strong spirit, such as brandy or vodka, along with other herbs and spices (Tonutti and Liddle, 2010). In addition, wormwood is the main ingredient of the legendary drink absinthe (Goud et al., 2015).

Wormwood is a popular remedy of traditional medicine and ensured its study and subsequent introduction into official medicine (Sharifi-Rad et al., 2022). In traditional European medicine, it has been used as an effective agent in gastrointestinal ailments and in the treatment of helminthiasis, anaemia, insomnia, bladder diseases, difficult-to-heal wounds, and fever (Szopa et al., 2020). It possesses anti-inflammatory, immunomodulatory, hepatoprotective, anti-helminthic, and antidepressant activity (Hashimi et al., 2019).

12.1.12 *Betula pendula* (birch)

Birch is a tree native to Europe, Asia, and North America. Birch grows as a tree or bush-like tree, up to 30 m high, with one or many stems. The bark is smooth and silvery white above, often black and fissured into rhomboid bosses at the base. Branches may be pendulous, spreading or ascending, but constant throughout the tree. Twigs are glabrous and brown, with pale warts, especially on the younger twigsn. Birch buds are acute, not sticky. They are collected in the appropriate vegetative period of the plant before opening, preferably in the early spring or before the end of winter. Leaves are coarsely and unequally double-serrated. Inflorescences develope as unisexual catkins; the male develop in summer, shedding pollen the following spring, a few days after female flowers have emerged. Female catkins are smaller and develop into fruits (Atkinson, 1992, Beck and Caudullo, 2016, Penkov et al., 2018, Vladimirov et al., 2019).

Birch, as a remedy in Serbian traditional medicine, is known as *bela breza* (the word is of Proto-Slavic origin, Russian *берёза*, Belarus *бяроза*, Ukrainian *береза*). Usually used are leaf buds (*Betulae gemmae*) and young birch leaves (*Betulae folium*), collected early in the spring specifically for treating hyperglycemia, kidney problems, prostate issues, and skin diseases (Tucakov, 2006). In addition, the infusion made from these plant parts is believed to have diuretic, anti-inflammatory, and antiseptic properties, which may help to alleviate these health conditions (Šavikin et al., 2013,

Jarić et al., 2015, Matejić et al., 2020). In the Serbian folk belief, birch has apotropaic properties; witches and evil spirits are afraid of it, and when the wedding guests leave home for church, they pass over the birch broom in the hope that it will keep spells at bay (Čajkanović, 2022).

It has been used for centuries in traditional medicine for various health benefits Bljajić et al., 2016. It is known that birch has exhibited antimicrobial, antioxidant, and anticancer properties, and is often used for treating diabetes and urinary tract disorders (Bljajić et al., 2016, Efthimiou et al., 2022). Moreover, ethnobotanical and ethnopharmacological use of the birch in many European countries is widespread. The leaf and bark have been known as colorant substances in decoctions, for dying linen and wool to yellow, green, brown, or grey, while bark has been applied for tanning leathers (Papp et al., 2014).

12.1.13 *CARLINA ACAULIS* (STEMLESS CARLINE THISTLE)

The stemless carline (Figure 12.2a) thistle is a perennial plant that grows in dry and rocky habitats. The rhizome is massive and vertically set in the ground. The leaves are elliptic-oblong, pinnatifid to pinnatisect, up to 30 cm long, and set in a rosette. The stem rarely reaches 15 cm in height, and usually, the plants are acaulescent. The capitula (25–50 mm diameter) are surrounded by silvery white or pale pink bracts. Flowering occurs from July to September. The flowers are hermaphrodites. The fruits are achenes 5 mm long (Trejgell et al., 2009).

In Serbian, this plant is known as *vilino sito* (literally fairie's sieve – after the fruits fall from the receptacle, a round-like structure with sieve-like holes remains; in Serbian belief, it is an amulet that attracts God's gratitude, brings health, understanding, wealth, and happiness into the home). The root of stemless carline thistle (*Carlinae radix*) extracted in autumn is used for medicinal purposes in Serbian folk medicine (Tucakov, 2006). The root tends to mold quickly, and must be dried immediately and stored in a dry place. The powdered root is used to make a gruel, which is then applied directly to the skin to treat a variety of skin complaints such as ulcers, acne, eczema, and wounds (Jarić et al., 2007, Thas, 2008).

The root of the stemless carline thistle contains several bioactive compounds, such as inulin, known for its anti-inflammatory, antimicrobial, and wound-healing properties. Furthermore, inulin is a carbohydrate that is slowly digested and absorbed by the body, which means it does not raise blood sugar levels as rapidly as other types of carbohydrates. Because of this, the root of the stemless carline thistle is also used as a food source and is recommended as a dietary supplement for people with diabetes (Tucakov, 2006, Strzemski et al., 2019).

12.1.14 *CASTANEA SATIVA* (CHESTNUT)

Chestnut, or sweet chestnut, is a medium-large deciduous tree native tree to Europe. The bark is brown-greyish and often has net-shaped venations with deep furrows or fissures. Leaves are oblong-lanceolate, with a dentate-crenate margin and a brighter green upper leaf surface. This species of tree is monoecious and flowers develop in late June to July. Male flowers are gathered in catkins whereas female flowers are usually positioned at the base of the male ones in the upper part of the current year's shoots. By autumn, the female flowers develop into spiny cupules (commonly called bur) containing 3–7 brownish nuts that are shed during October (Conedera et al., 2016, Krebs et al., 2019).

In Serbia, chestnut is commonly referred to as *pitomi kesten*. It is a popular tree in the country and is often used for fruit and wood (Jarić et al., 2015). In traditional medicine, various parts of the sweet chestnut tree are used to treat various ailments, such as respiratory problems, stomach issues, and skin conditions (Dajić Stevanović et al., 2014). The extract prepared with chopped fruit is kept in plum brandy for 40 days, then filtrated, is used as an anti-rheumatic remedy, as well as to control coughing (Jarić et al., 2007).

FIGURE 12.2   Plants used as wild edible in Serbia: a) *Carlina acaulis*, b) *Cicorium intybus*, c) *Cornus mas*, d) *Crataegus monogyna*, e) *Equsetum arvense*, f) *Filipendula vulgaris*, g) *Fragaria vesca*, h) *Galium verum*, i) *Gentiana lutea*. Photographs by the authors.

Many European folk medicines used chestnuts in the treatment of skin and soft tissue infections (Quave et al., 2015). In addition, the chestnut fruit is well known for its nutritional properties, namely its high concentration of carbohydrates (starch) and its low-fat content (Santos et al., 2022).

12.1.15   *Chenopodium album* (lamb's quarters)

Lamb's quarters is the principal weed species. It is a cosmopolitan, annual, broad-leaved herbaceous plant, yielding a large number of seeds (up to 70,000 seeds per plant) (Bana et al., 2022). The stem is striped with greyish to bluish-green and red. It tends to grow upright at first, reaching heights of 20–250 cm, but typically becomes recumbent after flowering (due to the weight of the foliage and seeds). Leaves are greyish green to green, sometimes after flowering reddish green; margin shallowly and irregularly serrate to dentate; apex acute; leaf blade rhombic to elliptic in basal and middle leaves and lanceolate to elongated-lanceolate almost completely in the upper part. The plant flowers from late May to October. The small flowers are radially symmetrical and grow in small cymes on a dense branched inflorescence (Bassett and Crompton, 1978, Grozeva, 2014).

In Serbia, lamb's quarters is called *bela pepeljuga* (literally ash-coloured because of grayish ash coating over the leaf) or *divlji spanać* (literally wild spinach – leaves are an acceptable substitute for spinach). It is a widely distributed plant in the country and is often used as a food source due to its high nutritional value (Anđelković et al., 2021). In traditional medicine, various parts of the plant are used to treat gastrointestinal issues, respiratory problems, abdominal pain, and skin conditions (Matejić et al., 2020).

Lamb's quarters is a valuable plant in herbal medicine around the world as well as a plant with significant nutritious benefits (Polito et al., 2016, Saini and Saini, 2020). It is used as an anthelmintic, antiphlogistic, antirheumatic, antidiarrheal, antioxidant, antimicrobial, contraceptive, laxative, odontalgic, and so on, as well as a popular green leafy vegetable (Saini and Saini, 2020).

12.1.16 *Cichorium intybus* (chicory)

Chicory (Figure 12.2b) is a plant native to Europe that is now widely distributed worldwide in temperate and semi-arid regions. Chicory is an erect perennial herb, with a rhizome that is light yellow from the outside and white from within, containing bitter milky juice. Radical and lower leaves are 7.5–15 cm long, while upper leaves are alternate, small, entire, and their bases clasp the stem. Heads are ligulate 2.5–3.8 cm. in diameter, terminal and solitary or axillary and clustered, sessile or on short, thick stalks. Flowers are white to light blue and lavender, toothed at the ends. There are two rows of involucral bracts; the inner is longer and erect in comparison to the outer bracts which are shorter and spreading. Flowering occurs from July to October (Das et al., 2016, Prasanna Kumar et al., 2020).

In the Serbian language, this plant is known as *vodopija* (it has a long, spindle-shaped root that reaches the water even when it is dry) or *ženetrga* (the name dates back to the time when this root was used to induce abortion mechanically). It is a common plant in the country, growing along roadsides and in fields. In Serbian traditional medicine, it has a long history of use: the dried crushed root (*Cichorii radix*) is made into decoctions for digestive upsets and to improve appetite; it reduces blood sugar and acts as diuretic and laxative (Jarić et al., 2007, Šavikin et al., 2013, Zlatković et al., 2014). The flowering plant (*Cichorii herba*) is less often used for the same medicinal purposes as the root.

The root is extracted in the autumn, when it is the thickest and richest in inulin. Interestingly, the root of chicory can also be used as a substitute for coffee. When roasted and ground, the root has a rich, nutty flavor similar to coffee but without caffeine. This makes it a popular choice for people wanting to reduce their caffeine intake or looking for a caffeine-free alternative to coffee (Tucakov, 2006, Khan and Khatoon, 2008). However, it is believed to have similar benefits to the root and may be used to treat digestive issues and other ailments in some traditional medicine practices (Khare, 2008).

12.1.17 *Cornus mas* (cornel)

Cornel or cornelian cherry (Figure 12.2c) is a bushy shrub or small tree producing olive-shaped red fruits. The trunk is straight and the bark is grey-brownish, peeling off in scaly flakes like crocodile skin. The young shoots are hairy grey-greenish, becoming hairless later. The leaves are opposite with a short stalk, oval, 3–5 cm wide and 6–8 cm long, with an entire margin that is shortly acuminate and supplied with visible parallel veins. They turn to mahogany red in autumn. The flowers are small, yellow, hermaphroditic, and clustered in umbels. They bloom before the leaves sprout (March). The fruit is a fleshy, bright red cherry-like drupe, which ripens in mid-late summer (Da Ronch et al., 2016).

Cornel in Serbian is known as *dren* (the word is of Proto-Slavic origin, Belarus *дзёран звычайны*, Ukrainian *дерён справжній*). In traditional medicine, it is used fresh or as dried fruits (*Corni fruits*) as astringent, against diarrhea, as immune system strengthening, and for treatment of anemia when prepared as decoction (Jarić et al., 2007, Zlatković et al., 2014). In addition to their medicinal uses, the fruits of the cornel plant are often used in culinary applications. They can be made into juices, marmalade, and jam, and are commonly used to produce a traditional Serbian brandy known as *drenjevača*. This brandy, by fermenting and distilling the fruits of the cornel plant, has a distinct flavor and is often consumed as a digestive or traditional remedy for colds and flu (Tucakov, 2006, Radovanović et al., 2013). According to popular belief, dogwood has an extraordinary healing power, which is what the proverb "healthy as a dogwood" means.

Cornel fruits, flowers, and leaves have been traditionally used against sore throat, measles, chicken pox, anemia, kidney and liver ailments, and digestive system disorders in many parts of the world (Çevik et al., 2022). Cornel fruits contain sugar and a mineral salt, polyphenols, monoterpenes, organic acids, and vitamins, and therefore are used for the treatment of many disorders, as well as a nutrient source, consumed dried or fresh as well as processed (Adams et al., 2007, Szczepaniak et al., 2019, Zeliha and Koca, 2021).

12.1.18 *Corylus avellana* (European filbert)

European filbert or common hazel is typically a shrub with a branched stem. The bark is grey with white and large spots. The leaves are deciduous, rounded, with a double serrate margin and hairy on both sides. The flowers appear in early spring, before the leaves; male catkins are usually grouped together and are yellowish-brown and up to 10 cm long, while female catkins are very small. The fruit is a nut, grouped in clusters of one to four together. Each nut is held in a short leafy involucre (husk) that encloses about half of the nut. The nut is roughly spherical, up to 2 cm long, and yellow-brown with a pale scar at the base (Enescu et al., 2016).

European filbert in the Serbian language is known as *leska* (the word is of Proto-Slavic origin, Rusian *лещина обыкновённая*, Belarus *ляшчына звычайная*, Ukrainian *ліщина звичайна*). In traditional medicine, leaf and bark (*Corylli avellana folium et cortex*) are used as vasoconstrictor remedies to treat venous inflammations and hemorrhoids, mainly externally, using balms and ointments. Additionally, the leaf and bark infusion can be used internally to treat diarrhea and heavy menstruation (Tucakov, 2006, Jarić et al., 2007).

Apart from their medicinal uses, the hazelnuts (*Corylli avellana fructus*) are commonly used as food and in confectionery. Hazelnuts are one of the most popular tree nuts on a worldwide basis, used as a nutritious food, with a high content of healthy lipids (Bottone and Cerulli, 2019). Furthermore, they are often used as an ingredient in various sweet treats such as chocolates and pastries. Additionally, hazelnuts can be processed to extract fatty oil (*Corylli avellana oleum*), which is used in the production of skin and hair preparations (Allegrini et al., 2022). On the other side, leaves and bark are used in folk medicine for the treatment of hemorrhoids, varicose veins, phlebitis, and lower members' oedema, as a consequence of its astringency, vasoprotective, and anti-edema properties (Oliveira et al., 2007).

12.1.19 *Crataegus* sp.

Hawthorn species (*Crataegus* sp.) (Figure 12.2d) are deciduous, most often thorny shrubs or low-growing trees. In Serbia, it is known as *glog* (the word is of Proto-Slavic origin, Belarus *глог*, ukraina *глід*). The leaves are alternate, lobed to pinnately cut, simple or double serrated, or with a whole rim. The flowers are small and arranged in rich clusters. Five calyx leaves are short, triangular, drooping, or remaining on the fruit. Five coronal leaves are round, usually white, rarely pink. The flowers develop at a time when the leaves have not reached average size. The fruit is pomum;

stone-shaped; red, black, or orange (Đordjević and Nikolić, 2021). The genus is very rich in species, varieties, and hybrids, and in Serbia's flora, eight species are noted, however, the most commonly used are *C. laevigata* [syn. *C. oxyacantha*] (red hawthorn or midland hawthorn) and *C. monogyna* (white hawthorn). In Slovenic, hawthorn is the most powerful apotropaic against vampires, witches, and evil demons (Čajkanović, 2022).

Red hawthorn in the Serbian language is known as *crveni glog*. In traditional medicine, the inflorescence with leaves that are collected in May are used as a mild sedative against insomnia and as a blood pressure–lowering agent. They are also known to reduce the tone of smooth muscles, specifically the uterus and intestines. The ripe fruits (*Crataegi folium cum flore*) that are collected during September–October (Tucakov, 2006) are used as berries and for the preparation of tincture, marmalade, jellies, juices, and jams.

Another similar species, white hawthorn, in Serbian *beli glog*, in traditional medicine is used as a leaf infusion for heart failure, against hypertension and atherosclerosis, and intestinal diseases (Šavikin et al., 2013, Zlatković et al., 2014). Additionally, the plant's fruit is a rich source of vitamin C (Živković et al., 2020).

Hawthorn leaf, flower, and fruit, as well as their preparations, have a long history in European countries since that time, predominantly in the prevention and treatment of heart failure (Đordjević and Nikolić, 2021). However, they have long been used for medicinal purposes such as digestive disorders, hyperlipidemia, and dyspnea, inducing diuresis, and preventing kidney stones (de Quadros et al., 2017).

12.1.20 *Equisetum arvense* (common horsetail)

Common horsetail (Figure 12.2e) is a perennial plant with a thin black horizontal root. The permanent underground part of the stem is black in color and, like a thread, it extends into depth and width. It is very resistant (Đurić et al., 2019). It forms two types of stems (a spring stem and a summer stem) and the part used for this plant is its infertile bases that contain chlorophyll particles (Wang et al., 2023).

Common horsetail in Serbia is known as *rastavić*. Traditional medicine uses the dried aerial part of the plant (*Equiseti herba*) for preparation infusions and tinctures, known to have diuretic properties, which help reduce fluid retention in the body. They are also believed to have anti-inflammatory and antioxidant effects, which may contribute to their therapeutic benefits when used for urogenital tract ailments such as kidney diseases and kidney stones, urinary infections, and fluid retention (Šavikin et al., 2013, Zlatković et al., 2014, Shikov et al., 2014).

Pharmacological studies showed that common horsetail possessed antioxidant, anticancer, antimicrobial, smooth muscle relaxant effects on the vessels and ileum, anticonvulsant, sedative, antianxiety, dermatological immunological, antinociceptive, anti-inflammatory, antidiabetic, diuretic, inhibition of platelet aggregation, promotion of osteoblastic response, anti-leishmanial, and many other effects (Asgarpanah and Roohi, 2012, Al-Snafi, 2017).

12.1.21 *Filipendula vulgaris* [syn. *F.* hexapetala] (dropwort, meadowsweet)

Dropwort (Figure 12.2f), also known as meadowsweet, grows wild in Europe, Asia, and northwest Africa, mainly on dry nonacidic grasslands and sunny slopes (Rubio et al., 2022). It is a perennial plant that produces short, tuberous rhizomes with clonal growth. The flower stalk is straight up to 80 cm high. It has double-odd-pinnate leaves, clustered in a rosette at the base of the shoot. Its flowers are small, cream-colored, and gathered in a paniculate inflorescence (Barabasz-Krasny et al., 2022).

In Serbian, dropwort is known as *suručica* or *medunika*. Traditional medicine uses the flowering tops and leaves of the plant (*Ulmariae summitas et folium*) for various medicinal purposes. One of the primary uses of dropwort is for lowering blood pressure. It is also known for its astringent and tonic properties, which can help strengthen the body and improve overall health. Additionally, drop-wort has been used for treating hepatitis (Jarić et al., 2007, Matejić et al., 2020).

Traditionally dropwort is used for fever, pain, inflammatory diseases (arthrosis, rheumatism, and arthritis), gastric disorders, liver dysfunction, and gout (Farzaneh et al., 2022). The results showed that this plant possesses anti-inflammatory activity, which supports the use for relieving pain in diseases (Samardžić et al., 2016).

12.1.22 *FRAGARIA VESCA* (WILD STRAWBERRY)

Wild strawberry (Figure 12.2g) is a perennial herb species with short stalks and rounded leaves. It occurs in light forests of Eurasia from the lowlands up to the mountains (Malinίková et al., 2013). The leaf is on a sparsely haired stem. Leaves are green, basal and palmate, compound, in groups of three. Leaflets are coarsely toothed, nearly hairless on the top surface, prominently veined, and oval to egg-shaped. The flowers are with five round to oval white petals with about 20 yellow stamens surrounding a yellow center. They are usually in clusters of two to five. The fruit is a small red straw-berry that is egg shaped to conic with tiny seeds raised on the surface (Igić et al., 2010).

In Serbia, wild strawberry is known as *šumska jagoda*. The leaves of the wild strawberry plant are commonly used for external application to treat hemorrhoids. The leaves are known for their astringent and anti-inflammatory properties, which help reduce inflammation and bleeding in the affected area. Wild strawberries are also used internally for treating digestive disorders such as diarrhea. The plant contains tannins, which have astringent properties that help reduce inflam-mation in the digestive tract. The leaves are often prepared as a tea or infusion, which is then consumed to relieve symptoms of diarrhea. In addition to their use for digestive disorders, wild strawberries are also used as a natural remedy for coughs (Jarić et al., 2007, Zlatković et al., 2014). The fresh fruit is used in skin care products due to their high vitamin C content. Vitamin C is known to have antioxidant properties that help in protecting the skin from damage caused by free radicals (Šavikin et al., 2013).

Wild strawberries have been grown in gardens at least since the time of the Romans (Liston et al., 2014). They have nutritional benefits, being a good source of vitamin C, folate, and potassium. They are also low in calories and high in fiber, making them a healthy addition to the diet (Alvarez-Suarez et al., 2014).

12.1.23 *GALIUM VERUM* (YELLOW BEDSTRAW, LADY'S BEDSTRAW)

Yellow bedstraw (Figure 12.2h) is common throughout Europe, North Africa, and Asia, but it can occur in North America. It is a perennial herbaceous plant, with elongated stems growing to 60–120 cm. The leaves are glossy and dark green, while flowers are golden yellow, grouped in many-flowered panicles. The aerial parts of this plant are collected during dry and sunny days of the blooming period. Its golden yellow flowers are present from June to September (Bradic et al., 2021).

In Serbian, yellow bedstraw is known as *ivanjsko cveće*. In traditional medicine, it has been used for various purposes; the upper half of the flowering plant (*Galii very herba*) is used to prepare infusions that are known for their diuretic and sedative properties (Tucakov, 2006). These infusions are often used to relieve symptoms associated with urinary tract infections, as well as to promote relaxation and improve sleep quality. In addition to its internal use, yellow bedstraw is also used externally to treat various skin conditions such as wounds, ulcers, acne, and burns. The plant is prepared as an infusion or ointment and applied directly to the affected area to promote healing and reduce inflammation (Jarić et al., 2007). Furthermore, Serbian herbalists often use this plant against malignant throat diseases (Jarić et al., 2015). This is the favorite flower of the Serbian people, it is dedicated to St. John, and wreaths are woven from it and sprinkled on houses to protect against evil spirits and diseases (Popović-Radović, 2010, Čajkanović, 2022).

Previous pharmacological studies have shown that the yellow bedstraw exhibits various medicinal properties, including antioxidant, central nervous system regulatory, antimicrobial, endocrine, anti-hemolytic, and cytotoxic effects (Lakić et al., 2010, Zabihi and Pashapour, 2022).

12.1.24 *Gentiana lutea* (yellow gentian)

Yellow gentian (Figure 12.2i) is native to the mountainous regions of Europe, including the Alps, the Pyrenees, the Carpathians, and the Balkans. It is also found in some parts of western Asia (Momčilović et al., 2001). However, it is endangered in its natural habitat in many countries due to overexploitation, so the interest in its cultivation is increasing due to increased demand (Radanović et al., 2016). It is a long-lived, rhizomatous plant. The stem rises 3 or 4 feet long or more, with a pair of wide lanceolate to elliptic leaves contrary to one another, at every joint. In summer, the plant produces inflorescences up to 120 cm, with yellow flowers grouped in pseudo-whorls with a corolla divided into five open lobes (Prakash et al., 2017, Dettori et al., 2018).

In Serbian, yellow gentian is known as *žuta lincura*. Its root (*Gentianae radix*) is used primarily as a digestive aid and is typically taken before meals. The root is commonly prepared in the form of an alcoholic tincture, wine, or bitter gentian brandy – *rakija od lincure* – in case of loss of appetite and digestive problems (Marković et al., 2021). It is also believed to have immune system strengthening properties, making it a popular natural remedy for combating fever and other illnesses (Jarić et al., 2015, Šavikin et al., 2013, Živković et al., 2020).

Traditionally, worldwide the yellow gentian has long-term use in traditional medicine as a stomachic, digestive, or bitter tonic to treat dyspepsia, gastric inefficiency in infants, catarrhal diarrhea, anaemia, malarial disease, and more (Kusšar et al., 2006, Prakash et al., 2017). Moreover, phytochemical investigations show presence of large number of secondary metabolites with promising bioactivities, such as iridoids and secoiridoids, flavonoids, triterpenoids, alkaloids and others (Yang et al., 2010).

12.1.25 *Glechoma hederacea* [syn. *Nepeta glechoma*] (ground ivy)

Ground ivy (Figure 12.3a) is native to Eurasia, however it is introduced and widespread in territories of the USA and Europe (Sipek et al., 2021). It is a perennial herb with creeping, rooting stems and ascending, flowering branches. Leaves are 6 to 76 mm long and kidney to heart-shaped with long petioles. Two to five flowers are borne in a series of whorls. Flowers are usually blue-violet with purple spots on the lower lip, but they are sometimes pink or white (Hutchings and Price, 1999).

In Serbia, ground ivy is known as *dobričica*. According to traditional medicine fresh plant juice (*Glechomae succus*) or dried aerial parts in flowers (*Glechomae herba*) collected between May and June can be used as an infusion to treat bronchitis, improve appetite, strengthen the immune system, and against diarrhea (Jarić et al., 2007, Janaćković et al., 2019). Ground ivy can also be used externally to treat various ailments. It is commonly used to treat inflammation of the nasal mucous membrane, as well as wounds and injuries. The herb is believed to have anti-inflammatory properties that can help reduce swelling and promote healing. To use ground ivy externally, a poultice or compress can be made by crushing the fresh leaves and applying them directly to the affected area (Tucakov, 2006, Popović et al., 2014).

Ground ivy has a long history of use in traditional medicine, but still, there is a lack of complete and scientifically based bioactivities (Šeremet et al., 2020). There are approved antioxidant, anticancer, antimicrobial, anti-inflammatory, and wound healing properties (Lee et al., 2018, Grabowska et al., 2022, Gwiazdowska et al., 2022, El-Aasr et al., 2023). Beside medicinal properties, ground ivy possesses potential use in foodstuffs (Milovanovic et al., 2010).

FIGURE 12.3 Plants used as wild edible in Serbia: a) *Glechoma hederacea*, b) *Glycyrrhiza glabra*, c) *Hedera helix*, d) *Helianthus tuberosus*, e) *Humulus lupulus*, f) *Hypericum perforatum*, g) *Leunurus cardiaca*, h) *Inula helenium*, i) *Linum usitatissimum*. Photographs by the authors.

12.1.26 *Glycyrrhiza glabra* (common liquorice)

Liquorice (Figure 12.3b) is a perennial herb native to the Mediterranean region, central to southern Russia and Asia Minor to Iran. It grows in subtropical climates in rich soil, and the root system is extensive with a main taproot and numerous runners. The main taproot, which is harvested for medicinal use, is soft, fibrous, and has a bright yellow interior. The plant grows as a shrub, up to 1 m in height, with pinnate leaves, with nine to 17 oval leaflets. The flowers are purple to pale whitish blue, produced in a loose inflorescence. The fruit is an oblong pod, containing several seeds (Sharma et al., 2018, Dastagir and Rizvi, 2016).

In Serbia, common liquorice is known as *slatki koren* or *sladić* (literally sweet root). For medicinal purposes, the root (*Liquiritiae radix*) of the plant is particularly valued for its therapeutic properties. One of the most important uses of liquorice root is as a mild expectorant. It is often included as an ingredient in pulmonary teas, where it can help to loosen phlegm and mucus from the respiratory tract. This makes it an effective treatment for conditions such as bronchitis, coughs, and colds (Petrović et al., 2022). In addition to its expectorant properties, liquorice root is also known to have anti-inflammatory properties. This makes it an effective treatment for a variety of conditions that are caused by inflammation, including gastritis and stomach ulcers. Another important use of liquorice root is as a laxative remedy. When combined with other herbs such as glossy buckthorn, caraway, aniseed, fennel, prunes, and flax seed, liquorice root can help to relieve constipation and promote regular bowel movements. This makes it an effective treatment for conditions such as irritable bowel syndrome and other digestive disorders. (Tucakov, 2006).

It has been well known in pharmacies for many years, and natural products have been a prime source for the treatment of many forms of ailments, many of which treatments are consumed daily with the diet (Sharma et al., 2018, Dastagir and Rizvi, 2016). Liquorice root can help to reduce inflammation in the mucous membranes of the stomach, which can reduce pain and discomfort associated with these conditions (Račková et al., 2007).

12.1.27 *Hedera helix* (English ivy)

English ivy or common ivy (Figure 12.3c) is a perennial evergreen herbaceous plant with natural distribution limited to Europe, North Africa, and West Asia, but it has spread worldwide in a large variety of habitat types (Diljkan et al., 2022). It can grow as a climbing plant, with aerial rootlets and matted pads which cling strongly to the substrate. English ivy has two types of leaves, with palmately five-lobed juvenile leaves on creeping and climbing stems, and unlobed cordate adult leaves on fertile flowering stems exposed to full sun. The flowers are produced from late summer until late autumn, individually small, in umbels, greenish-yellow, and very rich in nectar. The fruits are purple-black berries ripening in late winter (Igić et al., 2010). Ivy is used in love divination; girls put a wreath on their heads and see their perfect match in their dreams (Čajkanović, 2022).

In Serbia, English ivy is known as *bršljan* and is a popular ornamental plant that is often used to cover walls and buildings. In addition to its ornamental uses, common ivy has a long history of use in traditional medicine. External use involves tea prepared from leaves and applied as a compress on swollen tissue, and juice from the fresh leaves is applied directly onto the wound to treat wounds and ulcers (Jarić et al., 2007). Common ivy is also used topically to treat skin conditions such as eczema and psoriasis. One of the most important uses of common ivy in traditional medicine is as a treatment for respiratory conditions. The plant has expectorant properties, which can help to loosen mucus and phlegm from the respiratory tract. This makes it an effective treatment for conditions such as bronchitis, coughs, and colds when it is internally consumed as an infusion (Zlatković et al., 2014, Huntley and Ernst, 2000).

English ivy is classified as a conventional plant used as a medicinal product in the cure and prevention of upper respiratory tract inflammation and infection due to its pectolytic and bronchiolitis effects (Shokry et al., 2022). The plant has anti-inflammatory properties that can help to reduce swelling and redness, while its astringent properties can help to tighten and tone the skin (Happy et al., 2021).

12.1.28 *Helianthus tuberosus* (Jerusalem artichoke)

Jerusalem artichoke (Figure 12.3d) is a member of the sunflower family, a native plant of North America. It is a rhizomatous perennial herb that grows up to 3–4 m high, with an adventitious, fibrous root system that develops cord-like rhizomes, forming a fleshy tuber at the swollen apical part of the rhizome (Saengthongpinit and Sajjaanantakul, 2005, Pacanoski and Mehmeti, 2020). It is widely grown for its edible tubers, which have a sweet, nutty flavor and can be eaten raw or cooked. Leaves and stems are frost sensitive, but tubers survive and the plant can easily regrow from the tubers after winter. The leaves are simple, oval to lance-shaped, with coarsely toothed edges on well-developed petioles. The flower heads are bright yellow and resemble those of the cultivated sunflower but are only 3–5 cm in diameter (Swanton et al., 1992).

In Serbia, Jerusalem artichoke is known as *čičoćka*. In addition to its use as a food crop, Jerusalem artichoke is also used in traditional medicine to treat a variety of ailments, including diabetes, digestive problems, and skin conditions (Jarić et al., 2007, Jarić et al., 2015, Shariati et al., 2021).

Jerusalem artichoke can be used for human health and treatments can be extracted from its tubers, leaves, and flowers (Méndez-Yáñez et al., 2022). It contains inulins, dietary fiber, and as prebiotics it stimulates the growth of probiotic gut bacteria, mediates sugar and lipid metabolism, and stimulates the immune system (Dias et al., 2016). In addition, Jerusalem artichoke is considered to be one of the most promising multipurpose bioenergetic crops (Manokhina et al., 2022).

12.1.29 *HUMULUS LUPULUS* (HOP)

Hop (Figure 12.3e) is native to Europe and western Asia. However, it is now widely cultivated and can be found growing in many regions of the world (Naraine and Small, 2017). It is a perennial herbaceous plant up to 10 metres tall. The leaves of H. lupulus are dark green, heart-shaped with 3–5 lobes, and sharp-toothed with a smooth surface. The plant is dioecious, with male and female flowers on separate plants. The male flowers do not have petals, while the female flowers gather in a pendent, glandular, cone-like, light green inflorescence formed by persistent ovate and acute bracts, commonly called strobila (Igić et al., 2010, Nezi et al., 2022, Bektur et al., 2022).

In Serbia, hop is known as *hmelj* and is widely used in beer production. Dried female inflorescence (*Strobilus lupuli*) contains several metabolites used in brewing and pharmaceutics. In addition to its use in brewing, hop has a long history of use in traditional medicine. In Serbian traditional medicine, hop cones are commonly used as a sedative, for treating insomnia and anxiety. The plant is believed to have a calming effect on the nervous system, which can help to promote relaxation and improve sleep quality. The plant has anti-inflammatory and antiseptic properties that can help to reduce inflammation and fight bacterial infections. This makes it an effective treatment in the form of infusion for a variety of skin conditions (Jarić et al., 2015, Matejić et al., 2020).

Hops are traditionally used to relieve the symptoms of mental stress and treatment of sleep disorders; they stimulate the secretion of gastric juices and have a strong spasmolytic effect on intestinal smooth muscles (Bektur et al., 2022). Hop has strong therapeutic potential due to the presence of a wide range of bioactive molecules (Carbone and Gervasi, 2022). Apart from medicine, hops strobiles have been used as food flavoring to prepare spices, sauces, tobacco, and alcoholic beverages (Bektur et al., 2022).

12.1.30 *HYPERICUM PERFORATUM* (ST. JOHN'S WORT)

St. John's wort (Figure 12.3f) is characterized by very wide ecological amplitude and can grow under different environmental conditions (Đorđević, 2015). It is a perennial herbaceous plant with a strongly branched root system and a branched stem up to 100 cm high. The leaves are ovate, sessile, and bare with airy spots. Yellow flowers with black dots around the rim are collected in umbels from June to August (Igić et al., 2010).

St. John's wort in Serbian language is commonly referred as *kantarion*. The upper half of the plant in flower (*Hyperici herba*) is collected from May to September and is widely used in traditional medicine for its medicinal properties (Tucakov, 2006). Internally, St. John's wort is often consumed as an herbal tea for treating moderate depression and insomnia. The plant is believed to have a mood-enhancing effect, and can help to alleviate symptoms of anxiety and depression. It is also used to treat a range of gastrointestinal ailments, including stomach ulcers, liver and bile ailments, and jaundice. Additionally, St. John's wort is known for its immune system–strengthening properties (Matejić et al., 2020, Živković et al., 2020). Externally it is used in the form of infused oil or cream to treat a range of conditions, including burns, wounds, cuts, muscular pain, sciatica, and neuralgia (Jarić et al., 2007, Zlatković et al., 2014). According to Serbian folk belief, a plant harvested in the period from *Velika* (15.08.) to *Mala* (21.09.) *Gospoine*, so-called intermediate days, has a special healing effect and is worn as a charm (Čajkanović, 2022).

St. John's wort has a long tradition of use as a medicinal plant and is currently one of the most consumed medicinal plants in the world (Monteiro et al., 2022). The main biologically active compounds of St. John's wort are naphthodianthrones, primarily represented by hypericin and pseudo-hypericin, flavonoids and phloroglucinol derivatives (Bagdonaite et al., 2012). Its putative medicinal properties include wound-healing and diuretic, antibiotic, and antiviral effects, as well as antidepressive activity (Mennini and Gobbi, 2004, Wölfle, 2014). The plant has anti-inflammatory and analgesic properties, which can help to reduce pain and inflammation and promote healing, and therefore dermatological applications also have a long tradition (Phalen, 2012, Wölfle, 2014).

12.1.31 *INULA HELENIUM* (ELECAMPANE)

Elecampane (Figure 12.3h) is a perennial herbaceous plant that is native to Europe and western Asia. It is commonly found growing in meadows, along riverbanks, and in other damp, open areas (Shekhar et al., 2013). The root is thick, branchy and sticky, with a bitter taste. Leaves are large and toothed, with lower petiolate and the rest cover the stem; blades are ovoid, elliptical, or spear shaped, measuring 30 cm long and 12 cm wide. The leaves are green above with light, scattered hairs and whitish below due to a thick layer of wool. Flower heads are up to 5 cm (Qizi and Obidjon, 2021).

Elecampane in the Serbian language is known as *oman* (the word is of Proto-Slavic origin, Ukrainian *оман високий*) or *devesilje*. For medicinal purposes, tradition recommends using the root (*Inulae radix*) as an infusion or tincture for treating bronchitis, coughs, hay fever, asthma, tuberculosis, and pleurisy (Jarić et al., 2007). In addition, powdered root mixed with honey is used as a remedy for coughs (Šavikin et al., 2013). The flower of the elecampane plant is also used in traditional medicine to treat various conditions, including irregular menstrual bleeding, dry tinea infections and itchiness, and as a stomachic and anti-rheumatic remedy (Živković et al., 2020). Elecampane has a strong apotropaic power; it is worn as a charm against spells (Čajkanović, 2022).

Since ancient times, dried roots and rhizomes of elecampane have been used for the treatment of respiratory and dermal ailments (Kenny et al., 2022). Studies have shown that the roots of the elecampane plant possess a cytotoxic and antiproliferative effect on cancer cell lines, as well as anti-inflammatory, antioxidant, antibacterial, antifungal, and anthelmintic activities (Buza et al., 2020).

12.1.32 *JUGLANS REGIA* (WALNUT)

Walnut, commonly known as Persian or English walnut, is a species of tree native to the regions stretching from the Balkans to the Himalayas. It is a deciduous tree that can reach heights of up to 30 meters and has a broad crown with a diameter of up to 25 meters (Blondel and Aronson, 1999). The bark is silver-grey and smooth between deep, wide fissures. The leaves are 20–45 cm long, with five to nine leaflets, the ones from the apex being larger compared with those from the base of the leaf. Male flowers are single-stemmed catkins, while females are in clusters of three to nine, with or just after the leaves. The fruit ripens during hot summers and is a large, rounded nut of 4–5 cm (de Rigo et al., 2016).

Walnut in Serbian is known as *orah*. During June and July, young leaves (*Juglandis folium*) are collected and dried. Leaves infusion is taken internally as a digestive, tonic, and against constipation, while applied externally for cuts, grazes, and skin disorders such as eczema, herpes, and eruptive skin complaints (Jarić et al., 2007). Apart from leaves, rinds (*Juglandis fructus cortex*) are also a valuable traditional remedy infusion for treating diarrhea and anemia (Jarić et al., 2007), as well as pericarp (*Juglandis pericarpum*) heated with olive oil for treating inflammation of the ears (Šavikin et al., 2013). Unripe fruits (*Juglandis imaturi fructus*) are mixed with honey to treat hypothyrosis

(Šavikin et al., 2013). In folk medicine and witchcraft, the walnut plays an important role, and it is related to the "underworld" and demons (Čajkanović, 2022).

According to previous studies, walnut is a rich source of various phytochemicals and exhibits a range of properties, including antioxidant and antiproliferative activity (Anjum et al., 2017), anti-inflammatory activity (Fizeşan et al., 2021), antimicrobial activity (Delaviz et al., 2017), anticancer activity (Zhao et al., 2023), cardiovascular benefits (Anderson et al., 2001), and neuroprotective properties (Liu et al., 2019).

12.1.33 *JUNIPERUS COMMUNIS* (JUNIPER)

Juniper is a small evergreen tree or shrub that belongs to the Cupressaceae family. It is widely distributed throughout the northern hemisphere and can be found in many habitats, including woodlands, heaths, and rocky areas (Farjon and Filer, 2013). Leaves are needlelike, borne in whorls of three. The needles are sessile, and have one white band on the upper side. The fruit is cone-shaped, fleshy and berry-like, purple to black in colour, up to 1 cm, globose or longer than broad. They ripen in the second or third year (Enescu et al., 2016). It is typically harvested in the late summer or early autumn, when the berries are fully ripened and have turned a bluish-black color.

Juniper is known in Serbia as *kleka*. Fresh ripe juniper berries (*Juniperi fructus*) are too bitter to eat raw and are usually dried for further use to flavour meats, sauces, stuffings, and alcoholic beverages such as gin, liquors, bitters, and traditional Serbian juniper brandy – *klekovača* (Jarić et al., 2015). In addition, internally dried fruit infusion is used for kidney inflammation, poor digestion with wind, urethritis, and cystitis, and externally for rheumatic pain (Jarić et al., 2007, Šavikin et al., 2013).

The plant has a long history of use in traditional medicine and has been used for various medicinal purposes (Pirani et al., 2011). Modern science suggests that juniper has a range of pharmacological activities, including antimicrobial, diuretic, anti-inflammatory, antioxidant, and hypoglycemic activity (Cosentino et al., 2003, Simić et al., 2002, Raina et al., 2019, Al-Snafi et al., 2019).

12.1.34 *LAMIUM ALBUM* (WHITE DEAD-NETTLE)

White dead-nettle is native to Europe and western Asia. It has also been introduced to other parts of the world, including North America, where it is considered an invasive species in some regions (Luczaj et al., 2012). It is a perennial herb, with short, creeping rhizomes. Stems are simple, ascending to erect, sparsely hairy, and it grows 20 to 80 cm. On the cross-section the stem is four-angled. Leaves are opposite, petiolated, ovate to ovate-lanceolate, hairy, and net-veined, with pointed tips and coarsely toothed margins. The flowers are white and sessile, appearing from May to August in groups at the top of stems and branches in groups of three to nine from opposite leaf axils (Josifovic, 1975, Dénes et al., 2012).

White dead-nettle in Serbian is known as *bela mrtva kopriva*. The whole plant when flowering is collected as a remedy to regulate menstruation, against increased vaginal secretions, and for the treatment of kidneys and bladder disorders (Tucakov 2006). Fresh herb is also used as an antidote to snake bites (rub onto the bite having first pierced the skin with a sharp object) (Jarić et al., 2015).

Traditional medicinal uses of white dead-nettle have been reported to exhibit uterotonic, astringent, antispasmodic, and anti-inflammatory activities and therefore are utilized in menorrhagia, uterine hemorrhage, vaginal and cervical inflammation, and leukorrhea treatment (Yalçin and Kaya, 2006). In addition, a large number of phytochemicals are reported in this plant such as iridoid and secoiridoid glucosides, phenylpropanoids, flavonoids, essential oils, phytoecdysteroids, benzoxazinoids, and triterpene saponins, as well as megastigmenone and hemiterpene-type compounds (Morteza-Semnani et al., 2016). According to the literature, dead-nettle appears to have certain antioxidant, diuretic, and anti-inflammatory properties (Ríos et al., 2000, Salehi et al., 2019).

12.1.35 *Leonurus cardiaca* (motherwort)

Motherwort (Figure 12.3g) is native to Europe and Asia and has been introduced to North America and other parts of the world as an ornamental plant. It is a perennial herb that can grow up to 1.5 meters in height and has square stems that are covered with fine hairs (Wojtyniak et al., 2013). The leaves of motherwort are deeply lobed and toothed, with a hairy surface that is rough to the touch. The flowers are small and pink or purple, and grow in dense clusters at the end of the stems (Borna et al., 2017).

Motherwort in Serbia is known as *srdačica*. In traditional medicine, the aboveground parts (*Leonuri herba*) are collected during the flowering (from June to September). It is used internally as an infusion for treating heart complaints, calming effect on the heart, anaemia, and numerous problems associated with menstruation and urinary disorders (Jarić et al., 2007).

Several scientific studies have investigated the potential health benefits of motherwort, showing that extracts of motherwort may have anti-inflammatory effects and may help reduce oxidative stress and protect against certain diseases (Koshovyi et al., 2021). Yu et al., 2019 examined motherwort injection for preventing postpartum hemorrhage in women with vaginal delivery. Other studies have shown that motherwort may have cardiovascular benefits, including reducing blood pressure and heart rate, and improving circulation (Gompf, 2005, Arbeláez et al., 2018).

12.1.36 *Linum usitatissimum* (flax)

Flax (Figure 12.3i) is one of the oldest known cultivated plants in the world, with benefits from seeds and fibers (Arslanoğlu and Aytac, 2020). Flax is an annual herb, the stem is upright, thin, densely covered with leaves, and weakly branched at the top. Leaves are linear, lanceolate or ovate, attenuated at both ends, and acute at the apex. Flowers are small, blue, in terminal panicles in corymbose. Fruits are capsular with five cells containing 10–20 seeds that are compressed, ellipsoid, smooth, dark brown, and shining (Rashid et al., 2018).

Flax in the Serbian language is known as *lan*. Seeds (*Lini semen*) are applied as an infusion for the treatment of inflamed mucosa of the mouth, stomach and gut, and gastrointestinal inflammatory diseases (Živković et al., 2020). It is often used as a laxative (2–4 spoons of whole seeds) (Tucakov, 2006). Externally crushed seeds are used for the preparation of compress for treating skin ulcers and inflammations, while oil is used for the treatment of burns (Tucakov, 2006). In Serbian mythology, flax is often used in witchcraft because it is often associated with the "upper world", and sowing and all work related to flax is done only by women. (Čajkanović, 2022).

Since ancient times to the present day flax's use emanates from its seed and stem fiber (Liu et al., 2011, González and Deyholos, 2012). Flaxseed is obtaining considerable attention as a functional food due to the reported health benefits of its components (Kaithwas et al., 2011). Bioactive peptides derived from hydrolysed flaxseed protein have demonstrated protection against cardiovascular disease and antioxidant, opioid, bile acid binding, immunostimulating, mineral utilizing and antimicrobial activities (Marambe et al., 2008, Shim et al., 2014). Stems, which contain bast fibers characterized by respectable chemical composition, fineness, whiteness, and mechanical and sorption properties, have been widely used throughout history as materials for clothing items (Lazić et al., 2017).

12.1.37 *Lythrum salicaria* (purple loosestrife)

Purple loosestrife (Figure 12.4a) is an herbaceous perennial species native to Eurasia, present in North America, many regions of Canada, and Central Europe (Bastlová and Květ, 2002). It grows in wet and flooded places, so it can often be found along rivers, streams, and lakes (Srećković et al., 2020).

FIGURE 12.4 Plants used as wild edible in Serbia: a) *Lythrum salicaria*, b) *Marrubium vulgare*, c) *Matricaria chamomilla*, d) *Melissa officinalis*, e) *Mentha* sp., f) *Morus nigra*, g) *Origanum vulgare*, h) *Papaver rhoeas*, i) *Plantago major*. Photographs by the authors.

The aboveground shoots develop from winter buds formed on the rootstock in the previous year. The stem is erect, sparingly branched, 50–150 cm high, subglabrous to densely grey pubescent. Leaves are mostly opposite or in whorls of three, but the upper ones sometimes alternate, ovate to lanceolate-oblong, acute, sessile, truncate and semi-amplexicaul at the base. Flowers are trimorphic, in whorl-like cymes in the axils of small bracts, forming long, terminal spikes of reddish-purple, appearing during July and August (Olsson and Ågren, 2002, Piwowarski et al., 2015, Mattingly et al., 2023).

In the Serbian language, purple loosestrife is known as *vrbičica* or *potočnjak* (Tucakov, 2006). In traditional medicine, the top part of the plant in flowering (*Salicariae summitas*) is used for inflammatory diseases, gastrointestinal ailments, dysentery, and as astringent for external use (Matejić et al., 2020, Srećković et al., 2020).

Purple loosestrife pharmacological activity is mainly due to its phenolic compounds, primarily tannins (Humadi and Istudor, 2009). Scientific investigations of purple loosestrife demonstrated anti-inflammatory, anti-nociceptive, anti-hyperlipidemic, antiatherosclerotic, antioxidant, and anti-microbial effects (Tunalier et al., 2007, Guclu et al., 2014).

12.1.38 *MALUS SYLVESTRIS* (CRAB APPLE)

Crab apple is a woody wild apple species native to Western and Central Europe (Jacques et al., 2009). The trees can gain sizes up to 10 m, but frequently their growth habit reaches more shrubs than trees (Reim et al., 2012). Leaves are almost hairless, oval in shape and with small triangular teeth. The flowers are largely white, developing frompink buds. Fruits are less than 4 cm in diameter (McInerny and Gray, 2018). Crab apple was mainly used as a decorative plant because of its ornamental features, such as the color of its leaves and flowers (Kolarević et al., 2020).

Crab apple in Serbia is known as *šumska jabuka* and *kiselica*, where it is consumed fresh or as apple vinegar (Solieri and Giudici, 2009, Šarić-Kundalić et al., 2011). Decoction made from fruits acts as an immune system strengthener against colds, digestive disorders, and as an antihypertensive (Zlatković et al., 2014). It is used for weight loss and blood detoxification (vinegar mixed with water and honey) (Jarić et al., 2015).

Some studies have shown that wild apples possess more phenolic compounds (flavonol and phenolic acid) and higher antioxidant activity than cultivated apples (Iacopini et al., 2010, Jakobek and Barron, 2016, Mihailović et al., 2018, Astuti et al., 2022).

12.1.39 *MARRUBIUM VULGARE* (WHITE HOREHOUND)

White horehound (Figure 12.4b) is a perennial herb that grows naturally in the region between the Mediterranean Sea and Central Asia. It has a tough, woody, branched taproot or numerous fibrous lateral roots and numerous stems, which are quadrangular, erect, very downy, and from 20 to 100 cm high. Leaves are roundish, ovate, usually toothed, petiolate, veined, and hoary on the surface, and they are arranged in opposite pairs on a long stem. Inflorescences are formed in axils of upper leaves, with white flowers (Aćimović et al., 2021a). Leaves and tops have a musky odor and a pungent bitter and aromatic taste and are harvested just before full green color (Lodhi et al., 2017).

In Serbia, white horehound is known as *bela očajnica*. The aboveground parts collected during flowering (*Marrubii herba*) are used for medicinal purposes. As a bitter remedy, white horehound is used for treating gastrointestinal and respiratory disorders, as well as for treatment of menstrual difficulties, normalizing the menstrual cycle, female infertility, and against hemorrhoids (Zlatković et al., 2014, Jarić et al., 2015, Aćimović et al., 2020a).

Various uses of white horehound in folk medicine are mentioned, including as an expectorant, hypoglycemic, anti-thyphoid, antipyretic, antidiarrhetic, anti-icteric, diuretic, choleretic, and tonic, in bilious stimulation and for menstrual pains; it is especially known for anti-inflammatory activity (Zawiślak, 2012, El Abbouyi et al., 2013). White horehound is a valuable source of bioactive compounds and preparations with health-promoting effects: antioxidant, hepatoprotective, antiproliferative, anti-inflammatory, antidiabetic, and antimicrobial (Aćimović et al., 2020b, 2021b).

12.1.40 *MATRICARIA CHAMOMILLA* (CHAMOMILE)

Chamomile (Figure 12.4c) is an annual plant native to southern and Eastern Europe, and it can also be found in North Africa, Asia, North and South America, Australia, and New Zealand (Murti et al., 2012). The plant has a round, hollow, erect, branched stem, and grows to a height of 10–80 cm. The root is spindle-shaped, with roots only penetrating flatly into the soil. The leaves are feathery light green bipinnate to tripinnate. The flower heads are placed separately, daisy-like (Pirkhezri et al., 2010, Singh et al., 2011). It is a very important commercial crop in Serbia, with 345 ha it is the leading cultivated medicinal plant (Pavlić et al., 2023).

Chamomile in Serbian is known as *kamilica*. It is used internally as infusion for treating stomach ache, dyspepsia, and constipation; externally it is used as a tea for skin and mucous complaints (burns, wounds, ulcers), for vaginal douche, and steam inhalation for sinusitis (Jarić et al., 2007).

Around the world, chamomile is a highly favored and much used medicinal plant in traditional and official medicine (Singh et al., 2011). Chamomile has anti-inflammatory, antiseptic, antispasmodic, and mildly sudorific properties. It is mainly used as a tisane for stomach problems, diarrhea, nausea, urinary tract inflammation, and painful menstruation. Externally, it can be applied to wounds, skin eruptions, infections, hemorrhoids, and inflammation of the mouth, throat, and eyes. Chamomile oil is widely used in the perfumery, cosmetics, aromatherapy, and food industry (Singh et al., 2011, El Mihyaoui et al., 2022, Chauhan et al., 2022).

12.1.41 *MEDICAGO SATIVA* (ALFALFA)

Alfalfa is a forage crop distributed worldwide, with high agronomic and economic value, known for its exceptional persistence compared to other field crops (Ben-Laouane et al., 2020). Alfalfa is a perennial herbaceous legume with a strong taproot with many lateral roots. From buds near the soil surface (crown-produced buds) every spring develops between five and 15 stems up to 1 m in height. Tri- or multi-foliolate leaves form alternately on the stem, and the flowers are purple. It is a nectar source for honey production (Teuber et al., 1980).

Alfalfa in Serbian is known as *lucerka*. Young alfalfa leaves (*Medicaginis sativae herba*) are a rich source of nutrients and vitamins. It is often prepared in human nutrition as *medovina* – dry leaves boiled in water, cooled, and strained, then mixed with honey (Tucakov, 2006).

It has a long tradition of benefit in folk therapy in many parts of the world, with the most typical ethnomedical uses in treating digestive, vascular, reproductive, urinary tract, or respiratory system disorders (Wunsch, 2010). The plant displays antioxidant, anti-inflammatory, and antidiabetic properties (Karimi et al., 2013). Even though alfalfa extracts were widely used in traditional therapy to treat various conditions, including asthma, arthritis, diabetes, kidney, prostate, and bladder disorders, there is insufficient clinical evidence of their influence (Rafińska et al., 2017).

12.1.42 *MELISSA OFFICINALIS* (LEMON BALM)

Lemon balm (Figure 12.4d) is a perennial plant with natural distribution in central and western Asia and Europe, especially the Mediterranean basin; however, it is being naturalized worldwide and today is a commonly grown species in other parts of the globe (Taiwo et al., 2012). Lemon balm is a herbaceous, bushy plant species, with a hairy root system with numerous lateral roots. The stem is a square or quadrangular stem growing to a height of 50–150 cm. Leaves are hairy, ovate to cordate, and opposite decussate pairs. Small white flowers are arranged in pseudo whorls (Abdel-Naime et al., 2020). In Serbia it grows spontaneously, but it is also cultivated as an important industrial crop (approximately 150 ha in Serbia is the estimated area in 2021) (Pavlić et al., 2023).

Lemon balm in Serbian is known as *matičnjak*. Dried leaves for the preparation of infusion are taken for nervous anxiety, depression, tension headaches, and digestive disorders (Jarić et al., 2007, Šavikin et al., 2013). A lemony aroma justifies its common use as a spice, natural seasoning, or flavoring (Jovanovic et al., 2022).

Lemon balm has various applications such as a food additive, herbal tea, cosmetics ingredient, ornament, and medicament. It is commonly used for its anti-herpes, antiviral, anti-HIV, antioxidant, antimicrobial, anticancer and antitumor, anti-stress, anti-anxiolytic, antidepressant, anti-Alzheimer, anti-cardiovascular disease, memory improvement, concentration, and anti-inflammatory effects (Miraj et al., 2016). Traditionally it has been a remedy for treating catarrh, fever, flatulence, headaches, influenza and toothaches; it is known for sedative, antidepressant, antiviral, antibacterial, and antispasmodic effects (Boyadzhiev and Dimitrova, 2006, Aharizad et al., 2013).

12.1.43 MENTHA SP. (MINT)

Mentha sp. (Figure 12.4e) is native to Europe, western and central Asia, and northern and southern Africa (Huxley 1992). Species of the genus *Mentha* are very polymorphic and slanted to hybridization. Ten species are represented as wild growing in Serbian flora (Josifovic 1975), while peppermint (*M × piperita*) is a leading cultivated medicinal plant in Serbia with between 300 and 350 ha per year under cultivation (Pavlić et al., 2023). However, peppermint has been and remains one of the most widely used herbs for medicinal purposes (Shelepova et al., 2021).

Peppermint is a perennial herbaceous plant that became a natural hybrid between *M. viridis* and *M. aquatica* (Wei et al., 2023). The root is poorly developed and shallow, but it possesses an abundantly developed system of subterranean or epigeous stolons (Bokić et al., 2020). The plant is 30–100 cm tall, with reddish-purple stems, and is fragrant, toothed, and hairy on the underside leaves, with pinkish-purple flowers (Abbaszadeh et al., 2009). Peppermint, in Serbian *pitoma nana*, is cultivated because of dried leaves (*Menthae piperitae folium*) for treating insomnia, anxiety, and stomach disorders as infusion (Šavikin et al., 2013).

Among all wild species the most valuable in Serbia is pennyroyal (*M. pulegium*), in Serbian called *barska nana*, widely used against colds, as a digestive, and antispasmodic (Zlatković et al., 2014), followed by spearmint (*M. spicata*), called *divlja nana*, used against respiratory complaints, laryngitis, and women's illnesses (Jarić et al., 2015).

According to previously conducted studies mints posseses antioxindant potential, antimicrobial, antibacterial, and cytotoxic activities (Kanatt et al., 2007, Boukhebti et al., 2011, Zaidi and Dahiya, 2015, Alsaraf et al., 2021).

12.1.44 MORUS NIGRA (BLACK MULBERRY)

Black mulberry (Figure 12.4f) is a deciduous tree well adjusted to a vast region range, including Asia, Europe, and North and South America (Ahlawat et al., 2016). Black mulberry tree grows very rapidly, up to 10 m height, and the stem is very branched. Leaves are simple, alternate, thick, rough, dark green, broadly ovate, petiolate, and sometimes irregularly lobed. Flowers are unisexual spikes; the female flowers are borne in erect, cylindrical short capitate spikes, while the male flowers are in catkin-like spikes. Fruits are fleshy deep red drupes, cylindrical or ovoid (Arshad et al., 2018).

Black mulberry is in Serbian *crni dud*. In traditional medicine, leaves (*Mori nigrae folium*) are used for infusion preparation against obesity and diabetes, to rinse the mouth (inflammation of the throat and oral cavity), liver disease, abdominal pain, and constipation (Janaćković et al., 2019, Matejić et al., 2020). On the other hand, the fruits are rich in phenols and have a unique sour and refreshing taste, and are used as a food and for the preparation of syrups and marmelades (Tucakov, 2006, Miljković et al., 2014).

The black mulberry fruits and leaves possess antioxidant activity (Ercisli and Orhan, 2008, Sánchez-Salcedo et al., 2015), while its fresh juice has antimicrobial activity (Khalid et al., 2011) and anti-inflammatory properties (Chen et al., 2016). Also, mulberry fruits are used for the prevention and treatment of cardiovascular diseases (Jiang et al., 2011), to treat fever, protect the liver, strengthen the joints, facilitate discharge of urine, and lower blood pressure (Okatan et al., 2016).

12.1.45 ORCHIS MORIO [SYN. ANACAMPTIS MORIO] (GREEN-WINGED ORCHID)

The green-winged orchid has six subspecies located across Europe and, in some regions, the Mediterranean seaside zone (Bailarote et al., 2012). Tubers are produced each year to replace current tubers, the food reserves of which are used up in the flowering process. The plant produces its leaves in autumn (September–October), which remain green and functional throughout the winter and spring, and die down in mid-June after the plants have flowered (Wells et al., 1998). The plant during flowering has a height of 10–30 cm and a single stem with green basal leaves in the form of a rosette. The flowers are small and have a complex structure typical of orchids, a deep purple color with a three-lobed lip purple-brown color with a whitish border (Caputo et al., 1997).

In Serbia green-winged orchid is known as *kaćun*. For medicinal purposes, bulbs (*Salep tuber*) are used as a typical mucous medicine for treating inflammation of the intestinal and gastric mucosa accompanied by diarrhea, especially in children (Tucakov, 2006). Furthermore, a bulb cooked in milk or water with honey is a nutritious drink that is consumed as a delicacy. Especially in the past, dried bulbs were very popular as love talismans and aphrodisiacs (Pieroni, 2008).

There is a lack of research focusing on the phytochemistry and biological activities of orchids, and numerous traditional uses of these plants have not been scientifically verified (Bazzicalupo et al., 2023).

12.1.46 *Origanum vulgare* (oregano)

Oregano (Figure 12.4g) is native to the Mediterranean region, including Southern Europe and Western Asia. It is also widely cultivated in many other parts of the world for its culinary and medicinal uses (Stefanaki and van Andel, 2021). It is a perennial herb that develops creeping roots; branched woody stems; and opposite, petiolate, and hairy leaves. The leaves are spade-shaped, olive-green, and are covered with fine hairs called trichomes. Flowers are purple or pink, occuring in spikes forming a panicle with multiple branched stems growing from a central stalk (Alekseeva et al., 2020).

In the Serbian language oregano is known as *vranilova trava* or *vranilovka*, because this plant was used to obtain a black colour of wool with the addition of ferrous sulfate until the discovery of synthetic dyes (Tucakov, 2006). In the serbian tradition the plant collected during flowering (*Origani herba*) is used for the preparation of infusions for gastrointestinal ailments such as minor digestive upsets, insomnia, headache, colds, and bronchitis (Jarić et al., 2007).

Numerous studies have confirmed its strong antimicrobial (Sakkas and Papadopoulou 2017), antibacterial (Saeed and Tariq 2009, Özkalp et al., 2010, Aćimović et al., 2020a), antifungal (Tomić et al., 2023), antioxidant (Cervato et al., 2000, Karakaya et al., 2011), and anti-inflamatory activity (Avola et al., 2020).

12.1.47 *Papaver rhoeas* (common poppy, corn poppy)

The common poppy (Figure 12.4h) is native to many parts of Europe, and Asia, however, today it is a common cosmopolitan weed plant (Montefusco et al., 2015, Grauso et al., 2021). It is an erect herb usually around 20–80 cm high. Leaves vary greatly in shape and size; basal leaves form a rosette of pinnatipartite leaves with seven to nine lanceolate or elliptic segments and serrate or dentate margins, while apical leaves are smaller but more dissected. Flowers are solitary and red, fruits are ovoid capsules (Grauso et al., 2021).

Common poppy in Serbian is known as *bulka* or *turčinak*. The common poppy had various folk uses throughout history. In medicinal purposes dry petals (*Rhoeados flos*) are used as ingredient of pulmonary teas, especially for treating productive cough (Matejić et al., 2020), while syrup is made from fresh petals for pain relief (Pinke et al., 2022).

Poppy preparations have been used to treat a variety of ailments, including insomnia, anxiety, and diarrhea (Lal, 2022). The milky sap from the poppy plant has been used as a painkiller, sedative, and cough suppressant. It contains alkaloids such as morphine and codeine, which are pain-relieving and calming (Muhammad et al., 2021). For culinary purposes, poppy plant seeds are used in baking and cooking to flavor bread, cakes, and pastries, especially in European and Middle Eastern cuisine (Pushpangadan et al., 2012). On the other hand, the petals of the common poppy can be used to dye fabrics and yarns, producing a range of shades from pale pink to deep red (Abdel-Kareem, 2012).

12.1.48 *Plantago major* (common plantain)

Common plantain or broadleaf plantain (Figure 12.4i) is a common perennial weed with a wide range of distribution. It is native to central Europe, Asia, and North Africa, but it has since been introduced and naturalized in many other parts of the world (Zubair et al., 2011). It forms a rosette of leaves with parallel venation that have an ovate to elliptical shape with an acute apex and a smooth margin. On the top of the stem, brownish-green flowers with purple stamens form a compact spike, which produces a large number of seeds (Keivani et al., 2020).

Serbian traditional medicine uses fresh and dry leaves of common plantain (in Serbian *ženska bokvica*), or less frequently, the whole plant and root (*Plantaginis folium, herba et radix*) (Tucakov, 2006). The leaf is collected during flowering, and the root is extracted throughout the year. The fresh leaf is applied externally to wounds and ulcers, while dry leaf as infusion is applied internally as tea for treating digestive complaints such as diarrhea, spasms, or intestinal and stomach ulcer. Plantain leaf increases secretion in the lungs and thus facilitates expectoration. It is often used in combination with other plants (marshmallow flower, fennel fruits, and coltsfoot) (Jarić et al., 2007). Another similar species is ribwort plantain (*P. lanceolata*), in Serbian known as *muška bokvica*, which is used for the same purposes as common plantain, for the treatment of respiratory infections, as antitussive and expectorant, against cough and bronchitis – in the juice mixed with honey (Marković et al., 2022b).

Plantains have been used as natural remedies to treat a variety of ailments related to the respiratory and digestive systems, skin, and infections caused by microorganisms (Sanna, et al., 2022). Studies on the pharmacological activities showed analgesic, anti-inflammatory, and antibacterial potential (Guillén et al., 1997, Razik et al., 2012). They contain mucilage, tannins, iridoid glycosides, silicic and phenolic carboxylic acids, flavonoids, and minerals (Serag et al., 2010).

12.1.49 *Portulaca oleracea* (common purslane)

Common purslane or purslane is a plant species native to Asia and has since spread to many other parts of the world (Mitich, 1997). It is a cosmopolitan species and is considered to be one of the most aggressive weeds. Common purslane is an annual succulent herbaceous, creeping plant that forms a large biomass. The leaves are thick, obovate, and sessile. The flowers are small, yellow, and star-shaped, characterized by a prolonged flowering period (from July until the first frosts). The seeds are small, black, and with germination ability for up to 40 years (Anđelković et al., 2022).

In Serbia, common purslane is known as *tušt* or *ledena trava*. This plant is used fresh externally for various skin complaints, such as eczema, ulcers, and acne, and to give relief from insect bites (Sicari et al., 2018, Saulić 2022). It is an edible plant that is commonly used as a vegetable in salads, soups, and stews (Srivastava, et al., 2023).

Common purslane has been traditionally used as a medicinal plant (Kumari et al., 2016). It has anti-inflammatory and analgesic properties, anticancer and antioxidant activity, anti-hyperglycemic and anti-hyperlipidemic, renoprotective, and hepatoprotective effects, and is usually applied to treat a variety of ailments, such as digestive issues, fever, and inflammation (Mosaddegh et al., 2012, Iranshahy et al., 2017). In addition, purslane is a rich source of vitamins and minerals, protein, and carbohydrates, while the sour taste is due to malic acid (Uddin et al., 2014).

12.1.50 *Primula veris* [syn *P.* officinalis] (common cowslip)

Common cowslip (Figure 12.5a) is native to Europe and parts of western Asia (Brys and Jacquemyn, 2009). It is an herbaceous perennial plant that typically grows in meadows, grasslands, and open woods (Apel et al., 2017). The rhizome is short, stout, ascending, girthy at the apex with more or less fleshy scales formed by the bases of withered leaves, furnished with numerous fibrous roots. Leaves ovate to ovate-oblong, rugose with the veins impressed above and prominent below, pubescent or glabrescent on the upper surface, thinly greyish or whitish tomentose to glabrescent on the lower

FIGURE 12.5 Plants used as wild edible in Serbia: a) *Primula veris*, b) *Prunus spinosa*, c) *Pulmonaria officinalis*, d) *Robinia pseudoacacia*, e) *Rosa canina*, f) *Rubus fruticosus*, g) *Salvia officinalis*, h) *Sambucus nigra*, i) *Satureja kitaibelii*. Photographs by the authors.

surface. Scape originates among the leaf axils of rosettes, bearing one or two umbels each with one to 30 bright yellow flowers which bloom in the spring (Brys and Jacquemyn, 2009). Cowslip is also a popular garden plant and is cultivated for its attractive flowers (Hickey and King, 1997).

Common cowslip, in Serbian, is *jagorčevina* or *jaglika*. Root and rhizome (*Primulae rhizoma et radices*) extracted in autumn, washed, and dried, is used in Serbian traditional medicine as decoction as expectorant for treating respiratory conditions cough, colds, and bronchitis (Šavikin et al., 2013, Zlatković et al., 2014, Jarić et al., 2015). Sometimes the leaf and flower are used (*Primulae flos et folium*), but they contain significantly less saponides, approximately 2%, although they are a source of vitamin C (Tucakov, 2006). The leaves of the cowslip are also edible and have a slightly sweet and nutty flavor (Fernald et al., 1996).

Sometimes *P. vulgaris*, common primrose, flowers or root infusion are used against pneumonia, productive cough, strengthening of the heart, sedative, and insomnia (Janaćković et al., 2019, Matejić et al., 2020). However, this species in Serbia is critically endangered. A similar species is

P. elatior, oxlip. Cowslip has been used for medicinal purposes for centuries and is known for its anti-inflammatory and expectorant properties (Budniak et al., 2021).

Previous studies have identified that *Primula* sp. possesses anti-inflammatory, anti-microbial, anti-asthmatic, anti-viral, and anti-fungal effects (Başbülbül et al., 2008, Eliopoulos et al., 2022).

12.1.51 *PRUNUS* SP.

Prunus is a genus that belongs to deciduous, evergreen trees and shrubs native to the temperate regions of the Northern Hemisphere, with a large number of species worldwide (Balkrishan et al., 2021). The genus *Prunus* has seven autochthonous (wild) species in Serbia (Josifovic, 1975). However, a number of species from this genus are in cultivation for their edible seeds and edible fruits, particularly *P. amygdalis* (almonds), *P. domestica* (plums), *P. persica* (peaches), *P. cerasus* (cherries), *P. americana* (apricots), and others. (Poonam et al., 2011).

Wild cherry or sweet cherry (*P. avium*) is a medium-sized, fast-growing deciduous tree, which produces small, round, and juicy fruit that is commonly eaten fresh or used in desserts such as pies and cakes (Welk et al., 2016, Janick, 2005). In Serbia, it is commonly known as *divlja trešnja*, and it is a popular fruit often used in traditional Serbian cuisine to make jams, preserves, and brandy (Iličić et al., 2022). In addition to its culinary and commercial uses, wild cherry has a long history of use in traditional medicine in Serbia and is believed to have medicinal properties that can help with various health conditions such as coughs, colds, and fever (Pilipović et al., 2011). An infusion prepared from leaves and stalks is used for urinary tract ailments, fluid retention, and prostate issues (Šavikin et al., 2013, Zlatković et al., 2014, Matejić et al., 2020). The domesticated form cultivated for fruit is known as the sweet cherry (Russell, 2003).

Blackthorn (*P. spinosa*) (Figure 12.5b) is a spiny, deciduous shrub that produces small, purple, edible plums (Popescu and Caudullo, 2016). In Serbian it is known as *trnjina*. The flower is picked half-open (*Pruni spinosae flos*), the ripe fruit is harvested before the first frosts (*Pruni spinosae fruit*), and at the same time, the bark is peeled (*Pruni spinosae cortex*) (Tucakov, 2006). Fruits and bark infusion and alcoholic extract is used for high blood cholesterol and triglyceride levels (Šavikin et al., 2013), while fruits decoct a digestive stimulant (Zlatković et al., 2014). According to popular belief, there is no real spring until the blackthorn blooms (Čajkanović, 2022).

12.1.52 *PULMONARIA OFFICINALIS* (BLUE LUNGWORT)

Blue lungwort (Figure 12.5c) is native to Europe and Western Asia (Meeus et al., 2013). The species is a distylous, perennial rosette hemicryptophyte that occurs predominantly in the understory of broad-leaved, mixed, open woods rich in hornbeam, beech, and oak (Brys et al., 2008). The underground parts consist of a slowly creeping rhizome with adventitious roots. The leaves and flowering stalks are covered in hairs of varied length and stiffness, and stems and inflorescences are sparsely covered with glandular hairs. Flowers of blue lungwort are heterostylous, with distinct pin and thrum morphs. The corolla varies from purple, violet, or blue to shades of pink and red, or sometimes white. The plant reproduces both by sexual and vegetative means (Meeus et al., 2013).

Blue lungwort in the Serbian language is known as *plućnjak*. For medicinal purposes, the leaves or the whole plant are collected during flowering (*Pulmonariae folium et herba*). As infusion from blue lungwort is used for pulmonary problems, asthma, and tuberculosis (Jarić et al., 2015, Matejić et al., 2020, Mamedov and Ibadullayeva, 2022).

It has been used as a medicinal plant in folk and traditional medicine, while various studies have observed its biological activities, such as antioxidant, anti-inflammatory, anti-neurodegenerative, skin whitening, anticoagulant, antibacterial, anti-anemic, anticonvulsant, and wound-healing properties (Malinowska, 2013, Sorescu et al., 2020, Krzyżanowska-Kowalczyk et al., 2021, Chauhan et al., 2022).

12.1.53 *Quercus* sp. (oak)

Quercus species, also known as oak, represent an important genus of the Fagaceae family. It is widely distributed in temperate forests of the northern hemisphere and tropical climatic areas (Taib et al., 2020). The *Quercus* genus is comprised of around 450 species worldwide (Vinha et al., 2016), while ten species are defined in *Flora of Serbia* (Josifovic, 1975). The Serbian people have great respect for oak (Čajkanović, 2022).

For medicinal purposes the *Q. pedunculata* [syn. *Q. robur*] and *Q. sessiliflora* is used. Early in the spring before the vegetation starts or in the autumn when the leaves fall, the bark (*Quercus cortex*) is peeled from the young trees (Tucakov, 2006). Oak kernels were traditionally used in medicine, particularly roasted ones (*Quercus semen tostum*) as astringents, antidiarrheals, antidotes, bark decoction against pneumonia, antidiabetic, immune system strengthener, gargle for the throat, and externally as antirheumatic and for treating eczema (Tucakov, 2006, Jarić et al., 2015 Janaćković et al., 2019). Oak bark is often used in tea blends with flowers of the common daisy and nettle leaves (Josifovic, 1975). Roasted oak acorns (*Quercus semen tostum*) are used as a substitute for coffee (Rakić et al., 2018).

Oak bark can be obtained from 10- to 15-year-old oak trees, and it is widely used in traditional medicines around the world for treating gastric disorders because of its astringent effects, as well as antibacterial and wound-healing properties and many others (Hemmati et al., 2015, Bahmani et al., 2015). The bioactivity assay revealed antioxidant, antibacterial, antidiabetic, and anticancer activity in the oak bark extracts (Ştefănescu et al., 2022a).

12.1.54 *Robinia pseudoacacia* (black locust)

Black locust (Figure 12.5d) is a strong, fast-growing (10 m to 14 m high) and light-demanding tree species native to eastern North America (Nicolescu et al., 2020). The tree is usually bent-stemmed with greyish-brown to dark brown bark, becoming longitudinally fissured with age. The leaves are composed and pinnate, usually with a pair of spines at the base which persists on young shoots. The leaflets are commonly in two to 12 pairs, usually opposite, with an additional one at the end of the rachis. Leaf blades are oblong, elliptic, or ovate with an entire margin. Black locust is a monoecious species: the hermaphroditic, scented flowers have a white to cream corolla with yellow spots inside. The fruit is a legume, 5–10 cm long dark brown pods hanging in winter and containing four to ten seeds, mainly dispersed by gravity and wind (Sitzia et al., 2016).

This plant in Serbian is known as *bagrem*. In Serbian traditional medicine, black locust flowers are used, or, infrequently leaves (*Robiniae pseudoacaciae flos et folium*) (Tucakov, 2006). Flower infusion is used for treating colds, productive cough, bronchitis, and asthma, and for strengthening the immune system, while leaf infusion is used for stomach problems, as a cholagogue (Jarić et al., 2015, Janaćković et al., 2019, Matejić et al., 2020).

Black locust edible flowers are high in bioactive and volatile compounds, including a huge group of polyphenols compounds, quercetin, epigallocatechin, and ferulic acid (Hallmann, 2020). The flower should be picked immediately before opening, and the leaf while it is still young. In traditional cuisine, black locust and bristly locust flowers can be consumed as concentrates, syrups, flower flour, and fried in pancakes, similar to elderberry flowers (Bobiş et al., 2021).

12.1.55 *Rosa canina* (rosehip)

Rosehip (Figure 12.5e) is an ornamental plant with an erect prickly shrub (1–3 m high) native to Central Asian and Anatolian regions (Saricaoglu et al., 2019). It has fragrant pink or white flowers and is grown for decorative purposes in gardens and landscape design projects. Its branches are often curved or arched (Khazaei and Pazhouhi, 2020). The rosehip (or rose haw) is the pseudofruit of the rose plant. It is usually 1.5–2 cm long, bright red when mature, and glabrous. The hips overwinter on the plant (Chrubasik et al., 2008).

Rosehip in Serbian is known as *šipak* or *širurak*. Fresh or dried fruit (*Cygosbati fructus*) is used to make tea, jams, jellies, syrups, and even wine, but it is rarely consumed raw (Tomljenović et al., 2021, Miljković et al., 2022). Internally, fruit decoction is widely used for colds and influenza, and as mild astringents for intestinal disorders, especially diarrhea (Jarić et al., 2007).

Rosehip has a tart, slightly sweet flavor and contains vitamins C, B, K, and provitamin A (Mureşan et al., 2018, Amirova et al., 2023). It has gained importance in medical applications as a natural source of antioxidants (Tumbas et al., 2012, Roman et al., 2013, Ersoy et al., 2015). Various preparations of the rose hip have importance as probiotic, stool-regulating, and smooth muscle–relaxing actions, as well as the rose hip seed lipid-lowering, antiobese, and antiulcerogenic effects (Chrubasik et al., 2008).

12.1.56 *RUBUS* SP.

Rubus species are widely distributed in the warmer temperate zones of the Northern Hemisphere, and have been known since ancient times for their medicinal and nutraceutical properties (Rocabado et al., 2008, Kucharski et al., 2022). The genus *Rubus* is represented by over 700 species, while only ten species have been recorded as wildgrown in the flora of Serbia (Josifovic, 1975). However, some species from this genus have horticultural importance, such as *R. fruticosus* (blackberry), *R. ideaus* (red raspberry), *R. occidentalis* (black raspberry), and *R. chingii* (raspberry), with numerous varieties as industrial plants for commercialization as nutritious and pleasant-tasting fruits (Rocabado et al., 2008, Meng et al., 2022).

Blackberry (*R. fruticosus*) (Figure 12.5f) in Serbian is known as *divlja kupina*. Young blackberry leaves (*Rubi fruticosi folium*) are collected without petiole in April and May, and dried in a well-aerated place in the shade. In traditional medicine, blackberry folium infusion is used as an astringent and as a remedy against diarrhea, intestinal inflammation and stomach bleeding, and hemorrhoids (Živković et al., 2020). Ripe blackberry fruits act as a mild laxative. They are consumed fresh or used for the preparation of blackberry wine, syrup, juices, and preserves (Tucakov 2006, Jarić et al., 2015).

Wild raspberry (*R. ideaus*) in Serbian is known as *malina*. They are viewed as beneficial fruits due to their increased content of nutrients and phytochemicals, which carry considerably valuable effects on human health (Fotirić Akšić et al., 2022). Wild raspberry leaves (*Rubi idaei folium*) are used as astringent against diarrhea and for rinsing with mild infections of the oral cavity, as well as used as a substitute for Chinese tea (Tucakov, 2006). Fresh ripe fruit is used for the preparation of syrup, juices, preserves, and dried is used for flavouring fruit tea, mainly consumed for fatigue after cold (Šavikin et al., 2013).

The bioactive properties of *Rubus* species include antioxidant activity, antimicrobial activity, and anti-inflammatory activity, and they can vary depending on the specific species, as well as factors such as growing conditions and harvesting methods (Heinonen, 2007, Tosun et al., 2009, Viskelis et al., 2010, Fazio, 2013).

12.1.57 *RUMEX CRISPUS* (CURLY DOCK)

Curly dock is a perennial herbaceous plant that is native to Europe but has since spread throughout much of the world. It is often considered a weed and can be found in fields, gardens, and disturbed areas, but some of the members of genus *Rumex* include invasive weeds (Morris et al., 2009, Benoliel, 2011, Vasas et al., 2015). Curly dock is a perennial wild plant, which can reach a height of 30–100 cm. It has a well-developed root. The stem is bare and furrowed. The leaves are large and somewhat fleshy, on a stalk as long as the leaf blade (30 cm). The lower leaves are lanceolate and wrinkled, while the upper are linear. The flowers are bisexual, up to twice as short as the flower stalk. They are gathered in whorls, and these in loose panicles. The fruit is a nut, brown in colour, sharpened at the top. It blooms from June to August (Bektašević et al., 2022).

Curly dock has been used for centuries in traditional medicine for a variety of purposes promoting digestive and skin health (Lust, 2014, Ćebović et al., 2020). In Serbia, curly dock is known as *kiselica* or *štavelj*. Because of a high content of calcium oxalate it has pleasant sour taste; mainly used are young leaves as vegetable (Tucakov, 2006). Infusion from seed can be used as antidiarrheal (Jarić et al., 2015).

It is important to note that while curly dock has a long history of traditional use, scientific research on its effectiveness is limited (Park et al., 2018). Plants produce a large number of biologically important secondary metabolites such as anthraquinones, naphthalenes, stilbenoids, steroids, flavonoid glycosides, leucoanthocyanidins, and phenolic acids (Vasas et al., 2015). In some regions, the leaves of *Rumex* sp. are utilized as foods, mainly in the forms of sour soups (usually in milk), sauces, and salads, while roots have been used as gentle laxans (Vasas et al., 2015).

12.1.58   *Salvia* sp. (sage)

Fourteen species are defined in *Flora of Serbia* called with the common name *žalfija* (Josifovic, 1975). The best-known and most used is *S. officinalis* (Figure 12.5g) or common sage. It originates from southern Europe; the plant is 30 to 70 cm tall, with violet flowers that grow along the spines of the stem; the leaves are covered with tiny, thick hairs, have a silvery sheen, and spread a specific, slightly strong aroma. The leaves are collected in May and June before the plant blooms (Treben, 2020). However, this plant grows spontaneously in Serbian nature only in Sićevo Gorge (Southeast Serbia), therefore this plant has great commercial importance as cultivated (Aćimović et al., 2022b, Pavlić et al., 2023). Common sage leaves infusion is used for mouth and throat infections, wounds, and skin disease (Šavikin et al., 2013, Matejić et al., 2020). It is also used combined with birch, bearberry, and horsetail for urinary tract inflammations (Tucakov, 2006).

S. verticillata, lilac sage, is known in Serbian by the name *sjeruša* (Stanković et al., 2020). Aerial parts are used for the preparation of infusion, which acts as an expectorant, for disinfection of the oral cavity, and as cataplasm for healing wounds, cuts, thyroid gland, and cystic ovaries (Jarić et al., 2015, Matejić et al., 2020).

S. sclarea, clary sage, is known in Serbian as *muskatna žalfija*. Spikes in full flowering stadiums contain volatile oil with a characteristic scent, while dried seeds are rich in fatty oil, both highly valuable in the perfumery and cosmetic industry (Aćimović et al., 2022c). Traditionally, clary sage is applied as an aromatic compound in food and beverages, but used as an agent against gingivitis, stomatitis, and aphthae (Aćimović et al., 2018).

12.1.59   *Sambucus nigra* (elderberry)

Elderberry (Figure 12.5h) is a small tree widely distributed in Europe, North America, and some parts of Asia (Akbulut et al., 2009). Elderberry can typically grow around 3–4 meters in height. It has a rounded shape with compound and pinnate leaves. It produces small, creamy-white flowers in large, flat-topped clusters called umbels (Charlebois, 2007). Elderberry creates clusters of small, dark purple-to-black berries about 5–6 mm in diameter. The berries are edible when cooked but should not be consumed raw as they can cause digestive upset (Finn et al., 2008, Liu et al., 2022).

In Serbia, elderberry is known as *zova* or *bazga* (the word is of Proto-Slavic origin, Russian *бузина чёрная*, Belarus *бузіна чорная*, Ukrainian *бузина чорна*). It is called a demonic tree by Serbs, as well as by other peoples – fairies or the devil live on it (Čajkanović, 2022). Traditional medicine uses flowers (*Sambuci flos*) and ripe fruits (*Sambuci fructus*). The elderberry inflorescences are harvested as soon as the flowers begin to open, in sunny and dry weather. After drying in a well-aerated and dry place, the plant material should be light yellow, and the flowers should separate from the flower stalks (Tucakov, 2006). Flower infusion causes sweating and it is used for fever and cold; it also acts as a diuretic (Šavikin et al., 2013). Fruits are used for the preparation of jams, jellies, juices, and wines. Sometimes elderberry bark (*Sambuci cortex*) is strongly laxative in effect (purgative) (Tucakov, 2006).

In the literature, there is much evidence of elderberry health benefits, including boosting the immune system (Kronbichler et al., 2020), reducing inflammation (David et al., 2014), relieving cold and flu symptoms (Tiralongo et al., 2016), and supporting cardiovascular health (Farrell et al., 2015).

12.1.60 *Satureja* sp. (savory)

The genus *Satureja* counts about 30 species out of which in Serbia only nine may be found (Radanović et al., 2018). According to literature survey, *S. montana* and *S. hortensis* are the most common species (Tepe and Cilkiz, 2016). However, *S. kitaibeli* is endemic species for this part of Europe. All plants from this genus are native to southeastern Europe and western Asia, and is commonly grown for its culinary and medicinal uses (Jelenković et al., 2016). It is a semi-evergreen subshrub, with narrow, elongated leaves that are about 1–2 cm; a slightly hairy texture; a dark green color; and pink flowers that are arranged in clusters at the tips of the stems. The flowers bloom in the summer months and are attractive to bees and other pollinators (Perrino et al., 2021). In addition, all savory has a strong, spicy flavor and is often used as a culinary herb, where it is used to flavor meat dishes, stews, and soups (Šeregelj et al., 2022).

S. montana, commonly known as winter savory, or in Serbian *čubar*, is a perennial plant species that naturally grows in the sub-Mediterranean area. Winter savory is known for its strong, spicy flavor and is often used as a seasoning in Mediterranean cuisine, particularly in dishes featuring meat, fish, or vegetables (Green, 2006).

S. hortensis, commonly known as summer savory, or in Serbian *čubrica*, is an annual plant species. It is native to the Mediterranean region but is widely cultivated in many parts of the world. Summer savory has a milder and sweeter flavor compared to winter savory and is often used in lighter dishes such as salads, soups, and stews. The plant is also known for its medicinal properties and has been used in traditional medicine to treat a variety of conditions, including digestive problems, sore throat, and cough. In addition to its culinary and medicinal uses, summer savory is also used as an insect repellent in some cultures (Hassanzadeh et al., 2016).

S. kitaibelii (Figure 12.5i) is commonly known as Kitaibel's savory, or in Serbian *Rtanjski čaj*. It is used in traditional medicine in Serbia for its potential health benefits, which include digestive and respiratory support, herb infusion for immune strengthening against cold, as well as antioxidant and antimicrobial properties (Zlatković et al., 2014, Marković et al., 2022b).

The dried leaves of *Satureja* are used as a pleasant spice and food additive as well as an herbal tea. It has been reported that *Satureja* species extracts and fractions possess various therapeutic effects. They are antioxidants, control blood lipids, attenuate blood glucose, and inhibit lipid peroxidation (Babajafari et al., 2015). Recent studies have highlighted the antioxidant, antihyperglycemic, anti-inflammatory, and antimicrobial activity of Kitaibel's savory extracts (Aćimović et al., 2021a) and biological activity of its essential oil (Aćimović et al., 2022a).

12.1.61 *Silybum marianum* (milk thistle)

Milk thistle (Figure 12.6a), is a plant species native to the Mediterranean region but has been widely naturalized in many parts of the world, especially in Central Europe (Porwal et al., 2019). It is an annual or biennial herb that grows up to 2 m, with an erect stem. It is characterized by big prickly leaves, large purple flowering heads, and strong spinescent stems. Milk thistle is named for its milky veins on the leaves (Emadi et al., 2022).

In Serbian language, milk thistle is known as *gujina trava* or *blaženi čkalj*. In traditional medicine fruit is used (*Silybi fructus*), rarely leaves (*Silybi folium*) (Tucakov, 2006). Milk thistle has been

FIGURE 12.6 Plants used as wild edible in Serbia: a) *Sylibum marianum*, b) *Taraxacum officinale*, c) *Teucrium chamaedrys*, d) *Tilia* sp., e) *Thymus serpyllum*, f) *Tusilago farfara*, g) *Urtica dioica*, h) *Verbascum phlomoides*, i) *Viola odorata*. Photographs by the authors.

used for centuries as a traditional herbal remedy for various ailments, particularly seed decoction for liver cleansing, digestive system disorders (alleviates gallbladder attack, constipation), diabetes, headache, tonic, asthma, and hemorrhoids (Janaćković et al., 2019, Živković et al., 2020).

It has been traditionally used in Europe for more than 2000 years as a vegetable in salads, and the seeds used as a galactagogue for breastfeeding mothers, as well as for different diseases, especially because it has protective effects against different biological poisons (such as mycotoxins, snake venoms, and bacterial toxins) and chemical poisons (such as metals, fluoride, pesticides, cardiotoxic, neurotoxic, hepatotoxic, and nephrotoxic agents) (Emadi et al., 2022). It is believed to protect the liver from toxins and support its ability to regenerate damaged cells (Greenlee et al., 2007). Both in vitro and in vivo research have proved the antioxidant activity of silymarin from milk thistle and its ability to stimulate protein synthesis and cell regeneration; therefore, it is being used to treat toxic liver damage and chronic inflammatory liver diseases, and liver cirrhosis (Pradhan and Girish, 2006, Vargas-Mendoza et al., 2014, Abenavoli et al., 2018).

12.1.62 *Taraxacum officinale* (dandelion)

Dandelion (Figure 12.6b) is widely known as a weed and can be found in lawns, gardens, and along roadsides; it is native to Europe and Asia but now grows in many parts of the world (Stewart-Wade et al., 2002, Haragan, 2014). Dandelions usually bloom in the spring and summer, with yellow flowers about 2.5–5 cm in diameter. The leaves are green and deeply lobed, forming a rosette at the base of the plant. The stem is smooth and hollow, exuding a milky sap when broken (Bergen et al., 1990, González-Castejón et al., 2012).

In the Serbian language, dandelion is known as *maslačak*, and in traditional medicine, the root is highly valuable (*Taraxaci radix*). Infusion increases appetite and helps digestion (Tucakov, 2006). Leaves are collected early in spring, before flowering; eaten fresh as a vegetable, which eases problems with the liver and gallbladder. An infusion is prepared from dry leaves for gastrointestinal ailments, diuretics, expectorants, eczema, acne, and wounds (Šavikin et al., 2013, Zlatković et al., 2014).

Dandelion leaves, roots, and flowers have long been used in traditional medicine for potential health benefits (Fatima et al., 2018). They are a rich source of vitamins and minerals, including vitamins A, C, and K, as well as potassium, iron, and calcium (Li et al., 2022). The leaves are often used in salads and teas, and the roots are sometimes roasted and used as a coffee substitute (Siminiuc and Turcanu, 2021). Dandelion is also a natural diuretic and has been used to help improve liver function, digestion, and inflammation (Sweeney et al., 2005, Qureshi et al., 2017).

12.1.63 *Teucrium chamaedrys* (wall germander)

The genus *Teucrium* is a large and polymorphic genus distributed in mild climate zones, particularly in the Mediterranean basin and Central Asia (Candela et al., 2020). The genus is represented by about 300 species in the world and seven species in Serbia (Josifovic, 1975).

Wall germander (*C. chamaedrys*) (Figure 12.6c) is a small, evergreen plant native to the Mediterranean region and is often grown as an ornamental plant in gardens and landscapes (De Smet, 1997). The leaves are small, glossy, dark green, measuring only 1–2 cm in length, and have a slightly serrated edge. The plant produces small, pink-purple flowers in the summer, which are arranged in tight clusters at the tips of the stems (Ciocarlan et al., 2022).

Wall germander in Serbian is known as *podubica*. In traditional medicine, the dried aerial part of the plant (*Chamedryos herba*) is used internally as tea for gastrointestinal aliments (gallbladder problems and diarrhea), hemorrhoids, wounds, and increased vaginal discharge with no infections (Jarić et al., 2007, Šavikin et al., 2013, Živković et al., 2020).

A similar species is mountain germander, *T. montanum*, well-known in Serbia as *trava iva*. Internally as an infusion, mountain germander herb (*Teucrii montani herba*) is used for gastrointestinal and respiratory ailments and for immune system strengthening (Jarić et al., 2007, Šavikin et al., 2013, Zlatković et al., 2014).

Plants from genus *Teucrium* have been used as medicine since ancient times. Species of this genus have been widely implemented for their biological properties, including antimicrobial, anti-inflammatory, antispasmodic, insecticidal, antimalaria, and so on, for treating different ailments (Ulubelen et al., 2000, Candela et al., 2020).

12.1.64 *Thymus sp.* (thyme)

Thymus sp. (thyme) is a genus of perennial herbaceous plants native to the Mediterranean region and is widely cultivated around the world for its aromatic leaves, which are used as a culinary herb and for medicinal purposes (Kim et al., 2022). The genus *Thymus* is very complex and highly polymorphous, and in *Flora of Serbia* 31 species are listed, with many varieties (Josifovic, 1975, Dajić-Stevanović et al., 2008, Marković et al., 2020).

The most common species include *T. vulgaris* (common thyme), *T. serpyllum* (creeping thyme) (Figure 12.6e), and *T. pannonicus* (Hungarian or Eurasian thyme) (Maksimović et al., 2008). However, all species have characteristic aroma and serve as medicinal (*sensu lato*). In the Serbian language, creeping thyme is known as *majkina dušica*, and it is a favourite herbal tea, obtained mainly by collecting it from nature. On the other hand, thyme is known as *timijan*; it is a plant with commercial importance as it is cultivated mainly on a large scale as well as in yards and gardens (Pavlić et al., 2023). Furthermore, Hungarian thyme, known as *Panonski timijan*, is attributable to lemon scented (because of mixture of geranial and neral), and has significantly different essential oil composition in comparison to the previously mentioned species (where dominant compounds are thymol and carvacrol) (Pluhár et al., 2010).

The aerial part of *Thymus* spp. is frequently as an infusion for treating anxiety, gastrointestinal disorders, and respiratory problems (Šavikin et al., 2013). Thyme is known for its strong, fragrant flavor and aroma, and is often used in savory dishes such as soups, stews, and sauces, as well as in herbal teas and as a natural remedy for a variety of health conditions (Stefanaki and van Andel 2021). Thyme posesses antioxidative and antibacterial properties (Martins et al., 2015, Gavaric et al., 2015, Abramovic et al., 2018).

12.1.65 Tilia sp. (linden)

Tilia sp., commonly known as linden or lime (Figure 12.6d), is a genus of trees native to temperate regions of the Northern Hemisphere, often planted as ornamental trees in parks and gardens, and their wood is used in carpentry and furniture making (Savill, 2019). Four species are defined in *Flora of Serbia*, but those used for medicinal purposes are *T. cordata* (small-leaved lime, in Serbian *sitnolisna lipa*) and *T. platyphullos* (large-leaved lime in Serbian *krupnolisna lipa*) (Josifovic, 1975). Both species are very similar trees, but the first is the more common species in Europe, while the second extends farther south (Eaton et al., 2016). The trees are known for their distinctive heart-shaped leaves, fragrant flowers, and durable wood (Thomas, 2014, De Benedetti et al., 2022).

Linden in Serbian is known as *lipa*, and as remedy linden inflorescence (*Tiliae flos cum bracteis*) is collected. They have been used for centuries in traditional medicine to treat a range of conditions, such as anxiety, insomnia, and respiratory ailments (Rodriguez-Fragoso et al., 2008). Flower infusion is used for treating fever and cold, pneumonia, sore throat, immune system strengthening, and as a mild sedative against insomnia (Šavikin et al., 2013, Zlatković et al., 2014, Janaćković et al., 2019). Linden is a sacred tree among Serbs, and in general among Slavs, and it is a sin to cut it down (Čajkanović, 2022).

Since the Middle Ages, linden flowers have been used in Europe for the treatment of the common cold or to relieve symptoms of mental stress (Hake et al., 2022). Therapeutic use of lime flowers have various medicinal properties, including being a natural sedative, anti-inflammatory, and pain reliever (Poljšak and Glavač, 2021).

12.1.66 Tussilago farfara (coltsfoot)

Coltsfoot (Figure 12.6f) is native to Europe and parts of Asia and has been introduced to other parts of the world (Chen et al., 2021). Coltsfoot is a perenial plant, with long, creeping, white scaly rhizomes that branch and have fibrous roots. It blooms in early spring, before the leaves appear, and produces yellow flowers that resemble dandelions on a stem that grows up to 30 cm tall. Basal leaves develop after flowering stems and grow from the rhizomes in rosettes. The topside of leaves is hairless (glabrous) or with some hairs, with undersides more or less white and woolly (gray-tomentose) (Tilford, 1997, Stewart-Wade et al., 2002).

In the Serbian language coltsfoot is known as *podbel* (literally the underside of the leaf is white). Flowers (*Farfarae flos*) are collected before flowering, early in the spring, while leaves (*Farfarae folium*) are collected in late spring, after flowering. Wounds are to be bandaged with fresh leaves for fastening wounds and ulcers, and for vein inflammation, kneaded fresh leaves mixed with fresh sour cream are put onto the skin and wrapped in a linen cloth (Jarić et al., 2007). Dried leaves and flowers are used for preparation infusions for asthma, bronchitis, laryngitis, and other diseases accompanied by a strong cough (Šavikin et al., 2013).

Coltsfoot has a long history of medicinal use, particularly for respiratory conditions (Chen et al., 2021). Its leaves and flowers contain compounds that have been shown to have anti-inflammatory, expectorant, and soothing properties (Chanaj-Kaczmarek et al., 2013, Armutcu and Kucukbayrak, 2021). Since the coltsfoot leaves contain a significant amount of arginine, studies involve the prevention and adjunctive treatment of viral diseases (Chromchenkova et al., 2020).

12.1.67   *Urtica diorica* (nettle)

Nettle (Figure 12.6g) is originally from northern Europe, Asia, and parts of North Africa. However, it has now been widely introduced and naturalized worldwide (Jan et al., 2017). Nettle is an herbaceous perennial plant that can grow up to 1–2 meters in height. It has a square stem that is covered with fine hairs and is typically green or reddish-brown in color (Nkhabu et al., 2021). Nettle flowers from May to September, with small, green, dioecious flowers occuring as racemes in the axils of the upper leaves. Usually, the plant has either male or female flowers, in separate inflorescences (Ahmed KK and Parsuraman, 2014).

In the Serbian language nettle is known as *kopriva* or *žeža*. Young nettle leaves (*Urticae folium*) are collected in early spring (Tucakov, 2006). One of the traditional uses of nettle is as an infusion, which is used for anemia, heavy menstrual bleeding, jaundice, and stones in the urinary bladder. Stinging nettle is also commonly used externally, in the form of a spice, ointment, or tincture to treat a variety of conditions including sciatica, neuralgia, anemia, heavy menstrual bleeding, jaundice, stones in the urinary bladder and rheumatism (Randall, 2003, Jarić et al., 2007, Šavikin et al., 2013). Fresh young leaves are widely used as a vegetable green and for juice and syrup (with honey) (Tucakov, 2006).

The available literature suggests that nettle possesses antioxidant activity (Almasi et al., 2016, Knežević et al., 2019, Elez Garofulić et al., 2021), anti-aging activity (Bourgeois et al., 2016, Masłowski et al., 2022), antimicrobial activity (Knežević et al., 2015), and antiulcer and analgesic activities (Gülçin et al., 2004).

12.1.68   *Vaccinium myrtilis* (bilberry)

Bilberry is native to northern Europe, typically grows to a height of 30–60 cm, and produces dark blue, edible berries often used for culinary purposes known for their sweet, tangy flavor (Martinussen et al., 2008). The rhizomes are generally found at 15–20 cm from the surface and always are confined to the humus layers of a profil. The aerial shoots branch sympodially (primarily) and the buds overwinter. Leaves are ovate to lanceolate or broadly elliptic in shape, simple and alternate arrangement, and light green colour that turns red in autumn. Flowers are bell shaped, reddish or pink, and appear from April to June. The fruit is a blue-black or purple berry (Ritchie, 1956).

In the Serbian language, bilberry is known as *borovnica*, and in traditional medicine, leaf and fruit (*Myrtilli folium et fructus*) are used for medicinal purposes (Tucakov, 2006). The leaf is collected in the spring throughout the flowering period, and the fruit when at the fully ripe stage. Leaves applied internally as an infusion are used for treating varicose and thread veins, poor circulation, and diabetes, as well as acting as a diuretic. Dried fruits are used against diarrhea, while fresh fruits, rich in vitamin C, are used as berries in nutrition and for preparation of extracts, juices and wines, and as a remedy for anemia (Jarić et al., 2007).

Bilberry leaves have been used in the traditional medicine of different cultures and the berries are widely consumed as food (Tundis et al., 2021). Bilberry leaves contain a bewildering variety of polyphenolic compounds that are presumed to be responsible for the majority of its therapeutic effects such as anti-inflammatory and antimicrobial properties (Ben Lagha et al., 2015, Huang et al., 2018, Bell et al., 2021). Bilberry is therefore included in many supplements for diabetes, iron deficiency anemia, and many others (Ştefănescu et al., 2022b).

12.1.69 *Verbascum phlomoides* (orange mullein)

Orange mullein (Figure 12.6h) is native to Europe, but has been introduced to other parts of the world, including North America (Panchal et al., 2010, Georgiev et al., 2011). Orange mullein is a biennial plant that produces a basal rosette up to 60 cm in diameter in the first growth year which overwinters, and in the second season the plants form flower stalks with tall spikes of yellow-orange flowers, which bloom in mid to late summer. Leaves possess smooth edges with dense silvery hairs on both sides, giving the leaves a woolly appearance, and leaves on the flowering stem are alternate and become smaller and more pointed close to the top of the plant (Gross and Werner, 1978, Bercu et al., 2017).

In the Serbian language, orange mullein is known as *divizma* or *kraljevska sveća*. The flowers of orange mullein (*Verbasci flos*), that is, the petal with stamens, is collected throughout the summer (Tucakov, 2006). The plant has a long history of use in traditional medicine, and its leaves and flowers have been used to treat various respiratory and inflammatory conditions (Georgiev et al., 2011). In Serbian, this plant is known as *divizma*, and in traditional medicine it is used as an infusion for respiratory ailments (bronchitis, laryngitis, asthma, influenza, tuberculosis) and externally for rheumatic pain (Jarić et al., 2007).

Most of the common traditional uses of orange mullein are based on the anti-inflammatory action, and therefore it is widely used in folk medicine to treat pathologies related to the musculature and skeleton, and circulatory, digestive, and respiratory systems, as well as to treat infectious diseases and organ-sense illnesses (Blanco-Salas et al., 2021).

12.1.70 *Viola odorata* (sweet violet)

Sweet violet (Figure 12.6i) is native to Europe and Asia (Mittal et al., 2015). It is a small, herbaceous perennial plant, spreads with stolons, typically grows in shady woodland areas and along the edges of fields and hedgerows (Singh and Dhariwal, 2018). The plant is characterized by its fragrant purple or white flowers that bloom in the spring (Erdogan Orhan et al., 2015).

In Serbia, sweet violet is commonly known as *ljubičica* or *mirisna ljubičica*. The flowers of the sweet violet are edible and are often used as a garnish for salads or desserts. In Serbian traditional medicine sweet violet flowers are used as infusion for treating colds and respiratory diseases, as it aids the coughing of mucus (Jarić et al., 2015). The violet has its role in etymological magic: a woman gives it to her husband or a girl to a boyfriend so that he will always kiss her (Čajkanović, 2022).

The plant has a long history of use in traditional medicine and has been used to treat a variety of ailments, including respiratory conditions, headaches, and digestive problems (Lim, 2014). Sweet violet is also valued for its cosmetic properties and is used in some skin care products due to its moisturizing and anti-inflammatory effects (Gautam et al., 2012). Sweet violet leaves are valuable materials for the flavor and fragrance industry (Saint-Lary et al., 2014). In addition to its medicinal and cosmetic uses, sweet violet is also a popular garden plant grown for its attractive flowers (Ammarellou et al., 2021).

12.1.71 Conclusion

Wild edible plants have been integral to human diets since ancient times. They are used for food, medicinal purposes, animal nutrition, firewood, and fiber. In the past, wild edible plants were food for poor people and remedies for local people, but today they are delicacies and specific gastronomic offers of rural tourist destinations. In gene pools this plant material is very important for breeding. As the world modernizes, contemporary science has confirmed many biological activities of ancient plant varieties and traditional recipes. However, many traditional practices and knowledge have been lost, and the same fate has befallen traditional wisdom. The knowledge, which was once passed down from generation to generation, now only survives in the memories of the elderly, putting it in great danger of disappearing altogether. To prevent this loss, conducting further ethnobotanical research in the field and ethnographic archives is essential. In this chapter, over 70 wild edible plants of Serbia were reviewed, belonging to various families. These plants are widely used for different purposes and have great potential in the pharmaceutical industry. Wild edible plants remain an essential resource for human existence and play a crucial role in promoting sustainable living. This chapter aims to bridge the gap in the existing literature by analyzing information from various sources and presenting comprehensive data on the traditional use of wild edible plants in Serbia. By doing so, we hope to promote further research on forgotten useful plants as potential new food sources and prevent the loss of valuable knowledge.

REFERENCES

Abbaszadeh, Bohloul, Sayed Alireza Valadabadi, Hossein Aliabadi Farahani, and Hossein Hasanpour Darvishi. "Studying of essential oil variations in leaves of Mentha species." *African Journal of Plant Science* 3, no. 10 (2009): 217–221.

Abdel-Kareem, Omar. "History of dyes used in different historical periods of Egypt." *Research Journal of Textile and Apparel* 16, no. 4 (2012): 79–92.

Abdel-Naime, Waleed A., John R. Fahim, Mostafa A. Fouad, and Mohamed S. Kamel. "Botanical studies on the stem and root of Melissa officinalis L.(lemon balm)." *Journal of Advanced Biomedical and Pharmaceutical Sciences* 3, no. 4 (2020): 184–189.

Abenavoli, Ludovico, Angelo A. Izzo, Natasa Milić, Carla Cicala, Antonello Santini, and Raffaele Capasso. "Milk thistle (Silybum marianum): A concise overview on its chemistry, pharmacological, and nutraceutical uses in liver diseases." *Phytotherapy Research* 32, no. 11 (2018): 2202–2213.

Abramovic, Helena, Veronika Abram, Anja Cuk, Barbara Ceh, Sonja Smole-Mozina, Mateja Vidmar, Martin Pavlovic, and Natasa Poklar Ulrih. "Antioxidative and antibacterial properties of organically grown thyme (Thymus sp.) and basil (Ocimum basilicum L.)." *Turkish Journal of Agriculture and Forestry* 42, no. 3 (2018): 185–194.

Aceituno-Mata, Laura, Javier Tardío, and Manuel Pardo-de-Santayana. "The persistence of flavor: Past and present use of wild food plants in Sierra Norte de Madrid, Spain." *Frontiers in Sustainable Food Systems* 4 (2021): 610238.

Aćimović, Milica, Stefan Ivanović, Katarina Simić, Lato Pezo, Tijana Zeremski, Jelena Ovuka, and Vladimir Sikora. "Chemical characterization of Marrubium vulgare volatiles from Serbia." *Plants* 10, no. 3 (2021a): 600.

Aćimović, Milica, Katarina Jeremić, Nebojša Salaj, Neda Gavarić, Biljana Kiprovski, Vladimir Sikora, and Tijana Zeremski. "Marrubium vulgare L.: A phytochemical and pharmacological overview." *Molecules* 25, no. 12 (2020a): 2898.

Aćimović, Milica, Biljana Kiprovski, Milica Rat, Vladimir Sikora, Vera Popović, Anamarija Koren, and Milka Brdar-Jokanović. "Salvia sclarea: Chemical composition and biological activity." *Journal of Agronomy, Technology and Engineering Management (JATEM)* 1, no. 1 (2018): 18–28.

Aćimović, Milica G., Biljana Lj Lončar, Valtcho D. Jeliazkov, Lato L. Pezo, Jovana P. Ljujic, Ana R. Miljkovic, and Ljubodrag V. Vujisic. "Comparison of volatile compounds from clary sage (Salvia sclarea L.) verticillasters essential oil and hydrolate." *Journal of Essential Oil Bearing Plants* 25, no. 3 (2022b): 555–570.

Aćimović, Milica, Lato Pezo, Ivana Čabarkapa, Anika Trudić, Jovana Stanković Jeremić, Ana Varga, Biljana Lončar, Olja Šovljanski, and Vele Tešević. "Variation of Salvia officinalis L. essential oil and hydrolate composition and their antimicrobial activity." *Processes* 10, no. 8 (2022c): 1608.

Aćimović, Milica, Vanja Šeregelj, Olja Šovljanski, Vesna Tumbas Šaponjac, Jaroslava Švarc Gajić, Tanja Brezo-Borjan, and Lato Pezo. "In vitro antioxidant, antihyperglycemic, anti-inflammatory, and antimicrobial activity of Satureja kitaibelii Wierzb. ex Heuff. subcritical water extract." *Industrial Crops and Products* 169 (2021b): 113672.

Aćimović, Milica, Olja Šovljanski, Lato Pezo, Vanja Travičić, Ana Tomić, Valtcho D. Zheljazkov, Gordana Ćetković, Jaroslava Švarc-Gajić, Tanja Brezo-Borjan, and Ivana Sofrenić. "Variability in biological activities of Satureja montana subsp. montana and subsp. variegata based on different extraction methods." *Antibiotics* 11, no. 9 (2022a): 1235.

Aćimović, Milica, Miroslav Zorić, Valtcho D. Zheljazkov, Lato Pezo, Ivana Čabarkapa, Jovana Stanković Jeremić, and Mirjana Cvetković. "Chemical characterization and antibacterial activity of essential oil of medicinal plants from Eastern Serbia." *Molecules* 25, no. 22 (2020b): 5482.

Adams, Michael, Francine Gmünder, and Matthias Hamburger. "Plants traditionally used in age related brain disorders—A survey of ethnobotanical literature." *Journal of Ethnopharmacology* 113, no. 3 (2007): 363–381.

Ad'hiah, Ali H., Orooba NH Al-Bederi, and Khulood W. Al-Sammarrae. "Cytotoxic effects of Agrimonia eupatoria L. against cancer cell lines in vitro." *Journal of the Association of Arab Universities for Basic and Applied Sciences* 14, no. 1 (2013): 87–92.

Ahamad, Javed. "A pharmacognostic review on Artemisia absinthium." *International Research Journal of Pharmacy* 10, no. 1 (2019): 25–31.

Aharizad, Saeid, Mohammad Hasan Rahimi, Mahmoud Toorchi, and Nasser Mohebalipour. "Assessment of relationship between effective traits on yield and citral content of lemon balm (Melissa officinalis L.) populations using path analysis." *Indian Journal of Science and Technology* 6, no. 5 (2013): 4447–4452.

Ahlawat, Timur, N. L. Patel, Roshni Agnihotri, C. R. Patel, and Yatin Tandel. "Black mulberry (Morus nigra)." *Underutilized fruit Crops: Importance and Cultivation* (2016): 195–212.

Ahmed, K. K. M., and Subramani Parsuraman. "Urtica dioica L.,(Urticaceae): A stinging nettle." *Systematic Reviews in Pharmacy* 5, no. 1 (2014): 6–8.

Akbulut, Mustafa, Sezai Ercisli, and Murat Tosun. "Physico-chemical characteristics of some wild grown European elderberry (Sambucus nigra L.) genotypes." *Pharmacognosy Magazine* 5, no. 20 (2009): 320–323.

Alekseeva, Marina, Tzvetelina Zagorcheva, Ivan Atanassov, and Krasimir Rusanov. "Origanum vulgare L.-a review on genetic diversity, cultivation, biological activities and perspectives for molecular breeding." *Bulgarian Journal of Agricultural Science* 26, no. 6 (2020).

Allegrini, Agnese, Pietro Salvaneschi, Bartolomeo Schirone, Kevin Cianfaglione, and Alessandro Di Michele. "Multipurpose plant species and circular economy: Corylus avellana L. as a study case." *Frontiers in Bioscience-Landmark* 27, no. 1 (2022): 11.

Almasi, Hadi, Mohsen Zandi, Sara Beigzadeh, Sara Haghju, and Nazila Mehrnow. "Chitosan films incorporated with nettle (Urtica Dioica L.) extract-loaded nanoliposomes: II. Antioxidant activity and release properties." *Journal of microencapsulation* 33, no. 5 (2016): 449–459.

Alsaraf, Shahad, Zainab Hadi, Md Jawaid Akhtar, and Shah Alam Khan. "Chemical profiling, cytotoxic and antioxidant activity of volatile oil isolated from the mint (Mentha spicata L.) grown in Oman." *Biocatalysis and Agricultural Biotechnology* 34 (2021): 102034.

Al-Snafi, Ali Esmail. "The pharmacological and therapeutic importance of Agrimonia eupatoria-A review." *Asian Journal of Pharmaceutical Science and Technology* 5, no. 2 (2015): 112–117.

Al-Snafi, Ali Esmail. "The pharmacology of Equisetum arvense-A review." *IOSR Journal of Pharmacy* 7, no. 2 (2017): 31–42.

Al-Snafi, Ali Esmail, Wajdy J. Majid, Tayseer Ali Talab, and A. A. Al-Battat. "Medicinal plants with antidiabetic effects-An overview (Part 1)." *IOSR Journal of Pharmacy* 9, no. 3 (2019): 9–46.

Alvarez-Suarez, Jose M., Luca Mazzoni, Tamara Y. Forbes-Hernandez, Massimiliano Gasparrini, Silvia Sabbadini, and Francesca Giampieri. "The effects of pre-harvest and post-harvest factors on the nutritional quality of strawberry fruits: A review." *Journal of Berry Research* 4, no. 1 (2014): 1–10.

Amirova, Noila, Dildora Qulmaxamatova, Komila Bebitova, Foziljon Saitkulov, and Khasan Nasimov. "Technology of creating cool beverages rich in vitamins based on rose hip fruit." *Theoretical Aspects in the Formation of Pedagogical Sciences* 2, no. 5 (2023): 169–172.

Ammarellou, Ali, Justyna Zabicka, Aneta Słomka, Jerzy Bohdanowicz, Thomas Marcussen, and Elżbieta Kuta. "Seasonal and simultaneous cleistogamy in rostrate violets (Viola, subsect. Rostratae, Violaceae)." *Plants* 10, no. 10 (2021): 2147.

Anđelković, Ana, Slađana Popović, Milica Živković, Dušanka Cvijanović, Maja Novković, Dragana Marisavljević, Danijela Pavlovic, and Snežana Radulović. "Catchment area, environmental variables and habitat type as predictors of the distribution and abundance of Portulaca oleracea L. in the riparian areas of Serbia." *Acta Agriculturae Serbica* 27, no. 53 (2022): 9–15.

Anđelković, Ana, Goran Tmušić, Dragana Marisavljević, Mladen Marković, Dušanka Cvijanović, Goran Anačkov, Snežana Radulović, and Danijela Pavlović. "Distribution of economically important weed species in the riparian and roadside vegetation of Serbia." *Acta Herbologica* 30, no. 1 (2021): 51–64.

Anderson, Koren J., Suzanne S. Teuber, Alayne Gobeille, Peader Cremin, Andrew L. Waterhouse, and Francene M. Steinberg. "Walnut polyphenolics inhibit in vitro human plasma and LDL oxidation." *The Journal of Nutrition* 131, no. 11 (2001): 2837–2842.

Andritoiu, Calin Vasile, Corina Elena Andriescu, Constanta Ibanescu, Cristina Lungu, Bianca Ivanescu, Laurian Vlase, Cornel Havarneanu, and Marcel Popa. "Effects and characterization of some topical ointments based on vegetal extracts on incision, excision, and thermal wound models." *Molecules* 25, no. 22 (2020): 5356.

Anjum, Syed, Adil Gani, Mudasir Ahmad, Asima Shah, F. A. Masoodi, Yasir Shah, and Asir Gani. "Antioxidant and antiproliferative activity of walnut extract (Juglans regia L.) processed by different methods and identification of compounds using GC/MS and LC/MS technique." *Journal of Food Processing and Preservation* 41, no. 1 (2017): e12756.

Apel, Lysanne, Dietmar R. Kammerer, Florian C. Stintzing, and Otmar Spring. "Comparative metabolite profiling of triterpenoid saponins and flavonoids in flower color mutations of Primula veris L." *International Journal of Molecular Sciences* 18, no. 1 (2017): 153.

Arab, Atefeh, Rezvan Yazdianâ, Hasan Rezaei-Seresht, Melika Ehtesham-Gharaee, and Fatemeh Soltani. "Evaluation of Neuroprotective Effect of Althaea Officinalis Flower Aqueous and Methanolic Extracts against H2O2-Induced Oxidative Stress in PC12 Cells: Protective activity of Althaea Officinalis against oxidative stress." *Iranian Journal of Pharmaceutical Sciences* 13, no. 1 (2017): 49–56.

Arbeláez, Luisa F. González, Alejandro Ciocci Pardo, Juliana C. Fantinelli, Guillermo R. Schinella, Susana M. Mosca, and José-Luis Ríos. "Cardioprotection and natural polyphenols: An update of clinical and experimental studies." *Food & Function* 9, no. 12 (2018): 6129–6145.

Armutcu, Ferah, and Abdulkadir Kucukbayrak. "Herbal Remedies for Respiratory Tract Infections." *Infectious Diseases* 5 (2021): 46.

Arshad, Sadia, Rafia Rehman, Ayesha Mushtaq, and Abdul Qayyum. "Morphology, chemical composition and medicinal properties of Morus nigra L. A review." *IJCBS* 13 (2018): 100–103.

Arslanoğlu, Sf, and S. Aytac. "The important of flax (Linum usitatissimum L.) in terms of health." *International Journal of Life Sciences and Biotechnology* 3, no. 1 (2020): 95–107.

Asgarpanah, Jinous, and Elnaz Roohi. "Phytochemistry and pharmacological properties of Equisetum arvense L." *Journal of Medicinal Plants Research* 6, no. 21 (2012): 3689–3693.

Astuti, Engrid Juni, Agustin Rafikayanti, SeptianaTita Purwanto, Chenchen Puspitasari, Wenni Fista Rika, and Yeni Yesica. "The Antioxidant Activity Comparison of Malus sylvestris Mill and Its Processed Products." *Food Science and Technology Journal (Foodscitech)* (2022): 46–52.

Atkinson, M. D. "Betula pendula Roth (B. verrucosa Ehrh.) and B. pubescens Ehrh." *Journal of Ecology* 80, no. 4 (1992): 837–870.

Avola, Rosanna, Giuseppe Granata, Corrada Geraci, Edoardo Napoli, Adriana Carol Eleonora Graziano, and Venera Cardile. "Oregano (Origanum vulgare L.) essential oil provides anti-inflammatory activity and facilitates wound healing in a human keratinocytes cell model." *Food and Chemical Toxicology* 144 (2020): 111586.

Babajafari, Siavash, Farzad Nikaein, Seyed Mohammad Mazloomi, Mohammad Javad Zibaeenejad, and Arman Zargaran. "A review of the benefits of Satureja species on metabolic syndrome and their possible mechanisms of action." *Journal of Evidence-based Complementary & Alternative Medicine* 20, no. 3 (2015): 212–223.

Bagdonaite, Edita, Valdimaras Janulis, Liudas Ivanauskas, and Juozas Labokas. "Between species diversity of Hypericum perforatum and H. maculatum by the content of bioactive compounds." *Natural Product Communications* 7, no. 2 (2012): 1934578X1200700220.

Bahmani, Mahmoud, S. H. Forouzan, Ezatollah Fazeli-Moghadam, Mahmoud Rafieian-Kopaei, Ahmad Adineh, and S. H. Saberianpour. "Oak (Quercus branti): An overview." *Journal of Chemical and Pharmaceutical Research* 7, no. 1 (2015): 634–9.

Bailarote, Bruno Cachapa, Bart Lievens, and Hans Jacquemyn. "Does mycorrhizal specificity affect orchid decline and rarity?" *American Journal of Botany* 99, no. 10 (2012): 1655–1665.

Balakumbahan, Ramachandran, K. Rajamani, and K. Kumanan. "Acorus calamus: An overview." *Journal of Medicinal Plants Research* 4, no. 25 (2010): 2740–2745.

Balemie, Kebu, and Fassil Kebebew. "Ethnobotanical study of wild edible plants in Derashe and Kucha Districts, South Ethiopia." *Journal of Ethnobiology and Ethnomedicine* 2 (2006): 1–9.

Balkrishan, Acharya, Samriti Tanwar, and Uday Bhan Prajapati. "Medicinal and nutritional aspect of genus Prunus L. with phytoetymology." *International Journal of Unani and Integrative Medicine* 5 (2021): 24–27.

Bana, Ram Swaroop, Vipin Kumar, Seema Sangwan, Teekam Singh, Annu Kumari, Sachin Dhanda, Rakesh Dawar, Samarth Godara, and Vijay Singh. "Seed germination ecology of Chenopodium album and Chenopodium murale." *Biology* 11, no. 11 (2022): 1599.

Barabasz-Krasny, Beata, Katarzyna Możdżeń, Agnieszka Tatoj, Katarzyna Rożek, Peiman Zandi, Ewald Schnug, and Alina Stachurska-Swakoń. "Ecophysiological parameters of medicinal plant Filipendula vulgaris in diverse habitat conditions." *Biology* 11, no. 8 (2022): 1198.

Başbülbül, Gamze, Ali Özmen, Halil Biyik, and Özge Şen. "Antimitotic and antibacterial effects of the Primula veris L. flower extracts." *Caryologia* 61, no. 1 (2008): 88–91.

Bassett, I. J., and C. W. Crompton. "The biology of canadian weeds.: 32 Chenopodium album L." *Canadian Journal of Plant Science* 58, no. 4 (1978): 1061–1072.

Bastlová, Daša, and Jan Květ. "Differences in dry weight partitioning and flowering phenology between native and non-native plants of purple loosestrife (Lythrum salicaria L.)." *Flora-Morphology, Distribution, Functional Ecology of Plants* 197, no. 5 (2002): 332–340.

Bazzicalupo, Miriam, Jacopo Calevo, Antonella Smeriglio, and Laura Cornara. "Traditional, Therapeutic Uses and Phytochemistry of Terrestrial European Orchids and Implications for Conservation." *Plants* 12, no. 2 (2023): 257.

Beck, Pieter, Giovanni Caudullo, Daniele de Rigo, and Willy Tinner. "Betula pendula, Betula pubescens and other birches in Europe: Distribution, habitat, usage and threats." (2016): 70–73.

Bektašević, Mejra, Melisa Oraščanin, and Edina Šertović. "Biological activity and food potential of plants Rumex crispus L. and Rumex obtusifolius L.–a review." *Technologica Acta-Scientific/Professional Journal of Chemistry and Technology* 15, no. 1 (2022): 61–67.

Bektur, Zehra, Methiye Mancak Karakus, and Ufuk Koca Caliskan. "Which Humulus lupulus drug samples meet the European Pharmacopoeia criteria: Cultivated, obtained from the herbalists or online shopping sites in Turkey?" *Journal of Research in Pharmacy* 26, no. 5 (2022).

Bell, Surelys Ramos, Luis Guillermo Hernández Montiel, Ramsés Ramón González Estrada, and Porfirio Gutiérrez Martínez. "Main diseases in postharvest blueberries, conventional and eco-friendly control methods: A review." *LWT* 149 (2021): 112046.

Ben Lagha, Amel, Stéphanie Dudonné, Yves Desjardins, and Daniel Grenier. "Wild blueberry (Vaccinium angustifolium Ait.) polyphenols target Fusobacterium nucleatum and the host inflammatory response: Potential innovative molecules for treating periodontal diseases." *Journal of Agricultural and Food Chemistry* 63, no. 31 (2015): 6999–7008.

Ben-Laouane, Raja, Marouane Baslam, Mohamed Ait-El-Mokhtar, Mohamed Anli, Abderrahim Boutasknit, Youssef Ait-Rahou, Salma Toubali et al. "Potential of native arbuscular mycorrhizal fungi, rhizobia, and/or green compost as alfalfa (Medicago sativa) enhancers under salinity." *Microorganisms* 8, no. 11 (2020): 1695.

Benoliel, Doug. *Northwest Foraging: The Classic Guide to Edible Plants of the Pacific Northwest.* Skipstone, 2011.

Bercu, R., A. Bavaru, and R. Popoviciu. "Contribution to the anatomy of some angiosperms flower parts with therapeutic value." *Annals of the Romanian Society for Cell Biology* 21, no. 2 (2017): 51–58.

Bergen, Peter, James R. Moyer, and Gerald C. Kozub. "Dandelion (Taraxacum officinale) use by cattle grazing on irrigated pasture." *Weed Technology* 4, no. 2 (1990): 258–263.

Blagojević, Jovana, Jelena Vukojević, Borko Ivanović, and Žarko Ivanović. "Characterization of Alternaria species associated with leaf spot disease of Armoracia rusticana in Serbia." *Plant Disease* 104, no. 5 (2020): 1378–1389.

Blanco-Salas, Jose, Lorena Gutierrez-Garcia, Juana Labrador-Moreno, and Trinidad Ruiz-Tellez. "Wild plants potentially used in human food in the Protected Area 'Sierra Grande de Hornachos' of Extremadura (Spain)." *Sustainability* 11, no. 2 (2019): 456.

Blanco-Salas, José, María P. Hortigón-Vinagre, Diana Morales-Jadán, and Trinidad Ruiz-Téllez. "Searching for Scientific Explanations for the Uses of Spanish Folk Medicine: A Review on the Case of Mullein (Verbascum, Scrophulariaceae)." *Biology* 10, no. 7 (2021): 618.

Bljajić, Kristina, Nina Šoštarić, Roberta Petlevski, Lovorka Vujić, Andrea Brajković, Barbara Fumić, Isabel Saraiva de Carvalho, and Marijana Zovko Končić. "Effect of Betula pendula leaf extract on α-glucosidase and glutathione level in glucose-induced oxidative stress." *Evidence-Based Complementary and Alternative Medicine* 2016 (2016).

Blondel, Jacques, and James Aronson. *Biology and wildlife of the Mediterranean region.* Oxford University Press, USA, 1999.

Bobiş, Otilia, Victoriţa Bonta, Mihaiela Cornea-Cipcigan, Gulzar Ahmad Nayik, and Daniel Severus Dezmirean. "Bioactive molecules for discriminating Robinia and helianthus honey: High-performance liquid chromatography–electron spray ionization–mass spectrometry polyphenolic profile and physico-chemical determinations." *Molecules* 26, no. 15 (2021): 4433.

Bogdanov, Natalija, Vesna Rodić, and Matteo Vittuari. "Structural change and transition in the agricultural sector: Experience of Serbia." *Communist and Post-Communist Studies* 50, no. 4 (2017): 319–330.

Bokić, Bojana S., Milica M. Rat, Nebojša V. Kladar, Goran T. Anačkov, and Biljana N. Božin. "Chemical diversity of volatile compounds of mints from southern part of Pannonian plain and Balkan Peninsula–New data." *Chemistry & Biodiversity* 17, no. 8 (2020): e2000211.

Boroja, T., V. Mihailović, J. Katanić, S.-P. Pan, S. Nikles, P. Imbimbo, D. M. Monti, N. Stanković, M. S. Stanković, and R. Bauer. "The biological activities of roots and aerial parts of Alchemilla vulgaris L." *South African Journal of Botany* 116 (2018): 175–184.

Borna, F., Nazeri, V., Shokrpour, M. and Ghaziani, F., 2017. Study on response to water deficit stress in Motherwort (Leonurus cardiaca) ecotypes using stress tolerance indices, dry matter and essential oil content. *Iranian Journal of Horticultural Science* 48, no. 4: 1001–1012.

Bottone, Alfredo, Antonietta Cerulli, Gilda D'Urso, Milena Masullo, Paola Montoro, Assunta Napolitano, and Sonia Piacente. "Plant specialized metabolites in hazelnut (Corylus avellana) kernel and byproducts: An update on chemistry, biological activity, and analytical aspects." *Planta Medica* 85, no. 11/12 (2019): 840–855.

Boukhebti, Habiba, Adel Nadjib Chaker, Hani Belhadj, Farida Sahli, Messaoud Ramdhani, Hocine Laouer, and Daoud Harzallah. "Chemical composition and antibacterial activity of Mentha pulegium L. and Mentha spicata L. essential oils." *Der Pharmacia Lettre* 3, no. 4 (2011): 267–275.

Boudreau, A., Richard, A.J., Harvey, I. and Stephens, J.M., 2022. Artemisia scoparia and metabolic health: untapped potential of an ancient remedy for modern use. *Frontiers in endocrInology*, 12: 727061.

Bourgeois, Capucine, Émilie A. Leclerc, Cyrielle Corbin, Joël Doussot, Valérie Serrano, Jean-Raymond Vanier, Jean-Marc Seigneuret et al. "Nettle (Urtica dioica L.) as a source of antioxidant and anti-aging phytochemicals for cosmetic applications." *Comptes Rendus Chimie* 19, no. 9 (2016): 1090–1100.

Boyadzhiev, L. and Dimitrova, V. Extraction and liquid membrane preconcentration of rosmarinic acid from lemon balm (Melissa officinalis L.). *Separation Science and Technology* 41, no. 5 (2006): 877–886.

Bradic, Jovana, Anica Petkovic, and Marina Tomovic. "Phytochemical and Pharmacological Properties of Some Species of the Genus." *Serbian Journal of Experimental and Clinical Research* 22, no. 3 (2021): 187–193.

Brys, Rein, and Hans Jacquemyn. "Biological flora of the British Isles: Primula veris L." *Journal of Ecology* 97, no. 3 (2009): 581–600.

Brys, Rein, Hans Jacquemyn, Martin Hermy, and Tom Beeckman. "Pollen deposition rates and the functioning of distyly in the perennial Pulmonaria officinalis (Boraginaceae)." *Plant Systematics and Evolution* 273 (2008): 1–12.

Budniak, Liliia, Marjana Vasenda, and Liudmyla Slobodianiuk. "Determination of flavonoids and hydroxycinnamic acids in tablets with thick extract of Primula denticulata SMITH." *PharmacologyOnLine* 2 (2021): 1244–1253.

Buza, Victoria, Maria-Cătălina Matei Lațiu, and Laura-Cristina Ștefănuț. "Inula helenium: A literature review on ethnomedical uses, bioactive compounds and pharmacological activities." (2020).

Čajkanović, Veselin. *Rečnik srpskih narodnih verovanja o biljkama.* Talija izdavaštvo, Niš.2022. [in Serbian].

Ćalić-Dragosavac, Dušica, Snežana Zdravković-Korać, Katarina Šavikin-Fodulović, Ljiljana Radojević, and Branka Vinterhalter. "Determination of escin content in androgenic embryos and hairy root culture of Aesculus hippocastanum." *Pharmaceutical Biology* 48, no. 5 (2010): 563–567.

Candela, Rossella Gagliano, Sergio Rosselli, Maurizio Bruno, and Gianfranco Fontana. "A review of the phytochemistry, traditional uses and biological activities of the essential oils of genus Teucrium." *Planta Medica* 87, no. 06 (2020): 432–479.

Caputo, Paolo, Serena Aceto, Salvatore Cozzolino, and Roberto Nazzaro. "Morphological and molecular characterization of a natural hybrid between Orchis laxiflora and O. morio (Orchidaceae)." *Plant Systematics and Evolution* 205 (1997): 147–155.

Carbone, K. and Gervasi, F., 2022. An updated review of the genus Humulus: A valuable source of bioactive compounds for health and disease prevention. *Plants* 11, no. 24: 3434.

Ćebović, Tatjana, Dunja Jakovljević, Zoran Maksimović, Snezana Djordjevic, Sanja Jakovljević, and Dragana Četojević-Simin. "Antioxidant and cytotoxic activities of curly dock (Rumex crispus L., Polygonaceae) fruit extract." *Vojnosanitetski pregled* 77, no. 3 (2020).

Cervato, Giovanna, Marta Carabelli, Silvia Gervasio, Andrea Cittera, Roberta Cazzola, and Benvenuto Cestaro. "Antioxbdant properties of oregano (Origanum vulgare) leaf extracts." *Journal of Food Biochemistry* 24, no. 6 (2000): 453–465.

Çevik, Can Kerem, Kevser Taban Akça, And Ipek Suntar. "Cornelian Cherry (Cornus mas L.): Insight into its phytochemistry and bioactivity." *Journal of Research in Pharmacy* 26, no. 6 (2022).

Chanaj-Kaczmarek, Justyna, Małgorzata Wojcińska, and Irena Matławska. "Phenolics in the leaves." *Herba Polonica* 59, no. 1 (2013): 35–43.

Chandler, R. F., S. N. Hooper, and Mh J. Harvey. "Ethnobotany and phytochemistry of yarrow, Achillea mille-folium, Compositae." *Economic Botany* 36 (1982): 203–223.

Changizi Ashtiyani, Parvin Yarmohammady, Nasser Hosseini, and Majid Ramazani. "The Effect of Althaea officinalis. L Root Alcoholic Extract on Blood Sugar Level and Lipid Profiles of Streptozotocin Induced-Diabetic Rats." *Iranian Journal of Endocrinology and Metabolism* 17, no. 3 (2015): 238–250.

Charlebois, D. "Elderberry as a medicinal plant." *Issues in new crops and new uses.* ASHS Press, Alexandria, VA (2007): 284–292.

Chauhan, Shweta, Varun Jaiswal, Yeong-Im Cho, and Hae-Jeung Lee. "Biological activities and phytochemi-cals of lungworts (Genus Pulmonaria) focusing on pulmonaria officinalis." *Applied Sciences* 12, no. 13 (2022): 6678.

Chen, Hu, Junsong Pu, Dan Liu, Wansha Yu, Yunying Shao, Guangwei Yang, Zhonghuai Xiang, and Ningjia He. "Anti-inflammatory and antinociceptive properties of flavonoids from the fruits of black mulberry (Morus nigra L.)." *PloS one* 11, no. 4 (2016): e0153080.

Chen, Shujuan, Lin Dong, Hongfeng Quan, Xirong Zhou, Jiahua Ma, Wenxin Xia, Hao Zhou, and Xueyan Fu. "A review of the ethnobotanical value, phytochemistry, pharmacology, toxicity and quality control of Tussilago farfara L.(coltsfoot)." *Journal of Ethnopharmacology* 267 (2021): 113478.

Cheng, Zhuo, Xiaoping Lu, Fengke Lin, Abid Naeem, and Chunlin Long. "Ethnobotanical study on wild edible plants used by Dulong people in northwestern Yunnan, China." *Journal of Ethnobiology and Ethnomedicine* 18, no. 1 (2022): 1–21.

Chromchenkova E.P., Bokov D.O., Bessonov V.V., Samylina I.A., Kakhramanova S.D., Chevidaev V.V., Sokhin D.M., Balobanova N.P., Evgrafov A.A., Krasnyuk I.I., Kudashkina N.V., Galiakhmetova E.K., Marakhova A.I., Moiseev D.V. "Coltsfoot leaves (Tussilago farfara L.) A promising source of essential amino acids." *Systematic Reviews in Pharmacy* 11, no. 6 (2020).

Chrubasik, Cosima, Basil D. Roufogalis, Ulf Müller-Ladner, and Sigrun Chrubasik. "A systematic review on the Rosa canina effect and efficacy profiles." *Phytotherapy Research: An International Journal Devoted to Pharmacological and Toxicological Evaluation of Natural Product Derivatives* 22, no. 6 (2008): 725–733.

Ciocarlan, Alexandru, Ion Dragalin, Aculina Aricu, Lucian Lupascu, Nina Ciocarlan, Konstantin Vergel, Octavian G. Duliu, Gergana Hristozova, and Inga Zinicovscaia. "Chemical profile, elemental compo-sition, and antimicrobial activity of plants of the Teucrium (Lamiaceae) genus growing in Moldova." *Agronomy* 12, no. 4 (2022): 772.

Conedera, Marco, Willy Tinner, Patrik Krebs, Daniele de Rigo, and Giovanni Caudullo. "Castanea sativa in Europe: Distribution, habitat, usage and threats." In San-Miguel-Ayanz, J., de Rigo, D., Caudullo, G., Houston Durrant, T., Mauri, A. (eds.) *European Atlas of Forest Tree Species.* Luxembourg: Publication Office of the European Union (2016): 78–79.

Cosentino, Sofia, Andrea Barra, Barbara Pisano, Maddalena Cabizza, Filippo Maria Pirisi, and Francesca Palmas. "Composition and antimicrobial properties of Sardinian Juniperus essential oils against foodborne pathogens and spoilage microorganisms." *Journal of Food Protection* 66, no. 7 (2003): 1288–1291.

Da Ronch, F., G. Caudullo, T. Houston Durrant, and D. De Rigo. "Cornus mas in Europe: Distribution, habi-tat, usage and threats." *European atlas of forest tree species. Publication Office of the European Union, Luxembourg* (2016): 82–83.

Dajić Stevanović, Zora, Milica Petrović, and Svetlana Aćić. "Ethnobotanical knowledge and traditional use of plants in Serbia in relation to sustainable rural development." *Ethnobotany and Biocultural Diversities in the Balkans: Perspectives on sustainable rural development and reconciliation* (2014): 229–252.

Dajić-Stevanović, Zora, Ivan Šoštarić, Petar D. Marin, Danilo Stojanović, and Mihailo Ristić. "Population vari-ability in Thymus glabrescens Willd. From Serbia: Morphology, anatomy and essential oil composition." *Archives of Biological Sciences* 60, no. 3 (2008): 475–483.

Das, Bhrigu Kumar, AHM Viswanatha Swamy, Basavaraj C. Koti, and Pramod C. Gadad. "Experimental evi-dence for use of Acorus calamus (asarone) for cancer chemoprevention." *Heliyon* 5, no. 5 (2019).

Das, Sneha, Neeru Vasudeva, and Sunil Sharma. "Cichorium intybus: A concise report on its ethnome-dicinal, botanical, and phytopharmacological aspects." *Drug Development and Therapeutics* 7, no. 1 (2016): 1–1.

Dastagir, Ghulam, and Muhammad Afzal Rizvi. "Glycyrrhiza glabra L.(Liquorice)." *Pakistan Journal of Pharmaceutical Sciences* 29, no. 5 (2016).

David, Luminita, Bianca Moldovan, Adriana Vulcu, Liliana Olenic, Maria Perde-Schrepler, Eva Fischer-Fodor, Adrian Florea et al. "Green synthesis, characterization and anti-inflammatory activity of silver nanoparticles using European black elderberry fruits extract." *Colloids and Surfaces B: Biointerfaces* 122 (2014): 767–777.

De Benedetti, Claudia, Natalia Gerasimenko, Cesare Ravazzi, and Donatella Magri. "History of Tilia in Europe since the Eemian: Past distribution patterns." *Review of Palaeobotany and Palynology* (2022): 104778.

de Quadros, Ana Paula Oliveira, Dania Elisa Christofoletti Mazzeo, Maria Aparecida Marin-Morales, Fábio Ferreira Perazzo, Paulo Cesar Pires Rosa, and Edson Luis Maistro. "Fruit extract of the medicinal plant Crataegus oxyacantha exerts genotoxic and mutagenic effects in cultured cells." *Journal of Toxicology and Environmental Health, Part A* 80, no. 3 (2017): 161–170.

de Rigo, Daniele, Cristian Mihai Enescu, Tracy Houston Durrant, Willy Tinner, and Giovanni Caudullo. "Juglans regia in Europe: Distribution, habitat, usage and threats." (2016): 103.

De Smet, P. A. G. M. "Notes added in proof." In *Adverse effects of herbal drugs*, pp. 229–240. Berlin, Heidelberg: Springer, 1997.

Delaviz, Hamdollah, Jamshid Mohammadi, Ghasem Ghalamfarsa, Bahram Mohammadi, and Naser Farhadi. "A review study on phytochemistry and pharmacology applications of Juglans regia plant." *Pharmacognosy Reviews* 11, no. 22 (2017): 145.

Dell'Annunziata, Federica, Stefania Cometa, Roberta Della Marca, Francesco Busto, Veronica Folliero, Gianluigi Franci, Massimiliano Galdiero, Elvira De Giglio, and Anna De Filippis. "In vitro antibacterial and anti-inflammatory activity of arctostaphylos uva-ursi leaf extract against Cutibacterium acnes." *Pharmaceutics* 14, no. 9 (2022): 1952.

Dénes, Andrea, Nóra Papp, Dániel Babai, Bálint Czúcz, and Zsolt Molnár. "Wild plants used for food by Hungarian ethnic groups living in the Carpathian Basin." *Acta societatis botanicorum Poloniae* 81, no. 4 (2012).

Dettori, Caterina Angela, Laura Serreli, Alba Cuena Lombraña, Mauro Fois, Elena Tamburini, Marco Porceddu, Giuseppe Fenu, Donatella Cogoni, and Gianluigi Bacchetta. "The genetic structure and diversity of Gentiana lutea subsp. lutea (Gentianaceae) in Sardinia: Further insights for its conservation planning." *Caryologia* 71, no. 4 (2018): 489–496.

Dias, Nildo S., Jorge FS Ferreira, Xuan Liu, and Donald L. Suarez. "Jerusalem artichoke (Helianthus tuberosus, L.) maintains high inulin, tuber yield, and antioxidant capacity under moderately-saline irrigation waters." *Industrial Crops and Products* 94 (2016): 1009–1024.

Diljkan, Maja, Siniša Škondrić, Dino Hasanagić, Mirjana Žabić, Ljiljana Topalić-Trivunović, Carlos Raúl Jiménez-Gallardo, and Biljana Kukavica. "The antioxidant response of Hedera helix leaves to seasonal temperature variations." *Botanica Serbica* 46, no. 2 (2022): 295–309.

Đorđević, Aleksandra S. "Chemical composition of Hypericum perforatum L. essential oil." *Advanced Technologies* 4, no. 1 (2015): 64–68.

Đordjević, Sofija, and Nada Ćujić Nikolić. "Hawthorn (Crataegus spp.) from botanical source to phytopreparations." *Lekovite sirovine* 41 (2021): 63–71.

Đurić, Milena, Jelena Mladenović, Ljiljana Bošković-Rakočević, Gordana Šekularac, Duško Brković, and Nenad Pavlović. "Use of different types of extracts as biostimulators in organic agriculture." *Acta Agriculturae Serbica* 24, no. 47 (2019): 27–39.

Eaton, E., G. Caudullo, and D. De Rigo. "Tilia cordata, Tilia platyphyllos and other limes in Europe: Distribution, habitat, usage and threats." *European Atlas of Forest Tree Species* (2016): 184–185.

Efthimiou, Ioanna, Dimitris Vlastos, Vassilios Triantafyllidis, Antonios Eleftherianos, and Maria Antonopoulou. "Investigation of the Genotoxicological Profile of Aqueous Betula pendula Extracts." *Plants* 11, no. 20 (2022): 2673.

El Abbouyi, Ahmed, Said El Khyari, Rabia Eddoha, and Najoie Filaliâ. "Antiâ€'inflammatory effect of hydromethanolic extract from Marrubium vulgare Lamiaceae on leukocytes oxidative metabolism: An in vitro and in vivo studies." *International Journal of Green Pharmacy (IJGP)* 7, no. 3 (2013).

El-Aasr, Mona, Toshihiro Nohara, Tsuyushi Ikeda, Sally E. Abu-Risha, Engy Elekhnawy, Haytham O. Tawfik, Nagwa Shoeib, and Ghada Attia. "LC-MS/MS metabolomics profiling of Glechoma hederacea L. methanolic extract; in vitro antimicrobial and in vivo with in silico wound healing studies on Staphylococcus aureus infected rat skin wound." *Natural Product Research* 37, no. 10 (2023): 1730–1734.

El Mihyaoui, A., Esteves da Silva, J.C., Charfi, S., Candela Castillo, M.E., Lamarti, A. and Arnao, M.B., 2022. Chamomile (Matricaria chamomilla L.): A review of ethnomedicinal use, phytochemistry and pharmacological uses. *Life* 12, no. 4: 479.

Elez Garofulić, Ivona, Valentina Malin, Maja Repajić, Zoran Zorić, Sandra Pedisić, Meta Sterniša, Sonja Smole Možina, and Verica Dragović-Uzelac. "Phenolic profile, antioxidant capacity and antimicrobial activity of nettle leaves extracts obtained by advanced extraction techniques." *Molecules* 26, no. 20 (2021): 6153.

Eliopoulos, Aristides G., Apostolis Angelis, Anastasia Liakakou, and Leandros A. Skaltsounis. "In Vitro Anti-Influenza Virus Activity of Non-Polar Primula veris subsp. veris Extract." *Pharmaceuticals* 15, no. 12 (2022): 1513.

Emadi, Seyyed Amir, Mahboobeh Ghasemzadeh Rahbardar, Soghra Mehri, and Hossein Hosseinzadeh. "A review of therapeutic potentials of milk thistle (Silybum marianum L.) and its main constituent, silymarin, on cancer, and their related patents." *Iranian Journal of Basic Medical Sciences* 25, no. 10 (2022): 1166.

Enescu, C. M., T. Houston Durrant, G. Caudullo, and D. de Rigo. "Juniperus communis in Europe: Distribution, habitat, usage and threats." *European Atlas of Forest Tree Species; San-Miguel-Ayanz, J., de Rigo, D., Caudullo, G., Houston Durrant, T., Mauri, A., Eds* (2016): e01d2de.

Ercisli, Sezai, and Emine Orhan. "Some physico-chemical characteristics of black mulberry (Morus nigra L.) genotypes from Northeast Anatolia region of Turkey." *Scientia Horticulturae* 116, no. 1 (2008): 41–46.

Erdogan Orhan, Ilkay, Fatma Sezer Senol, Sinem Aslan Erdem, Erdem, I. Irem Tatli, Murat Kartal, and Sevket Alp. "Tyrosinase and cholinesterase inhibitory potential and flavonoid characterization of Viola odorata L.(Sweet Violet)." *Phytotherapy Research* 29, no. 9 (2015): 1304–1310.

Ersoy, Nilda, Yavuz Bagci, Hamdi Zenginbal, Merve Salman Ozen, and Ayse Yalcin Elidemir. "Antioxidant properties of Rosehip fruit types (Rosa canina sp.) selected from Bolu-Turkey." *International Journal of Science and Knowledge* 4, no. 1 (2015): 51–59.

Ezberci, F. and Ünal, E. "Aesculus Hippocastanum (Aescin, Horse Chestnut) in the Management of Hemorrhoidal Disease: Review". *Turkish Journal of Colorectal* 28, (2018): 54–57.

Farjon, Aljos, and Denis Filer (eds.). "Global and Trans-Continental Distributions." In Aljos Farjon and Denis Filer (eds). *An Atlas of the World's Conifers*, pp. 11–33. Brill, 2013.

Farrell, Nicholas, Gregory Norris, Sang Gil Lee, Ock K. Chun, and Christopher N. Blesso. "Anthocyanin-rich black elderberry extract improves markers of HDL function and reduces aortic cholesterol in hyperlipidemic mice." *Food & Function* 6, no. 4 (2015): 1278–1287.

Farzaneh, Avishan, Abbas Hadjiakhoondi, Mahnaz Khanavi, Azadeh Manayi, and Roodabeh Bahram Soltani. "Filipendula ulmaria (L.) Maxim.(Meadowsweet): A Review of Traditional Uses, Phytochemistry and Pharmacology." *Research Journal of Pharmacognosy* 9, no. 3 (2022): 85–106.

Fatima T., Bashir O., Naseer B., Hussain S.Z. (2018). Dandelion: Phytochemistry and clinical potential. *Journal of Medicinal Plants Studies*, 6(2):198–202.

Fazio, Alessia, Pierluigi Plastina, Jocelijn Meijerink, Renger F. Witkamp, and Bartolo Gabriele. "Comparative analyses of seeds of wild fruits of Rubus and Sambucus species from Southern Italy: Fatty acid composition of the oil, total phenolic content, antioxidant and anti-inflammatory properties of the methanolic extracts." *Food Chemistry* 140, no. 4 (2013): 817–824.

Fernald, Merritt Lyndon, Alfred Charles Kinsey, and Reed Clark Rollins. *Edible wild plants of eastern North America*. Courier Corporation, 1996.

Finn, Chad E., Andrew L. Thomas, Patrick L. Byers, and Sedat Serçe. "Evaluation of American (Sambucus canadensis) and European (S. nigra) elderberry genotypes grown in diverse environments and implications for cultivar development." *HortScience* 43, no. 5 (2008): 1385–1391.

Fizeşan, Ionel, Marius Emil Rusu, Carmen Georgiu, Anca Pop, Maria-Georgia Ştefan, Dana-Maria Muntean, Simona Mirel, Oliviu Vostinaru, Béla Kiss, and Daniela-Saveta Popa. "Antitussive, antioxidant, and anti-inflammatory effects of a walnut (Juglans regia L.) septum extract rich in bioactive compounds." *Antioxidants* 10, no. 1 (2021): 119.

Fotirić Akšić, Milica, Milica Nešović, Ivanka Ćirić, Živoslav Tešić, Lato Pezo, Tomislav Tosti, Uroš Gašić, Biljana Dojčinović, Biljana Lončar, and Mekjell Meland. "Chemical fruit profiles of different raspberry cultivars grown in specific Norwegian agroclimatic conditions." *Horticulturae* 8, no. 9 (2022): 765.

Gautam, Shiv Shanker, Navneet, and Sanjay Kumar. "The antibacterial and phytochemical aspects of Viola odorata Linn. extracts against respiratory tract pathogens." *Proceedings of the National Academy of Sciences, India Section B: Biological Sciences* 82 (2012): 567–572.

Gavaric, Neda, Sonja Smole Mozina, Nebojša Kladar, and Biljana Bozin. "Chemical profile, antioxidant and antibacterial activity of thyme and oregano essential oils, thymol and carvacrol and their possible synergism." *Journal of Essential Oil Bearing Plants* 18, no. 4 (2015): 1013–1021.

Georgiev, Milen I., Kashif Ali, Kalina Alipieva, Robert Verpoorte, and Young Hae Choi. "Metabolic differentiations and classification of Verbascum species by NMR-based metabolomics." *Phytochemistry* 72, no. 16 (2011): 2045–2051.

Ghedira, K., P. Goetz, and R. Jeune. "Alchemilla vulgaris L.: Alchémille (Rosaceae)." *Phytotherapie-Heidelberg* 10, no. 4 (2012): 263.

Gompf, Rebecca E. "Nutritional and herbal therapies in the treatment of heart disease in cats and dogs." *Journal of the American Animal Hospital Association* 41, no. 6 (2005): 355–367.

González, Leonardo Galindo, and Michael K. Deyholos. "Identification, characterization and distribution of transposable elements in the flax (Linum usitatissimum L.) genome." *BMC Genomics* 13 (2012): 1–17.

González-Castejón, Marta, Francesco Visioli, and Arantxa Rodriguez-Casado. "Diverse biological activities of dandelion." *Nutrition Reviews* 70, no. 9 (2012): 534–547.

Gordanić, Stefan, Dragoja Radanović, Sandra Vuković, Stefan Kolašinac, Sofija Kilibarda, Tatjana Marković, Đorđe Moravčević, and Aleksandar Ž. Kostić. "Phytochemical characterization and antioxidant potential of Allium ursinum L. cultivated on different soil types-a preliminary study." *Emirates Journal of Food and Agriculture* 34, no. 11 (2022): 904–914.

Gordanić, Stefan, Aleksandar Simić, Dragoja Radanović, Tatjana Marković, Snežana Mrđan, Sandra Vuković, Vladimir Filipović, Sara Mikić, and Đorđe Moravčević. "Morphological definition populations of Allium ursinum L. from the western part of the Republic of Serbia." *AGRORES* 2021, no. 10 (2021): 104.

Goud, Busineni Jayasimha, V. Dwarakanath, and B. K. Swamy. "A review on history, controversy, traditional use, ethnobotany, phytochemistry and pharmacology of Artemisia absinthium Linn." *International Journal of Advanced Research in Engineering and Applied Sciences* 4, no. 5 (2015): 77–107.

Grabowska, Karolina, Kinga Amanowicz, Paweł Paśko, Irma Podolak, and Agnieszka Galanty. "Optimization of the extraction procedure for the phenolic-rich Glechoma hederacea L. herb and evaluation of its cytotoxic and antioxidant potential." *Plants* 11, no. 17 (2022): 2217.

Grasgruber, Pavel, Bojan Mašanović, Stipan Prce, Stevo Popović, Fitim Arifi, Duško Bjelica, Dominik Bokůvka et al. "Mapping the mountains of giants: Anthropometric data from the western balkans reveal a nucleus of extraordinary physical stature in Europe." *Biology* 11, no. 5 (2022): 786.

Grauso, Laura, Bruna de Falco, Riccardo Motti, and Virginia Lanzotti. "Corn poppy, Papaver rhoeas L.: A critical review of its botany, phytochemistry and pharmacology." *Phytochemistry Reviews* 20 (2021): 227–248.

Green, Aliza. *Field Guide to Herbs & Spices: How to Identify, Select, and Use Virtually Every Seasoning on the Market*. Quirk books, 2006.

Greenlee, Heather, Kathy Abascal, Eric Yarnell, and Elena Ladas. "Clinical applications of Silybum marianum in oncology." *Integrative Cancer Therapies* 6, no. 2 (2007): 158–165.

Gross, Katherine L., and Patricia A. Werner. "The biology of Canadian weeds.: 28. Verbascum thapsus L. and V. blattaria L." *Canadian Journal of Plant Science* 58, no. 2 (1978): 401–413.

Grozeva, Neli. "A comparative morphological characteristics of Chenopodium album L., C. missouriense Aellen and C. probstii Aellen." *Türk Tarım ve Doğa Bilimleri Dergisi* 1, no. Özel Sayı-2 (2014): 1949–1954.

Guclu, Ertugrul, Hayriye Genc, Mustafa Zengin, and Oguz Karabay. "Antibacterial activity of Lythrum salicaria against multidrug-resistant Acinetobacter baumannii and Pseudomonas aeruginosa." *Annual Research & Review in Biology* (2014): 1099–1105.

Guillén Núñez, María Elena, José Artur da Silva Emim, Caden Souccar, and Antonio José Lapa. "Analgesic and anti-inflammatory activities of the aqueous extract of Plantago major L." *International Journal of Pharmacognosy* 35, no. 2 (1997): 99–104.

Gülçin, Ilhami, Ö. İrfan Küfrevioğlu, Münir Oktay, and Mehmet Emin Büyükokuroğlu. "Antioxidant, antimicrobial, antiulcer and analgesic activities of nettle (Urtica dioica L.)." *Journal of Ethnopharmacology* 90, no. 2–3 (2004): 205–215.

Guna, Gowher. "Theraprutic value of arctium lappa linn-a review." *Asian Journal of Pharmaceutical and Clinical Research* 12, no. 7 (2019): 53–9.

Gwiazdowska, Daniela, Pascaline Aimee Uwineza, Szymon Frąk, Krzysztof Juś, Katarzyna Marchwińska, Romuald Gwiazdowski, and Agnieszka Waśkiewicz. "Antioxidant, antimicrobial and antibiofilm properties of Glechoma hederacea extracts obtained by supercritical fluid extraction, using different extraction conditions." *Applied Sciences* 12, no. 7 (2022): 3572.

Hake, Alexander, Nico Symma, Stefan Esch, Andreas Hensel, and Martina Düfer. "Alkaloids from lime flower (Tiliae flos) exert spasmodic activity on murine airway smooth muscle involving acetylcholinesterase." *Planta Medica* 88, no. 08 (2022): 639–649.

Hallmann, Ewelina. "Quantitative and qualitative identification of bioactive compounds in edible flowers of black and bristly locust and their antioxidant activity." *Biomolecules* 10, no. 12 (2020): 1603.

Happy, Afroza Akter, Ferdoushi Jahan, and Md Abdul Momen. "Essential oils: Magical ingredients for skin care. J." *Plant Science* 9, no. 2 (2021): 54.

Haragan, Patricia Dalton. *Weeds of Kentucky and adjacent states: A field guide*. University Press of Kentucky, 2014.

Hashimi, Ayshah, Mantasha Binth Siraj, Yasmeen Ahmed, A. Siddiqui, and Umar Jahangir. "One for All–Artemisia absinthium (Afsanteen) "A Potent Unani Drug"." *Tang (Humanitas Medicine)* 9, no. 4 (2019): 1–9.

Hassanzadeh, Mohammad K., Zahra Tayarani Najaran, Maryam Nasery, and Seyed Ahmad Emami. "Summer savory (Satureja hortensis L.) oils." In *Essential Oils in Food Preservation, Flavor and Safety*, pp. 757–764. Academic Press, 2016.

Heinonen, Marina. "Antioxidant activity and antimicrobial effect of berry phenolics–a Finnish perspective." *Molecular Nutrition & Food Research* 51, no. 6 (2007): 684–691.

Hemmati, Ali Asghar, Gholamreza Houshmand, Mohammad Nemati, Mohammad Bahadoram, Nozar Dorestan, Mohammad Reza Rashidi-Nooshabadi, and Hamidreza Zargar. "Wound healing effects of persian Oak (Quercus brantii) ointment in rats." *Jundishapur Journal of Natural Pharmaceutical Products* 10, no. 4 (2015).

Hickey, Michael, and Clive King. *Common families of flowering plants*. Cambridge University Press, 1997.

Hong, Jungha, and Nazim S. Gruda. "The potential of introduction of Asian vegetables in Europe." *Horticulturae* 6, no. 3 (2020): 38.

Horikawa, Kazumi, Takami Mohri, Yoshito Tanaka, and Hiroshi Tokiwa. "Moderate inhibition of mutagenicity and carcinogenicity of benzo [a] pyrene, 1, 6-dinitropyrene and 3, 9-dinitrofluoranthene by Chinese medicinal herbs." *Mutagenesis* 9, no. 6 (1994): 523–526.

Huang, Wuyang, Zheng Yan, Dajing Li, Yanhong Ma, Jianzhong Zhou, and Zhongquan Sui. "Antioxidant and anti-inflammatory effects of blueberry anthocyanins on high glucose-induced human retinal capillary endothelial cells." *Oxidative Medicine and Cellular Longevity* (2018).

Humadi, S.S. and Istudor, V., 2009. Lythrum salicaria (purple loosestrife). Medicinal use, extraction and identification of its total phenolic compounds. *Farmacia* 57, no. 2: 192–200.

Huntley, A., and E. Ernst. "Herbal medicines for asthma: A systematic review." *Thorax* 55, no. 11 (2000): 925.

Hutchings, Michael J., and Elizabeth AC Price. "Glechoma hederacea L.(Nepeta glechoma Benth., N. hederacea (L.) Trev.)." *Journal of Ecology* (1999): 347–364.

Iacopini, Patrizia, Fabiano Camangi, Agostino Stefani, and Luca Sebastiani. "Antiradical potential of ancient Italian apple varieties of Malus× domestica Borkh. in a peroxynitrite-induced oxidative process." *Journal of Food Composition and Analysis* 23, no. 6 (2010): 518–524.

Ibdah, Mwafaq, Shada Hino, Bhagwat Nawade, Mosaab Yahyaa, Tejas C. Bosamia, and Liora Shaltiel-Harpaz. "Identification and characterization of three nearly identical linalool/nerolidol synthase from Acorus calamus." *Phytochemistry* 202 (2022): 113318.

Igić R., Vukov D., Božin B., Orlović S. Medicinal plants: Natural resources of Vojvodina. Vrelo Novi Sad. (2010).

Iličić, Renata, Aleksandra Jelušić, Sanja Marković, Goran Barać, Ferenc Bagi, and Tatjana Popović. "Pseudomonas cerasi, the new wild cherry pathogen in Serbia and the potential use of recG helicase in bacterial identification." *Annals of Applied Biology* 180, no. 1 (2022): 140–150.

Iranshahy, Milad, Behjat Javadi, Mehrdad Iranshahi, Seyedeh Pardis Jahanbakhsh, Saman Mahyari, Faezeh Vahdati Hassani, and Gholamreza Karimi. "A review of traditional uses, phytochemistry and pharmacology of Portulaca oleracea L." *Journal of Ethnopharmacology* 205 (2017): 158–172.

Ivanova, Antoaneta, Bozhanka Mikhova, Hristo Najdenski, Iva Tsvetkova, and Ivanka Kostova. "Chemical composition and antimicrobial activity of wild garlic Allium ursinum of Bulgarian origin." *Natural Product Communications* 4, no. 8 (2009): 1934578X0900400808.

Jacques, Dominique, Kristine Vandermijnsbrugge, Sébastein Lemaire, Adriana Antofie, and Marc Lateur. "Natural distribution and variability of wild apple (Malus sylvestris) in Belgium." *Belgian Journal of Botany* (2009): 39–49.

Jafari-Sales, Abolfazl, Behboud Jafari, Javad Sayyahi, and Tahereh Zohoori-Bonab. "Evaluation of antibacterial activity of ethanolic extract of malva neglecta and althaea officinalis l. On antibiotic-resistant strains of staphylococcus aureus." *Journal of Biology and Today's World* 4, no. 2 (2015): 58–62.

Jakobek, Lidija, and Andrew R. Barron. "Ancient apple varieties from Croatia as a source of bioactive polyphenolic compounds." *Journal of Food Composition and Analysis* 45 (2016): 9–15.

Jan, Khan Nadiya, Khan Zarafshan, and Sukhcharn Singh. "Stinging nettle (Urtica dioica L.): A reservoir of nutrition and bioactive components with great functional potential." *Journal of Food Measurement and Characterization* 11 (2017): 423–433.

Janaćković, Pedja, Milan Gavrilović, Jasmina Savić, Petar D. Marin, and Zora Dajić Stevanović. "Traditional knowledge on plant use from Negotin Krajina (Eastern Serbia): An ethnobotanical study." (2019).

Janick, Jules. "The origins of fruits, fruit growing, and fruit breeding." *Plant Breeding Reviews* 25, no. 1 (2005): 255–320.

Jarić, Snežana, Marina Mačukanović-Jocić, Lola Djurdjević, Miroslava Mitrović, Olga Kostić, Branko Karadžić, and Pavle Pavlović. "An ethnobotanical survey of traditionally used plants on Suva planina mountain (south-eastern Serbia)." *Journal of Ethnopharmacology* 175 (2015): 93–108.

Jarić, Snežana, Zorica Popović, Marina Mačukanović-Jocić, Lola Djurdjević, Miroslava Mijatović, Branko Karadžić, Miroslava Mitrović, and Pavle Pavlović. "An ethnobotanical study on the usage of wild medicinal herbs from Kopaonik Mountain (Central Serbia)." *Journal of Ethnopharmacology* 111, no. 1 (2007): 160–175.

Jelača, Sanja, Zora Dajić-Stevanović, Nenad Vuković, Stefan Kolašinac, Antoaneta Trendafilova, Paraskev Nedialkov, Miroslava Stanković et al. "Beyond traditional use of Alchemilla vulgaris: Genoprotective and antitumor activity in vitro." *Molecules* 27, no. 23 (2022): 8113.

Jelenković, Ljiljana, Predrag Jelenković, and Ljubo Pejanović. "The economic possibilities and perspectives of aromatic and medicinal herbs (Satureja kitaibelii)." *Економика пољопривреде* 63, no. 2 (2016): 375–388.

Jiang, H., L. Xu, J. C. Liu, and X. Z. Huang. "Research progress on active ingredients and pharmacological functions of black mulberry (Morus nigra L.)." *Science of Sericulture* 1 (2011): 20.

Jovanović, Aleksandra A., Milica Mosurović, Branko Bugarski, Petar Batinić, Natalija Čutović, Stefan Gordanić, and Tatjana Marković. "Melissa officinalis extracts obtained using maceration, ultrasoundand microwave-assisted extractions: Chemical composition, antioxidant capacity, and physical characteristics." *Lekovite sirovine* 42 (2022): 51–59.

Josifovic, M., (Ed.), *Flora SR Srbije (Flora of the Republic of Serbia) 7*, Serbian Academy of Science and Art, Belgrade, 1975 (in Serbian)

Kaithwas, Gaurav, Alok Mukherjee, A. K. Chaurasia, and Dipak K. Majumdar. "Antiinflammatory, analgesic and antipyretic activities of Linum usitatissimum L. (flaxseed/linseed) fixed oil." *Indian Journal of Experimental Biology* 49, no. 12 (2011): 932–938.

Kanak, Sebastian, Barbara Krzemińska, Rafał Celiński, Magdalena Bakalczuk, and Katarzyna Dos Santos Szewczyk. "Phenolic Composition and Antioxidant Activity of Alchemilla Species." *Plants* 11, no. 20 (2022): 2709.

Kanatt, Sweetie R., Ramesh Chander, and Arun Sharma. "Antioxidant potential of mint (Mentha spicata L.) in radiation-processed lamb meat." *Food Chemistry* 100, no. 2 (2007): 451–458.

Karakaya, Sibel, Sedef Nehir El, Nural Karagözlü, and Serpil Şahin. "Antioxidant and antimicrobial activities of essential oils obtained from oregano (Origanum vulgare ssp. hirtum) by using different extraction methods." *Journal of Medicinal Food* 14, no. 6 (2011): 645–652.

Karimi, E., Oskoueian, E., Oskoueian, A., Omidvar, V., Hendra, R. and Nazeran, H., 2013. Insight into the functional and medicinal properties of Medicago sativa (Alfalfa) leaves extract. *Journal of Medicinal Plants Research* 7, no. 7: 290–297.

Keivani, Mahnaz, Iraj Mehregan, and Dirk C. Albach. "Genetic diversity and population structure of Plantago major (Plantaginaceae) in Iran." *The Iranian Journal of Botany* 26, no. 2 (2020): 111–124.

Kenny, Ciara-Ruth, Anna Stojakowska, Ambrose Furey, and Brigid Lucey. "From monographs to chromatograms: The antimicrobial potential of Inula helenium L.(Elecampane) naturalised in Ireland." *Molecules* 27, no. 4 (2022): 1406.

Khalid, Nauman, Sardar Atiq Fawad, and Iftikhar Ahmed. "Antimicrobial activity, phytochemical profile and trace minerals of black mulberry (Morus nigra L.) fresh juice." *Pakistan Journal of Botany* 43, no. 6 (2011): 91–96.

Khan, Sher Wali, and Surayya Khatoon. "Ethnobotanical studies on some useful herbs of Haramosh and Bugrote valleys in Gilgit, northern areas of Pakistan." *Pakistan Journal of Botany* 40, no. 1 (2008): 43.

Khare, Chandrama P. *Indian medicinal plants: An illustrated dictionary*. Springer Science & Business Media, 2008.

Khazaei, M. K. M. R., and M. Pazhouhi. "An overview of therapeutic potentials of Rosa canina: A traditionally valuable herb." *WCRJ* 7 (2020): e1580.

Kim, Minju, Kandhasamy Sowndhararajan, and Songmun Kim. "The chemical composition and biological activities of essential oil from Korean native thyme Bak-Ri-Hyang (Thymus quinquecostatus Celak.)." *Molecules* 27, no. 13 (2022): 4251.

Knežević, Violeta, Lato Pezo, Biljana Lončar, Milica Nićetin, Vladimir Filipović, Stanislava Gorjanović, Desanka Sužnjević, and Tatjana Kuljanin. "Osmotic treatment of nettle leaves-optimization of kinetics and antioxidant activity." *Journal on Processing and Energy in Agriculture* 19, no. 4 (2015): 175–178.

Knežević, Violeta, Lato Pezo, Biljana Lončar, Vladimir Filipović, Milica Nićetin, Stanislava Gorjanović, and Danijela Šuput. "Antioxidant capacity of nettle leaves during osmotic treatment." *Periodica Polytechnica-Chemical Engineering* 63, no. 3 (2019): 491–498.

Kojičić, Ksenija, Dejan Pljevljakušić, Marija Marković, Dragoljub Miladinović, Biljana Nikolić, Olivera Papović, Ljubinko Rakonjac, and Snežana Cupara. "Traditional use of antitumor plants in the Pirot County of Southeastern Serbia." *Macedonian Pharmaceutical Bulletin* no. 68 (Suppl 2) (2022):197 – 198.

Kolarević, Ana, Dragana Stojiljković, Sandra Dinić, and Ivana Nešić. "Effect of application of emulsions with standardized extract of wild apple fruit (Malus sylvestris (L.) Mill., Rosaceae) on biophysical skin parameters: An in vivo study." *Acta Medica Medianae* 59, no. 3 (2020): 48–55.

Koshovyi, Oleh, Ain Raal, Igor Kireyev, Nadiya Tryshchuk, Tetiana Ilina, Yevhen Romanenko, Sergiy M. Kovalenko, and Natalya Bunyatyan. "Phytochemical and psychotropic research of motherwort (Leonurus cardiaca L.) modified dry extracts." *Plants* 10, no. 2 (2021): 230.

Krebs, Patrik, Gianni Boris Pezzatti, Giorgia Beffa, Willy Tinner, and Marco Conedera. "Revising the sweet chestnut (Castanea sativa Mill.) refugia history of the last glacial period with extended pollen and mac-rofossil evidence." *Quaternary Science Reviews* 206 (2019): 111–128.

Kronbichler, Andreas, Maria Effenberger, Michael Eisenhut, Keum Hwa Lee, and Jae Il Shin. "Seven recom-mendations to rescue the patients and reduce the mortality from COVID-19 infection: An immunological point of view." *Autoimmunity Reviews* 19, no. 7 (2020): 102570.

Krzyżanowska-Kowalczyk, Justyna, Mariusz Kowalczyk, Michał B. Ponczek, Łukasz Pecio, Paweł Nowak, and Joanna Kolodziejczyk-Czepas. "Pulmonaria obscura and pulmonaria officinalis extracts as mitiga-tors of peroxynitrite-induced oxidative stress and cyclooxygenase-2 inhibitors–in vitro and in silico stud-ies." *Molecules* 26, no. 3 (2021): 631.

Kucharski, Łukasz, Krystyna Cybulska, Edyta Kucharska, Anna Nowak, Robert Pełech, and Adam Klimowicz. "Biologically Active Preparations from the Leaves of Wild Plant Species of the Genus Rubus." *Molecules* 27, no. 17 (2022): 5486.

Kumar, Amit. "Medicinal properties of Acorus calamus." *Journal of Drug Delivery and Therapeutics* 3, no. 3 (2013): 143–144.

Kumari, Jyanti, and C. T. N. Singh. "Portulaca oleracea L. sp. Pl a small herb of religious and medicinal sig-nificance." *Journal of Medicinal Plants* 4, no. 4 (2016): 196–197.

Kuššar, A., A. Zupančič, M. Šentjurc, and D. Baričevič. "Free radical scavenging activities of yellow gentian (Gentiana lutea L.) measured by electron spin resonance." *Human & Experimental Toxicology* 25, no. 10 (2006): 599–604.

Kwon, Dur Han, Hyuk Yun Kwon, Hyun Jung Kim, Eun Joo Chang, Man Bae Kim, Seung Kew Yoon, Eun Young Song et al. "Inhibition of hepatitis B virus by an aqueous extract of Agrimonia eupatoria L." *Phytotherapy Research: An International Journal Devoted to Pharmacological and Toxicological Evaluation of Natural Product Derivatives* 19, no. 4 (2005): 355–358.

Lakić, Neda S., Neda M. Mimica-Dukić, Jelena M. Isak, and Biljana N. Božin. "Antioxidant properties of Galium verum L.(Rubiaceae) extracts." *Central European Journal of Biology* 5 (2010): 331–337.

Lal, R. K. "The opium poppy (Papaver somniferum L.): Historical perspectives recapitulate and induced muta-tion towards latex less, low alkaloids in capsule husk mutant: A review." *The Journal of Medicinal Plants Studies* 10, no. 3 Part A (2022): 19–29.

Lazić, Biljana D., Svjetlana Janjić, Tatjana Rijavec, and Mirjana Kostić. "Effect of chemical treatments on the chemical composition and properties of flax fibers." *Journal of the Serbian Chemical Society* 82, no. 1 (2017): 83–97.

Lee, Jin-Young, Dan-Hee Yoo, Yong-Seong Jeong, Sung-Hyun Joo, and Jung-Woo Chae. "Verification of Anti-inflammatory Activities of the Ethanol Extracts of Glechoma hederacea var. longituba in RAW 264.7 Cells." *Journal of Life Science* 28, no. 4 (2018): 429–434.

Lee, Ki Yong, Lim Hwang, Eun Ju Jeong, Seung Hyun Kim, Young Choong Kim, and Sang Hyun Sung. "Effect of neuroprotective flavonoids of Agrimonia eupatoria on glutamate-induced oxidative injury to HT22 hippocampal cells." *Bioscience, Biotechnology, and Biochemistry* 74, no. 8 (2010): 1704–1706.

Li, Yanni, Yilun Chen, and Dongxiao Sun-Waterhouse. "The potential of dandelion in the fight against gastro-intestinal diseases: A review." *Journal of Ethnopharmacology* 293 (2022): 115272.

Lim T.K. "Viola odorata." In *Edible Medicinal and Non Medicinal Plants: Volume 8, Flowers*, pp. 795–807. Dordrecht: Springer Netherlands, 2014.

Liston, Aaron, Richard Cronn, and Tia-Lynn Ashman. "Fragaria: A genus with deep historical roots and ripe for evolutionary and ecological insights." *American Journal of Botany* 101, no. 10 (2014): 1686–1699.

Liu, Dan, Xiao-Qin He, Ding-Tao Wu, Hua-Bin Li, Yi-Bin Feng, Liang Zou, and Ren-You Gan. "Elderberry (Sambucus nigra L.): Bioactive compounds, health functions, and applications." *Journal of Agricultural and Food Chemistry* 70, no. 14 (2022): 4202–4220.

Liu, Fei-Hu, Xia Chen, Bo Long, Rui-Yan Shuai, and Chun-Lin Long. "Historical and botanical evidence of distribution, cultivation and utilization of Linum usitatissimum L.(flax) in China." *Vegetation History and Archaeobotany* 20 (2011): 561–566.

Liu, Mingchuan, Shengjie Yang, Jinping Yang, Yita Lee, Junping Kou, and Chaojih Wang. "Neuroprotective and memory-enhancing effects of antioxidant peptide from walnut (Juglans regia L.) protein hydroly-sates." *Natural Product Communications* 14, no. 7 (2019): 1934578X19865838.

Lodhi, Santram, Gautam Prakash Vadnere, Vimal Kant Sharma, and Md Rageeb Usman. "Marrubium vulgare L.: A review on phytochemical and pharmacological aspects." *Journal of Intercultural Ethnopharmacology* 6, no. 4 (2017): 429–452.

Luczaj, Lukasz, Andrea Pieroni, Javier Tardío, Manuel Pardo-de-Santayana, Renata Sõukand, Ingvar Svanberg, and Raivo Kalle. "Wild food plant use in 21 st century Europe, the disapperance of old traditions and the search for new ciusines involving wild edibles." *Acta societatis botanicorum poloniae* 81, no. 4 (2012).

Lust, John. *The herb book: The most complete catalog of herbs ever published.* Courier Corporation, 2014.

Maksimović, Zoran, Marina Milenković, Dragana Vučićević, and Mihailo Ristić. "Chemical composition and antimicrobial activity of Thymus pannonicus All.(Lamiaceae) essential oil." *Open Life Sciences* 3, no. 2 (2008): 149–154.

Maliníková, Erika, Ján Kukla, Margita Kuklová, and Mária Balážová. "Altitudinal variation of plant traits: Morphological characteristics in Fragaria vesca L.(Rosaceae)." *Annals of Forest Research* 56, no. 1 (2013): 79–89.

Malinowska, Paulina. "Effect of flavonoids content on antioxidant activity of commercial cosmetic plant extracts." *Herba Polonica* 59, no. 3 (2013): 63–75.

Mamedov, Nazim A., And Sayyara C. Ibadullayeva. "The Correlation Between Color Of Flowers And Healing Abilities Of Medicinal Plants." *Acta Botanica Caucasica* (2022): 12.

Manokhina, Aleksandra A., Alexey S. Dorokhov, Tamara P. Kobozeva, Tatiana N. Fomina, and Oksana A. Starovoitova. "Varietal characteristics of jerusalem artichoke as a high nutritional value crop for her-bivorous animal husbandry." *Applied Sciences* 12, no. 9 (2022): 4507.

Marambe, P. W. M. L. H. K., P. J. Shand, and J. P. D. Wanasundara. "An in-vitro investigation of selected bio-logical activities of hydrolysed flaxseed (Linum usitatissimum L.) proteins." *Journal of the American Oil Chemists' Society* 85 (2008): 1155–1164.

Marković, Marija, Dejan Pljevljakušić, Jelena Matejić, Biljana Nikolić, Mirjana Smiljić, Gorica Đelić, Olivera Papović, Mrđan Đokić, and Vesna Stankov-Jovanović. "The plants traditionally used for the treatment of respiratory infections in the Balkan Peninsula (Southeast Europe)." *Lekovite sirovine* 42 (2022b): 68–88.

Marković, Marija, Dejan Pljevljakušić, Nebojša Menković, Jelena Matejić, Olivera Papović, and Vesna Stankov-Jovanović. "Traditional knowledge on the medicinal use of plants from genus Gentiana in the Pirot County (Serbia)." *Lekovite sirovine* 41 (2021): 46–53.

Marković, Marija, Dejan Pljevljakušić, Biljana Nikolić, Ljubinko Rakonjac, and Vesna Stankov-Jovanović. "Ethnomedicinal application of species from genus Thymus in the Pirot County (Southeastern Serbia)." *Lekovite sirovine* 40 (2020): 27–32.

Marković, Marija, Dejan Pljevljakušić, Olivera Papović, and Jovanović Vesna Stankov. "Ethnopharmacological use of burdock ('Arctium lappa') in the Pirot County." *Pirotski zbornik* 47 (2022a): 133–142.

Martins, Natália, Lillian Barros, Celestino Santos-Buelga, Sónia Silva, Mariana Henriques, and Isabel CFR Ferreira. "Decoction, infusion and hydroalcoholic extract of cultivated thyme: Antioxidant and antibacte-rial activities, and phenolic characterisation." *Food Chemistry* 167 (2015): 131–137.

Martinussen, Inger, Rolf Nestby, and Arnfinn Nes. "Potential of the European wild blueberry (Vaccinium myr-tillus) for cultivation and industrial exploitation in Norway." In *IX International Vaccinium Symposium 810*, pp. 211–216. 2008.

Masłowski, Marcin, Andrii Aleksieiev, Justyna Miedzianowska, Magdalena Efenberger-Szmechtyk, and Krzysztof Strzelec. "Antioxidant and anti–aging activity of freeze–dried alcohol–water extracts from common nettle (Urtica dioica L.) and peppermint (Mentha piperita L.) in elastomer vulcanizates." *Polymers* 14, no. 7 (2022): 1460.

Matejić, Jelena S., Nikola Stefanović, Milan Ivković, Nemanja Živanović, Petar D. Marin, and Ana M. Džamić. "Traditional uses of autochthonous medicinal and ritual plants and other remedies for health in Eastern and South-Eastern Serbia." *Journal of Ethnopharmacology* 261 (2020): 113186.

Mattingly, Kali Z., Brenna N. Braasch, and Stephen M. Hovick. "Greater flowering and response to flooding in Lythrum virgatum than L. salicaria (purple loosestrife)." *AoB Plants* 15, no. 2 (2023): plad009.

McInerny, C. J., and R. Gray. "Largest wild crab apple (Malus sylvestris (Linnaeus) Mill.) in Scotland on the shores of Loch Lomond." *Glasgow Naturalist (online 2018)* 27, no. Part 1 (2018): 71–73.

Meeus, Sofie, Rein Brys, Olivier Honnay, and Hans Jacquemyn. "Biological flora of the british isles: Pulmonaria officinalis." *Journal of Ecology* 101, no. 5 (2013): 1353–1368.

Méndez-Yáñez, Angela, Patricio Ramos, and Luis Morales-Quintana. "Human Health Benefits through Daily Consumption of Jerusalem Artichoke (Helianthus tuberosus L.) Tubers." *Horticulturae* 8, no. 7 (2022): 620.

Meng, Qinglin, Hakim Manghwar, and Weiming Hu. "Study on supergenus Rubus L.: Edible, medicinal, and phylogenetic characterization." *Plants* 11, no. 9 (2022): 1211.

Menković, Nebojša R., and Vanja M. Tadić. "Comprehensive combined chemical and pharmacognostic approach in the investigation of Montenegrin flora, with emphasis on endemic species: Past performance and future potential." *Lekovite sirovine* 41 (2021): 106–125.

Mennini, Tiziana, and Marco Gobbi. "The antidepressant mechanism of Hypericum perforatum." *Life Sciences* 75, no. 9 (2004): 1021–1027.

Mihailović, Nevena R., Vladimir B. Mihailović, Samo Kreft, Andrija R. Ćirić, Ljubinka G. Joksović, and Predrag T. Đurđević. "Analysis of phenolics in the peel and pulp of wild apples (Malus sylvestris (L.) Mill.)." *Journal of Food Composition and Analysis* 67 (2018): 1–9.

Miljković, Vojkan, Milica Nešić, Jelena Mrmošanin, Ivana Gajić, Bojana Miladinović, and Dušica Stojanović. "Rosa canina L. fruit and jam made of it–natural food colors E160a and E160d content and antioxidant capacity." *Facta Universitatis, Series: Physics, Chemistry and Technology* (2022): 079–086.

Miljković, Vojkan M., Goran S. Nikolić, Ljubiša B. Nikolić, and Biljana B. Arsić. "Morus species through centuries in pharmacy and as food." *Savremene tehnologije* 3, no. 2 (2014): 111–115.

Milovanovic, Mirjana, Dusan Zivkovic, and Biljana Vucelic-Radovic. "Antioxidant effects of Glechoma hederacea as a food additive." *Natural Product Communications* 5, no. 1 (2010): 1934578X1000500116.

Miraj, Sepide, Niloufar Azizi, and Sara Kiani. "A review of chemical components and pharmacological effects of Melissa officinalis L." *Der Pharmacia Lettre* 8, no. 6 (2016): 229–37.

Miskoska-Milevska, Elizabeta, Angelina Stamatoska, and Suzana Jordanovska. "Traditional uses of wild edible plants in the Republic of North Macedonia." *Phytologia Balcanica* 26 (2020): 155–162.

Mitich, Larry W. "Common purslane (Portulaca oleracea)." *Weed Technology* 11, no. 2 (1997): 394–397.

Mittal, Payal, Vikas Gupta, Manish Goswami, Nishant Thakur, and Praveen Bansal. "Phytochemical and pharmacological potential of viola odorata." *International Journal of Pharmacognosy* 2, no. 5 (2015): 215–20.

Momčilović, I., D. Grubiršić, and M. Nešković. "Transgenic Gentiana species (Gentian)." *Transgenic Crops III* (2001): 123–138.

Monteiro, Maria-do-Céu, Alberto CP Dias, Daniela Costa, António Almeida-Dias, and Maria Begoña Criado. "Hypericum perforatum and Its potential antiplatelet effect." *Healthcare* 10, no. 9 (2022): 1774.

Montefusco, A., Semitaio, G., Marrese, P.P., Iurlaro, A., De Caroli, M., Piro, G., Dalessandro, G. and Lenucci, M.S., 2015. Antioxidants in varieties of chicory (Cichorium intybus L.) and wild poppy (Papaver rhoeas L.) of Southern Italy. *Journal of Chemistry* (2015).

Morris, L. R., F. A. Baker, C. Morris, and R. J. Ryel. "Phytolith types and type-frequencies in native and introduced species of the sagebrush steppe and pinyon–juniper woodlands of the Great Basin, USA." *Review of Palaeobotany and Palynology* 157, no. 3–4 (2009): 339–357.

Morteza-Semnani K., Saeedi M., Akbarzadeh M. "Chemical composition of the essential oil of the flowering aerial parts of *Lamium album* L." *Journal of Essential Oil Bearing Plants* 19, no. 3 (2016): 773–777.

Mosaddegh, Mahmoud, Farzaneh Naghibi, Hamid Moazzeni, Atefeh Pirani, and Somayeh Esmaeili. "Ethnobotanical survey of herbal remedies traditionally used in Kohghiluyeh va Boyer Ahmad province of Iran." *Journal of Ethnopharmacology* 141, no. 1 (2012): 80–95.

Motley, Timothy J. "The ethnobotany of sweet flag, Acorus calamus (Araceae)." *Economic Botany* 48, no. 4 (1994): 397–412.

Motti, Riccardo. "Wild edible plants: A challenge for future diet and health." *Plants* 11, no. 3 (2022): 344.

Mueller, Monika, Stefanie Hobiger, and Alois Jungbauer. "Anti-inflammatory activity of extracts from fruits, herbs and spices." *Food Chemistry* 122, no. 4 (2010): 987–996.

Muhammad, Aleem, Aqsa Akhtar, Sadia Aslam, Rao Sanaullah Khan, Zaheer Ahmed, and Nauman Khalid. "Review on physicochemical, medicinal and nutraceutical properties of poppy seeds: A potential functional food ingredient." *Functional Foods in Health and Disease* 11, no. 10 (2021): 522–547.

Mureşan, Elena Andruţa, Romina Alina Vlaic, Vlad Mureşan, Constantin Gheorghe Cerbu, Simona Chiş, Melinda Fogoraşi, Simona Man, and Sevastiţa Muste. "Conversion of rosehip, cranberry, goji and coconut fruits into a functional tea rich in biologically active principles." *Hop and Medicinal Plants* 26, no. 1/2 (2018): 122–130.

Murti, Krishna, Mayank A. Panchal, Vipul Gajera, and Jinal Solanki. "Pharmacological properties of Matricaria recutita: A review." *Pharmacologia* 3, no. 8 (2012): 348–351.

Muruzović, Mirjana Ž., Katarina G. Mladenović, Olgica D. Stefanović, Sava M. Vasić, and Ljiljana R. Čomić. "Extracts of Agrimonia eupatoria L. as sources of biologically active compounds and evaluation of their antioxidant, antimicrobial, and antibiofilm activities." *Journal of Food and Drug Analysis* 24, no. 3 (2016): 539–547.

Muruzović, Mirjana Ž., Katarina G. Mladenović, Olgica D. Stefanović, Petrović Tanja D. Žugić, and Ljiljana R. Čomić. "In vitro interaction between Agrimonia eupatoria L.: Extracts and antibiotic." *Kragujevac Journal of Science* 39 (2017): 157–164.

Naraine, Steve GU, and Ernest Small. "Germplasm sources of protective glandular leaf trichomes of hop (Humulus lupulus)." *Genetic Resources and Crop Evolution* 64 (2017): 1491–1497.

Nasir, Omaima. "Protective effect of Acorus calamus on kidney and liver functions in healthy mice." *Saudi Journal of Biological Sciences* 28, no. 5 (2021): 2701–2708.

Nezi, Paola, Vittoria Cicaloni, Laura Tinti, Laura Salvini, Matteo Iannone, Sara Vitalini, and Stefania Garzoli. "Metabolomic and proteomic profile of dried hop inflorescences (Humulus lupulus L. cv. Chinook and cv. Cascade) by SPME-GC-MS and UPLC-MS-MS." *Separations* 9, no. 8 (2022): 204.

Nićetin, M., V. Filipović, J. Filipović, B. Lončar, B. Cvetković, V. Knežević, and D. Šuput. "Osmotic dehydration of wild garlic in sucrose–salt solution." *Acta Universitatis Sapientiae, Alimentaria* 15, no. 1 (2022): 27–39.

Nicolescu, Valeriu-Norocel, Károly Rédei, William L. Mason, Torsten Vor, Elisabeth Pöetzelsberger, Jean-Charles Bastien, Robert Brus et al. "Ecology, growth and management of black locust (Robinia pseudoacacia L.), a non-native species integrated into European forests." *Journal of Forestry Research* 31 (2020): 1081–1101.

Nkhabu, Khothatso, Mpho Liphoto, Takalimane Ntahane, and Katleho Senoko. "Genetic Diversity of Stinging Nettle (Urtica dioica) by Agro Morphological Markers." *European Journal of Botany, Plant Sciences and Phytology* 6, no. 1 (2021): 51–68.

Okatan, Volkan, Mehmet Polat, and Mehmet Atilla Aşkin. "Some physico-chemical characteristics of black mulberry (Morus nigra L.) in Bitlis." *Scientific Papers-Series B, Horticulture* 60 (2016): 27–30.

Oliveira, Ivo, Anabela Sousa, Patrícia Valentão, Paula B. Andrade, Isabel CFR Ferreira, Federico Ferreres, Albino Bento, Rosa Seabra, Letícia Estevinho, and José Alberto Pereira. "Hazel (Corylus avellana L.) leaves as source of antimicrobial and antioxidative compounds." *Food Chemistry* 105, no. 3 (2007): 1018–1025.

Olsson, Katarina, and Jon Ågren. "Latitudinal population differentiation in phenology, life history and flower morphology in the perennial herb Lythrum salicaria." *Journal of Evolutionary Biology* 15, no. 6 (2002): 983–996.

Özkalp, Birol, Fatih Sevgi, Mustafa Özcan, and Mehmet Musa Özcan. "The antibacterial activity of essential oil of oregano (Origanum vulgare L.)." *Journal of Food, Agriculture and Environment* 8, no. 2 (2010): 6–8.

Pacanoski, Zvonko, and Arben Mehmeti. "The first report of the invasive alien weed Jerusalem artichoke (Helianthus tuberosus L.) in the Republic of North Macedonia." *Agricultural and Forest Meteorology* 66 (2020): 115–127.

Panchal, Mayank A., Krishna Murti, and Vijay Lambole. "Pharmacological properties of Verbascum thapsus— A review." *International Journal of Pharmaceutical Sciences Review and Research* 5, no. 2 (2010): 73–7.

Papp, Nóra, Dóra Czégényi, Anita Hegedus, Tamás Morschhauser, Cassandra L. Quave, Kevin Cianfaglione, and Andrea Pieroni. "The uses of betula pendula roth among Hungarian csángós and székelys in Transylvania, Romania." *Acta Societatis Botanicorum Poloniae* 83, no. 2 (2014): 113–122.

Park, Eui Seong, Gyl Hoon Song, Seung Min Lee, Tae Young Kim, and Kun Young Park. "Increased anti-inflammatory effects of processed curly dock (Rumex crispus L.) in ex vivo LPS-induced mice splenocytes." *Journal of the Korean Society of Food Science and Nutrition* 47, no. 5 (2018): 599–604.

Pavlić, Branimir, Milica Aćimović, Aleksandra Sknepnek, Dunja Miletić, Živan Mrkonjić, Aleksandra Cvetanović Kljakić, Jelena Jerković et al. "Sustainable raw materials for efficient valorization and recovery of bioactive compounds." *Industrial Crops and Products* 193 (2023): 116167.

Pavlović, Dragana R., Milica Veljković, Nikola M. Stojanović, Marija Gočmanac-Ignjatović, Tatjana Mihailov-Krstev, Suzana Branković, Dušan Sokolović, Mirjana Marčetić, Niko Radulović, and Mirjana Radenković. "Influence of different wild-garlic (Allium ursinum) extracts on the gastrointestinal system: Spasmolytic, antimicrobial and antioxidant properties." *Journal of Pharmacy and Pharmacology* 69, no. 9 (2017): 1208–1218.

Pawera, Lukas, Ali Khomsan, Ervizal AM Zuhud, Danny Hunter, Amy Ickowitz, and Zbynek Polesny. "Wild food plants and trends in their use: From knowledge and perceptions to drivers of change in West Sumatra, Indonesia." *Foods* 9, no. 9 (2020): 1240.

Pejatović, T., D. Samardžić, and S. Krivokapić. "Antioxidative properities of a traditional tincture and several leaf extracts of Allium ursinum L.(collected in Montenegro and Bosnia and Herzegovina)." *Journal of Materials and Environmental Sciences* 8, no. 6 (2017): 1929–1934.

Perrino, Enrico V., Francesca Valerio, Shaima Jallali, Antonio Trani, and Giuseppe N. Mezzapesa. "Ecological and biological properties of Satureja cuneifolia Ten. and Thymus spinulosus Ten.: Two wild officinal species of conservation concern in Apulia (Italy). A preliminary survey." *Plants* 10, no. 9 (2021): 1952.

Petrović, Bojana, Predrag Vukomanović, Vera Popović, Ljubica Šarčević-Todosijević, Marko Burić, Milica Nikolić, and Snežana Đorđević. "Significance and efficacy of triterpene saponin herbal drugs with expectorant action in cough therapy." *Agriculture and Forestry* 68, no. 3 (2022): 221–239.

Penkov, Dimitar, Velichka Andonova, Delian Delev, Ilia Kostadinov, and Margarita Kassarova. "Antioxidant activity of dry birch (Betula Pendula) leaves extract." *Folia Med (Plovdiv)* 60, no. 4 (2018): 571–9.

Phalen, Kathleen F. *Wellness East & West: Achieving Optimum Health through Integrative Medicine.* Tuttle Publishing, 2012.

Pieroni, Andrea. "Local plant resources in the ethnobotany of Theth, a village in the Northern Albanian Alps." *Genetic Resources and Crop Evolution* 55 (2008): 1197–1214.

Pilipović, Andrej, Saša Orlović, Srđan Stojnić, Vladislava Galović, and Miroslav Marković. "Inventarization of wild cherry (Prunus avium) genefond in Serbia in the aim of directed genetic potential utilization." *Topola* 187–188 (2011): 53–63.

Pinke, Gyula, Viktória Kapcsándi, and Bálint Czúcz. "Iconic arable weeds: The significance of corn poppy (Papaver rhoeas), cornflower (Centaurea cyanus), and field larkspur (Delphinium consolida) in Hungarian ethnobotanical and cultural heritage." *Plants* 12, no. 1 (2022): 84.

Pirani, Atefeh, Hamid Moazzeni, Shahab Mirinejad, Farzaneh Naghibi, and Mahmoud Mosaddegh. "Ethnobotany of Juniperus excelsa M. Bieb.(Cupressaceae) in Iran." *Ethnobotany Research and Applications* 9 (2011): 335–341.

Pirkhezri, M., M. Hassani, and J. Hadian. "Genetic diversity in different populations of Matricaria chamomilla l. Growing in southwest of Iran, based on morphological." *Research Journal of Medicinal Plant* 4, no. 1 (2010): 1–13.

Piwowarski, Jakub P., Sebastian Granica, and Anna K. Kiss. "Lythrum salicaria L.—Underestimated medicinal plant from European traditional medicine. A review." *Journal of Ethnopharmacology* 170 (2015): 226–250.

Pleşca, Ioana Maria, and Tatiana Blaga. "Documenting Achillea L. genus using herbarium records." *Scientific Study and Research – Biology* 28, no. 1 (2019): 20–28.

Pluhár, Zsuzsanna, Szilvia Sárosi, Adrienn Pintér, and Hella Simkó. "Essential oil polymorphism of wild growing Hungarian thyme (Thymus pannonicus) populations in the Carpathian Basin." *Natural Product Communications* 5, no. 10 (2010): 1934578X1000501034.

Pluta-Kubica, Agnieszka, Dorota Najgebauer-Lejko, Jacek Domagała, Jana Štefániková, and Jozef Golian. "The Effect of Cow Breed and Wild Garlic Leaves (Allium ursinum L.) on the Sensory Quality, Volatile Compounds, and Physical Properties of Unripened Soft Rennet-Curd Cheese." *Foods* 11, no. 24 (2022): 3948.

Polito, Letizia, Massimo Bortolotti, Stefania Maiello, Maria Giulia Battelli, and Andrea Bolognesi. "Plants producing ribosome-inactivating proteins in traditional medicine." *Molecules* 21, no. 11 (2016): 1560.

Poljšak, Nina, and Nina Kočevar Glavač. "Tilia sp. Seed Oil—Composition, antioxidant activity and potential use." *Applied Sciences* 11, no. 11 (2021): 4932.

Poonam, V., G. Kumar, C. S. Reddy, R. Jain, S. K. Sharma, A. K. Prasad, and V. S. Parmar. "Chemical constituents of the genus Prunus and their medicinal properties." *Current Medicinal Chemistry* 18, no. 25 (2011): 3758–3824.

Popescu, I., and G. Caudullo. "Prunus spinosa in Europe: Distribution, habitat, usage and threats." *U: European Atlas of Forest Tree Species,(San-Miguel-Ayanz, J., de Rigo, D., Caudullo, G., Houston Durrant, T., Mauri, A. ured.), Luxembourg, str* 145 (2016).

Popović, Zorica, Miroslava Smiljanić, Miroslav Kostić, Predrag Nikić, and Snežana Janković. "Wild flora and its usage in traditional phytotherapy (Deliblato Sands, Serbia, South East Europe)." (2014).

Popović-Radović, M. "Mitološki rečnik biljnog sveta – mitske priče o bilju" Građevinska knjiga d.o.o. Beograd (2010) [in Serbian]

Porwal, Omji, Muath Sheet Mohammed Ameen, Esra T. Anwer, Subasini Uthirapathy, Javed Ahamad, and Amani Tahsin. "Silybum marianum (Milk Thistle): Review on Its chemistry, morphology, ethno medical uses, phytochemistry and pharmacological activities." *Journal of Drug Delivery and Therapeutics* 9, no. 5 (2019): 199–206.

Pradhan, S.C. and Girish, C., 2006. Hepatoprotective herbal drug, silymarin from experimental pharmacology to clinical medicine. *Indian Journal of Medical Research* 124, no. 5: 491–504.

Prakash, Om, Ruchi Singh, Saroj Kumar, Shweta Srivastava, and Akash Ved. "Gentiana lutea Linn.(yellow gentian): A comprehensive." *Journal of Ayurvedic and Herbal Medicine* 3 (2017): 175–81.

Prasanna Kumar C.N., Shruthi, R. Kannan, and U. V. Babu. "Botanical pharmacognosy of Cichorium intybus seeds." *Ornamental and Medicinal Plants* no. 4 (3–4) (2020): 4–10.

Pushpangadan, P., V. George, and S. P. Singh. "Poppy." In *Handbook of herbs and spices*, pp. 437–448. Woodhead Publishing, 2012.

Qizi, Rasulova Mamurahon Obidjon. "Study of the chemical composition of the inula helenium plant." *European Scholar Journal* no. 2 (2021):59–61.

Quave, Cassandra L., James T. Lyles, Jeffery S. Kavanaugh, Kate Nelson, Corey P. Parlet, Heidi A. Crosby, Kristopher P. Heilmann, and Alexander R. Horswill. "Castanea sativa (European Chestnut) leaf extracts rich in ursene and oleanene derivatives block Staphylococcus aureus virulence and pathogenesis without detectable resistance." *PLoS One* 10, no. 8 (2015): e0136486.

Qureshi, S., S. Adil, M. E. Abd El-Hack, M. Alagawany, and M. R. Farag. "Beneficial uses of dandelion herb (Taraxacum officinale) in poultry nutrition." *World's Poultry Science Journal* 73, no. 3 (2017): 591–602.

Račková, Lucia, Viera Jančinová, Margita Petríková, Katarína Drábiková, Radomír Nosáľ, Milan Štefek, Daniela Košťálová, Naďa Prónayová, and Mária Kováčová. "Mechanism of anti-inflammatory action of liquorice extract and glycyrrhizin." *Natural Product Research* 21, no. 14 (2007): 1234–1241.

Radanović, Dragoja, Tatjana Marković, Jovica Vasin, and Dušana Banjac. "The efficiency of using different mulch films in the cultivation of yellow gentian (Gentiana lutea L.) in Serbia." *Ratarstvo i povrtarstvo* 53, no. 1 (2016): 30–37.

Radanović, Dragoja, Ana Matković, Rada Đurović-Pejčev, Tatjana Marković, Vladimir Filipović, Snežana Mrđan, and Jovica Vasin. "Preliminary results of winter savory (Satureja montana L.) cultivated under permeable mulch film in dry farming conditions of South Banat." *Lekovite sirovine* 38 (2018): 51–57.

Radovanović, Blaga C., S. M. Anđelković, Aleksandra B. Radovanović, and Marko Z. Anđelković. "Antioxidant and antimicrobial activity of polyphenol extracts from wild berry fruits grown in southeast Serbia." *Tropical Journal of Pharmaceutical Research* 12, no. 5 (2013): 813–819.

Radovanović, Katarina, Neda Gavarić, Jaroslava Švarc-Gajić, Tanja Brezo-Borjan, Bojan Zlatković, Biljana Lončar, and Milica Aćimović. "Subcritical water extraction as an effective technique for the isolation of phenolic compounds of Achillea species." *Processes* 11, no. 1 (2022): 86.

Radovanović, Katarina, Neda Gavarić, and Milica Aćimović. "Anti-Inflammatory Properties of Plants from Serbian Traditional Medicine." *Life* 13, no. 4 (2023): 874.

Radulović, Niko, Polina Blagojević, and Radosav Palić. "Comparative study of the leaf volatiles of Arctostaphylos uva-ursi (L.) Spreng. and Vaccinium vitis-idaea L.(Ericaceae)." *Molecules* 15, no. 9 (2010): 6168–6185.

Radulovic, Niko, Bojan Zlatković, Radosav Palic, and Gordana Stojanovic. "Chemotaxonomic significance of the Balkan Achillea volatiles." *Natural Product Communications* 2, no. 4 (2007): 1934578X0700200417.

Rafińska, Katarzyna, Paweł Pomastowski, Olga Wrona, Ryszard Górecki, and Bogusław Buszewski. "Medicago sativa as a source of secondary metabolites for agriculture and pharmaceutical industry." *Phytochemistry Letters* 20 (2017): 520–539.

Raina, Rajinder, Pawan K. Verma, Rajinder Peshin, and Harpreet Kour. "Potential of Juniperus communis L as a nutraceutical in human and veterinary medicine." *Heliyon* 5, no. 8 (2019): e02376.

Rakić, Sveto, Jelena Kukic-Markovic, Silvana Petrovic, Vele Teševic, Snežana Jankovic, and Dragan Povrcnovic. "Oak kernels—volatile constituents and coffee-like beverages." *Journal of Agricultural Science* 10, no. 5 (2018): 117.

Rančić, Dragana, Slobodan Dražić, Zora Dajić-Stevanović, and Radenko Radošević. "Anatomical features of the marshmallow (Althaea officinalis L.) root." *Journal of Scientific Agricultural Research* 70, no. 4 (2009): 51–60.

Randall, Colin. "Historical and modern uses of Urtica." In *Urtica*, pp. 28–40. CRC Press, 2003.

Rankovic, Marina, Milos Krivokapic, Jovana Bradic, Anica Petkovic, Vladimir Zivkovic, Jasmina Sretenovic, Nevena Jeremic et al. "New insight into the cardioprotective effects of Allium ursinum L. extract against myocardial ischemia-reperfusion injury." *Frontiers in Physiology* 12 (2021): 690696.

Rashid, Nahida, Pervaiz Ahmad Dar, Hakeem Naseer Ahmad, and Shameem Ahmad Rather. "Alsi (Linum Usitatissimum Linn.): A potential multifaceted Unani drug." *Journal of Pharmacognosy and Phytochemistry* 7, no. 5 (2018): 3294–3300.

Ravazzi, C., and Giovanni Caudullo. "Aesculus hippocastanum in Europe: Distribution, habitat, usage and threats." *European Atlas of Forest Tree Species. Publication Office of the European Union, Luxemburg* (2016): 60.

Razik, Abd, Basma Monjd, Hiba Ali Hasan, and Muna Khalil Murtadha. "The study of antibacterial activity of Plantago major and Ceratonia siliqua." *The Iraqi Postgraduate Medical Journal* 11, no. 1 (2012): 130–5.

Reim, Stefanie, Anke Proft, Simone Heinz, and Monika Höfer. "Diversity of the European indigenous wild apple Malus sylvestris (L.) MILL. in the East Ore Mountains (Osterzgebirge), Germany: I. Morphological characterization." *Genetic Resources and Crop Evolution* 59 (2012): 1101–1114.

Ríos, J. L., M. C. Recio, S. Maáñez, and R. M. Giner. "Natural triterpenoids as anti-inflammatory agents." *Studies in Natural Products Chemistry* 22 (2000): 93–143.

Ritchie, J. C. "Vaccinium myrtillus L." *Journal of Ecology* 44, no. 1 (1956): 291–299.

Rocabado, Guillermo Omar, Luis Miguel Bedoya, María José Abad, and Paulina Bermejo. "Rubus-a review of its phytochemical and pharmacological profile." *Natural Product Communications* 3, no. 3 (2008): 1934578X0800300319.

Rodriguez-Fragoso, Lourdes, Jorge Reyes-Esparza, Scott W. Burchiel, Dea Herrera-Ruiz, and Eliseo Torres. "Risks and benefits of commonly used herbal medicines in Mexico." *Toxicology and Applied Pharmacology* 227, no. 1 (2008): 125–135.

Roman, Ioana, Andreea Stănilă, and Sorin Stănilă. "Bioactive compounds and antioxidant activity of Rosa canina L. biotypes from spontaneous flora of Transylvania." *Chemistry Central Journal* 7, no. 1 (2013): 1–10.

Rubio, Laura, del Carmen Valiño, Jesús Expósito, Marta Lores, and Carmen Garcia-Jares. "Sourcing New Ingredients for Organic Cosmetics: Phytochemicals of Filipendula vulgaris Flower Extracts." *Cosmetics* 9, no. 6 (2022): 132.

Russell, K. "EUFORGEN Technical Guidelines for genetic conservation and use for wild cherry (Prunus avium)". *Bioversity International*, 2003.

Saeed, Sabahat, and Perween Tariq. "Antibacterial activity of oregano (Origanum vulgare Linn.) against gram positive bacteria." *Pakistan Journal of Pharmaceutical Sciences* 22, no. 4 (2009): 421–425.

Saeidnia, S., A. R. Gohari, N. Mokhber-Dezfuli, and Fumiyuki Kiuchi. "A review on phytochemistry and medicinal properties of the genus Achillea." *DARU: Journal of Faculty of Pharmacy, Tehran University of Medical Sciences* 19, no. 3 (2011): 173.

Saengthongpinit, Wanpen, and Tanaboon Sajjaanantakul. "Influence of harvest time and storage temperature on characteristics of inulin from Jerusalem artichoke (Helianthus tuberosus L.) tubers." *Postharvest Biology and Technology* 37, no. 1 (2005): 93–100.

Saini, Shilpa, and Kamal Kant Saini. "Chenopodium album Linn: An outlook on weed cum nutritional vegetable along with medicinal properties." *Emergent Life Sciences Research* 6 (2020): 28–33.

Saint-Lary, Laure, Céline Roy, Jean-Philippe Paris, Pascal Tournayre, Jean-Louis Berdagué, Olivier P. Thomas, and Xavier Fernandez. "Volatile compounds of Viola odorata absolutes: Identification of odorant active markers to distinguish plants originating from France and Egypt." *Chemistry & Biodiversity* 11, no. 6 (2014): 843–860.

Sakkas, Hercules, and Chrissanthy Papadopoulou. "Antimicrobial activity of basil, oregano, and thyme essential oils." *Journal of Microbiolog and Biotechnology* 27, no. 3 (2017): 429–438.

Salehi, Bahare, Lorene Armstrong, Antonio Rescigno, Balakyz Yeskaliyeva, Gulnaz Seitimova, Ahmet Beyatli, Jugreet Sharmeen et al. "Lamium plants—A comprehensive review on health benefits and biological activities." *Molecules* 24, no. 10 (2019): 1913.

Salama, M., and Shaimaa G. Salama. "Arctium lappa L.(Asteraceae); a new invasive highly specific medicinal plant growing in Egypt." *Pyrex Journal of Plant and Agricultural Research* 2, no. 2 (2016): 44–53.

Samardžić, Stevan, Maja Tomić, Uroš Pecikoza, Radica Stepanović-Petrović, and Zoran Maksimović. "Antihyperalgesic activity of Filipendula ulmaria (L.) Maxim. and Filipendula vulgaris Moench in a rat model of inflammation." *Journal of Ethnopharmacology* 193 (2016): 652–656.

Sampliner, Danielle, and Allison Miller. "Ethnobotany of horseradish (Armoracia rusticana, Brassicaceae) and its wild relatives (Armoracia spp.): Reproductive biology and local uses in their native ranges." *Economic Botany* 63 (2009): 303–313.

Sánchez-Salcedo, Eva M., Pedro Mena, Cristina García-Viguera, Francisca Hernández, and Juan José Martínez. "(Poly) phenolic compounds and antioxidant activity of white (Morus alba) and black (Morus nigra) mulberry leaves: Their potential for new products rich in phytochemicals." *Journal of Functional Foods* 18 (2015): 1039–1046.

Sanna, Federico, Giovanna Piluzza, Giuseppe Campesi, Maria Giovanna Molinu, Giovanni Antonio Re, and Leonardo Sulas. "Antioxidant Contents in a Mediterranean Population of Plantago lanceolata L. Exploited for Quarry Reclamation Interventions." *Plants* 11, no. 6 (2022): 791.

Santos, Maria João, Teresa Pinto, and Alice Vilela. "Sweet chestnut (Castanea sativa Mill.) nutritional and phenolic composition interactions with chestnut flavor physiology." *Foods* 11, no. 24 (2022): 4052.

Saricaoglu, F.T., Atalar, I., Yilmaz, V.A., Odabas, H.I. and Gul, O., 2019. Application of multi pass high pressure homogenization to improve stability, physical and bioactive properties of rosehip (Rosa canina L.) nectar. *Food Chemistry* 282: 67–75.

Saulić, Markola. "Rezerve semena korovskih biljaka u zemljištu u zavisnisti od plodoreda i đubrenja." *Универзитет у Београду*, PhD thesis (2022).

Savić, Dragiša, Goran Anačkov, and Pal Boža. "New chorological data for flora of the Pannonian region of Serbia." *Central European Journal of Biology* 3 (2008): 461–470.

Šavikin, Katarina, Gordana Zdunić, Nebojša Menković, Jelena Živković, Nada Ćujić, Milena Tereščenko, and Dubravka Bigović. "Ethnobotanical study on traditional use of medicinal plants in South-Western Serbia, Zlatibor district." *Journal of Ethnopharmacology* 146, no. 3 (2013): 803–810.

Savill, Peter S. *The silviculture of trees used in British forestry.* CABI, 2019.

Serag, M. M. S., Amina Z. Abo El-Naga, and Mahitab El-Ramal. "Ecological Study On Plantago Major L. In The Different Habitats Of Damietta." *Journal of Plant Production* 1, no. 7 (2010): 1007–1031.

Šeregelj, Vanja, Olja Šovljanski, Jaroslava Švarc-Gajić, Teodora Cvanić, Aleksandra Ranitović, Jelena Vulić, and Milica Aćimović. "Modern green approaches for obtaining Satureja kitaibelii Wierzb. ex Heuff extracts with enhanced biological activity." *Journal of the Serbian Chemical Society* 87, no. 12 (2022): 1359–1365.

Šeremet, Danijela, Aleksandra Vojvodić Cebin, Ana Mandura, Ivana Žepić, Ksenija Marković, and Draženka Komes. "An insight into the chemical composition of ground ivy (Glechoma hederacea L.) by means of macrocomponent analysis and fractionation of phenolic compounds." *Hrvatski časopis za prehrambenu tehnologiju, biotehnologiju i nutricionizam* 15, no. 3–4 (2020): 133–138.

Shaheen, Shabnum, Mushtaq Ahmad, and Nidaa Haroon. *Edible Wild Plants: An alternative approach to food security.* Cham, Switzerland: Springer International Publishing, 2017.

Shamilov, Arnold A., Valentina N. Bubenchikova, Maxim V. Chernikov, Dmitryi I. Pozdnyakov, Ekaterina R. Garsiya, and Mikhail V. Larsky. "Bearberry (Arctostaphylos uva-ursi (L.) Spreng.): Chemical content and pharmacological activity." *Journal of Excipients and Food Chemicals* 12, no. 3 (2021): 49–66.

Shariati, Mohammad Ali, Muhammad Usman Khan, Lukas Hleba, Carolina Krebs de Souza, Zhaiyk Tokhtarov, Sergei Terentev, Sergey Konovalov et al. "Topinambur (the Jerusalem artichoke): Nutritional value and its application in food products: An updated treatise." *Journal of Microbiology, Biotechnology and Food Sciences* 10, no. 6 (2021): e4737–e4737.

Sharifi-Rad, Javad, Jesús Herrera-Bravo, Prabhakar Semwal, Sakshi Painuli, Himani Badoni, Shahira M. Ezzat, Mai M. Farid et al. "Artemisia spp.: An update on its chemical composition, pharmacological and toxicological profiles." *Oxidative Medicine and Cellular Longevity* 2022 (2022): 5628601.

Sharma, Varsha, Akshay Katiyar, and R. C. Agrawal. "Glycyrrhiza glabra: Chemistry and pharmacological activity." *Sweeteners* (2018): 87.

Shekhar, Shweta, Arun K. Pandey, and Arne A. Anderberg. "The genus Inula (Asteraceae) in India." *Rheedea* 23, no. 2 (2013): 113–127.

Shelepova, Olga V., Ekaterina V. Tkacheva, and Elena V. Golosova. "The history of the introduction of peppermint (Mentha× piperita L.) in Imperial Russia." In *BIO Web of Conferences*, vol. 38, p. 00115. EDP Sciences, 2021.

Shikov, Alexander N., Olga N. Pozharitskaya, Valery G. Makarov, Hildebert Wagner, Rob Verpoorte, and Michael Heinrich. "Medicinal plants of the Russian Pharmacopoeia; their history and applications." *Journal of Ethnopharmacology* 154, no. 3 (2014): 481–536.

Shim, Youn Young, Bo Gui, Paul G. Arnison, Yong Wang, and Martin JT Reaney. "Flaxseed (Linum usitatissimum L.) bioactive compounds and peptide nomenclature: A review." *Trends in Food Science & Technology* 38, no. 1 (2014): 5–20.

Shokry, Aya A., Riham A. El-Shiekh, Gehan Kamel, Alaa F. Bakr, and Amer Ramadan. "Bioactive phenolics fraction of Hedera helix L.(common ivy leaf) standardized extract ameliorates LPS-induced acute lung injury in the mouse model through the inhibition of proinflammatory cytokines and oxidative stress." *Heliyon* 8, no. 5 (2022): e09477.

Sicari, Vincenzo, Monica Rosa Loizzo, Rosa Tundis, Antonio Mincione, and Teresa Maria Pellicano. "Portulaca oleracea L.(Purslane) extracts display antioxidant and hypoglycaemic effects." *Applied Botany and Food Quality* 91, no. 1 (2018): 39–46.

Simić, Ana M., Marina Soković, Jelena B. Vukojević, Mihailo S. Ristić, Jovanović Slavica M. Grujić, and Petar D. Marin. "Antimicrobial activity of essential oil of Juniperus communis L." *Lekovite sirovine* 22 (2002): 25–30.

Siminiuc, Rodica, and Dinu Turcanu. "Study of Edible Spontaneous Herbs in the Republic of Moldova for Ensuring a Sustainable Food System." *Food and Nutrition Sciences* 12 (2021): 703–718.

Singh, A., and S. Dhariwal. "Traditional uses, antimicrobial potential, Pharmacological properties and Phytochemistry of Viola odorata: A Mini Review." *International Journal of Phytopharmacology* 7 (2018): 103–105.

Singh, Ompal, Zakia Khanam, Neelam Misra, and Manoj Kumar Srivastava. "Chamomile (Matricaria chamomilla L.): An overview." *Pharmacognosy Reviews* 5, no. 9 (2011): 82.

Sipek, Mirjana, Aleksandra Perčin, Željka Zgorelec, and Nina Sajna. "Morphological plasticity and ecophysiological response of ground ivy (Glechoma hederacea, Lamiaceae) in contrasting natural habitats within its native range." *Plant Biosystems-An International Journal Dealing with all Aspects of Plant Biology* 155, no. 1 (2021): 136–147.

Sitzia, Tommaso, Arne Cierjacks, Daniele de Rigo, and Giovanni Caudullo. "Robinia pseudoacacia in Europe: Distribution, habitat, usage and threats." *European Atlas of Forest Tree Species* (2016): 166–167.

Šarić-Kundalić, B., Dobeš, C., Klatte-Asselmeyer, V. and Saukel, J., 2011. Ethnobotanical survey of traditionally used plants in human therapy of east, north and north-east Bosnia and Herzegovina. *Journal of Ethnopharmacology* 133, no. 3: 1051–1076.

Škobić, Slađana, Mirjana D. Marčetić, Tatjana Kundaković-Vasović, and Jovan Crnobarac. "Nitrogen fertilization and the essential oils profile of the rhizomes of different sweet flag populations (Acorus calamus L.)." *Industrial Crops and Products* 142 (2019): 111871.

Sobolewska, Danuta, Irma Podolak, and Justyna Makowska-Wąs. "Allium ursinum: Botanical, phytochemical and pharmacological overview." *Phytochemistry Reviews* 14 (2015): 81–97.

Solieri, Lisa, and Paolo Giudici. "Vinegars of the World." In *Vinegars of the World*, pp. 1–16. Milano: Springer Milan, 2009.

Sorescu, A. A., A. Nuta, M. Grigore, E. R. Andrei, G. I. Radu, and L. Iancu. "Antioxidant activity of environmentally-friendly noble metallic nanoparticles." In *Journal of Physics: Conference Series*, vol. 1426, no. 1, p. 012046. IOP Publishing, 2020.

Srećković, Nikola, Jelena S. Katanić Stanković, Sanja Matić, Nevena R. Mihailović, Paola Imbimbo, Daria Maria Monti, and Vladimir Mihailović. "Lythrum salicaria L.(Lythraceae) as a promising source of phenolic compounds in the modulation of oxidative stress: Comparison between aerial parts and root extracts." *Industrial Crops and Products* 155 (2020): 112781.

Srivastava, Rajani, Vineet Srivastava, and Ajeet Singh. "Multipurpose benefits of an underexplored species purslane (Portulaca oleracea L.): A critical review." *Environmental Management* 72, no. 2 (2023): 309–320.

Stanković, Jelena S. Katanić, Nikola Srećković, Danijela Mišić, Uroš Gašić, Paola Imbimbo, Daria Maria Monti, and Vladimir Mihailović. "Bioactivity, biocompatibility and phytochemical assessment of lilac sage, Salvia verticillata L.(Lamiaceae)-A plant rich in rosmarinic acid." *Industrial Crops and Products* 143 (2020): 111932.

Stefanaki, Anastasia, and Tinde van Andel. "Mediterranean aromatic herbs and their culinary use." In *Aromatic Herbs in Food*, pp. 93–121. Academic Press, 2021.

Ştefănescu, Ruxandra, Cristina Nicoleta Ciurea, Anca Delia Mare, Adrian Man, Adrian Nisca, Alexandru Nicolescu, Andrei Mocan, Mihai Babotă, Năstaca-Alina Coman, and Corneliu Tanase. "Quercus Robur Older Bark—A Source of Polyphenolic Extracts with Biological Activities." *Applied Sciences* 12, no. 22 (2022b): 11738.

Ştefănescu, Ruxandra, Eszter Laczkó-Zöld, Bianca-Eugenia Ősz, and Camil-Eugen Vari. "An Updated Systematic Review of Vaccinium myrtillus Leaves: Phytochemistry and Pharmacology." *Pharmaceutics* 15, no. 1 (2022a): 16.

Stewart-Wade, S. M., S. Neumann, L. L. Collins, and G. J. Boland. "The biology of Canadian weeds. 117. Taraxacum officinale GH Weber ex Wiggers." *Canadian Journal of Plant Science* 82, no. 4 (2002): 825–853.

Strzemski, Maciej, Magdalena Wójciak-Kosior, Ireneusz Sowa, Daniel Załuski, and Rob Verpoorte. "Historical and traditional medical applications of Carlina acaulis L.-A critical ethnopharmacological review." *Journal of Ethnopharmacology* 239 (2019): 111842.

Swanton, C. J., D. R. Clements, M. J. Moore, and P. B. Cavers. "The biology of Canadian weeds. 101. Helianthus tuberosus L." *Canadian Journal of Plant Science* 72, no. 4 (1992): 1367–1382.

Sweeney, Brooke, Mamta Vora, Catherine Ulbricht, and Ethan Basch. "Evidence-based systematic review of dandelion (Taraxacum officinale) by natural standard research collaboration." *Journal of Herbal Pharmacotherapy* 5, no. 1 (2005): 79–93.

Szczepaniak, Oskar M., Joanna Kobus-Cisowska, Weronika Kusek, and Monika Przeor. "Functional properties of Cornelian cherry (Cornus mas L.): A comprehensive review." *European Food Research and Technology* 245, no. 10 (2019): 2071–2087.

Szopa, Agnieszka, Joanna Pajor, Paweł Klin, Agnieszka Rzepiela, Hosam O. Elansary, Fahed A. Al-Mana, Mohamed A. Mattar, and Halina Ekiert. "Artemisia absinthium L.—Importance in the history of medicine, the latest advances in phytochemistry and therapeutical, cosmetological and culinary uses." *Plants* 9, no. 9 (2020): 1063.

Tadić, Vanja Milija, Nemanja Krgović, and Ana Žugić. "Lady's mantle (Alchemilla vulgaris L., Rosaceae): A review of traditional uses, phytochemical profile, and biological properties." *Lekovite sirovine* 40 (2020): 66–74.

Taib, Mehdi, Yassine Rezzak, Lahboub Bouyazza, and Badiaa Lyoussi. "Medicinal uses, phytochemistry, and pharmacological activities of Quercus species." *Evidence-based Complementary and Alternative Medicine* 2020 (2020): 1920683.

Taiwo, Adefunmilayo E., Franco B. Leite, Greice M. Lucena, Marilia Barros, Dâmaris Silveira, Mônica V. Silva, and Vania M. Ferreira. "Anxiolytic and antidepressant-like effects of Melissa officinalis (lemon balm) extract in rats: Influence of administration and gender." *Indian Journal of Pharmacology* 44, no. 2 (2012): 189.

Tepe, Bektas, and Mustafa Cilkiz. "A pharmacological and phytochemical overview on Satureja." *Pharmaceutical Biology* 54, no. 3 (2016): 375–412.

Teuber, L. R., M. C. Albertsen, D. K. Barnes, and G. H. Heichel. "Structure of floral nectaries of alfalfa (Medicago sativa L.) in relation to nectar production." *American Journal of Botany* 67, no. 4 (1980): 433–439.

Thas, J. Joseph. "Siddha medicine—background and principles and the application for skin diseases." *Clinics in Dermatology* 26, no. 1 (2008): 62–78.

Thomas, Peter. *Trees: Their natural history*. Cambridge University Press, 2014.

Tilford, Gregory L. *Edible and medicinal plants of the West*. Mountain Press Publishing, 1997.

Tiralongo, Evelin, Shirley S. Wee, and Rodney A. Lea. "Elderberry supplementation reduces cold duration and symptoms in air-travellers: A randomized, double-blind placebo-controlled clinical trial." *Nutrients* 8, no. 4 (2016): 182.

Tobyn, Graeme, Alison Denham, and Margaret Whitelegg. *The Western Herbal Tradition E-Book: The Western Herbal Tradition E-Book*. Elsevier Health Sciences, 2010.

Tomić, Ana, Olja Šovljanski, Višnja Nikolić, Lato Pezo, Milica Aćimović, Mirjana Cvetković, Jovana Stanojev, Nebojša Kuzmanović, and Siniša Markov. "Screening of Antifungal Activity of Essential Oils in Controlling Biocontamination of Historical Papers in Archives." *Antibiotics* 12, no. 1 (2023): 103.

Tomljenović, Nikola, Tomislav Jemrić, and Marko Vuković. "Variability in pomological traits of dog rose (Rosa canina L.) under the ecological conditions of the Republic of Croatia." *Acta Agriculturae Serbica* 26, no. 51 (2021): 41–47.

Tomšik, Alena, Branimir Pavlić, Jelena Vladić, Milica Ramić, Ján Brindza, and Senka Vidović. "Optimization of ultrasound-assisted extraction of bioactive compounds from wild garlic (Allium ursinum L.)." *Ultrasonics Sonochemistry* 29 (2016): 502–511.

Tonutti, Ivan, and Peter Liddle. "Aromatic plants in alcoholic beverages. A review." *Flavour and Fragrance Journal* 25, no. 5 (2010): 341–350.

Tosun, Murat, S. Ercisli, Hüseyin, Karlidag, and M. Sengul. "Characterization of red raspberry (Rubus idaeus L.) genotypes for their physicochemical properties." *Journal of Food Science* 74, no. 7 (2009): C575–C579.

Treben, M. *Zdravlje iz Bozije apoteke*, 2020, Edicija.

Trejgell, Alina, Marlena Bednarek, and Andrzej Tretyn. "Micropropagation of Carlina acaulis L." *Acta Biologica Cracoviensia Series Botanica* 51, no. 1 (2009): 97–103.

Tucakov, J. *Lečenje biljem: fitoterapija*. 2006, Beograd. [in Serbian]

Tumbas, Vesna T., Jasna M. Čanadanović Brunet, Dragana D. Četojević Simin, Gordana S. Ćetković, Sonja M. Đilas, and Lars Gille. "Effect of rosehip (Rosa canina L.) phytochemicals on stable free radicals and human cancer cells." *Journal of the Science of Food and Agriculture* 92, no. 6 (2012): 1273–1281.

Tunalier, Zeynep, Müberra Koşar, Esra Küpeli, İhsan Çaliş, and K. Hüsnü Can Başer. "Antioxidant, anti-inflammatory, anti-nociceptive activities and composition of Lythrum salicaria L. extracts." *Journal of Ethnopharmacology* 110, no. 3 (2007): 539–547.

Tundis, Rosa, Maria C. Tenuta, Monica R. Loizzo, Marco Bonesi, Federica Finetti, Lorenza Trabalzini, and Brigitte Deguin. "Vaccinium species (Ericaceae): From chemical composition to bio-functional activities." *Applied Sciences* 11, no. 12 (2021): 5655.

Turner, Nancy J., Łukasz Jakub Łuczaj, Paola Migliorini, Andrea Pieroni, Angelo Leandro Dreon, Linda Enrica Sacchetti, and Maurizio G. Paoletti. "Edible and tended wild plants, traditional ecological knowledge and agroecology." *Critical Reviews in Plant Sciences* 30, no. 1–2 (2011): 198–225.

Tutin, T. G. "Allium Ursinum L." *Journal of Ecology* 45, no. 3 (1957): 1003–1010.

Uddin, Md Kamal, Abdul Shukor Juraimi, Md Sabir Hossain, Most Nahar, Altaf Un, Md Eaqub Ali, and M. M. Rahman. "Purslane weed (Portulaca oleracea): A prospective plant source of nutrition, omega-3 fatty acid, and antioxidant attributes." *The Scientific World Journal* 2014 (2014).

Ulubelen, Ayhan, Güolati Topu, and Ufuk Sönmez. "Chemical and biological evaluation of genus Teucrium." *Studies in Natural Products Chemistry* 23 (2000): 591–648.

Uprety, Yadav, Ram C. Poudel, Krishna K. Shrestha, Sangeeta Rajbhandary, Narendra N. Tiwari, Uttam B. Shrestha, and Hugo Asselin. "Diversity of use and local knowledge of wild edible plant resources in Nepal." *Journal of Ethnobiology and Ethnomedicine* 8 (2012): 1–15.

Vargas-Mendoza, Nancy, Eduardo Madrigal-Santillán, Ángel Morales-González, Jaime Esquivel-Soto, Cesar Esquivel-Chirino, Manuel García-Luna y González-Rubio, Juan A. Gayosso-de-Lucio, and José A. Morales-González. "Hepatoprotective effect of silymarin." *World Journal of Hepatology* 6, no. 3 (2014): 144.

Vasas, Andrea, Orsolya Orbán-Gyapai, and Judit Hohmann. "The Genus Rumex: Review of traditional uses, phytochemistry and pharmacology." *Journal of Ethnopharmacology* 175 (2015): 198–228.

Vašková, J., A. Fejerčáková, G. Mojžišová, L. Vaško, and P. Patlevič. "Antioxidant potential of Aesculus hippocastanum extract and escin against reactive oxygen and nitrogen species." *European Review for Medical & Pharmacological Sciences* 19, no. 5 (2015): 879–886.

Vinha, Ana F., João CM Barreira, Anabela SG Costa, and M. Beatriz PP Oliveira. "A new age for Quercus spp. fruits: Review on nutritional and phytochemical composition and related biological activities of acorns." *Comprehensive Reviews in Food Science and Food Safety* 15, no. 6 (2016): 947–981.

Viskelis, Pranas, Marina Rubinskiene, Ramunė Bobinaite, and Edita Dambrauskiene. "Bioactive compounds and antioxidant activity of small fruits in Lithuania." *Journal of Food, Agriculture and Environment* 8, no. 3–4 (2010): 259–263.

Vladimirov, Marijana S., Vesna D. Nikolić, Ljiljana P. Stanojević, Ljubiša B. Nikolić, and Ana D. Tačić. "Common birch (Betula pendula Roth.): Chemical composition and biological activity of isolates." *Advanced Technologies* 8, no. 1 (2019): 65–77.

Vlaisavljević, Sanja, Sanja Jelača, Gökhan Zengin, Neda Mimica-Dukić, Sanja Berežni, Milorad Miljić, and Zora Dajić Stevanović. "Alchemilla vulgaris agg.(Lady's mantle) from central Balkan: Antioxidant, anti-cancer and enzyme inhibition properties." *RSC Advances* 9, no. 64 (2019): 37474–37483.

Voća, Sandra, Jana Šic Žlabur, Sanja Fabek Uher, Marija Peša, Nevena Opačić, and Sanja Radman. "Neglected potential of wild garlic (Allium ursinum L.)—Specialized metabolites content and antioxidant capacity of wild populations in relation to location and plant phenophase." *Horticulturae* 8, no. 1 (2021): 24.

Vučić, Dragana M., Miroslav R. Petković, Branka B. Rodić-Grabovac, Sava M. Vasić, and Ljiljana R. Čomić. "In vitro efficacy of extracts of Arctostaphylos uva-ursi L. on clinical isolated Escherichia coli and Enterococcus faecalis strains." *Kragujevac Journal of Science* 35 (2013): 107–113.

Vujic, Zorica, Dragana Novovic, Ivana Arsic, and Dusanka Antic. "Optimization of the extraction process of escin from dried seeds of Aesculus hippocastanum L. by Derringer's desirability function." *Journal of Animal and Plant Sciences* 17, no. 1 (2013): 2514–2521.

Walters, Stuart Alan. "Horseradish: A neglected and underutilized plant species for improving human health." *Horticulturae* 7, no. 7 (2021): 167.

Wang, Lei, Luojun Zhang, Guangtao Zheng, Haiping Luo, Attalla F. El-Kott, and Ayman E. El-Kenawy. "Equisetum arvense L aqueous extract: A novel chemotherapeutic supplement for treatment of human colon carcinoma." *Archives of Medical Science: AMS* 19, no. 5 (2023): 1472.

Wei, Hao, Shuai Kong, Vanitha Jayaraman, Dhivya Selvaraj, Prabhakaran Soundararajan, and Abinaya Manivannan. "Mentha arvensis and Mentha× piperita-Vital Herbs with Myriads of Pharmaceutical Benefits." *Horticulturae* 9, no. 2 (2023): 224.

Welk, E., D. De Rigo, and G. Caudullo. "Prunus avium in Europe: Distribution, habitat, usage and threats." *European Atlas of Forest Tree Species. Publ. Off. EU, Luxembourg* 3 (2016): 140–141.

Wells, T. C. E., P. Rothery, Ruth Cox, and S. Bamford. "Flowering dynamics of Orchis morio L. and Herminium monorchis (L.) R. Br. at two sites in eastern England." *Botanical Journal of the Linnean Society* 126, no. 1–2 (1998): 39–48.

Wojtyniak, Katarzyna, Marcin Szymański, and Irena Matławska. "Leonurus cardiaca L.(motherwort): A review of its phytochemistry and pharmacology." *Phytotherapy Research* 27, no. 8 (2013): 1115–1120.

Wölfle, Ute, Günter Seelinger, and Christoph Schempp. "Topical application of St. John's wort (Hypericum perforatum)." *Planta Medica* 80, no. 02/03 (2014): 109–120.

Wunsch, M., 2010. Characterization of Fusarium Oxysporum and Phoma Sclerotioides, Pathogens of Birdsfoot Trefoil and Alfalfa.

Yalçin, Funda Nuray, and Duygu Kaya. "Ethnobotany, pharmacology and phytochemistry of the genus Lamium (Lamiaceae)." *FABAD Journal of Pharmaceutical Sciences* 31, no. 1 (2006): 43.

Yang, Jun-Li, Lei-Lei Liu, and Yan-Ping Shi. "Phytochemicals and biological activities of Gentiana species." *Natural Product Communications* 5, no. 4 (2010): 1934578X1000500432.

Yoon, Seong-Jin, Eun-Ji Koh, Chang-Soo Kim, Ok-Pyo Zee, Jong-Hwan Kwak, Won-Jin Jeong, Jee-Hyun Kim, and Sun-Mee Lee. "Agrimonia eupatoria protects against chronic ethanol-induced liver injury in rats." *Food and Chemical Toxicology* 50, no. 7 (2012): 2335–2341.

Young, Jane P., Anna Rallings, P. Michael Rutherford, and Annie L. Booth. "The Effect of NaCl and CMA on the Growth and Morphology of Arctostaphylos uva-ursi (Kinnikinnick)." *Journal of Botany* 2012 (2012): 789879.

Yu, Jiajie, Yujia Cai, Guanyue Su, and Youping Li. "Motherwort injection for preventing postpartum hemorrhage in women with vaginal delivery: A systematic review and meta-analysis of randomized evidence." *Evidence-based Complementary and Alternative Medicine* 2019 (2019).

Zabihi, A., and S. Pashapour. "Therapeutic effects of the Galium verum." *Food Therapy and Health Care*. 4, no. 3 (2022).

Zabihi, Mohsen, Arefeh Shojaeemehr, Ali Mohammad Ranjbar, Mohammadreza Rashidi Nooshabadi, Fatemeh Shishehbor, and Vahid Ramezani. "Althaea officinalis L. Extract Heals Skin Wounds in Second-Degree Burns in Mice." *Jundishapur Journal of Natural Pharmaceutical Products* 18, no. 1 (2023).

Zaidi, S., and P. Dahiya. "In vitro antimicrobial activity, phytochemical analysis and total phenolic content of essential oil from Mentha spicata and Mentha piperita." *International Food Research Journal* 22, no. 6 (2015): 2440.

Zawiślak, Grażyna. "The chemical composition of the essential oil of Marrubium vulgare L. from Poland." *Farmacia* 60, no. 2 (2012): 287–292.

Zeliha, Kaya, and İlkay Koca. "Health benefits of cornelian cherry (Cornus mas L.)." *Middle Black Sea Journal of Health Science* 7, no. 1 (2021): 154–162.

Zhao, Chengye, Xunjia Qian, Minni Qin, Xinyang Sun, Qingqing Yu, Jianyu Liu, Qing Zhu, and Andong Wang. "Juglans mandshurica Maximowicz as a traditional medicine: Review of its phytochemistry and pharmacological activity in East Asia." *Journal of Pharmacy and Pharmacology* 75, no. 1 (2023): 33–48.

Živković, Jelena, Milan Ilić, Katarina Šavikin, Gordana Zdunić, Aleksandra Ilić, and Dejan Stojković. "Traditional use of medicinal plants in South-Eastern Serbia (Pčinja District): Ethnopharmacological investigation on the current status and comparison with half a century old data." *Frontiers in Pharmacology* 11 (2020): 1020.

Zlatković, Bojan K., Stefan S. Bogosavljević, Aleksandar R. Radivojević, and Mila A. Pavlović. "Traditional use of the native medicinal plant resource of Mt. Rtanj (Eastern Serbia): Ethnobotanical evaluation and comparison." *Journal of Ethnopharmacology* 151, no. 1 (2014): 704–713.

Zubair, Muhammad, Hilde Nybom, Christina Lindholm, and Kimmo Rumpunen. "Major polyphenols in aerial organs of greater plantain (Plantago major L.), and effects of drying temperature on polyphenol contents in the leaves." *Scientia Horticulturae* 128, no. 4 (2011): 523–529.

13 Nutritional, Pharmaceutical, and Antinutritional Properties of Wild Edible Plants

What Have We Learned from the Mediterranean Diet

Alice Nabatanzi

13.1 NUTRITIONAL COMPOSITION OF WILD EDIBLE PLANTS

Developing countries are more prone to malnutrition due to the lack of nutrient-rich foods coupled with food insecurity. Leveraging underutilized WEPs is the best way to overcome malnutrition. In developing countries, approximately one billion people rely on WEPs (Taylor et al. 2015). This is majorly because WEPs are considered to be highly nutritious. Additionally, WEPs are critical during food shortages, especially for rural populations, children, and women. The nutritional quality of WEPs has been reported to be superior to domesticated varieties (Gupta et al. 2005). Furthermore, WEPs contain protein, fiber, carbohydrates, lipids, and micronutrients required in the human diet (Datta et al. 2019). Dietary fiber in WEPs is important for diabetic patients and reduces blood cholesterol (Adepoju and Oyewole 2008). Proteins are body-building materials that regulate fluid and acid–base balance in addition to making up the highest percentage of enzymes, antibodies, and hormones. They also act as transporters in the body and provide energy (Nabatanzi et al. 2022). Animal proteins are of superior quality as opposed to plant protein, thus supplementation of plant protein is a prerequisite (Champe, Harvey, and Ferrier 2005). Therefore, plant and animal foods should be consumed in combination to ensure maximum value and avert protein-energy malnutrition (Okello et al. 2017). Minerals regulate metabolic activity in the human body (Gopalan, Sastri, and Balasubramanian 2004). Calcium builds strong and healthy bones, thus aiding the functioning of cardiac muscles (Seal 2016). Magnesium and potassium work in combination to maintain a healthy nervous system (Devi and Jekendra 2014). Magnesium also helps to prevent bleeding disorders, muscle degeneration, and impaired spermatogenesis (Chaturvedi, Shrivastava, and Upreti, 2004). Manganese plays a role in the formation of hemoglobin and activates phosphoenolpyruvate carboxykinase and glutamine synthetase (Indrayan et al. 2005). Food quality is defined by the concentration of macro- and micronutrients in a particular food (Seal 2016). Thus the nutritional qualities of WEPs are in most cases superior to domesticated plants. The superiority of WEPs over conventional crop plants has been reported (Seal 2016).

13.2 ANTINUTRITIONAL FACTORS

Despite the nutritional value of WEPs, they also contain antinutrients which greatly affect their nutritional quality. These antinutrients include oxalates, protease inhibitors, amylase inhibitors, antivitamin factors, phytates, tannins, alkaloids, saponins, tannins, phytic acid, gossypol, lectins, metal binding ingredients, and goitrogens (Gupta et al. 2005). Antinutrients evolve as plants defend themselves against enemies and greatly reduce food quality (Shanthakumari and Britto 2008). Plants

DOI: 10.1201/9781003395935-13

acquire antinutrients from fertilizers, pesticides, and several naturally occurring chemicals (Igile 1996). Several studies have reported the nature of some of the antinutrients and the adverse effects they cause on the health of humans and animals.

Saponins are found in more than 100 plant families. Triterpenoid saponins are usually in soybeans, peanuts, chickpeas, lentils, sunflower seeds, spinach leaves, tea leaves, sugar beets, and allium species. Steroid saponins are present in oats, tomato seeds, asparagus, and yam (Tessa, Papadopoulou, and Osbourn 2014). Saponins cause hemolysis upon interaction with the cholesterol group of erythrocytes (Fleck et al. 2019) and inhibit activities of amylase, glucosidase, trypsin, chymotrypsin, and lipase (Ercan and El 2016). They also impair gastrointestinal digestion (Amin, Hanna, and Mohamed, 2011). Furthermore, saponins lead to a reduction in liver cholesterol, nutrient, and intestinal absorption (Kregiel et al. 2017) and form complexes that interfere with sterol activity (Cheeke 1971). Phytates in plants concentrate mainly in legumes, peanuts, cereals, and oil seeds (García-Estepa et al. 1999). The vegetarian diet culture mainly consumed by people in developing countries contributes to the high phytate ingestion levels. Enzymatic activity is necessary for protein degradation in the small intestine and the stomach is hindered by phytic acid (Kies et al. 2006). Additionally, the bioavailability of minerals in infants, pregnant and lactating women is hindered by phytic acid on consumption of cereal-based foods (Al Hasan et al. 2016). Phytic acid forms complexes with zinc, iron, magnesium, and calcium, thereby reducing their bioavailability. Phytic acid also causes mineral deficiencies due to its chelating property, thereby considering it the most effective antinutrient (Gutiérrez, Kalantar-Zadeh, and Mehrotra 2018). Tannins are either gallotannins and ellagitannins or proanthocyanidins. They form tannin-protein complexes, thereby reducing the digestibility of proteins (Raes et al. 2014). Though tannins can be found in sorghum and barley, they concentrate in beverages, berry fruits, cocoa beans, and pomegranate (Morzelle et al. 2019). Tannins also cause the inactivation of many digestive enzymes (Joye 2019).

Enzyme inhibitors occur in plant foods and slow down the catalytic effects of enzymes, thereby affecting the bioavailability of nutrients, especially in legumes. Proteinases improve the functional properties of various protein molecules (Salas, Dittz, and Torres 2018). Protease inhibitors affect signal initiation, inflammatory response, and blood coagulation (Gomes et al. 2011). Cereal seeds contain plant serpins also known as "suicide inhibitors" (Habib and Fazili 2007). Serpins inhibit trypsin and chymotrypsin enzyme activities (Dahl, Rasmussen, and Hejgaard, 1996). Protease inhibitors also limit enzyme activity by forming protein–protein interactions (Otlewski et al. 2005). Low levels of mineral bioavailability are reported in legumes due to high levels of protease inhibitors, α-amylase inhibitors, and lectins (Yasmin et al. 2008). Cereals have lesser amounts of digestive inhibitors, as compared to other crops (Nikmaram et al. 2017). Furthermore, protease inhibitors affect enzyme activity within the gastrointestinal tract of animals (Nørgaard et al. 2019). In soybeans, disulfide bridges are absent on the Kunitz trypsin inhibitor and Bowman-Birk inhibitor and thus are not readily inactivated by heat treatment (van der Ven et al. 2005). Inhibitors of α-amylase activity increase carbohydrate digestion time, thereby reducing the glucose absorption rate and affecting normal postprandial plasma glucose level (Bhutkar 2018). Seeds rich in trypsin inhibitors increase cholecystokinin (CCK), causing a reduction in food intake and body weight (Serquiz et al. 2016). Trypsin inhibitors in human diets lead to decreased growth rate and also cause pancreatic hyperplasia (Adeyemo and Onilude 2013). However, alpha-amylase, alpha-glucosidase, and lipase aid in the prevention of obesity and type 2 diabetes (Li and Tsao 2019).

Lectins and hemagglutinins found in raw cereals and legumes cause agglutination by attaching to red blood cells (Lagarda-Diaz, Guzman-Partida, and Vazquez-Moreno 2017). On consumption of foods containing lectins the hydrolytic functions and transport of the enterocyte are impaired (Hamid et al. 2013). Phytohemagglutinin is found in kidney beans and consists of two diverse subunits (Krupa 2008). Kidney bean phytohemagglutinin upregulates the function and metabolism of the whole gastrointestinal tract (Lajolo and Genovese 2002). In another study, phytohemagglutinins increased CCK hormone release, thereby enhancing the growth of the rat pancreas (Bardocz et al. 1995). Lectins bind to intestinal epithelial cells, causing damage in the intestinal tract and allowing bacteria to come into contact with the bloodstream (Herzig et al. 1997). Cyanide ions interfere with essential amino acids, inhibit enzyme systems, and cause acute toxicity, neuropathy, and death Neurological disorders are caused by alkaloids (Muramoto 2017). *Solanum* spp. contain solanine and chaconine which are hemolytically active and toxic to humans (Ohsaka, Hayashi and Sawai 1976). Saponins from *Bulbostermma paniculatum* and *Pentapamax leschenaultii* contain saponins that affect human spermatozoa by damaging the sperm plasma membrane (Aletor 1993). Trypsin causes pancreatic enlargement and growth depression (Aletor 1991). Phytates and oxalates make calcium, iron, magnesium, and zinc unavailable due to the formation of complexes (Pant et al. 1988). This in turn causes muscle weakness, paralysis, and gastrointestinal tract irritation (Su and Guo 1986). Furthermore, oxalates cause nephrotic lesions in the kidney (Aletor and Fetuga 1987; Nelson et al. 1968).

13.3 ANTINUTRIENTS IN WILD EDIBLE PLANTS

Several researchers have studied the levels of antinutrients in different WEPS. The phytate content of *Alocacia indica* is 312.4 mg/100 g; *Asparagus officinalis* 340.8 mg/100 g; *Portulaca oleracia* 823.6 mg/100 g; *Momordica dioicia* 284.2 mg/100 g; *Eulophia ochreata* 255.6 mg/100 g; *Solanum indicum* 695.8 mg/100 g; *Cordia myxa* 248.0 mg/100 g; and *Chlorophytum comosum* 468.8 mg/100 g (Blood and Radostits 1989). Jones, Hunt and King (1997), reported phytic acid values of 323.6 mg/100 g (*Portulaca oleracia*); 695.8 mg/100 g (*Solanum indicum*), 255.6 mg/100 g (*Eulophia*). Phytate content in *Cassipourea congoensis* was (2.57 ± 0.41%) (Ojiyako and Igwe 2008: Nkafamiya et al. 2006). Studies have reported the level of phytic acid in raw cowpea seeds (*Sesquipedalis*) as 4.25 mg/100 g, in boiling seed 2.85 mg/100 g, in roasted 2.05 mg/100 g and autoclaved seeds 3.95 mg/100 g (Nkafamiya et al. 2006). The phytic acid level in *Corhorus olitorius* was 1.71% (Ndlovu and Afolayan 2008).

Nkafamiya et al. (2007) observed 9.16 ± 0.13% oxalate in seeds of *Gemlina arborea*. *Ficus asperifolia* oxalates (3.78 ± 0.28 mg/100 g) were also analyzed (Udensi, Ekwu, and Isinguzo 2007). Adepoju 2009, worked on *Mordii whytii* fruits saponins (1.82 ± 0.08 mg/100 g). Asyaz et al. 2010, analyzed *Citrullus colocynthis* saponins 0.52 mg/100 g. Hess et al. 2003, studied *Sapindus saponaria, Enterolobium cyclocarpum*, and *Pithecellobium saman*. They observed 120 mg/g DM saponin from *Sapindus saponaria*; 19 mg/g DM saponin from *Enterolobium cyclocarpum*; and 17 mg/g DM saponin from *Pithecellobium saman*. The tannin content of *Dioscorea bulbifera* var. *vera* was 2.55 ± 0.07 mg/100 g (Arinathan, Mohan, and Maruthupandian 2009). Tannin values of *Boscia foetida* and *Grewia retinervis* ranged from 0.26 % to 9.5 % respectively (Aganga and Adogla-Bessa 1999). *Dolichos lablab* seeds contained 85 ± 0.01 mg/100 g tannins (Ramakrishna, Jhansi and Rao 2006). *Cassipourea congoensis* fruits were rich in oxalate (11.40 ± 1.50 mg/100 g), phytate (2.57 ± 0.41 mg/100 g) and saponin (8.16 ± 0.21 mg/100 g), and *Nuclea latifolia* fruit contained tannins (4.62 ± 0.14 mg/100 g) (Nkafamiya et al. 2006). *Cassipourea congoensis* seeds contained oxalates (10.21 ± 1.11 mg/100 g), tannins (2.84 ± 0.12 mg/100 g), and saponins (7.17 ± 0.18 mg/100 g) while *Balanites aegytiaca* was rich in phytates (2.18 ± 0.14 mg/100 g) (Nkafamiya et al. 2010).

Umaru et al. (2007) studied *Sclerocarya birrea* (3.56 ± 0.54%) and *Haematostaphis barteri* (3.30 ± 0.10%), *Balanites aegyptiaca* (16.01 ± 0.02%), *Detarium microcarpum* (12.10 ± 0.05%, and *Parkia biglobosa* (12.23 ± 0.46%) saponin content. The tannin content of *B. aegyptiaca* (7.40 ± 0.14%), *Hyphaena thebaica* (6.39 ± 0.5%), and *Borassus aethiopum* (5.90 ± 0.13%) was also studied. They noticed that the highest levels of oxalates were found in *Zizyphus spinachristi* (16.20 ± 2.12%), *Zizyphus mauritiana* (15.50 ± 1.50%), and *Balanite aegyptiaca* (14.50 ± 2.08%).. Aberoumand and Deokule (2009) studied the phytic acid content of *Cordia* Ojoyako and Igwe (2008) analyzed the oxalate (0.58 ± 0.12%), phytate (0.11 ± 0.02%), and tannin (0.02 ± 0.05%) content of tomato fruit *dichotoma* fruit, which was 248 mg/100g. Osabor et al. (2009) profiled the oxalate and phytic acid content of *Treculia africana*, which were 3.01 ± 0.11 mg/100g and 0.76 ± 0.01 mg/100g respectively. *Glycosmis pentaphylla* is rich in oxalic acid (1.64 ± 0.052 g/100g) and *Ficus racemosa* is rich in phytic acid (0.7 ± 0.2 g/100g) (Ilelaboye and Pikuda 2009).

Llelaboye and pikuda (2009) studied *Trichosanthe cucumerina* and *Citrillues vulgaris* seeds. The highest levels of tannins (25.30 mg/100g) and saponins (2097 mg/100g) were found in *Trichosanthe cucumerina* seeds whereas the highest levels of oxalic and phytic acid were found in *Citrillues vulgaris* seeds ((40.65 mg/100g) and (20.12 mg/100g) respectively). Bello et al. (2008) found the highest levels of tannins in *Cola millenii* mesocarp 1.33 ± 0.06 mg/g, phytic acid in *Strychnos innocua* seed (6.65 ± 0.60 mg/g), and oxalic acid in *Strychnos innocua* juice (1.17 ± 0.10 g/100g). Adepoju 2009, worked on the fruits of *Spondias mombin*, *Dialium guineense*, and *Mordii whytii*. The highest levels of phytates (1.64 ± 0.04 mg/100g) and saponins (1.82 ± 0.08 mg/100g) were found in *M. whytii* fruits while oxalates (1.88 ± 0.06 mg/100g) and tannins (2.41 ± 0.02 mg/100g) were found in *S. mombin* fruits. High tannin content reduces the digestibility of protein (Murwan, Saifeldin, and Ishag 2010). Amoo, Finnie, and Van Staden 2009, reported *Mucuna preta* seeds being rich in oxalate (8.31 ± 0.03 mg/g), phytic acid 85.47 ± 0.62%, and tannin 10.30 ± 1.15 mg/g.

13.4 FOOD PROCESSING STRATEGIES USED TO REMOVE ANTINUTRIENTS IN FOOD

Heat treatment, soaking, milling, de-branning, roasting, cooking, germination, and fermentation could reduce or remove antinutrients in foods.

Whereas milling removes phytic acid, lectins, and tannins in the bran of grains it also removes important minerals (Gupta et al. 2013: Chowdhury and Punia 1997). Cooking time is reduced and endogenous phytases are enhanced by soaking nuts and grains. This improves the digestibility and nutritional quality of food (Kumari 2018). Additionally, during soaking the antinutrients leach into water, in turn reducing water-soluble vitamins and minerals in grains and legumes (Kruger et al. 2014). Furthermore, exogenous phytase enzymes are enhanced during soaking (Vashishth, Ram, and Beniwal 2017). In another study, soaking reduced phytate content at 45°C and 65°C (Greiner and Konietzny 2006). The mineral concentration and protein availability increased in soaked beans with a reduction in phytic acid levels (Abdoulaye, Brou, and Chen 2011). In another study, when the soaking time of chickpeas was increased from 2 to 12 h the phytic acid concentration decreased by 47.45 to 55.71% (Ertaş and Türker 2012).

The autoclaving method increases acidity and activates phytase (Ertop and Bektaş 2018) as soaking and cooking decrease phytic content in legume grains (Vadivel and Biesalski 2011). Boiling reduces tannins and trypsin inhibitors (Patterson, Curran, and Der 2016). Cooking and soaking decrease the phytic acid concentration in legumes (Vadivel and Biesalski 2011). Kürşat and Elgün (2014) reported that autoclave and microwave application reduced the phytic acid content of whole wheat bread. Another study reported a reduction in tannin content due to pressure cooking, leading to improved black gram protein digestibility (Zia-Ur and Shah). A reduction in the oxalate content of taro leaves by 47% when boiled in water for 40 minutes was reported (Savage and Mårtensson 2010). Roasting decreased trypsin inhibitors in soybean meal (Vagadia, Vanga, and Raghavan 2017). Phytase is activated through germination, leading to a decreased phytic acid concentration in food and thus reducing the antinutritional content of cereals (Oghbaei and Prakash 2016). After the malting of millet samples for 72 h and 96 h, it was found that phytic acid content was reduced by 23.95 and 45.3%, respectively (Abdoulaye, Kouakou, and Chen 2010). Azeke et al. (2010) observed that the phytate content of cereal grains reduced significantly after 10 days of germination. According to Zhang et al. (2015), germination increased phenolic, flavonoid, and tannin contents in germinated buckwheat. A study reported an increase in the isoflavone profile of soybean due to the activation of β-glucosidases by germination (De Camargo and da Silva Lima 2019).

Fermentation is useful in reducing bacterial contamination in foods, thus fermented millet products are probiotics for treating diarrhea in children (Nduti et al. 2016). The pH maintained by fermentation in cereals degrades metal cation complexes, decreasing the phytic acid and thus enhancing the nutritional quality of cereals (Gibson and Benson 2005). Fermentation also increases lysine, methionine, and tryptophan, improving the nutritional value of grains (Mohapatra et al. 2019). During lactic acid fermentation, tannin levels are reduced, leading to increased absorption of iron (Ray and

Didier 2014). Destruction of some antinutrients may not always be achieved (Aykroyd, Doughty, and Walker 1982); for example, phytates are not destroyed by cooking (Ranhotra, Loewe and Puyat 1974). Korte (1973) also observed that cooking did not destroy haemagglutinin in ground beans and cereals and this caused diarrhea. In developing countries, the majority of households are poverty stricken, thus they resort to using energy sources like kerosene, gas, and little firewood. Thus, due to the prohibitive energy costs, when preparing beans they do not process them to the recommended level to destroy antinutrients. As a result of this, people end up eating cooked toxic beans (Pusztai et al. 1993). A high occurrence of nitrates is in fresh vegetables (Walker 1975).

13.5 NUTRACEUTICAL SIGNIFICANCE OF WILD EDIBLE PLANTS

During cellular metabolism, reactive oxygen species (ROS) and reactive nitrogen species (RNS) are generated. These are attributed to causing aging; cellular injury; cancer; neurodegenerative, cardiovascular, and renal disorders; Parkinson's disease; arthritis; Alzheimer's disease; stroke; and chronic inflammatory diseases (Szabo et al. 2010). Production of ROS is also attributed to pollution, drugs, smoke, ionizing radicals, stresses, synthetic pesticides, and spicy foods (Agrawal, Kulkarni, and Sharma 2011). Catalase, superoxide dismutase, and glutathione reductase scavenge free radicals, thereby helping maintain cellular functions (Kurutas, 2015), thus, the need to search for antioxidants from natural sources (Tadhani, Patel and Subhash 2007). Vitamin C scavenges superoxide anion (O^{2-}), hydroxyl (OH^-), and singlet oxygen (O_2) (Loganayaki, Siddhuraju and Manian 2010). It also synthesizes tocopherol and is a cofactor for hydroxylases (Prasad et al. 2010). Plant-based antioxidants have been reported to have fewer side effects. Antioxidant phytochemicals are abundant in fresh fruits and vegetables (Agbor, Vinson, and Donnelly 2014). The free radical scavenging ability of phenols is due to the presence of the hydroxyl group (Parajuli et al. 2019). Terpenoids have antihepatotoxic, antimicrobial, and antiseptic properties (Setchell, and Cassidy 1999). Saponins are cough suppressants and possess hemolytic activity (Okwu 2004) as tannins possess anti-inflammatory, diuretic, and antimicrobial properties (Okwu 2005).

People around the world have become aware of the nutraceutical importance of WEPs. The phytochemicals in WEPs are responsible for their anti-cancer effects in human chronic myeloid leukemia K562 cells (Liu et al. 2010). The phenolic compounds in genus *Satureja* are responsible for their anti-inflammatory, antispasmodic, antidiarrheal, antioxidant, antifungal, antiviral, and antibacterial activities (Aqil and Ahmad 2007). Fifty-nine wild edible vegetables (WEVs) used in the treatment of several ailments have been identified from the Harnai Balochistan district in Pakistan (Tareen et al. 2016). In South Africa, 14 WEVs have been reported to treat more than 52 diseases (Naik et al. 2017). Eighteen WEVs are used as medicinal food in Southern Shan State, Myanmar (Shin et al. 2018). *Berberis chitria* contains berberin in its stem and roots and is known to cure eye and ear diseases, diabetes, fever, jaundice, stomach disorders, and skin diseases (Palai, 2022). *Capsella bursa-pastoris* cures edema caused by nephritis and hypertension (Song et al. 2007). *Stellaria aquatica* shoots and leaves have high iron content and constitute the diet of Indians (Sharma and Arora 2012). *Rheum emodi* is known as the wonder drug because of its extensive medicinal uses (Ahmad et al. 2015). *S. oleraceus, S. arvensis, S. asper, S. uliginosus, S. brachyotus*, and *S. lingianus* from China have revealed broad-spectrum antibacterial activities against *E. coli, S. enteric, V. parahaemolyticus*, and *S. aureus* (Xia et al. 2011). Twenty-six species of WEVs from China, Japan, Thailand, and Yemen have been reported to have antimicrobial activity against *B. cereus, S. aureus, L. monocytogenes, E. coli*, and *S. infantis*.

In India, WEVs from Odisha possessed broad-spectrum activity against eight food-borne pathogens (Panda, Padhi, and Mohanty 2011). In addition, *Ceropegia thwaitesii* methanolic extract showed excellent antimicrobial activity (Muthukrishnan et al. 2018). Amino acids help in the transportation of nutrients and prevention of illness, and their deficiency results in decreased immunity, digestive problems, and slowed growth in children. The leaves of *Centella asiatica* contained histidine, lysine, isoleucine, and phenylalanine with high levels of palmitic acid (Nwachukwu 2019). Furthermore, *C. asiatica* was found to be high in steroids, phenols, proanthocyanin, and rutin. Steroids help in wound recovery and improve athletes' performance.

13.6   WHY THE MEDITERRANEAN DIET?

The Mediterranean diet is composed of vegetables, fruits, legumes, nuts, cereals, fish, and seafood. Thus it promotes good health and long life and is recommended in the prevention of noncommunicable diseases. Furthermore, the Mediterranean diet promotes a healthy immune system as every green edible plant contains 52 health-promoting phytochemicals. Climate, growing conditions, cultivar properties, plant growth and development stages, and harvesting time affect the phytochemicals in plants. Stress conditions during plant growth and development increase the content of antioxidants in plants. Contaminated water, inorganic fertilizers, metal-based pesticides, industrial emissions, harvesting processes, and activities during storage and/or sale[5] lead to the accumulation of heavy metals in WEPs.

13.7   CONCLUSION

Frequently adding WEPs to your diet reduces the risk of developing cardiovascular and degenerative diseases. Plant-based diets also help reduce functionality loss with increasing age. Furthermore, WEPs are the best source of natural antioxidants. Unfortunately, indigenous knowledge of WEPs has been lost in the majority of local communities around the world. This is majorly due to urbanization and the adoption of Western culture, especially by Africans. Furthermore, there is a continued loss of indigenous knowledge custodians due to the high prevalence of noncommunicable diseases and the lack of proper medical care, especially in marginalized communities. This calls for the necessity to document and investigate the nutritional and pharmaceutical significance of WEPs. Additionally, there is a need to advocate for the adoption of the Mediterranean among underdeveloped nations where the majority of the young population lives a sedentary lifestyle. For sustainability and as a conservation strategy, there is a need to develop appropriate propagation methods for the different WEPs.

REFERENCES

Aberoumand, A, and SS Deokule. 2009. "Quantification of Protein and Total Nitrogen in Some Plant Foods of Iran." *Nigerian Food Journal* 27 (1). doi:10.4314/nifoj.v27i1.47461

Adepoju, O. 2009. "Proximate composition and micronutrient potentials of three locally available wild fruits in Nigeria." *African Journal of Agricultural Research* 4. 887–892.

Adepoju, O., and O. Oyewole. 2008. "Nutritional importance and micronutrient potentials of two non-conventional indigenous green leafy vegetables from Nigeria." *Medwell On line Agricultural Journal* 3 (5), 3. 362–365.

Adeyemo, S. M., and A. A. Onilude. 2013. "Enzymatic Reduction of Anti-Nutritional Factors in Fermenting Soybeans by Lactobacillus Plantarum Isolates from Fermenting Cereals." *Nigerian Food Journal* 31 (2): 84–90. doi:10.1016/s0189-7241(15)30080-1

Aganga, A. A. and T. Adogla-Bessa. 1999. "Dry matter degradation, tannin and crude protein contents of some indigenous browse plants of Botswana." *Archivos de Zootecnia* 48: 79–83.

Agbor, A., Gabriel Joe A. Vinson, and Patrick E. Donnelly. 2014. "Folin-Ciocalteau Reagent for Polyphenolic Assay." *International Journal of Food Science, Nutrition and Dietetics*, 147–156. doi:10.19070/2326-3350-1400028

Agrawal, Shalu, Giriraj T. Kulkarni, and V. N. Sharma. 2011. "A Comparative Study on the Antioxidant Activity of Methanolic Extracts of Terminalia Paniculata and Madhuca Longifolia." *Free Radicals and Antioxidants* 1 (4): 62–68. doi:10.5530/ax.2011.4.10

Ahmad, S. B., S. Bhat, I. Hussain, J. Parrah, S. P. Ahmad, M. Manzoor ur Rahman. 2015. "The Use of a Rabbit Model to Evaluate the Influence of Age on Excision Wound Healing." *Journal of Life Sciences Research* 2, 72–75.

Abdoulaye, C., K. Brou, and J. Chen, 2011. "Phytic Acid in Cereal Grains: Structure, Healthy or Harmful Ways to Reduce Phytic Acid in Cereal Grains and Their Effects on Nutritional Quality." *American Journal of Plant Nutrition and Fertilization Technology* 1, 1–22. doi: 10.3923/ajpnft.2011.1.22

Aletor, V. A. 1991. "Anti-nutritional factors in some Nigerian feedstuffs, herbage byproducts, crop residues and browse plants". A monograph prepared for the Presidential task force on alternative formulation of livestock feeds; Product Development, Quality Evaluation and Health Implications.

Aletor, V. A. 1993. "Allelochemicals in Plant Foods and Feeding Stuffs. Part I. Nutritional, Biochemical and Physiopathological Aspects in Animal Production." *Veterinary Human Toxicology* 35 (1): 57–67.

Aletor, V. A., and B. L. Fetuga. 1987. "Pancreatic and Intestinal Amylase (EC 3.2.1.1) in the Rat Fed Lima Bean Haemagglutinin Extract. II. Evidence of Impaired Dietary Starch Utilization." *Journal of Animal Physiology and Animal Nutrition* 57 (1–5): 113–117. doi:10.1111/j.1439-0396.1987.tb00016.x

Amin, Hala Abdel Salam, Atef G. Hanna, and Sayeda Saleh Mohamed. 2011. "Comparative Studies of Acidic and Enzymatic Hydrolysis for Production of Soyasapogenols from Soybean Saponin." *Biocatalysis and Biotransformation* 29 (6): 311–319. doi:10.3109/10242422.2011.632479

Amoo, S. O., J. F. Finnie, and J. Van Staden. 2009. "In Vitro Pharmacological Evaluation of Three Barleria Species." *Journal of Ethnopharmacology* 121 (2): 274–277. doi:10.1016/j.jep.2008.10.035

Aqil, F., and Ahmad, I. 2007. "Antibacterial Properties of Traditionally Used Indian Medicinalplants." *Methods and Findings in Experimental and Clinical Pharmacology* 29 (2): 79. doi:10.1358/mf.2007.29.2.1075347

Arinathan, V., V. Mohan, and A. Maruthupandian. 2009. "Little Known Wild Edible Seeds of Western Ghats, Tamil Nadu." *Journal of Non-Timber Forest Products* 16 (2): 119–124. doi:10.54207/bsmps2000-2009-37fe1i

Asyaz, S., F. U. Khan, I. Hussain, M. A. Khan, I. U. Khan. 2010. "Evaluation of chemical analysis profile of Citrullus colocynthis growing in Southern area of Khyber Pukhtunkhwa, Pakistan." *Journal of World Applied Sciences* 10: 402–405.

Aykroyd, W. R., Doughty, J. and Walker, A. 1982. "Legumes in Human Nutrition." Food and Agricultural Organization of the United Nations (FAO), Food and Nutrition Paper. No.20, FAO. Rome.

Azeke, M. A., S. J. Egielewa, M. Ugunushe Eigbogbo, and G. I. Ihimire. 2010. "Effect of Germination on the Phytase Activity, Phytate and Total Phosphorus Contents of Rice (Oryza Sativa), Maize (Zea Mays), Millet (Panicum Miliaceum), Sorghum (Sorghum Bicolor) and Wheat (Triticum Aestivum)." *Journal of Food Science and Technology* 48 (6): 724–729. doi:10.1007/s13197-010-0186-y

Bardocz, S., G. Grant, S. W. Ewen, T. J. Duguid, D. S. Brown, K. Englyst, and A. Pusztai. 1995. "Reversible Effect of Phytohaemagglutinin on the Growth and Metabolism of Rat Gastrointestinal Tract." *Gut* 37 (3): 353–360. doi:10.1136/gut.37.3.353

Bello, M. O., Falade, O. S., Adewusi, S. R. A. and Olawore, N. O. 2008. "Studies on the Chemical Compositions and Anti-nutrients of Some Lesser Known Nigeria Fruits." *African Journal of Biotechnology* 7 (21): 3972–3979.

Bhutkar, Mangesh A. 2018. "'In Vitro Studies on Alpha Amylase Inhibitory Activity of Some Indigenous Plants.'" *Modern Applications in Pharmacy & Pharmacology* 1 (4). doi:10.31031/mapp.2018.01.000518

Blood, D. C. and O. M. Radostits. 1989. *Veterinary Medicine*, 7th Edition, Balliere Tindall, London, pp. 589–630.

Champe, Pamela C., Richard A. Harvey, and Denise R. Ferrier. 2005. *Biochemistry*. Lippincott Williams & Wilkins. http://books.google

Chaturvedi, U. C., Richa Shrivastava, and R. K. Upreti. 2004. "Viral Infections and Trace Elements: A Complex Interaction." *Current Science* 87 (11): 1536–54. http://www.jstor.org/stable/24109032

Cheeke, P. R. 1971. "Nutritional and Physiological Implications of Saponins: A Review." *Canadian Journal of Animal Science* 51 (3): 621–632. doi:10.4141/cjas71-082

Chowdhury, S., and D. Punia. 1997. "Nutrient and Antinutrient Composition of Pearl Millet Grains as Affected by Milling and Baking." *Food / Nahrung* 41 (2): 105–107. doi:10.1002/food.19970410210

Datta, S., B. K. Sinha, S. Bhattachaeje and T. Seal 2019. "Nutritional Composition, Mineral Content, Antioxidant Activity and Quantitative Estimation of Water Soluble Vitamins and Phenolics by RP-HPLC in Some Lesser Used Wild Edible Plants." *Heliyon* 5: 1431. doi:10.1016/j.heliyon.2019.e01431

Dahl, S. W., S. K. Rasmussen, and J. Hejgaard. 1996. "Heterologous Expression of Three Plant Serpins with Distinct Inhibitory Specificities." *Journal of Biological Chemistry* 271 (41): 25083–25088. doi:10.1074/jbc.271.41.25083

De Camargo, A. C., and R. da Silva Lima. 2019. "A Perspective on Phenolic Compounds, Their Potential Health Benefits, and International Regulations: The Revised Brazilian Normative on Food Supplements." *Journal of Food Bioactives* 7 (September). doi:10.31665/jfb.2019.7193

Devi, M. N. and S. S. Jekendra. 2014. "Antioxidant, Mineral and Phytochemical Composition of Clerodendrum Colebrookianum Walp, a Well Known Home Remedy Herbal for Humankind." *World Journal of Pharmaceutical Research* 3 (2): 2667–2676.

Ercan, Pınar, and Sedef Nehir El. 2016. "Inhibitory Effects of Chickpea and Tribulus Terrestris on Lipase, α-Amylase and α-Glucosidase." *Food Chemistry* 205: 163–169. doi:10.1016/j.foodchem.2016.03.012

Ertaş, Nilgün, and Selman Türker. 2012. "Bulgur Processes Increase Nutrition Value: Possible Role in in-Vitro Protein Digestability, Phytic Acid, Trypsin Inhibitor Activity and Mineral Bioavailability." *Journal of Food Science and Technology* 51 (7): 1401–1405. doi:10.1007/s13197-012-0638-7

Ertop, M. H., and M. Bektaş. 2018. "Enhancement of Bioavailable Micronutrients and Reduction of Antinutrients in Foods with Some Processes." *Food and Health* 4 (3): 159–165.

Fleck, Juliane Deise, Andresa Heemann Betti, Francini Pereira Da Silva, Eduardo Artur Troian, Cristina Olivaro, Fernando Ferreira, and Simone Gasparin Verza. 2019. "Saponins from Quillaja Saponaria and Quillaja Brasiliensis: Particular Chemical Characteristics and Biological Activities." *Molecules* 24 (1). doi:10.3390/molecules24010171

García-Estepa, Rosa Mạ, Eduardo Guerra-Hernández, and Belén García-Villanova. 1999. "Phytic Acid Content in Milled Cereal Products and Breads." *Food Research International* 32 (3): 217–221. doi:10.1016/s0963-9969(99)00092-7

Gibson, L., and G. Benson. 2005. "Origin, history, and uses of soybean (Glycine max)." Iowa State University, Department of Agronomy.

Gomes, T. R. Marco, Maria L. Oliva, Miriam T. P. Lopes, and Carlos E. Salas. 2011. "Plant proteinases and inhibitors: An overview of biological function and pharmacological activity." *Current Protein & Peptide Science* 12 (5): 417–436. doi:10.2174/138920311796391089

Gopalan, C., and B. V. R. Sastri, S. C. Balasubramanian. 2004. Nutritive Value of Indian Foods. National Institute of Nutrition, Indian Council of Medical Research, Hyderabad, India. 2–58.

Greiner, R. and U. Konietzny. 2006. "Phytase for Food Application." *Food Technology and Biotechnology* 44, 125–140.

Gupta, Raj Kishor, Shivraj Singh Gangoliya, and Nand Kumar Singh. 2013. "Reduction of Phytic Acid and Enhancement of Bioavailable Micronutrients in Food Grains." *Journal of Food Science and Technology* 52 (2): 676–684. doi:10.1007/s13197-013-0978-y

Gupta, Sheetal, A. Jyothi Lakshmi, M. N. Manjunath, and Jamuna Prakash. 2005. "Analysis of Nutrient and Antinutrient Content of Underutilized Green Leafy Vegetables." *LWT - Food Science and Technology* 38 (4): 339–345. doi:10.1016/j.lwt.2004.06.012

Gutiérrez, Orlando M., Kamyar Kalantar-Zadeh, and Rajnish Mehrotra. 2018. *Clinical Aspects of Natural and Added Phosphorus in Foods.* Springer. http://books.google

Habib, H. and K. M. Fazili. 2007. "Plant Protease Inhibitors: A Defense Strategy in Plants". *Biotechnology and Molecular Biology Reviews* 2 (3): 68–85.

Hamid, R., Masood, A., Wani, I. H., Rafiq, S. 2013. Lectins: "Proteins with Diverse Applications." *Journal of Applied Pharmaceutical Science* 3 (4): S93–S103.

Al Hasan, Syed Mahfuz, Mahedi Hassan, Sonjoy Saha, Mominul Islam, Masum Billah, and Shimul Islam. 2016. "Dietary Phytate Intake Inhibits the Bioavailability of Iron and Calcium in the Diets of Pregnant Women in Rural Bangladesh: A Cross-Sectional Study." *BMC Nutrition* 2 (1). doi:10.1186/s40795-016-0064-8

Herzig, K.-H., S. Bardocz, G. Grant, R. Nustede, U. R. Fölsch, and A. Pusztai. 1997. "Red Kidney Bean Lectin Is a Potent Cholecystokinin Releasing Stimulus in the Rat Inducing Pancreatic Growth." *Gut* 41 (3): 333–338. doi:10.1136/gut.41.3.333

Hess, H. D., M. Kreuzer, T. E. Díaz, C. E. Lascano, J. E. Carulla, Carla R. Soliva, and Andrea Machmüller. 2003. "Saponin Rich Tropical Fruits Affect Fermentation and Methanogenesis in Faunated and Defaunated Rumen Fluid." *Animal Feed Science and Technology* 109 (1–4): 79–94. doi:10.1016/s0377-8401(03)00212-8

Igile, G. O. 1996. Phytochemical and Biological studies on some constituents of *Vernonia amygdalina* (compositae) leaves. Ph.D thesis, Department of Biochemistry, University of Ibadan, Nigeria.

Ilelaboye, N. O. A., and O. O. Pikuda. 2009. "Determination of Minerals and Anti-Nutritional Factors of Some Lesser-Known Crop Seeds." *Pakistan Journal of Nutrition* 8 (10): 1652–1656. doi:10.3923/pjn.2009.1652.1656

Indrayan, A. K., Sudeep Sharma, Deepak Durgapal, Neeraj Kumar, and Manoj Kumar. 2005. "Determination of Nutritive Value and Analysis of Mineral Elements for Some Medicinally Valued Plants from Uttaranchal." *Current Science* 89 (7): 1252–55. http://www.jstor.org/stable/24110980

Jones, T. C., Hunt, R. D., and N. W. King. 1997. *Veterinary pathology*, 6th Ed. Baltimore, MD; London: Williams & Wilkins.

Joye, Iris. 2019. "Protein Digestibility of Cereal Products." *Foods* 8 (6): 199. doi:10.3390/foods8060199

Kies, Arie K., Leon H. De Jonge, Paul A. Kemme, and Age W. Jongbloed. 2006. "Interaction between Protein, Phytate, and Microbial Phytase. In Vitro Studies." *Journal of Agricultural and Food Chemistry* 54 (5): 1753–1758. doi:10.1021/jf0518554

Kürşat, D. M. and A. Elgün. 2014. "Comparison of Autoclave, Microwave, IR and UV-C stabilization of Whole Wheat Flour Branny Fractions upon the Nutritional Properties of Whole Wheat Bread." *Journal of Food Science and Technology* 51 (1):59–66. doi: 10.1007/s13197-011-0475-0

Korte, R. 1973. "Processed Plant Foods in High Humid Tropics." *Ecological Food Nutrition*, (1): 303–306.

Kregiel, D., J. Berlowska, I. Witońska, H. Antolak, C. Proestos, M. Babic, L. Babic and B. Zhang. 2017. Saponin-Based, Biological-Active Surfactants from Plants. In *Application and characterization of surfactants* (pp. 183–205). 10.5772/68062.

Kruger, Johanita, André Oelofse, and John R. N. Taylor. 2014. "Effects of Aqueous Soaking on the Phytate and Mineral Contents and Phytate: Mineral Ratios of Wholegrain Normal Sorghum and Maize and Low Phytate Sorghum." *International Journal of Food Sciences and Nutrition* 65 (5): 539–546. doi:10.3109/09637486.2014.886182

Krupa, U. 2008. "Main Nutritional and Antinutritional Compounds of Bean Seeds – A Review." *Polish Journal of Food and Nutrition Sciences* 58 (2): 149–155.

Kumari, S. 2018. "The effect of soaking almonds and hazelnuts on Phytate and mineral concentrations." Doctoral dissertation, University of Otago.

Kurutas, Ergul Belge. 2015. "The Importance of Antioxidants Which Play the Role in Cellular Response against Oxidative/Nitrosative Stress: Current State." *Nutrition Journal* 15 (1). doi:10.1186/s12937-016-0186-5

Lagarda-Diaz, Irlanda, Ana Guzman-Partida, and Luz Vazquez-Moreno. 2017. "Legume Lectins: Proteins with Diverse Applications." *International Journal of Molecular Sciences* 18 (6). doi:10.3390/ijms18061242

Lajolo, Franco M., and Maria Inés Genovese. 2002. "Nutritional Significance of Lectins and Enzyme Inhibitors from Legumes." *Journal of Agricultural and Food Chemistry* 50 (22): 6592–6598. doi:10.1021/jf020191k

Li, Li, and Rong Tsao. 2019. "UF-LC-DAD-MSn for Discovering Enzyme Inhibitors for Nutraceuticals and Functional Foods." *Journal of Food Bioactives* 7 (September). doi:10.31665/jfb.2019.7195

Liu, Yan-li, Li-hua Tang, Zhong-qin Liang, Ben-gang You, and Shi-lin Yang. 2010. "Growth Inhibitory and Apoptosis Inducing by Effects of Total Flavonoids from Lysimachia Clethroides Duby in Human Chronic Myeloid Leukemia K562 Cells." *Journal of Ethnopharmacology* 131 (1): 1–9. doi:10.1016/j.jep.2010.04.008

Loganayaki, Nataraj, Perumal Siddhuraju, and Sellamuthu Manian. 2010. "Antioxidant Activity of Two Traditional Indian Vegetables: Solanum Nigrum L. and Solanum Torvum L." *Food Science and Biotechnology* 19 (1): 121–127. doi:10.1007/s10068-010-0017-y

Mohapatra, Debabandya, Avinash Singh Patel, Abhijit Kar, Sumedha S. Deshpande, and Manoj Kumar Tripathi. 2019. "Effect of Different Processing Conditions on Proximate Composition, Anti-Oxidants, Anti-Nutrients and Amino Acid Profile of Grain Sorghum." *Food Chemistry* 271 (January): 129–135. doi:10.1016/j.foodchem.2018.07.196

Morzelle, Maressa Caldeira, Jocelem Mastrodi Salgado, Adna Prado Massarioli, Patricia Bachiega, Alessandro de Oliveira Rios, Severino Matias Alencar, Andres R. Schwember, and Adriano Costa de Camargo. 2019. "Potential Benefits of Phenolics from Pomegranate Pulp and Peel in Alzheimer's Disease: Antioxidant Activity and Inhibition of Acetylcholinesterase." *Journal of Food Bioactives* 5 (March). doi:10.31665/jfb.2019.5181

Moses, T., K. K. Papadopoulou, and A. Osbourn. 2014. "Metabolic and Functional Diversity of Saponins, Biosynthetic Intermediates and Semi-synthetic Derivatives." *Critical Reviews in Biochemistry and Molecular Biology* 49 (6): 439–462. doi:10.3109/10409238.2014.953628

Muramoto, Koji. 2017. "Lectins as Bioactive Proteins in Foods and Feeds." *Food Science and Technology Research* 23 (4): 487–494. doi:10.3136/fstr.23.487

Murwan Sabahel Khier, K. M., H. A. Saifeldin, and K. E. A. Ishag. 2010. "Effect of Maturity Stage on Protein Fractionation, In Vitro Protein Digestibility and Anti-nutrition Factors in Pineapple (Ananas comosis) Fruit Grown in Southern Sudan." *African Journal of Food Science* 4 (8): 550–552.

Muthukrishnan, S., T. Senthil Kumar, A. Gangaprasad, F. Maggi, and M. V. Rao. 2018. "Phytochemical Analysis, Antioxidant and Antimicrobial Activity of Wild and in Vitro Derived Plants of Ceropegia Thwaitesii Hook – An Endemic Species from Western Ghats, India." *Journal of Genetic Engineering and Biotechnology* 16 (2): 621–630. doi:10.1016/j.jgeb.2018.06.003

Nabatanzi, A., A. J. N. Kazibwe, I. Nakalembe, A. Nabubuya, G. Tumwine, B. N. Kungu, and J. D. Kabasa. 2022. "Nutraceutical and Antinutritional Properties of Wild Edible Plants Consumed by Pregnant Women and School-Age Children (6-12 Years) in Najjembe Sub-County, Buikwe District, Uganda." *African Journal of Food, Agriculture, Nutrition and Development* 22 (115): 21990–22016. doi:10.18697/ajfand.115.20925

Naik, R., Sneha D. Borkar, S. Bhat, and R. Acharya. 2017. "Therapeutic Potential of Wild Edible Vegetables - A Review." *Journal of Ayurveda and Integrated Medical Sciences (JAIMS)* 2 (6). doi:10.21760/jaims.v2i06.10930

Ndlovu, J., and A. J. Afolayan. 2008. "Nutritional Analysis of the South African Wild Vegetable Corchorus Olitorius L." *Asian Journal of Plant Sciences* 7 (6): 615–618. doi:10.3923/ajps.2008.615.618

Nduti, N., A. McMillan, S. Seney, M. Sumarah, P. Njeru, M. Mwaniki, and G. Reid. 2016. "Investigating Probiotic Yoghurt to Reduce an Aflatoxin B1 Biomarker among School Children in Eastern Kenya: Preliminary Study." *International Dairy Journal* 63 (December): 124–129. doi:10.1016/j.idairyj.2016.07.014

Nelson, T. S., J. J. McGillivray, T. R. Shieh, R. J. Wodzinski, and J. H. Ware. 1968. "Effect of Phytate on the Calcium Requirement of Chicks." *Poultry Science* 47 (6): 1985–1989. doi:10.3382/ps.0471985

Nikmaram, Nooshin, Sze Ying Leong, Mohamed Koubaa, Zhenzhou Zhu, Francisco J. Barba, Ralf Greiner, Indrawati Oey, and Shahin Roohinejad. 2017. "Effect of Extrusion on the Anti-Nutritional Factors of Food Products: An Overview." *Food Control* 79 (September): 62–73. doi:10.1016/j.foodcont.2017.03.027

Nkafamiya, I. I., A. J. Manji, U. U. Modibbo and H. A. Umaru. 2006. "Biochemical Evaluation of *Casspourea congoensis* (Tunti) and *Nuclea latifloia* (Luzzi). Fruits." *African Journal Biotechnol* 6 (19): 2461–2463.

Nkafamiya, I. I., U. U. Modibbo, A. J. Manji, and D. Haggai. 2007. "Nutrient Content of Seeds of Some Wild Plants." *African Journal of Biotechnology* 6 (14): 1665–1669.

Nkafamiya, I. I., S. A. Osemeahon, U. U. Modibbo, and A. Aminu. 2010. "Nutritional Status of Non-conventional Leafy Vegetables, *Ficus asperifolia* and *Ficus sycomorus*. " *African Journal of Food Science* 4 (3), 104–108. doi:10.5897/AJFS.9000206

Nørgaard, Jan Værum, Navodita Malla, Giuseppe Dionisio, Claus Krogh Madsen, Dan Pettersson, Helle Nygaard Lærke, Rasmus L. Hjortshøj, and Henrik Brinch-Pedersen. 2019. "Exogenous Xylanase or Protease for Pigs Fed Barley Cultivars with High or Low Enzyme Inhibitors." *Animal Feed Science and Technology* 248 (February): 59–66. doi:10.1016/j.anifeedsci.2018.12.005

Nwachukwu, C. N. 2019. "Nutrient, Phytochemical and Anti-Nutrient Evaluation of *Jatropha Tanjorensis* Leaf (Hospital Too Far)." *Journal of Agriculture and Food Sciences* 16 (2): 36. doi:10.4314/jafs.v16i2.4

Oghbaei, M., and J. Prakash. 2016. "Effect of Primary Processing of Cereals and Legumes on Its Nutritional Quality: A Comprehensive Review." Edited by Fatih Yildiz. *Cogent Food & Agriculture* 2 (1). doi:10.1080/23311932.2015.1136015

Ohsaka, Akira, Kyozo Hayashi, and Yoshio Sawai. 1976. *Animal, Plant, and Microbial Toxins*. Springer. http://books.google.ie/books

Ojiyako, O. A. and C. Igwe. 2008. "The Nutritive, Antinutritive and Hepatotoxic Properties of *Trichosanthes anguina* (snake tomato) Fruits from Nigeria." *Pakistan Journal of Nutrition* 7 (1): 85–89.

Okello, Jaspher, John B. L. Okullo, Gerald Eilu, Philip Nyeko, and Joseph Obua. 2017. "Mineral Composition of Tamarindus Indica LINN (Tamarind) Pulp and Seeds from Different Agro-ecological Zones of Uganda." *Food Science & Nutrition* 5 (5): 959–966. doi:10.1002/fsn3.490

Okwu, D. E. 2004. "Phytochemicals and Vitamin Contents of Indigenous Species of South East Nigeria." *Journal of Sustainable Agriculture and Environment* 6 (2): 140–147.

Okwu, D. E. 2005. "Phytochemicals, Vitamins and Mineral Contents of Two Nigerian Medicinal Plants." *International Journal of Molecular Medicine and Advance Sciences* 1 (4): 375–381.

Osabor, V., D. A. Ogar, P. Okafor, E. Egbung. 2009. "Profile of the African Bread Fruit (*Treculia africana*)." *Pakistan Journal of Nutrition*. 8. doi:10.3923/pjn.2009.1005.1008

Otlewski, Jacek, Filip Jelen, Malgorzata Zakrzewska, and Arkadiusz Oleksy. 2005. "The Many Faces of Protease–Protein Inhibitor Interaction." *The EMBO Journal* 24 (7): 1303–1310. doi:10.1038/sj.emboj.7600611

Palai S. 2022. "Berberine," in *Nutraceuticals and health care*, San Diego, CA: Elsevier Academic Press, 359–368.

Panda, SujogyaK, L. P. Padhi, and G. Mohanty. 2011. "Antibacterial Activities and Phytochemical Analysis of Cassia Fistula (Linn.) Leaf." *Journal of Advanced Pharmaceutical Technology & Research* 2 (1): 62. doi:10.4103/2231-4040.79814

Pant, G., M. Panwar, D. Negi, and M. Rawat. 1988. "Spermicidal Activity and Chemical Analysis ofPentapanax Leschenaultii." *Planta Medica* 54 (5): 477–477. doi:10.1055/s-2006-962517

Parajuli, Sabina, Nirmala Tilija Pun, Shraddha Parajuli, and Nirmala Jamarkattel-Pandit. 2019. "Antioxidant Activity, Total Phenol and Flavanoid Contents in Some Selected Medicinal Plants of Nepal." *Journal of Health and Allied Sciences* 2 (1): 27–31. doi:10.37107/jhas.71

Patterson, C. A., J. Curran, and T. Der. 2016. "Effect of Processing on Antinutrient Compounds in Pulses." *Cereal Chemistry* 94 (1): 2–10. doi:10.1094/cchem-05-16-0144-fi

Prasad, K. Nagendra, Lye Yee Chew, Hock Eng Khoo, Kin Weng Kong, Azrina Azlan, and Amin Ismail. 2010. "Antioxidant Capacities of Peel, Pulp, and Seed Fractions of Canarium odontophyllumMiq. Fruit." *Journal of Biomedicine and Biotechnology* 2010: 1–8. doi:10.1155/2010/871379

Pusztai, A., G. Grant, J. C. Stewart, S. Bardocz, S. W. B. Ewen, and A. M. R. Gatehouse. 1993. "Nutritional Evaluation of RAZ-2, a New Phaseolus Vulgaris Bean Cultivar Containing High Levels of the Natural Insecticidal Protein Arcelin 1." *Journal of Agricultural and Food Chemistry* 41 (3): 436–440. doi:10.1021/jf00027a017.

Raes, Katleen, Dries Knockaert, Karin Struijs, and John Van Camp. 2014. "Role of Processing on Bioaccessibility of Minerals: Influence of Localization of Minerals and Anti-Nutritional Factors in the Plant." *Trends in Food Science & Technology* 37 (1): 32–41. doi:10.1016/j.tifs.2014.02.002

Ramakrishna, V., P. Jhansi, and P. Rao. 2006. "Anti-Nutritional Factors During Germination in Indian Bean (Dolichos lablab L.) Seeds.". *World Journal of Dairy and Food Sciences* 1 (1): 6–11.

Ranhotra, G. S., R. J. Loewe, and L. V. Puyat. 1974. "Phytic Acid In Soy And Its Hydrolysis During Breadmaking." *Journal of Food Science* 39 (5): 1023–1025. doi:10.1111/j.1365-2621.1974.tb07301.x

Ray, R. C., and M. Didier. (Eds.). 2014. *Microorganisms and fermentation of traditional foods*. CRC Press.

Salas, C. E., Dittz, D., and M. J. Torres. 2018. Plant Proteolytic Enzymes: Their Role as Natural Pharmacophores. In *Biotechnological applications of plant proteolytic enzymes* (pp. 107–127). Cham: Springer.

Savage, G. P., and L. Mårtensson. 2010. "Comparison of the Estimates of the Oxalate Content of Taro Leaves and Corms and a Selection of Indian Vegetables Following Hot Water, Hot Acid and in Vitro Extraction Methods." *Journal of Food Composition and Analysis* 23 (1): 113–117. doi:10.1016/j.jfca.2009.07.001

Seal, T.. 2016. "Quantitative HPLC Analysis of Phenolic Acids, Flavonoids and Ascorbic Acid in Four Different Solvent Extracts of Two Wild Edible Leaves, Sonchus Arvensis and Oenanthe Linearis of North-Eastern Region in India." *Journal of Applied Pharmaceutical Science*, 157–166. doi:10.7324/japs.2016.60225

Serquiz, Alexandre C., Richele J. A. Machado, Raphael P. Serquiz, Vanessa C. O. Lima, Fabiana Maria C. de Carvalho, Marcella A. A. Carneiro, Bruna L. L. Maciel, Adriana F. Uchôa, Elizeu A. Santos, and Ana H. A. Morais. 2016. "Supplementation with a New Trypsin Inhibitor from Peanut Is Associated with Reduced Fasting Glucose, Weight Control, and Increased Plasma CCK Secretion in an Animal Model." *Journal of Enzyme Inhibition and Medicinal Chemistry* 31 (6): 1261–1269. doi:10.3109/14756366.2015.1103236

Setchell, Kenneth D. R., and Aedin Cassidy. 1999. "Dietary Isoflavones: Biological Effects and Relevance to Human Health." *The Journal of Nutrition* 129 (3): 758S–767S. doi:10.1093/jn/129.3.758s

Shanthakumari, S. and Mohan V. J. Britto. 2008. Nutritional Evaluation and Elimination of Toxic Principles in Wild Yam (Dioscorea spp.). *Tropical and Subtropical Agroecosystems* 8: 319–325.

Sharma A., and D. Arora. 2012. "Phytochemical and pharmacological potential of genus stellaria: a review." *Journal of Pharmacy Research* 5 (7), 3591–3596.

Shin, T., K. Fujikawa, A. Zaw Moe, and H. Uchiyama. 2018. "Traditional Knowledge of Wild Edible Plants with Special Emphasis on Medicinal Uses in Southern Shan State, Myanmar." *Journal of Ethnobiology and Ethnomedicine* 14 (1). doi:10.1186/s13002-018-0248-1

Song N., W. Xu, H. Guan, X. Liu, Y. Wang, X. Nie. 2007. "Several Flavonoids from Capsella Bursa-pastoris (L.) Medic." *Asian Journal of Traditional Medicines* 2 (6), 218–222.

Su, H. and R. Guo. 1986. "Inhibition of acrosine activity of human spermatozoa by saponins of Bulbostermma paniculatum Xtian Yike Daxue Xuebae 7, 225." In Chem. Abstr (Vol. 1008, p. 49459).

Szabo, M. R., D. Radu, S. Gavrilas, D. Chambre, and C. Iditoiu. 2010. "Antioxidant and Antimicrobial Properties of Selected Spice Extracts." *International Journal of Food Properties* 13 (3): 535–545. doi:10.1080/10942910802713149

Tadhani, M. B., V. H. Patel, and Rema Subhash. 2007. "In Vitro Antioxidant Activities of Stevia Rebaudiana Leaves and Callus." *Journal of Food Composition and Analysis* 20 (3–4): 323–329. doi:10.1016/j.jfca.2006.08.004

Tareen, N., Saeed-Ur-Rehman, A. Mushtaq, S. Zabta and B. Tahira. 2016. "Ethnomedicinal utilization of wild edible vegetables in district harnai of balochistan Province-Pakistan." *Pakistan Journal of Botany* 48: 1159–1171.

Taylor, M., R. Kambuou, G. H. Lyons, D. Hunter, E. H. Morgan, A. Quartermain, N. Robert, J. Paofa, A. Lorens, and V. C. Tuia. 2015. "Realizing the Potential of Indigenous Vegetables through Improved Germplasm Information and Seed Systems." *Acta Horticulturae*, no. 1102 (September): 29–42. doi:10.17660/actahortic.2015.1102.3

Tessa, M., K. K. Papadopoulou, and A. Osbourn. 2014. "Metabolic and Functional Diversity of Saponins, Biosynthetic Intermediates and Semi-Synthetic Derivatives." *Critical Reviews in Biochemistry and Molecular Biology* 49 (6): 439–462. doi:10.3109/10409238.2014.953628

Udensi, E. A., F. C. Ekwu, and J. N. Isinguzo. 2007. "Antinutrient Factors of Vegetable Cowpea (Sesquipedalis) Seeds During Thermal Processing." *Pakistan Journal of Nutrition* 6 (2): 194–197. doi:10.3923/pjn.2007.194.197

Umaru, H. A., R. Adamu, D. Dahiru and M. S. Nadro. 2007. *African Journal of. Biotechnology.* 6 (16), 1935–1938.

Vadivel, Vellingiri, and Hans K. Biesalski. 2011. "Effect of Certain Indigenous Processing Methods on the Bioactive Compounds of Ten Different Wild Type Legume Grains." *Journal of Food Science and Technology* 49 (6): 673–684. doi:10.1007/s13197-010-0223-x

Vagadia, B. H., S. Vanga, and V. Raghavan. 2017. "Inactivation Methods of Soybean Trypsin Inhibitor – A Review." *Trends in Food Science & Technology* 64 (June): 115–125. doi:10.1016/j.tifs.2017.02.003.

Vashishth, Amit, Sewa Ram, and Vikas Beniwal. 2017. "Cereal Phytases and Their Importance in Improvement of Micronutrients Bioavailability." *3 Biotech* 7 (1). doi:10.1007/s13205-017-0698-5

van der Ven, Cornelly, Ariette M. Matser, and Robert W. van den Berg. 2005. "Inactivation of Soybean Trypsin Inhibitors and Lipoxygenase by High-Pressure Processing." *Journal of Agricultural and Food Chemistry* 53 (4): 1087–1092. doi:10.1021/jf048577d

Walker, Ronald. 1975. "Naturally Occuring Nitrate/Nitrite in Foods." *Journal of the Science of Food and Agriculture* 26 (11): 1735–1742. doi:10.1002/jsfa.2740261116

Xia, Dao-Zong, Xin-Fen Yu, Zhuo-Ying Zhu, and Zhuang-Dan Zou. 2011. "Antioxidant and Antibacterial Activity of Six Edible Wild Plants (Sonchusspp.) in China." *Natural Product Research* 25 (20): 1893–1901. doi:10.1080/14786419.2010.534093

Yasmin, Azra, Aurang Zeb, Abdul Wajid Khalil, Gholam Mohi-ud-Din Paracha, and Amal Badshah Khattak. 2008. "Effect of processing on anti-nutritional factors of red kidney bean (Phaseolus Vulgaris) Grains." *Food and Bioprocess Technology* 1 (4): 415–419. doi:10.1007/s11947-008-0125-3

Zhang, G., X. Zhicun, G. Yuanyuan, X. Huang, Y. Zou, and T. Yang. 2015. "Effects of Germination on the Nutritional Properties, Phenolic Profiles, and Antioxidant Activities of Buckwheat." *Journal of Food Science* 80 (5). doi:10.1111/1750-3841.12830

14 *Parkia speciosa* Hassk. *A Medicinal Plant from the Wild*

Yusof Kamisah and Hawa Nordin Siti

14.1 INTRODUCTION

The wilderness is a rich source of medicinal plants, which humans across the globe have been using since ancient times. Many native communities around the world still depend on the biodiversity of local plants for their well-being. Even today, the wilderness continues to provide a broad spectrum of medicinal plants for various diseases. For example, *Salvia fruticose* Mill. from the Shouf Biosphere Reserve in Lebanon is used for colic and gastrointestinal problems (Hani et al. 2022), *Piper umbellatum* L. from Chapada Diamantina National Park in Northeastern Brazil is used for liver problems (de Medeiros et al. 2021), and *Parkia speciosa* Hassk. from the forests of the Malaysian peninsula is used for diabetes (Azliza et al. 2012).

P. speciosa, also known as bitter bean or stink bean (Lim 2012), is synonymously identified as *P. harbesonii* Elmer, *P. roxburghii* G. Don (Mansf), *P. biglobosa* sensu auct., *P. macrocarpa* Miq., *P. brunonis* R. Grah. ex Wall., *Mimosa pedunculata* Hunter, and *Inga pyriformis* Jungh. (Lim 2012). It has several local names depending on its geographical location (Table 14.1) (Chhikara et al. 2018; Tunsaringkarn et al. 2012; Kamisah et al. 2013). The plant has many pharmacological properties, including antihypertensive (Kamisah et al. 2017), antidiabetic (Gao et al. 2023), and antihypertrophic effects in cardiac cells (Mustafa et al. 2023). Potential medicinal properties in plants are sought to develop medicines that can save lives.

14.2 GEOGRAPHICAL DISTRIBUTION AND AGROECOLOGY

P. speciosa grows well in Southeast Asia, including Indonesia, Thailand, Malaysia, and Singapore (Figure 14.1). It thrives wild in tropical lowland rainforests in loamy, sandy, and podzolic soils and on riverbanks up to 1,400 m above sea level (Lim 2012). However, today, due to its economic benefit, the plant is cultivated in rural residential areas. It can reach up to 40 m tall and a meter in trunk width (Figure 14.2). It is a perennial flowering plant classified under the family Fabaceae or Leguminosae. Its leaves are alternate and bipinnate (Saleh et al. 2021a), and it bears long, flat fruits, called pods, in bunches (Figure 14.3). Each pod contains up to 20 bright green, edible seeds that have a pungent smell (Kamisah et al. 2013), which is how the plant got its name.

TABLE 14.1
***Parkia speciosa* Local Names**

Local Name	Country
Sataw, sator, sato,	Thailand
Petè, petai, peuteuy	Indonesia
Petai	Malaysia, Singapore
U'pang	Philippines
Yongchak, yongchaa	India

FIGURE 14.1 Geographical distribution of *Parkia speciosa*. Figure constructed using the platform of Mind the Graph (www.mindthegraph.com).

FIGURE 14.2 *Parkia speciosa* plants with bunches of fruits. Photograph by the authors.

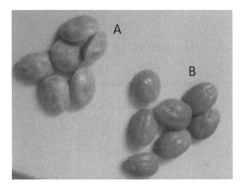

FIGURE 14.3 The seeds of *Parkia speciosa* with (A) and without (B) its coats. Photograph by the authors.

14.3 NUTRITIONAL VALUE

P. speciosa fruits are considered vegetables. The pods, including the seeds, have high nutritional value. The seeds are usually consumed without the pods, either eaten raw, stir-fried with shrimps or anchovies, or pickled in brine (Ong et al. 2011a). The seeds are high in carbohydrates, protein, fat, and minerals, including iron, calcium, phosphorus, and magnesium (Kamisah et al. 2013) (Table 14.2). Moreover, they contain vitamins C, E, and B1 (Ching and Mohamed 2001; Maisuthisakul et al. 2008).

The seeds and seed coats contain a high amount of tannin, a type of polyphenol, compared with other vegetables (Tunsaringkarn et al. 2012). However, eating foods high in tannins is unhealthy for children as tannins can reduce their protein digestion (Gilani et al. 2012). Therefore, the seeds are not advisable to be consumed in large amounts by children.

14.4 USE AS A TRADITIONAL MEDICINE

Different parts of the *P. speciosa* plant are employed in folk medicinal practice. The most common use of the plants is to treat the manifestation of diabetes mellitus (Ong et al. 2011b; Azliza et al.

TABLE 14.2

Nutritional Composition of *Parkia speciosa* Fruits (Kamisah et al. 2013) Parts Per Million (ppm)

Nutrition	Content (per 100 g)
Energy (kcal)	91.0–441.5
Carbohydrate (g)	13.2–52.9
Protein (g)	6.0–27.5
Fiber (g)	1.7–2.0
Fat (g)	1.6–13.3
Ash (g)	1.2–4.6
Calcium (mg)	108.0–265.1
Manganese (ppm)	42.0
Zinc (ppm)	8.2
Copper (ppm)	36.7
Phosphorus (mg)	115.0
Potassium (mg)	341.0
Ascorbic acid (vitamin C) (mg)	19.3
Thiamine (vitamin B1)	0.28
Tocopherol (vitamin E) (mg)	4.15

TABLE 14.3

Traditional Medicinal Use of *Parkia speciosa*

Parts	Indication of Traditional Use	Method of Consumption	Country	Reference
Leaves	Cough	Pounded with uncooked rice and applied to the neck	Malaysia	Ong et al. (2011b)
	Dermatitis	Pounded and rubbed	Indonesia	Roosita et al. (2008)
	Diabetes mellitus	Decocted and drunk	Thailand	Pranprawit (2019)
Seeds	Diabetes mellitus	Boiled or eaten raw	Malaysia, Thailand, and Singapore	Azliza et al. (2012); Siew et al. (2014)
	Kidney problems	Cooked	Malaysia	Samuel et al. (2010)
	To maintain body temperature	Eaten raw	Indonesia	Diliarosta et al. (2022)
	To increase appetite	Eaten raw	Indonesia	Batoro and Siswanto (2017)
Roots	Skin problems	Decocted and applied	Thailand	Srisawat et al. (2016)
	Hypertension and diabetes mellitus	Decocted and drunk	Malaysia	Ong et al. (2011b)
Empty pods	Diabetes mellitus	Decocted and drunk	Thailand	Pranprawit (2019)
	Poisoning due to overeaten of its seeds	Eaten raw	Indonesia	Fitrianti and Partasasmita (2020)

2012; Pranprawit 2019); leaves, seeds, roots, and pods are harvested for this purpose (Table 14.3). The roots and leaves are also taken to heal skin problems (Roosita et al. 2008; Srisawat et al. 2016). Pounded leaves are taken to alleviate cough (Ong et al. 2011c), and the seeds can increase appetite and help to reduce kidney problems (Batoro and Siswanto 2017; Samuel et al. 2010). Root decoction can be used as a remedy for hypertension (Ong et al. 2011b; Eswani et al. 2010). Some of these medicinal benefits have been proven pharmacologically in laboratories. Plants with traditional medicinal uses are an imperative source of modern drugs (Newman and Cragg 2020).

14.5 PHYTOCHEMICAL COMPOSITION

P. speciosa contains many secondary metabolites. The major metabolites are flavonoids, phenolics, and terpenoids, which are detected in almost all parts of the plant (Saleh et al. 2021b). However, phytochemicals have only been detected in the pods and seeds of the plant. The pods contain mostly polyphenolics, while other phytochemical compounds, such as terpenoids, steroids, and cyclic polysulfide, are found in the seeds (Table 14.4). Polyphenolics have several biological activities, including anti-inflammatory and antioxidant (Nani et al. 2021). Trithiolane is the main cyclic polysulfide in the seeds (Liang et al. 2017). It can release hydrogen sulfide, which has a hypotensive effect (Morales-Loredo et al. 2019). The polysulfide is responsible for the unique smell of the seeds (Asikin et al. 2018; Frérot et al. 2008). Boiling the seeds results in the formation of thiazolidine-4-carboxylic (Suvachittanont and Jaranchavanapet 2000), which has anticancer effects (Chen et al. 2008). Other compounds possess additional pharmacological activities, which will be discussed in subsequent subtopics. The molecular structures of major phytochemical compounds in *P. speciosa* are depicted in Figure 14.4.

TABLE 14.4

Major Phytochemicals Found in Pods and Seeds of *P. speciosa*

Type	Parts	Phytochemicals	Reference
Polyphenolics	Pod	Apigenin	Ko et al. (2014); Kamisah et al. (2017)
		Caffeic acid	
		Caftaric acid	
		Catechin	
		Chlorogenic acid	
		Cinnamic acid	
		p-Coumaric acid	
		Didymin	
		Ellagic acid	
		Epicatechin	
		Ferulic acid	
		Gallic acid	
		Hydroxybenzoic acid	
		Kaempferol	
		Malvidin	
		Myricetin	
		Nobiletin	
		Primulin	
		Punicalin	
		Rutin	
		Tangeritin	
		Theaflavin gallate	
		Quercetin	
		Vanillic acid	
Terpenoid	Seeds	Lupeol	Mohd Azizi et al. (2008)
Steroid	Seeds	Campesterol	Mohd Azizi et al. (2008); Rahman et al.
		Stigmasterol	(2012); Jamaluddin et al. (1994);
		Stigmastenone	Jamaluddin et al. (1995)
		Stigmastadienol	
		β-Sitosterol	
Cyclic polysulfide	Seeds	Dimethyl tetrasulfide	Mohd Azizi et al. (2008); Frérot et al.
		Dimethyl trithiolane	(2008); Gmelin et al. (1981); Tocmo
		Dithiabutane	et al. (2016); Salman et al. (2006);
		Dithiapentane	Liang et al. (2017)
		Hexathiolnane	
		Lenthionine	
		Trithiaheptane	
		Trithiahexane	
		Trithiane	
		Trithiolane	
		Tetrathiane	
		Tetrathiepane	
		Tetrathiocane	

(Continued)

TABLE 14.4 (Continued)
Major Phytochemicals Found in Pods and Seeds of *P. speciosa*

Type	Parts	Phytochemicals	Reference
Ester	Seeds	Tetradecyl acetate	Rahman et al. (2012); Salman et al.
		Methyl linoleate	(2006)
		Ethyl linoleate	
		Butyl palmitate	
		Ethyl palmitate	
		Methyl palmitate	
		Methyl laurate	
		Dodecyl acrylate	
		Methyl hexadecanoate	
		Butyl stearate	
		Propanoic acid, 3,30-thiobis-didodecyl ester	
		Linoleaidic acid methyl ester	
Alcohol	Seeds	Hexadecatetraenol	Mohd Azizi et al. (2008); Salman et al.
		Ethyl-nonanol	(2006)
		Tridecanol	
Ketone	Seeds	Nonadecanone	Mohd Azizi et al. (2008); Salman et al.
		Pyrrolidinone	(2006)
		Cyclodecanone	
Alkane	Seeds	Tetradecane	Rahman et al. (2012)
Aldehyde	Seeds	Decenal	Mohd Azizi et al. (2008); Salman et al.
		Cyclo-decanone-decadienal	(2006); Rahman et al. (2012); Frérot
		Pentanal	et al. (2008)
		Tetradecanal	
		Pentadecanal	
		Hexadecanal	
Fatty acid	Seeds	Arachidonic acid	Mohd Azizi et al. (2008); Salman et al.
		Linoleic acid	(2006); Rahman et al. (2012)
		Squalene	
		Lauric acid	
		Stearic acid	
		Stearoic acid	
		Eicosanic acid	
		Oleic acid	
		Palmitic acid	
		Myristic acid	
		Undecanoic acid	
		Stearolic acid	
		Hydnocarpic acid	
		Eicosanoic acid	
		Elaidic acid	
Chromanol	Seeds	Tocopherol	Mohd Azizi et al. (2008); Salman et al.
			(2006)

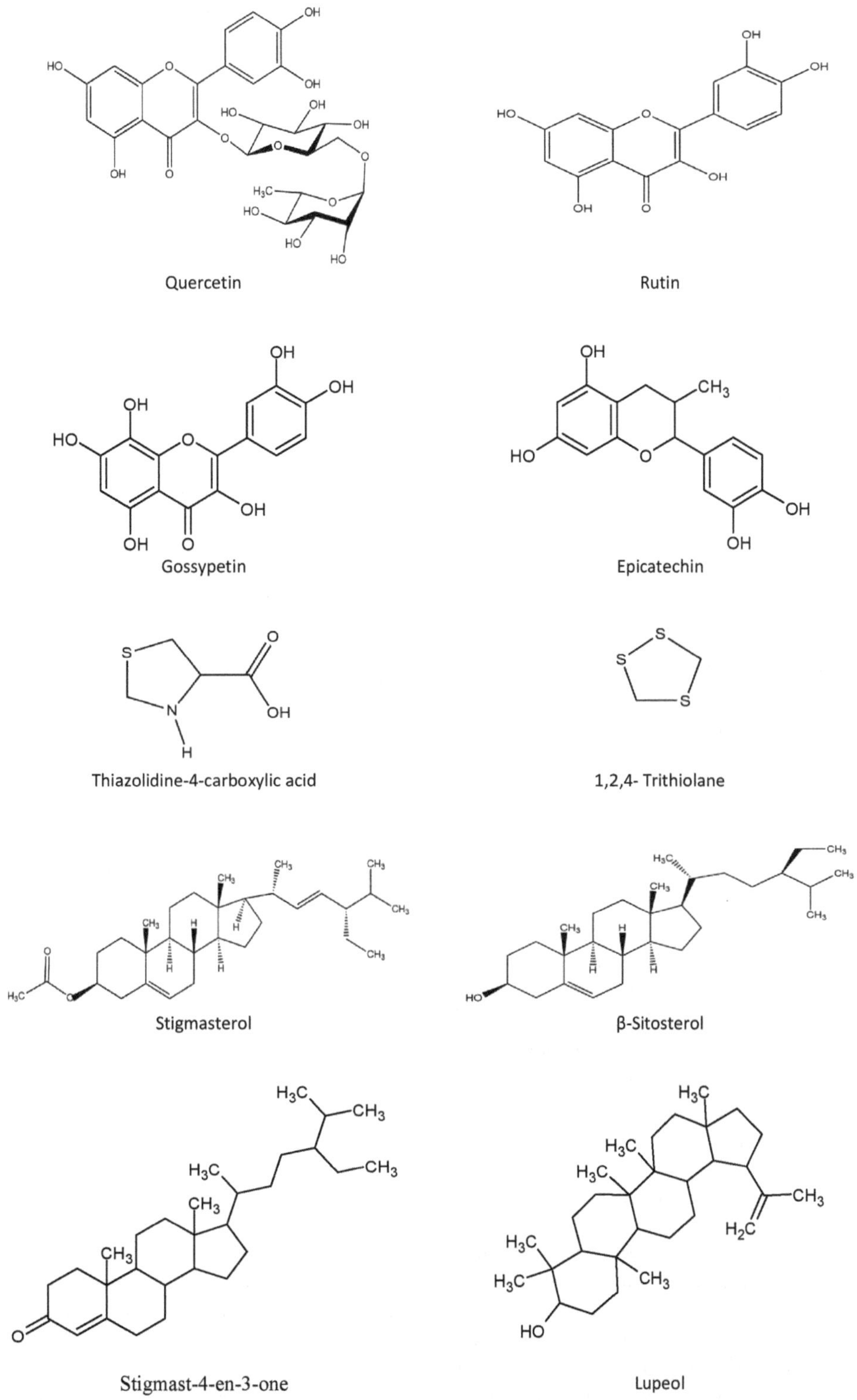

FIGURE 14.4 Major phytochemicals in *Parkia speciosa*. Chemical structures drawn by the authors using ChemSketch.

14.6 PHARMACOLOGICAL PROPERTIES

14.6.1 ANTIOXIDANT AND ANTI-INFLAMMATORY ACTIVITIES

Almost all plants possess antioxidant and anti-inflammatory activities. Antioxidant activity refers to the capability of a substance or plant extract to remove free radical – chemical compounds with unpaired electrons – and protect cells from attacks by free radicals, thereby alleviating oxidative stress. The anti-inflammatory property indicates the ability of a compound to reduce inflammation, usually by regulating immune cells and cytokine production. Inflammation and oxidative stress have a prominent part in the pathogenesis of almost all diseases.

Multiple studies have reported the antioxidant activity of *P. speciosa*. Almost all parts of the plant possess antioxidant activity. The combination of pod and seed extract exhibits a greater antioxidant activity than that of the seed or leaf extract (Zaini and Mustaffa 2017; Kamisah et al. 2013). In in vivo and in vitro studies, the plant extracts have been shown to diminish lipoperoxidation products and generation of reactive oxygen species (a type of free radical) by radical removal and iron-chelating activities (Siti et al. 2021b; Gui et al. 2019a; Al Batran et al. 2013; Ko et al. 2014). The extract suppresses oxidative stress by reducing NADPH oxidase, an enzyme that generates superoxide anion, and increasing antioxidant capacities, namely superoxide dismutase (Siti et al. 2021b), glutathione peroxidase, and catalase (Gao et al. 2023; Al Batran et al. 2013) activities, to combat free radicals.

Addition of powdered pod (200–800 mg/L) into frying oil was demonstrated to reduce the content of hepatic lipoperoxidation products in mice that consumed the oil (Ramadani 2012). The beneficial effect of the powdered pod was attributable to its high phenolic content (Ramadani 2012; Wonghirundecha et al. 2014), which reduced free radical formation in the oil during frying. Frying produces reactive oxygen species in vegetable oils which are harmful to health, such as causing elevation of blood pressure when chronically consumed (Ng et al. 2012).

P. speciosa extract exerts its anti-inflammatory activities by suppressing the activation of mitogen-activated protein kinase (MAPK) pathway. In hypertrophied cardiomyocytes, the empty pod ethanol extract and its ethyl acetate fraction attenuated MAPK subfamilies – c-Jun N-terminal protein kinase [JNK], extracellular signal-regulated protein kinase [ERK], and p38 kinase – activation (Siti et al. 2021b; Mustafa et al. 2023). The inhibition decreased the expression of downstream inflammatory biomarkers, including cyclooxygenase-2, vascular cell adhesion molecule, and inducible nitric oxide synthase in H9c2 cardiomyocyte cell and human umbilical vein endothelial cells. It also prevents nuclear factor kappa B (NF-κB) pathway activation. Upon inflammatory stimulation, inducible nitric oxide synthase produces nitric oxide, which is inhibited by *P. extract* pod extract (Gui et al. 2019b; Mustafa et al. 2018). In vivo, the combination of *P. speciosa* pods and seeds was reported to decrease swelling (edema) induced by lipopolysaccharide injection in rat paws. A reduction in total white cell count and inflammatory cell infiltration at the site of inflammation was noted (Norazlin et al. 2022).

The data suggest that the plant extract ameliorates pathological conditions, partly by diminishing oxidative stress and inflammation. Both activities are conferred by the existence of bioactive substances in the extract, including quercetin, rutin, and gallic acid.

14.6.2 ANTIDIABETIC ACTIVITY

Undesirable elevated fasting blood sugar (hyperglycemia) over 7 mmol/L is diagnosed as diabetes mellitus. It is categorized into types 1 and 2. Type 1 occurs due to the destruction of pancreatic β-cells, which are responsible for synthesizing and releasing insulin, leading to total insulin deficiency; type 2 is due to insulin resistance (American Diabetes Association 2015). This disease claimed approximately two million lives globally in 2019 (WHO 2023). It must be treated to prevent the complications of cardiovascular problems, kidney failure, and blindness. Based on the properties of *P. speciosa*, including its antioxidant and anti-inflammatory effects, this plant can be used to treat diabetic vasculopathy (a blood vessel disorder) (Azemi et al. 2022).

Two enzymes – α-glucosidase and α-amylase – participate in carbohydrate metabolism. The former is normally present in pancreatic juices, where it hydrolyzes carbohydrates to disaccharide

TABLE 14.5

Effects of *Parkia speciosa* Extracts and Its Isolated Compounds on Diabetes

Extract/Compound	Model	Dose and Mode of Administration	Findings	Reference
Seed aqueous/ethanol extract	In vitro	-	↓ α-amylase	Sonia et al. (2018)
Empty pod ethanol extract	In vitro	-	↓ α-glucosidase	Tunsaringkarn et al. (2008)
Empty pod aqueous and seed aqueous extracts	In vitro	-	↓ α-glucosidase Empty pod extract inhibits 15 times higher than seed extract	Tunsaringkarn et al. (2009); Chankhamjon et al. (2010)
Pod, leaf, and seed ethanol extract	In vitro	-	The rind extract had the lowest IC_{50} for α-glucosidase inhibition.	Fitrya et al. (2019)
Empty pod ethanol extract (from six locations)	In vitro	-	↓ α-glucosidase Different activities depending on location	Saleh et al. (2021a)
Empty pod ethanol extract	High-fat/ streptozotocin-induced diabetic rats	100 and 400 mg/kg/d (p.o.) for 30 days	↓ blood glucose ↑ insulin	Gao et al. (2023)
Seed chloroform and empty pod diethyl ether, petroleum ether, chloroform, dichloromethane, methanol, and ammoniacal chloroform extracts	Alloxan-induceddiabetic rats	50–1000 mg/kg (p.o.) once	Chloroform extract: ↓ blood glucose The activity in empty pods was lower than that of the seeds. No hypoglycemic effect observed with other extracts.	Jamaluddin and Mohamed (1993)
Combination of β-sitosterol (66%) and stigmasterol (34%) (empty pod + seed chloroform extract)	Alloxan-induced diabetic rats	100 mg/kg (p.o.) once	Combination: ↓ blood glucose Individual compounds: No hypoglycemic effect.	Jamaluddin et al. (1994)
Empty pod petroleum ether, chloroform, dichloromethane, ethyl acetate, ammoniacal chloroform, and methanol extracts and stigmast-4-en-3-one (chloroform extract)	Alloxan-induced diabetic rats	200 mg/kg (p.o.) once. Stigmast-4-en-3-one:100 mg/kg (p.o.) once	Chloroform extract: ↓ blood glucose Stigmast-4-en-3-one: ↓ blood glucose	Jamaluddin et al. (1995)

p.o., per oral; ↓, reduced; ↑, increased.

sugars, while the latter converts disaccharides into glucose before their absorption in the small intestines (Alqahtani et al. 2019). In in vitro studies, *P. speciosa* seed aqueous (water)/ethanol extract inhibited α-amylase activity (Sonia et al. 2018), while its seed aqueous and ethanol extracts suppressed the activity of α-glucosidase (Tunsaringkarn et al. 2008; Tunsaringkarn et al. 2009) (Table 14.5). Empty pod aqueous extract inhibits α-glucosidase 15-fold higher than that of seed

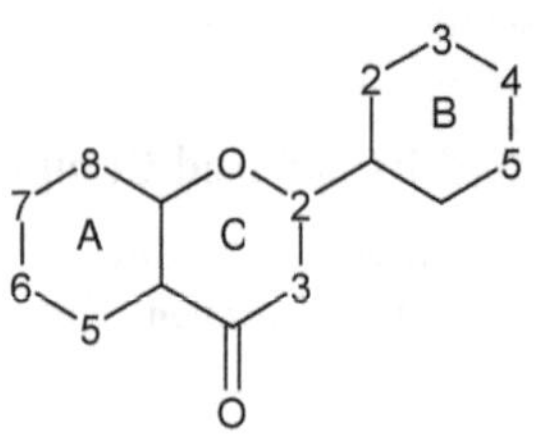

FIGURE 14.5   Backbone of flavonoid. Chemical structures drawn by the authors using ChemSketch.

extract (Chankhamjon et al. 2010; Tunsaringkarn et al. 2009). However, Fitrya et al. (2019) demonstrated that the pod extract had the best suppressive activity against α-glucosidase, followed by the leaf and seed extracts. The findings suggest that nearly all parts of the plant exhibit antidiabetic effects.

The α-glucosidase-suppressing activity of the empty pods varies according to the location of samplings, despite the presence of similar bioactive compounds (e.g., gallic acid, epigallocatechin, gossypetin, and myricitrin) in the pods (Saleh et al. 2021a). This discrepancy could be due to climatic factors in different geographical regions, such as solar radiation, temperature, humidity, and rainfall, which alter the synthesis of plant secondary products (Sampaio et al. 2016). Lim et al. (2022) reported specific structural criteria for flavonoids (a group of polyphenols) that determine their selective inhibition of α-glucosidase and α-amylase. For suppressing α-glucosidase, the existence of a covalent bond between C2 and C3 as well as hydroxyl groups at B3 and C3 is crucial, while for α-amylase-inhibiting activity, the existence of hydroxyl groups at A5 and B3 is pivotal (Figure 14.5). According to this requirement, only gossypetin fits the criteria for inhibiting α-glucosidase. A higher concentration of gossypetin present in the samples was associated with higher α-glucosidase-inhibiting activity (Saleh et al. 2021a).

Various extracts of the seeds and empty pods of *P. speciosa*, namely, chloroform, methanol, petroleum ether, dichloromethane, ammoniacal chloroform, diethyl ether, and ethyl acetate, were evaluated for hypoglycemic effects in rats. Only chloroform extract produced a prominent hypoglycemic effect (Jamaluddin and Mohamed 1993; Jamaluddin et al. 1994, 1995). Stigmast-4-en-3-one (Jamaluddin et al. 1995), stigmasterol, and β-sitosterol (Jamaluddin et al. 1994) were isolated from the chloroform extract. In diabetic rats, stigmast-4-en-3-one and the combination of stigmasterol and β-sitosterol decreased serum glucose (Jamaluddin et al. 1994, 1995). However, when β-sitosterol and stigmasterol were assessed individually, no reduction in blood glucose was noted (Jamaluddin et al. 1994), suggesting that the compounds act synergistically to exert positive effects. Both compounds increase the survival of pancreatic β cells by reducing apoptosis (programmed cell death) and oxidative stress. In addition, they promote insulin release by increasing the expression of glucose transporter protein 4 (Babu and Jayaraman 2020; Ward et al. 2017). This protein is the major glucose transporter to muscle and fat cells (Grunwald et al. 2022). The antidiabetic effect of stigmast-4-en-3-one was also described elsewhere (Alexander-Lindo et al. 2004). However, an in-depth study of the compound has not yet been conducted.

Collectively, the findings suggest that consumption of *P. speciosa* seeds or empty pods could wane postprandial blood sugar levels due to the suppression of both α-glucosidase and α-amylase. The protective effects of stigmast-4-en-3-one, stigmasterol, and β-sitosterol in diabetes should be probed further. The substances are promising candidates for antidiabetic agents.

14.6.3   HYPOCHOLESTEROLEMIC AND ANTIOBESITY ACTIVITIES

Studies investigating the impacts of *P. speciosa* extract on hyperlipidemia are still lacking; only three published studies were found. Empty pod and leaf extracts exhibited promising hypocholesterolemic effects (Table 14.6). The plant extract decreased low-density lipoprotein (LDL), triglyceride, and total cholesterol levels and elevated high-density lipoprotein (HDL) levels in experimental

TABLE 14.6
Effects of *Parkia speciosa* Extracts on Hyperlipidemia in Rats

Extract/ Compound	Model	Dose and Mode of Administration	Findings	Reference
Empty pod ethanol extract	Hypercholesterolemia induced by high-cholesterol diet in rats	300, 400, and 500 mg/kg/d, orally for 2 weeks	On days 28 and 35: ↓ TC	Tandi et al. (2020)
Leaf ethanol extract	Dyslipidemia induced by high-fat diet in rats	100, 200, and 400 mg/kg/d, orally for 2 weeks	All doses: ↓ TC ↑ HDL 200 and 400 mg/kg: ↓ TG ↓ LDL	Susilo et al. (2020)
Empty pod ethanol extract	Diabetes induced by streptozotocin and high-fat diet in rats	100 and 400 mg/kg/d, orally for 30 days	↓ TC ↑ HDL ↓ TG ↓ LDL	Gao et al. (2023)

TC, total cholesterol; HDL, high-density lipoprotein; TG, triglyceride; LDL, low-density lipoprotein; ↓, reduced; ↑, elevated.

hyperlipidemia (Tandi et al. 2020; Susilo et al. 2020), which is a desirable lipid profile. In addition, the protective effects of the plant were noted in diabetic rodents given a high-fat diet (Gao et al. 2023). However, the mechanisms of action of the hypocholesterolemic effects were not investigated; therefore, further studies should explore this aspect.

Obesity is a health problem that manifests as excessive fat accumulation in the body. It may increase the risk of acquiring other diseases such as diabetes and cardiovascular disease. Obesity develops when calorie intake outweighs its expenditure. Therefore, one of the strategies to curb obesity is to decrease calorie intake. In the human body, pancreatic lipase converts triglycerides into fatty acids and glycerol (de la Garza et al. 2011,) thereby reducing the absorption of lipids. An in vitro study reported that *P. speciosa* seed hydromethanolic extract at 500 µg/mL reduced pancreatic lipase activity by more than 89% (Sonia et al. 2018). This finding suggests that consumption of the plant seeds could be beneficial in treating obesity by reducing calorie intake in addition to its suppressive effects on α-glucosidase and α-amylase. The bioactive substances that are responsible for this activity should be isolated and investigated further.

14.6.4 ANTIHYPERTENSIVE ACTIVITY

High blood pressure, or hypertension, is confirmed when the systolic over diastolic blood pressure is measured to be greater than 130/80 mmHg on three different occasions. Several factors like genetics, age, and dietary lifestyle play a part in the pathogenesis of hypertension (Ng et al. 2014; Rossier et al. 2017). Untreated hypertension can be complicated by stroke, ventricular hypertrophy, and heart failure (Biswas et al. 2003).

Even though *P. speciosa* is used to reduce hypertension in folk medicine, its effects have not been extensively studied. The seed extract and hydrolyzed peptides from seeds were noted to inhibit the activity of angiotensin-converting enzyme (ACE) in vitro (Siow and Gan 2013; Khalid and Babji 2018) (Table 14.7). ACE metabolizes angiotensin I into angiotensin II (Ang II), which is a vasoconstrictor, leading to an elevation of blood pressure; therefore, suppression of ACE by the extract could presumably lower blood pressure. This was confirmed by a study on hypertensive rats that were administered *P. speciosa* empty pod extract. A reduction in systolic blood pressure and cardiac ACE activity was contemplated in the animals, and the decrease in plasma nitric oxide was restored (Kamisah et al. 2017). Nitric oxide mediates vasodilation, which decreases blood pressure

TABLE 14.7

Impacts of *Parkia speciosa* Extract on Vasoactive Substances and Blood Pressure

Extract/Compound	Model	Dose and Mode of Administration	Findings	Reference
Hydrolyzed peptides from seeds	In vitro	-	↓ ACE	Siow and Gan (2013)
Seeds	In vitro	-	↓ ACE	Khalid and Babji (2018)
Empty pod methanol extract	L-NAME-induced hypertension in rats	800 mg/kg/d, orally for 8 weeks	↓ systolic BP ↑ plasma NO ↓ cardiac ACE	Kamisah et al. (2017)

ACE, angiotensin-converting enzyme; L-NAME, N(G)-nitro-L-arginine methyl ester; BP, blood pressure; NO, nitric oxide; ↓, reduced; ↑, increased.

(Norsidah et al. 2013). The hydrogen sulfide released by the plant seeds (Tocmo et al. 2016) also possesses vasodilatory effects, like nitric oxide (Wu et al. 2018), which may contribute to its hypotensive activity.

Based on findings from a limited number of studies, *P. speciosa* seeds and empty pods can lower blood pressure, likely by weakening the activity of ACE, restoring the fall in plasma nitric oxide, and decreasing oxidative stress in the heart. Other comprehensive mechanistic actions of the antihypertensive activity of the plant remain vague.

14.6.5 ANTICARDIAC HYPERTROPHIC ACTIVITY

As previously mentioned, hypertension can lead to the development of cardiac hypertrophy (heart enlargement), which may lead to heart failure. One of the features of heart failure is a reduction in the blood-pumping efficiency of the heart. ACE activity in the heart is positively correlated with the size and thickness of cardiac muscle fibers (Kamisah et al. 2015), indicating the adverse brunt of ACE on cardiac function. Increased ACE activity increases the concentration of Ang II, which promotes cardiac hypertrophy (Siti et al. 2021a).

Only two studies examined the impacts of *P. speciosa* on cardiac hypertrophy in vitro (Table 14.8). The first study explored the impacts of empty pod ethanol extract on cell hypertrophy induced by Ang II in cardiomyocyte cells. The authors reported the beneficial effects of the extract in reducing cell hypertrophy and the B-type natriuretic peptide level, an indicator of cell enlargement. Quercetin and rutin were detected in the extract (Siti et al. 2021b). Another study examined the effects of empty pod ethanol extract and its fractions, namely ethyl acetate, hexane, chloroform, and methanol, using the same model. The authors found that the ethyl acetate fraction displayed the best antihypertrophic properties based on its ability to decrease BNP levels (Mustafa et al. 2023). In both studies, the protective effects of the extract likely functioned by suppressing the activation (phosphorylation) of MAPK signaling pathways – ERK, JNK, and p38 kinase – which are implicated in the development of cardiac hypertrophy (Siti et al. 2021b; Mustafa et al. 2023). The fraction (ethyl acetate) decreased the calcineurin expression, thereby inhibiting nuclear factor of activated T-cells C3 (NFATC3) activation. The sequence of events resulted in decreased formation of the NFATC3 and GATA–binding protein 4 (GATA4) complex (Mustafa et al. 2023). NFATC3 is a protein that has a role in immunity, and it is activated by calcineurin. GATA4 is a principal modulator of cardiac hypertrophy-related transcription factors (Yoon et al. 2020). These functions suggest that the plant extract inhibits cardiomyocyte hypertrophy by preventing immune-mediated inflammatory events. Nineteen compounds were putatively detected in the fraction, including picein, gossypetin, quercetin, herbacetin, and kaempferol (Mustafa et al. 2023). Further investigation is needed to confirm the bioactive components that are responsible for the observed effects.

TABLE 14.8

Impacts of *Parkia speciosa* on Cardiomyocyte Hypertrophy Induced by Exposure to Angiotensin II

Extract/Compound	Model	Concentration	Outcomes	Reference
Empty pod ethanol extract	Ang II-induced cardiomyocyte hypertrophy in vitro	12.5, 25, and 50 μg/mL	↓ cardiomyocyte size ↓ BNP ↓ p-ERK protein ↓ p-p38 protein ↓ p-JNK protein	Siti et al. (2021b)
Empty pod ethyl acetate fraction of ethanol extract	Ang II-induced cardiomyocyte hypertrophy in vitro	6.25, 12.5, and 25 μg/mL	↓ cardiomyocyte size ↓ BNP ↓ ANP gene ↓ protein content ↓ p-ERK protein ↓ p-p38 protein ↓ p-JNK protein ↓ p-GATA4 protein ↓ calcineurin protein ↓ NFATC3 gene	Mustafa et al. (2023)

ANP, atrial natriuretic peptide; Ang II, angiotensin II; BNP, B-type natriuretic peptide; p-GATA, phosphorylated GATA-binding protein 4; NFATC3, calcineurin-nuclear factor of activated T-cells C3; p-p38, phosphorylated p38 kinase; p-JNK, phosphorylated c-Jun N-terminal kinase; p-ERK, phosphorylated extracellular signal-related kinase; ↓, reduced; ↑, increased.

The empty pod extract of *P. speciosa* exhibits antihypertrophic activities by blocking MAPK and calcineurin-NFATC3 signaling pathways activation. It may also induce a decrease in cellular oxidative stress and inflammation. The bioactive compounds that possess these effects must be isolated and evaluated for their hypertrophy-lowering effects, which could be developed as antihypertrophic drugs.

14.6.6 ANTICANCER AND ANTIANGIOGENESIS

Consumption of *P. speciosa* raw seeds is believed to be related to a reduction in the incidence of esophageal cancer in Southern Thais (Vatanasapt et al. 2002); however, no clinical studies have been executed to date to examine this direct relationship. The effectiveness of the plant extract in cancer was exhibited in in vitro studies. The plant pod extracts were toxic to breast cancer cells, evidenced by inhibited cell growth or viability, and they were selectively non-toxic to normal cells. Moreover, the pod methanol extract suppressed the viability of colon and liver cancer cells (Aisha et al. 2009) (Table 14.9). Its ethyl acetate fraction selectively repressed the growth of breast cancer cells, with less cytotoxic impacts on colorectal cancer cells (Aisha et al. 2012).

To facilitate tumor growth, angiogenesis will be activated where new blood vessels are stemmed from existing blood vessels. If tumor angiogenesis is suppressed, tumor spread to other organs can be prevented. *P. speciosa* pod methanol extract exhibited antiangiogenic activity evidenced by a reduction in vascular endothelial growth factor expression in human umbilical vein endothelial cells (normal cells) (Aisha et al. 2012). Increased expression of the growth factor indicates new growth of blood vessels.

Activation of Epstein–Barr virus is linked to the formation of many cancers and autoimmune diseases (Kerr 2019). The pod extract inhibits virus activation (Murakami et al. 2000). Therefore, the pod extract indicates the ability of the extract to suppress tumor-promoting activity.

One of the processes in the development of cancer is mutagenicity, which refers to the ability of certain chemicals (mutagens), either in our food or environment, to induce permanent changes in our DNA (mutations). Chemicals or agents that could prevent this process are described as possessing antimutagenic properties. A study on Thai cuisine found that *P. speciosa* seeds exhibited antimutagenic effects when tested in a laboratory (Tangkanakul et al. 2011). A study reported that

TABLE 14.9

Screening of Anticancer Effects in *P. speciosa* Extracts

Extract/Compound	Model	Findings	Reference
Pod methanol extract	In vitro	Cytotoxic to HepG2 (liver cancer), HCT-116 (colorectal cancer), T47D (breast cancer) and MCF-7 (breast cancer) cells	Aisha et al. (2009)
Pod methanolic ethyl acetate fraction	In vitro	Cytotoxic to MCF-7 breast cancer cells	Aisha et al. (2012)
Pod methanol extract	In vitro	Weak inhibition of Epstein-Barr virus activation	Murakami et al. (2000)
Lectin from seeds	In vitro	↑ lymphocyte mitogenic activity ↑ incorporation of ^{3}H-thymidine into lymphocyte DNA	Suvachittanont and Jaranchavanapet (2000)
Seed homogenate	In vitro	Moderate antimutagenic via inhibition of Trp-P-1 induction in Ames test	Tangkanakul et al. (2011)

lectin (a carbohydrate-binding protein) isolated from the seeds increased lymphocyte (white blood cell) mitogenic activity, evident from increased introduction of ^{3}H-thymidine (a nucleoside in DNA) into lymphocyte DNA (Suvachittanont and Jaranchavanapet 2000). Lymphocytes are involved in our body's self-defense mechanism against diseases, including cancer. When mitogenicity increases, the division of lymphocytes increases, augmenting our immune system. Moreover, thiazolidine-4-carboxylic acid, which is formed after thermal treatment, such as boiling, may contribute to the anticancer property of the seeds (Suvachittanont et al. 1996). However, studies probing the impacts of *P. speciosa* components on tumor growth are still inadequate.

Collectively, *P. speciosa* extract may protect our body by boosting the immune response through its mitogenic properties and preventing the development of cancers via its antimutagenic activity. Nonetheless, in vivo studies are required to confirm its anticancer properties. Furthermore, its protective mechanisms at the molecular level should be pursued.

14.6.7 ANTIMICROBIAL ACTIVITY

Bacteria can cause various infections, such as pneumonia (lung infection), gastroenteritis (inflammation of the gastrointestinal tract), and meningitis (infection and inflammation of the membrane and fluid enclosing the brain and spinal cord). *P. speciosa* extracts were screened for their antibacterial effects, which are tabulated in Table 14.10. The raw seed extract is active against *Aeromonas hydrophila, Helicobacter pylori, Staphylococcus* species (*S. aureus, S. agalactiae,* and *S. anginosus*), and *Vibrio parahaemolyticus* (Musa et al. 2008; Sakunpak and Panichayupakaranant 2012; Uyub et al. 2010). Boiling broadened the seeds' spectrum of activity to further inhibit the growth of *Salmonella typhimurium* (Muhialdin et al. 2020). When the seeds were fermented, the activity was higher than that of boiled seeds, evidenced by inhibition of *Escherichia coli* and *Listeria monocytogenes* growth. The lower effect of the boiled extract is likely due to the denaturation of antimicrobial bioactive substances during heating. Seven unidentified peptides, which might have a role in the seeds' antibacterial properties, were detected in fermented seed extract (Muhialdin et al. 2020). Leaf aqueous extract also exhibits inhibitory activity against *S. aureus, Pseudomonas aeruginosa, Bacillus subtilis,* and *E. coli* (Ravichandran et al. 2019).

The plant pod and leaf extracts are effective against *S. aureus, E. coli, P. aeruginosa,* and *B. subtilis* (Ravichandran et al. 2019; Fatimah 2016; Hasim et al. 2015). With the advancements in nanotechnology, silver nanoparticles were developed as a carrier tool to deliver drugs to their targets (Fatimah 2016). The use of silver nanoparticles improved the antibacterial property of the plant leaf and pod extracts (Fatimah 2016; Ravichandran et al. 2019). The characteristics of nanoparticles – smaller particles with greater surface area – may improve the extracts' bacteria-killing (bactericidal) effects. Silver nanoparticles can invade bacterial cell walls and membranes, which may result in cell

TABLE 14.10

Common Bacterial Infections and Antibacterial Properties of *Parkia speciosa* Extracts

Bacteria	Common Infection	Antibacterial Effects of *P. speciosa*	Reference
Aeromonas hydrophila	Diarrhea in kids and individuals with weak immune system	+ (seed water extract)	Musa et al. (2008)
Bacillus subtilis	Not harmful to humans	+ (leaf water extract) + (leaf water extract silver nanoparticle) + (seed ethanol extract)	Ravichandran et al. (2019); Ghasemzadeh et al. (2018)
Citrobacter freundii	Urinary tract infection, diarrhea, and pneumonia	− (seed water extract)	Musa et al. (2008)
Edwardsiella tarda	Gastroenteritis	− (seed water extract)	Musa et al. (2008)
Escherichia coli	Gastroenteritis, urinary tract infection, meningitis in newborns	− (seed water extract) + (seed ethyl acetate extract) + (seed ethanol extract) + (fermented seed extract) + (leaf water extract) + (leaf water extract silver nanoparticle) + (pod water extract) + (pod water extract silver nanoparticle) + (pod ethyl acetate extract)	Musa et al. (2008); Sakunpak & Panichayupakaranant (2012); Muhialdin et al. (2020); Fatimah (2016); Hasim et al. (2015); Ghasemzadeh et al. (2018); Ravichandran et al. (2019)
Helicobacter pylori	Gastritis and duodenitis	+ (seed methanol, chloroform, and petroleum ether extracts)	Sakunpak & Panichayupakaranant (2012); Uyub et al. (2010)
Listeria monocytogenes	Listeriosis with symptoms of sepsis (a life-threatening body reaction to infection), meningitis, and encephalitis	+ (boiled seed extract) + (seed ethanol extract)	Muhialdin et al. (2020); Ghasemzadeh et al. (2018)
Pseudomonas aeruginosa	Pneumonia, urinary tract infection, gastrointestinal infection, and skin and soft tissue infections	+ (pod water extract) + (pod water extract silver nanoparticle) + (leaf water extract) + (leaf water extract silver nanoparticle) + (pod methanol extract) + (seed ethanol extract)	Fatimah (2016); Sonia et al. (2018); Ghasemzadeh et al. (2018); Ravichandran et al. (2019)
Salmonella typhi	Typhoid	− (seed ethyl acetate extract)	Sakunpak & Panichayupakaranant (2012)
Salmonella typhimurium	Gastroenteritis	− (seed ethyl acetate extract) + (boiled seed extract) + (seed ethanol extract)	Muhialdin et al. (2020); Ghasemzadeh et al. (2018)
Shigella sonnei	Shigellosis (fever, diarrhea, and abdominal pain	− (seed ethyl acetate extract)	Sakunpak & Panichayupakaranant (2012)

(Continued)

TABLE 14.10   (Continued)

Bacteria	Common Infection	Antibacterial Effects of *P. speciosa*	Reference
Staphylococcus aureus	Skin infection	+ (seed water extract) + (seed ethanol extract) + (leaf water extract + (leaf water extract silver nanoparticle) + (pod methanol extract) + (pod water extract) + (pod water extract silver nanoparticle) + (pod ethyl acetate extract)	Musa et al. (2008); Muhialdin et al. (2020), Sonia et al. (2018); Fatimah (2016); Hasim et al. (2015); Ghasemzadeh et al. (2018); Ravichandran et al. (2019)
Streptococcus agalactiae	Pneumonia, meningitis in newborns	+ (seed water extract)	Musa et al. (2008)
Streptococcus anginosus	Pharyngitis and bacteremia	+ (seed water extract)	Musa et al. (2008)
Vibrio alginolyticus	Otitis (ear infection), conjunctivitis (eye infection) and wound infection	− (seed water extract)	Musa et al. (2008)
Vibrio parahaemolyticus	Acute gastroenteritis, and wound infection	+ (seed water extract)	Musa et al. (2008)
Vibrio vulnificus	Acute gastroenteritis, necrotizing wound infection, and invasive sepsis (in individuals with compromised immune system)	− (seed water extract)	Musa et al. (2008)

+, antibacterial activity; −, no antibacterial activity.

death (Yin et al. 2020). Upon invasion, nanoparticles release their load (extract) into the bacteria, increasing the effectiveness of the antibacterial properties of the extract.

In addition to bacteria, bark methanol extract of *P. speciosa* demonstrated antifungal activity against *Gloeophyllum trabeum* but was ineffective against *Pycnoporus sanguineus* (Kawamura et al. 2011). A single study was conducted examining the antifungal properties of the plant; therefore, it is uncertain whether the plant has effective inhibitory activity against other fungi. Furthermore, the bark of the plant was screened for its antimalarial activity; however, no activity was detected (Leaman et al. 1995).

In summary, *P. speciosa* demonstrates antimicrobial activities against many bacteria. The seeds, leaves, and pods could be developed as cheaper, alternative antibiotics. However, additional studies are essential to explicate the antibacterial mechanisms of action for clinical use.

14.6.8   ANTIULCER ACTIVITY

Only two studies explored the influence of the plant extracts on gastric ulcer formation (Table 14.11). Al Batran et al. (2013) reported that a pretreatment dose of 50–400 mg/kg leaf ethanol extract dose-dependently diminished ethanol-induced ulcer area in rats. The extract reduced gastric acidity and increased mucus secretion (Al Batran et al. 2013), which produces a mucosal protective barrier that protects the gastric mucosa against injury (Iwabuchi et al. 2013). Neutrophil infiltration into the gastric mucosa was also decreased, evidenced by reduced gastric lesions in rats treated with the extract (Al Batran et al. 2013). Neutrophils are phagocytic white blood cells that have an essential role in

TABLE 14.11

Impacts of *Parkia speciosa* Extract on Gastric Ulcer in Animal Studies

Extract/ Compound	Model	Dose and Mode of Administration	Findings in Stomach	Reference
Leaf ethanol extract	Gastric ulcer induced by ethanol in rats	50, 100, 200, and 400 mg/ kg orally (pretreatment) once	↓ ulcer area ↓ acidity ↑ mucus secretion ↓ neutrophil infiltration ↓ apoptosis ↓ lipid peroxidation ↑ SOD ↑ GSH	Al Batran et al. (2013)
Seed ethanol extract	Gastric ulcer induced by indomethacin in rats	100, 200, and 400 mg/kg (pretreatment) for 14 days	↓ ulcer index ↓ mucosal lesion ↓ collagen ↓ fibrosis ↓ acidity	Maria et al. (2015)

GSH, reduced glutathione; SOD, superoxide dismutase; ↓, reduced; ↑, increased.

the early stage of inflammation. Decreased neutrophil infiltration induced by the extract indicates that only a low degree of gastric inflammation occurred. The positive effects were also related to a reduction in gastric apoptosis and lipid peroxidation (oxidative stress) and increased antioxidant status due to increased superoxide dismutase activity and glutathione content.

The beneficial impacts of *P. speciosa* seed ethanol extract against gastric ulcers have also been reported (Maria et al. 2015). Pretreatment with the extract for 14 days prior to gastric ulcer induction in rats protected the gastric mucosa by reducing gastric acidity and collagen content, thereby reducing fibrosis (scarring).

Studies have demonstrated the positive pretreatment effects of seed and leaf extracts on gastric ulcers. No study has explored the effects of the extract as a posttreatment that mimics the clinical condition. Furthermore, the protective effects were not extensively studied regarding their molecular mechanisms of action. However, the findings revealed the promising effects of the plant extracts.

14.6.9 OTHER PHARMACOLOGICAL ACTIVITIES

P. speciosa empty pod extract inhibits acetylcholinesterase enzyme in vitro (Rawa et al. 2019); however, this activity is not observed in its leaf extract. Acetylcholinesterase degrades acetylcholine, a neurotransmitter. A deficit in acetylcholine transmission is partly responsible for the pathogenesis of Alzheimer's disease, which is featured by impairment of memory and orientation to surroundings (George et al. 2022). Acetylcholinesterase inhibition by the extract may elevate acetylcholine levels in the brain and alleviate the symptoms of Alzheimer's disease.

Infection weakens the immune system, which often manifests as a reduction in white blood cell count. *P. speciosa* empty pod extract was shown to modulate the immune system in experimental animals. It elevates levels of white blood cells, such as neutrophils, leukocytes, monocytes, and CD4 in rats infected with *S. typhimurium* (Fitrya et al. 2020), enhancing the response to infection.

14.7 CONCLUSION AND SUGGESTIONS FOR FUTURE STUDIES

P. speciosa is a medicinal plant with proven pharmacological properties, including antihypertensive, antidiabetic, and antiulcer effects. Bioactive compounds from the plant should be isolated, and their mechanisms of action should be elucidated. Moreover, the biological properties of the plant and its bioactive substances ought to be studied clinically.

ACKNOWLEDGMENT

This study received a grant (GUP-2022-038) from the Universiti Kebangsaan Malaysia.

REFERENCES

Aisha, A. F., Abu-Salah, K. M., Alrokayan, S. A., Ismail, Z., & Abdul Majid, A. M. S. Evaluation of antiangiogenic and antoxidant properties of *Parkia speciosa* Hassk extract. *Pakistan Journal of Pharmaceutical Sciences* 2012, 25: 7–14.

Aisha, A. F., Abu-Salah, K. M., Darwis, Y., & Abdul Majid, A. Screening of antiangiogenic activity of some tropical plants by rat aorta ring assay. *International Journal of Pharmacology* 2009, 5: 370–376. doi: 10.3923/ijp.2009.370.376.

Alexander-Lindo, R. L., Morrison, E. Y., Nair, M. G. 2004. Hypoglycaemic effect of stigmast-4-en-3-one and its corresponding alcohol from the bark of *Anacardium occidentale* (cashew). *Phytotherapy Research* 2004 18: 403–07. doi: 10.1002/ptr.1459.

Al Batran, R., Al-Bayaty, F., Jamil Al-Obaidi, M. M., Abdualkader, A. M., Hadi, H. A., Ali, H. M., & Abdulla, M. A. In vivo antioxidant and antiulcer activity of *Parkia speciosa* ethanolic leaf extract against ethanol-induced gastric ulcer in rats. *PloS One* 2013, 8: e64751. doi: 10.1371/journal.pone.0064751.

Alqahtani, A. S., Hidayathulla, S., Rehman, M. T., ElGamal, A. A., Al-Massarani, S., Razmovski-Naumovski, V., Alqahtani, M. S., El Dib, R. A., & AlAjmi, M. F. Alpha-amylase and alpha-glucosidase enzyme inhibition and antioxidant potential of 3-oxolupenal and katononic acid isolated from *Nuxia oppositifolia*. *Biomolecules* 2019, 10: 61. doi: 10.3390/biom10010061.

American Diabetes Association. (2) Classification and diagnosis of diabetes. *Diabetes Care* 2015, 38: S8–S16.

Asikin, Y., Taira, E., & Wada, K. Alterations in the morphological, sugar composition, and volatile flavor properties of petai (*Parkia speciosa* Hassk.) seed during ripening. *Food Research International* 2018, 106: 647–53. doi: 10.1016/j.foodres.2018.01.044.

Azemi, A. K., Nordin, M. L., Hambali, K. A., Noralidin, N. A., Mokhtar, S. S., & Rasool, A. H. G. 2022. Phytochemical contents and pharmacological potential of *Parkia speciosa* Hassk. for diabetic vasculopathy: A review. *Antioxidants* 11: 431. doi: 10.3390/antiox11020431.

Azliza, M., Ong, H., Vikineswary, S., Noorlidah, A., & Haron, N. Ethno-medicinal resources used by the Temuan in Ulu Kuang Village. *Studies on Ethno-Medicine* 2012, 6: 17–22. doi:10.1080/09735070.2012.11886415.

Babu, S., & Jayaraman, S. An update on β-sitosterol: A potential herbal nutraceutical for diabetic management. *Biomedicine & Pharmacotherapy* 2020, 131: 110702.

Batoro, J., & Siswanto, D. Ethnomedicinal survey of plants used by local society in Poncokusumo district, Malang, East Java Province, Indonesia. *Asian Journal of Medical and Biological Research* 2017, 3: 158–167. doi:10.3329/AJMBR.V3I2.3356.

Biswas, S., Dastidar, D. G., Roy, K. S., Pal, S. K., Biswas, T. K., & Ganguly, S. B. Complications of hypertension as encountered by primary care physician. *Journal of Indian Medical Association* 2003, 101: 257–259.

Chankhamjon, K., Petsom, A., Sawasdipuksa, N., & Sangvanich, P. Hemagglutinating activity of proteins from *Parkia speciosa* seeds. *Pharmaceutical Biology* 2010, 48: 81–88. doi: 10.3109/13880200903046195.

Chen, J., Wang, Z., Lu, Y., Dalton, J. T., Miller, D. D., & Li, W. Synthesis and antiproliferative activity of imidazole and imidazoline analogs for melanoma. *Bioorganic & Medicinal Chemistry Letters* 2008, 18: 3183–3187. doi: 10.1016/j.bmcl.2008.04.073.

Chhikara, N., Devi, H. R., Jaglan, S., Sharma, P., Gupta, P., & Panghal, A. Bioactive compounds, food applications and health benefits of *Parkia speciosa* (stinky beans): A review. *Agriculture & Food Security* 2018, 7: 1–9. doi: 10.1186/s40066-018-0197-x.

Ching, L. S., & Mohamed, S. Alpha-tocopherol content in 62 edible tropical plants. *Journal of Agricultural and Food Chemistry* 2001, 49: 3101–3105. doi: 10.1021/jf000891u.

de la Garza, A. L., Milagro, F. I., Boque, N., Campión, J., & Martínez, J. A. Natural inhibitors of pancreatic lipase as new players in obesity treatment. *Planta Medica* 2011, 77: 773–785. doi: 10.1055/s-0030-1270924.

de Medeiros, P. M., Figueiredo, K. F., Gonçalves, P. H. S., Caetano, R. A., Santos, É. M. D. C., Dos Santos, G. M. C., Barbosa, D. M., de Paula, M., & Mapeli, A. M. Wild plants and the food-medicine continuum-an ethnobotanical survey in Chapada Diamantina (Northeastern Brazil). *Journal of Ethnobiology and Ethnomedicine* 2021, 17, 37. doi: 10.1186/s13002-021-00463-y.

Diliarosta, S., Sari, M. P., Ramadhani, R., & Efendi, A. Ethnomedicine study on medicinal plants used by communities in West Sumatera, Indonesia. In: El-Shemy, H. (ed). *Medicinal Plants from Nature*. London: IntechOpen; 2022, pp. 1–11. doi: 10.5772/intechopen.96810.

Eswani, N., Kudus, K. A., Nazre, M., Noor, A. G. A., & Ali, M. Medicinal plant diversity and vegetation analysis of logged over hill forest of Tekai Tembeling Forest Reserve, Jerantut, Pahang. *Journal of Agricultural Sciences* 2010, 2, 189.

Fatimah, I. Green synthesis of silver nanoparticles using extract of *Parkia speciosa* Hassk pods assisted by microwave irradiation. *Journal of Advanced Research* 2016, 7: 961–969.

Fitrianti, T., & Partasasmita, R. Medicinal plants of Cintaratu Village, Pangandaran, West Java. In: Prosiding Seminar Nasional Masyarakat Biodiversitas Indonesia. 2020, 6, 625–634. doi: 10.13057/psnmbi/m060124.

Fitrya, F., Amriani, A., Novita, R. P., Elfita, & Setiorini, D. Immunomodulatory effect of *Parkia speciosa* Hassk. pods extract on rat induced by *Salmonella typhimurium*. *Journal of Pharmacy & Pharmacognosy Research* 2020, 8: 457–465.

Fitrya, Annisa A, Nikita S, & Ranna C. Alpha glukosidase inhibitory test and total phenolic content of ethanol extract of *Parkia speciosa* plant. *Science and Technology Indonesia* 2019, 4: 1–4. doi: 10.26554/sti.2019.4.1.1-4.

Frérot, E., Velluz, A., Bagnoud, A., & Delort, E. Analysis of the volatile constituents of cooked petai beans (*Parkia speciosa*) using high-resolution GC/ToF–MS. *Flavour and Fragrance Journal* 2008, 23: 434–440.

Gao, L., Zhang, W., Yang, L., Fan, H., & Olatunji, O. J. Stink bean (*Parkia speciosa*) empty pod: a potent natural antidiabetic agent for the prevention of pancreatic and hepatorenal dysfunction in high fat diet/streptozotocin-induced type 2 diabetes in rats. *Archives of Physiology and Biochemistry* 2023, 129: 261–267. doi: 10.1080/13813455.2021.1876733.

George, G., Koyiparambath, V. P., Sukumaran, S., Nair, A. S., Pappachan, L. K., Al-Sehemi, A. G., Kim, H., & Mathew, B. Structural modifications on chalcone framework for developing new class of cholinesterase inhibitors. *International Journal of Molecular Sciences* 2022, 23: 3121. doi: 10.3390/ijms23063121.

Ghasemzadeh, A., Jaafar, H. Z., Bukhori, M. F. M., Rahmat, M. H., & Rahmat, A. Assessment and comparison of phytochemical constituents and biological activities of bitter bean (*Parkia speciosa* Hassk.) collected from different locations in Malaysia. *Chemistry Central Journal* 2018, 12: 1–9. doi: 10.1186/s13065-018-0377-6.

Gilani, G. S., Xiao, C. W., & Cockell, K. A. Impact of antinutritional factors in food proteins on the digestibility of protein and the bioavailability of amino acids and on protein quality. *British Journal of Nutrition* 2012, 108: S315–S332. doi: 10.1017/S0007114512002371.

Gmelin, R., Susilo, R., & Fenwick, G. Cyclic polysulphides from *Parkia speciosa*. *Phytochemistry* 1981, 20: 2521–2523. doi: 10.1016/0031-9422(81)83085-3.

Grunwald, S. A., Haafke, S., Grieben, U., Kassner, U., Steinhagen-Thiessen, E., & Spuler, S. Statins aggravate the risk of insulin resistance in human muscle. *International Journal of Molecular Sciences* 2022, 23: 2398. doi: 10.3390/ijms23042398.

Gui, J., Mustafa, N. H., Jalil, J., Jubri, Z., & Kamisah, Y. Modulation of NOX4 and MAPK signaling pathways by *Parkia speciosa* empty pods in H9c2 cardiomyocytes exposed to H_2O_2. *Indian Journal of Pharmaceutical Sciences* 2019a, 81: 1029–1035. doi: 10.36468/pharmaceutical-sciences.600.

Gui, J., Jalil, J., Jubri, Z., & Kamisah, Y. *Parkia speciosa* empty pod extract exerts anti-inflammatory properties by modulating NFκB and MAPK pathways in cardiomyocytes exposed to tumor necrosis factor-α. *Cytotechnology* 2019b, 71: 79–89. doi: 10.1007/s10616-018-0267-8.

Hani N, Baydoun S, Nasser H, Ulian T, & Arnold-Apostolides N. Ethnobotanical survey of medicinal wild plants in the Shouf Biosphere Reserve, Lebanon. *Journal of Ethnobiology and Ethnomedicine* 2022, 18, 73. doi: 10.1186/s13002-022-00568-y.

Hasim, D., Faridah, N., & Kurniawati, D. A. Antibacterial activity of *Parkia speciosa* Hassk. peel to *Escherichia coli* and *Staphylococcus aureus* bacteria. *Journal of Chemical and Pharmaceutical Research* 2015, 7: 239–243.

Iwabuchi, T., Iijima, K., Ara, N., Koike, T., Shinkai, H., Ichikawa, T., Kamata, Y., Ishihara, K., & Shimosegawa, T. Increased gastric mucus secretion alleviates non-steroidal anti-inflammatory drug-induced abdominal pain. *Tohoku Journal of Experimental Medicine* 2013, 231: 29–36. doi: 10.1620/tjem.231.29.

Jamaluddin, F., & Mohamed, S. Hypoglycemic effect of extracts of petai papan (*Parkia speciosa*, Hassk). *Pertanika Journal of Tropical Agricultural Science* 1993, 16: 161–165.

Jamaluddin, F., Mohamed, S., & Lajis, M. N. Hypoglycaemic effect of *Parkia speciosa* seeds due to the synergistic action of β-sitosterol and stigmasterol. *Food Chemistry* 1994, 49: 339–345. doi: 10.1016/0308-8146(94)90002-7.

Jamaluddin, F., Mohamed, S., & Lajis, M. N. Hypoglycaemic effect of stigmast-4-en-3-one, from *Parkia speciosa* empty pods. *Food Chemistry* 1995, 54: 9–13. doi: 10.1016/0308-8146(95)92656-5.

Kamisah, Y., Othman, F., Qodriyah, H. M. S., & Jaarin, K. *Parkia speciosa* Hassk.: A potential phytomedicine. *Evidence-Based Complementary and Alternative Medicine*, 2013, 2013: 709028. doi: 10.1155/2013/709028.

Kamisah, Y., Periyah, V., Lee, K. T., Noor-Izwan, N., Nurul-Hamizah, A., Nurul-Iman, B. S., Subermaniam, K., Jaarin, K., Azman, A., Faizah, O., & Qodriyah, H. M. Cardioprotective effect of virgin coconut oil in heated palm oil diet-induced hypertensive rats. *Pharmaceutical Biology* 2015, 53:1243–1249. doi: 10.3109/13880209.2014.971383.

Kamisah, Y., Zuhair, J. S. F., Juliana, A. H., & Jaarin, K. *Parkia speciosa* empty pod prevents hypertension and cardiac damage in rats given N (G)-nitro-l-arginine methyl ester. *Biomedicine & Pharmacotherapy* 2017, 96: 291–298. doi: 10.1016/j.biopha.2017.09.095.

Kawamura, F., Ramle, S. F. M., Sulaiman, O., Hashim, R., & Ohara, S. Antioxidant and antifungal activities of extracts from 15 selected hardwood species of Malaysian timber. *European Journal of Wood and Wood Products* 2011, 69, 207–212. doi: 10.1007/s00107-010-0413-2.

Kerr, J. R. Epstein-Barr virus (EBV) reactivation and therapeutic inhibitors. *Journal of Clinical Pathology* 2019, 72: 651–658. doi: 10.1136/jclinpath-2019-205822.

Khalid, N. M., & Babji, A. S. Antioxidative and antihypertensive activities of selected Malaysian ulam (salad), vegetables and herbs. *Journal of Food Research* 2018, 7: 27. doi: 10.5539/jfr.v7n3p27.

Ko, H. J., Ang, L. H., & Ng, L. T. Antioxidant activities and polyphenolic constituents of bitter bean *Parkia speciosa*. *International Journal of Food Properties* 2014, 17: 1977–1986. doi: 10.1080/10942912.2013.775152.

Leaman, D. J., Arnason, J. T., Yusuf, R., Sangat-Roemantyo, H., Soedjito, H., Angerhofer, C. K., & Pezzuto, J. M. Malaria remedies of the Kenyah of the Apo Kayan, East Kalimantan, Indonesian Borneo: A quantitative assessment of local consensus as an indicator of biological efficacy. *Journal of Ethnopharmacology* 1995, 49: 1–16. doi: 10.1016/0378-8741(95)01289-3.

Liang, D., Bian, J., Deng, L. W., & Huang, D. Cyclic polysulphide 1,2,4-trithiolane from stinky bean (Parkia speciosa seeds) is a slow releasing hydrogen sulphide (H2S) donor. *Journal of Functional Foods* 2017, 35: 197–204. doi: 10.1016/j.jff.2017.05.040.

Lim, J., Ferruzzi, M. G., & Hamaker, B. R. Structural requirements of flavonoids for the selective inhibition of α-amylase versus α-glucosidase. *Food Chemistry* 2022, 370: 130981. doi: 10.1016/j.foodchem.2021.130981.

Lim T. K. *Parkia speciosa* In: *Edible Medicinal And Non-Medicinal Plants: Volume 2, Fruits*, Springer Science+Business Media B. V. 2012; pp. 798–803. doi: 10.1007/978-94-007-1764-0_90.

Maisuthisakul, P., Pasuk, S., & Ritthiruangdej, P. Relationship between antioxidant properties and chemical composition of some Thai plants. *Journal of Food Composition and Analysis* 2008, 21: 229–240.

Maria, M. S., Devarakonda, S., Kumar, A. T., & Balakrishnan, N. Anti-ulcer activity of ethanol extract of *Parkia speciosa* against indomethacin induced peptic ulcer in albino rats. *International Journal of Pharmaceutical Sciences and Research* 2015, 6: 895–902. doi: 10.13040/IJPSR.0975-8232.6(2). 895-902.

Mohd Azizi, C. Y., Salman, Z., Nik Norulain, N., & Mohd Omar, A. Extraction and identification of compounds from *Parkia speciosa* seeds by supercritical carbon dioxide. *Journal of Chemical and Natural Resources Engineering* 2008, 2: 153–163.

Morales-Loredo, H., Barrera, A., Garcia, J. M., Pace, C. E., Naik, J. S., Gonzalez Bosc, L. V., & Kanagy, N. L. Hydrogen sulfide regulation of renal and mesenteric blood flow. *American Journal of Physiology-Heart and Circulatory Physiology* 2019, 317, H1157–H1165. doi: 10.1152/ajpheart.00303.2019.

Muhialdin, B. J., Rani, N. F. A., & Hussin, A. S. M. Identification of antioxidant and antibacterial activities for the bioactive peptides generated from bitter beans (*Parkia speciosa*) via boiling and fermentation processes. *LWT* 2020, 131: 109776. doi: 10.1016/j.lwt.2020.109776.

Murakami, A., Ali, A. M., Mat-Salleh, K., Koshimizu, K., & Ohigashi, H. Screening for the in vitro anti-tumor-promoting activities of edible plants from Malaysia. *Bioscience, Biotechnology and Biochemistry* 2000, 64: 9–16. doi: 10.1271/bbb.64.9.

Musa, N., Wei, L. S., Seng, C. T., Wee, W., & Leong, L. K. Potential of edible plants as remedies of systemic bacterial disease infection in cultured fish. *Global Journal of Pharmacology* 2008, 2: 31–36.

Mustafa, N. H., Jalil, J., Saleh, M. S. M., Zainalabidin, S., Asmadi, A. Y., & Kamisah, Y. *Parkia speciosa* Hassk. Empty pod extract prevents cardiomyocyte hypertrophy by inhibiting MAPK and calcineurin-NFATC3 signaling pathways. *Life* 2023, 13:43. doi: 10.3390/life13010043.

Mustafa, N. H., Ugusman, A., Jalil, J., & Kamisah, Y. Anti-inflammatory property of *Parkia speciosa* empty pod extract in human umbilical vein endothelial cells. *Journal of Applied Pharmaceutical Science* 2018, 8: 152–158. doi: 10.7324/JAPS.2018.8123.

Nani, A., Murtaza, B., Sayed Khan, A., Khan, N. A., & Hichami, A. Antioxidant and anti-inflammatory potential of polyphenols contained in Mediterranean diet in obesity: Molecular mechanisms. *Molecules* 2021, 26: 985. doi: 10.3390/molecules26040985.

Newman, D. J., & Cragg, G. M. Natural products as sources of new drugs over the nearly four decades from 01/1981 to 09/2019. *Journal of Natural Products* 2020, 83, 770–803. doi: 10.1021/acs.jnatprod.9b01285.

Ng, C. Y., Kamisah, Y., Faizah, O., & Jaarin, K. The role of repeatedly heated soybean oil in the development of hypertension in rats: Association with vascular inflammation. *International Journal of Experimental Pathology* 2012, 93: 377–87. doi: 10.1111/j.1365-2613.2012.00839.x.

Ng, C. Y., Leong X. F., Masbah, N., Adam, S. K., Kamisah, Y., Jaarin, K. Reprint of Heated vegetable oils and cardiovascular disease risk factors. *Vascular Pharmacology* 2014, 62: 38–46. doi: 10.1016/j.vph.2014.05.003.

Norazlin, Y., Usamah, N., Salamah, H., Alif, A., Qayyum, M., Hazilawati, H., Mazlina, M., & Hezmee, M. Anti-inflammatory activity of combined pods and seed extract of Parkia speciosa on lipopolysaccharide-induced paw edema in rats. *Comparative Clinical Pathology* 2022, 31: 787–796. doi: 10.1007/s00580-022-03380-y.

Norsidah, K. Z., Asmadi, A. Y., Azizi, A., Faizah, O., & Kamisah, Y. Palm tocotrienol-rich fraction improves vascular proatherosclerotic changes in hyperhomocysteinemic rats. *Evidence-Based Complementary and Alternative Medicine* 2013, 2013: 976967. doi: 10.1155/2013/976967.

Ong, H. C., Ahmad, N., & Milow, P. Traditional medicinal plants used by the Temuan villagers in Kampung Tering, Negeri Sembilan, Malaysia. *Studies on Ethno-Medicine* 2011b, 5: 169–173. doi: 10.1080/09735070.2011.11886406.

Ong, H. C., Chua, S., & Milow, P. Ethno-medicinal plants used by the Temuan villagers in Kampung Jeram Kedah, Negeri Sembilan, Malaysia. *Studies on Ethno-Medicine* 2011a, 5: 95–100. doi: 10.1080/09735070.2011.11886395.

Ong, H. C., Zuki, R. M., & Milow, P. Traditional knowledge of medicinal plants among the Malay villagers in Kampung Mak Kemas, Terengganu, Malaysia. *Studies on Ethno-Medicine* 2011c, 5: 175–85. doi: 10.1080/09735070.2011.11886407.

Pranprawit, A. Local vegetables traditionally used for reducing hyperglycemia in Surat Thani province, Thailand. *International Journal of Applied Pharmaceutics* 2019, 11: 3–6. doi: 10.22159/ijap.2019.v11s3.M0008.

Rahman, N. N. N. A., Zhari, S., Sarker, M. Z. I., Ferdosh, S., Yunus, M. A. C., & Kadir, M. O. A. Profile of *Parkia speciosa* Hassk metabolites extracted with SFE using FTIR-PCA method. *Journal of the Chinese Chemical Society* 2012, 59: 507–514. doi: 10.1002/jccs.201100104.

Ramadani, G. Pengaruh ekstrak kulit petai (*Parkia speciosa*) sebagai antioksidan alami pada pemakaian minyak goreng deep frying terhadap kadar MDA hepar mencit (*Mus musculus*). *Saintika Medika* 2012, 8: 28–34. doi: 10.22219/sm.v8i1.4096.

Ravichandran, V., Vasanthi, S., Shalini, S., Shah, S. A. A., Tripathy, M., & Paliwal, N. Green synthesis, characterization, antibacterial, antioxidant and photocatalytic activity of *Parkia speciosa* leaves extract mediated silver nanoparticles. *Results in Physics* 2019, 15, 102565. doi: 10.1016/j.rinp.2019.102565.

Rawa, M. S. A., Hassan, Z., Murugaiyah, V., Nogawa, T., & Wahab, H. A. Anti-cholinesterase potential of diverse botanical families from Malaysia: Evaluation of crude extracts and fractions from liquid-liquid extraction and acid-base fractionation. *Journal of Ethnopharmacology* 2019, 245: 112160. doi: 10.1016/j.jep.2019.112160.

Roosita, K., Kusharto, C. M., Sekiyama, M., Fachrurozi, Y., & Ohtsuka, R. Medicinal plants used by the villagers of a Sundanese community in West Java, Indonesia. *Journal of Ethnopharmacology* 2008, 115: 72–81. doi: 10.1016/j.jep.2007.09.010.

Rossier, B. C., Bochud, M., & Devuyst, O. The Hypertension Pandemic: An Evolutionary Perspective. *Physiology* 2017, 32: 112–125. doi: 10.1152/physiol.00026.2016.

Sakunpak, A., & Panichayupakaranant, P. Antibacterial activity of Thai edible plants against gastrointestinal pathogenic bacteria and isolation of a new broad spectrum antibacterial polyisoprenylated benzophenone, chamuangone. *Food Chemistry* 2012, 130: 826–831. doi: 10.1016/j.foodchem.2011.07.088.

Saleh, M. S. M., Jalil, J., Mustafa, N. H., Ramli, F. F., Asmadi, A. Y., & Kamisah, Y. UPLC-MS-based metabolomics profiling for α-glucosidase inhibiting property of *Parkia speciosa* pods. *Life* 2021a, 11: 78. doi: 10.3390/life11020078.

Saleh, M. S. M., Jalil, J., Zainalabidin, S., Asmadi, A. Y., Mustafa, N. H., & Kamisah, Y. Genus *Parkia*: Phytochemical, medicinal uses, and pharmacological properties. *International Journal of Molecular Sciences* 2021b, 22: 618. doi: 10.3390/ijms22020618.

Salman, Z., Mohd Azizi, C., Nik Norulaini, N., & Mohd Omar, A. Gas chromatography/time-of-flight mass spectrometry for identification of compounds from *Parkia speciosa* seeds extracted by supercritical carbon dioxide. In: *Proceedings of the First International Conference on Natural Resources Engineering & Technology* 2006, pp. 112–120.

Sampaio, B. L., Edrada-Ebel, R., & Da Costa, F. B. Effect of the environment on the secondary metabolic pro-file of *Tithonia diversifolia*: A model for environmental metabolomics of plants. *Scientific Reports* 2016, 6: 29265. doi: 10.1038/srep29265.

Samuel, A. J. S. J., Kalusalingam, A., Chellappan, D. K., Gopinath, R., Radhamani, S., Husain, H. A., Muruganandham, V., & Promwichit, P. Ethnomedical survey of plants used by the Orang Asli in Kampung Bawong, Perak, West Malaysia. *Journal of Ethnobiology and Ethnomedicine* 2010, 6: 5. doi: 10.1186/1746-4269-6-5.

Siew, Y. Y., Zareisedehizadeh, S., Seetoh, W. G., Neo, S. Y., Tan, C. H., & Koh, H. L. Ethnobotanical survey of usage of fresh medicinal plants in Singapore. *Journal of Ethnopharmacology* 2014, 155: 1450–1466. doi: 10.1016/j.jep.2014.07.024.

Siow, H. L., & Gan, C. Y. Extraction of antioxidative and antihypertensive bioactive peptides from *Parkia speciosa* seeds. *Food Chemistry* 2013, 141: 3435–3442. doi: 10.1016/j.foodchem.2013.06.030.

Siti, H. N., Jalil, J., Asmadi, A. Y., & Kamisah, Y. Rutin modulates MAPK pathway differently from quercetin in angiotensin II-induced H9c2 cardiomyocyte hypertrophy. *International Journal of Molecular Sciences* 2021a, 22: 5063. doi: 10.3390/ijms22105063.

Siti, H. N., Jalil, J., Asmadi, A. Y., & Kamisah, Y. *Parkia speciosa* Hassk. Empty pod extract alleviates angio-tensin II-induced cardiomyocyte hypertrophy in H9c2 cells by modulating the Ang II/ROS/NO Axis and MAPK pathway. *Frontiers in Pharmacology* 2021b, 12: 741623. doi: 10.3389/fphar.2021.741623.

Sonia, N., Dsouza, M. R., & Alisha. Pharmacological evaluation of *Parkia speciosa* Hassk. for antioxidant, anti-inflammatory, anti-diabetic and anti-microbial activities in vitro. *International Journal of Life Sciences* 2018, A11: 49–59.

Srisawat, T., Suvarnasingh, A., & Maneenoon, K. Traditional medicinal plants notably used to treat skin dis-orders nearby Khao Luang mountain hills region, Nakhon si Thammarat, Southern Thailand. *Journal of Herbs, Spices & Medicinal Plants* 2016, 22: 35–56. doi: 10.1080/10496475.2015.1018472.

Susilo, J., Astuti, A. W., & Larasati, D. Efficacy of Petai (*Parkia speciosa*, Hassk.) leaf extract as an anti-dyslipidemic herb in *Rattus norvegicus* induced by high-fat feed. *Ad-Dawaa. Journal of Pharmaceutical Sciences* 2020, 3: 47–55. doi: 10.24252/djps.v3i1.13761.

Suvachittanont, W., & Jaranchavanapet, P. Mitogenic effect of *Parkia speciosa* seed lectin on human lympho-cytes. *Planta Medica* 2000, 66: 699–704. doi: 10.1055/s-2000-9565.

Suvachittanont, W., Kurashima, Y., Esumi, H., & Tsuda, M. Formation of thiazolidine-4-carboxylic acid (thio-proline), an effective nitrite-trapping agent in human body *in Parkia speciosa* seeds and other edible leguminous seeds in Thailand. *Food Chemistry* 1996, 55: 359–363. doi: 10.1016/0308-8146(95)00132-8.

Tandi, J., Handayani, T. W., Tandebia, M., & Wijaya, J. A. Effect of *Parkia speciosa* Hassk peels extract on total cholesterol levels of hypercholesterolemia rats. *Indian Journal of Forensic Medicine & Toxicology* 2020, 14: 2988–2992. doi: 10.37506/ijfmt.v14i4.12045.

Tangkanakul, P., Trakoontivakorn, G., Saengprakai, J., Auttaviboonkul, P., Niyomwit, B., Lowvitoon, N., & Nakahara, K. Antioxidant capacity and antimutagenicity of thermal processed Thai foods. *Japan Agricultural Research Quarterly* 2011, 45: 211–218. doi: 10.6090/jarq.45.211.

Tocmo, R., Liang, D., Wang, C., Poh, J., & Huang, D. Organosulfide profile and hydrogen sulfide-releasing capacity of stinky bean (*Parkia speciosa*) oil: Effects of pH and extraction methods. *Food Chemistry* 2016, 190: 1123–1129. doi: 10.1016/j.foodchem.2015.06.072.

Tunsaringkarn, T., Rungsiyothin, A., & Ruangrungsi, N. α-Glucosidase inhibitory activity of Thai mimosa-ceous plant extracts. *Journal of Health Research* 2008, 22: 29–33.

Tunsaringkarn, T., Soogarun, S., Rungsiyothin, A., & Palasuwan, A. Inhibitory activity of Heinz body induc-tion in vitro antioxidant model and tannin concentration of Thai mimosaceous plant extracts. *Journal of Medicinal Plants Research* 2012, 6: 4096–4101. doi: 10.5897/JMPR12.304.

Tunsaringkarn, T., Rungsiyothin, A., & Ruangrungsi, N. α-Glucosidase inhibitory activity of water soluble extract from Thai mimosaceous plants. *Public Health Journal of Burapha University* 2009, 4: 54–63.

Uyub, A. M., Nwachukwu, I. N., Azlan, A. A., & Fariza, S. S. In-vitro antibacterial activity and cytotoxicity of selected medicinal plant extracts from Penang Island Malaysia on metronidazole-resistant-*Helicobacter pylori* and some pathogenic bacteria. *Ethnobotany Research and Applications* 2010, 8: 95–106.

Vatanasapt, V., Sriamporn, S., & Vatanasapt, P. Cancer control in Thailand. *Japanese Journal of Clinical Oncology* 2002, 32: S82–S91. doi: 10.1093/jjco/hye134.

Ward, M. G., Li, G., Barbosa-Lorenzi, V. C., & Hao, M. Stigmasterol prevents glucolipotoxicity induced defects in glucose-stimulated insulin secretion. *Scientific Reports* 2017, 7: 1–13. doi: 10.1038/s41598-017-10209-0.

World Health Organization (WHO). Diabetes, 2023. https://www.who.int/news-room/fact-sheets/detail/diabetes. Accessed on false7 April 2023.

Wonghirundecha, S., Benjakul, S., & Sumpavapol, P. Total phenolic content, antioxidant and antimicrobial activities of stink bean (*Parkia speciosa* Hassk.) pod extracts. *Songklanakarin Journal of Science & Technology* 2014, 36: 300–308.

Wu, D., Hu, Q., & Zhu, D. An update on hydrogen sulfide and nitric oxide interactions in the cardiovascular system. *Oxidative Medicine and Cellular Longevity* 2018, 2018: 4579140. doi: 10.1155/2018/4579140.

Yin, I. X., Zhang, J., Zhao, I. S., Mei, M. L., Li, Q., & Chu, C. H. The antibacterial mechanism of silver nanoparticles and its application in dentistry. *International Journal of Nanomedicine* 2020, 15: 2555–2562. doi: 10.2147/IJN.S246764.

Yoon, J. J., Son, C. O., Kim, H. Y., Han, B. H., Lee, Y. J., Lee, H. S., & Kang, D. G. Betulinic acid protects DOX-triggered cardiomyocyte hypertrophy response through the GATA4/Calcineurin/NFAT pathway. *Molecules* 2020, 26, 53. doi: 10.3390/molecules26010053.

Zaini, N., & Mustaffa, F. *Parkia speciosa* as valuable, miracle of nature. *Asian Journal of Medicine and Health* 2017, 2: 1–9. doi: 10.9734/AJMAH/2017/30997.

Index

Pages in *italics* refer to figures and pages in **bold** refer to tables.